NEUROMETHODS

Series Editor
Wolfgang Walz
University of Saskatchewan
Saskatoon, Canada

For further volumes:
http://www.springer.com/series/7657

Neuromethods publishes cutting-edge methods and protocols in all areas of neuroscience as well as translational neurological and mental research. Each volume in the series offers tested laboratory protocols, step-by-step methods for reproducible lab experiments and addresses methodological controversies and pitfalls in order to aid neuroscientists in experimentation. *Neuromethods* focuses on traditional and emerging topics with wide-ranging implications to brain function, such as electrophysiology, neuroimaging, behavioral analysis, genomics, neurodegeneration, translational research and clinical trials. *Neuromethods* provides investigators and trainees with highly useful compendiums of key strategies and approaches for successful research in animal and human brain function including translational "bench to bedside" approaches to mental and neurological diseases.

Extracellular Vesicles

Edited by

Amit K. Srivastava

Department of Medicine, Cardeza Foundation for Hematologic Research Sidney Kimmel Medical College, Thomas Jefferson University, Philadelphia, PA, USA

Humana Press

Editor
Amit K. Srivastava
Department of Medicine, Cardeza
Foundation for Hematologic Research
Sidney Kimmel Medical College
Thomas Jefferson University
Philadelphia, PA, USA

ISSN 0893-2336 ISSN 1940-6045 (electronic)
Neuromethods
ISBN 978-1-0716-4904-6 ISBN 978-1-0716-4905-3 (eBook)
https://doi.org/10.1007/978-1-0716-4905-3

This Humana imprint is published by the registered company Springer Science+Business Media, LLC, part of Springer Nature.
The registered company address is: 1 New York Plaza, New York, NY 10004, U.S.A.

Preface to the Series

Experimental life sciences have two basic foundations: concepts and tools. The Neuromethods series focuses on the tools and techniques unique to the investigation of the nervous system and excitable cells. It will not, however, shortchange the concept side of things as care has been taken to integrate these tools within the context of the concepts and questions under investigation. In this way, the series is unique in that it not only collects protocols but also includes theoretical background information and critiques which led to the methods and their development. Thus, it gives the reader a better understanding of the origin of the techniques and their potential future development. The Neuromethods publishing program strikes a balance between recent and exciting developments like those concerning new animal models of disease, imaging, in vivo methods, and more established techniques, including, for example, immunocytochemistry and electrophysiological technologies. New trainees in neurosciences still need a sound footing in these older methods in order to apply a critical approach to their results.

Under the guidance of its founders, Alan Boulton and Glen Baker, the Neuromethods series has been a success since its first volume published through Humana Press in 1985. The series continues to flourish through many changes over the years. It is now published under the umbrella of Springer Protocols. While methods involving brain research have changed a lot since the series started, the publishing environment and technology have changed even more radically. Neuromethods has the distinct layout and style of the Springer Protocols program, designed specifically for readability and ease of reference in a laboratory setting.

The careful application of methods is potentially the most important step in the process of scientific inquiry. In the past, new methodologies led the way in developing new disciplines in the biological and medical sciences. For example, Physiology emerged out of Anatomy in the nineteenth century by harnessing new methods based on the newly discovered phenomenon of electricity. Nowadays, the relationships between disciplines and methods are more complex. Methods are now widely shared between disciplines and research areas. New developments in electronic publishing make it possible for scientists that encounter new methods to quickly find sources of information electronically. The design of individual volumes and chapters in this series takes this new access technology into account. Springer Protocols makes it possible to download single protocols separately. In addition, Springer makes its print-on-demand technology available globally. A print copy can therefore be acquired quickly and for a competitive price anywhere in the world.

Saskatoon, Canada ***Wolfgang Walz***

Preface

Extracellular vesicles (EVs) have steadily moved from the periphery of biological curiosity to the center of serious scientific inquiry. Once seen as cellular cast-offs, they are now recognized as vital participants in intercellular communication, with implications for diagnostics, therapeutics, and our fundamental understanding of physiology and disease. The field has made impressive strides, but it also continues to face technical and conceptual challenges, many of which arise from the sheer complexity of working with EVs.

This textbook offers a comprehensive collection of protocols that reflect the current state of the art in EV research, drawing on the expertise of investigators across disciplines and institutions. With contributions from 72 authors spanning 16 countries across 6 continents, the chapters cover a wide array of experimental approaches, from foundational isolation and enrichment techniques to cutting-edge analytical tools and therapeutic applications.

Protocols include tangential flow filtration for large-scale isolation, magnetic bead-based separation, and optimized methods for recovering EVs from diverse biological sources such as blood, brain tissue, urine, red blood cells, and mesenchymal stem cells. Specialized procedures detail the capture of EV surface proteins, electron tomography-based structural analysis, and innovative strategies to load small RNAs into EVs. In addition, contributors present protocols for characterizing EV uptake, procoagulant potential, and for performing advanced molecular profiling using nanoparticle tracking analysis and mass spectrometry.

Recognizing the complexity of working with EVs, this volume also addresses practical considerations such as optimized storage buffers, quality control, and troubleshooting strategies. A dedicated chapter explores bioinformatics tools and databases to support data analysis and interpretation, acknowledging the growing importance of computational approaches in this data-rich field.

The goal of this volume is to serve as a practical and accessible reference for both new and experienced researchers. Whether working with clinical samples, developing therapeutic EVs, or studying their basic biology, investigators will find detailed protocols accompanied by critical commentary on experimental design, controls, and interpretation.

I am deeply grateful to the contributing authors for their clarity, rigor, and collaborative spirit. Their collective work highlights the international and interdisciplinary nature of EV science and provides a strong foundation to support continued innovation and discovery in this rapidly evolving field.

Philadelphia, PA, USA *Amit K. Srivastava*

Contents

Contributors

AMRITHA ACHUTHKUMAR • *Departments of Pharmacology, Physiology, and Cancer Biology, Thomas Jefferson University, Philadelphia, PA, USA*
KARINA A. BARREIRO • *Institute for Molecular Medicine Finland FIMM, University of Helsinki, Helsinki, Finland*
ZENIA BARRETTO • *Department of Medicine, Cardeza Foundation for Hematologic Research, Sidney Kimmel Medical College, Thomas Jefferson University, Philadelphia, PA, USA*
JENS BERNDTSSON • *Centre for Cellular Imaging Core Facility, Sahlgrenska Academy, University of Gothenburg, Gothenburg, Sweden*
SEYEDSHAHABADIN BIDOKI • *Department of Neurology, David Geffen School of Medicine, University of California, Los Angeles, CA, USA*
GAL BITAN • *Department of Neurology, David Geffen School of Medicine, University of California, Los Angeles, CA, USA; Molecular Biology Institute, University of California, Los Angeles, CA, USA; Brain Research Institute, University of California, Los Angeles, CA, USA*
CRISTIAN BODO • *Systems Oncology Group, Champalimaud Research, Champalimaud Centre for the Unknown, Lisbon, Portugal*
DYLAN BURGER • *University of Ottawa, Ottawa, Canada*
JESSICA C. CARDENAS • *Department of Surgery, Division of GI, Trauma, and Endocrine Surgery, School of Medicine, University of Colorado Denver, Aurora, CO, USA*
MINGHAO CHEN • *Center for Soft Nanoscience and Institute of Medical Physics and Biophysics, University of Muenster, Münster, Germany; Department of Structural Biochemistry, Max Planck Institute of Molecular Physiology, Dortmund, Germany*
SAI V. CHITTI • *Department of Biochemistry, La Trobe Institute for Molecular Science, La Trobe University, Melbourne, VIC, Australia*
ALISA MORSS CLYNE • *Fischell Department of Bioengineering, University of Maryland, College Park, MD, USA*
BRUNO COSTA-SILVA • *Systems Oncology Group, Champalimaud Research, Champalimaud Centre for the Unknown, Lisbon, Portugal*
ROSSELLA CRESCITELLI • *Sahlgrenska Center for Cancer Research and Wallenberg Centre for Molecular and Translational Medicine, Department of Surgery, Institute of Clinical Sciences, Sahlgrenska Academy, University of Gothenburg, Gothenburg, Sweden*
JULIANO C. DA SILVEIRA • *Department of Veterinary Medicine, College of Animal Science and Food Engineering, University of São Paulo, Pirassununga, SP, Brazil*
CAMILA AZZOLIN DE SOUZA • *Department of Veterinary Medicine, College of Animal Science and Food Engineering, University of São Paulo, Pirassununga, SP, Brazil*
UTA ERDBRÜGGER • *University of Virginia, Charlottesville, VA, USA*
SAMIR EL ANDALOUSSI • *Department of Laboratory Medicine, Division of Biomolecular and Cellular Medicine, Karolinska Institutet, Stockholm, Sweden; Department of Cellular Therapy and Allogeneic Stem Cell Transplantation (CAST), Karolinska University Hospital Huddinge and Karolinska Comprehensive Cancer Center, Stockholm, Sweden; Karolinska ATMP Center, ANA Futura, Huddinge, Sweden*
PAMALI FONSEKA • *Department of Biochemistry, La Trobe Institute for Molecular Science, La Trobe University, Melbourne, VIC, Australia*

Ella Fouhy • *UCD Conway SPHERE Research Group, Conway Institute, University College Dublin, Dublin, Ireland; School of Biomolecular and Biomedical Science, University College Dublin, Dublin, Ireland*

Ahmed Gad • *Animal Reproduction and Biotechnology Laboratory (ARBL), Department of Biomedical Sciences, College of Veterinary Medicine and Biomedical Sciences, Colorado State University, Fort Collins, CO, USA; Department of Animal Production, Faculty of Agriculture, Cairo University, Giza, Egypt*

Christos Gatsogiannis • *Center for Soft Nanoscience and Institute of Medical Physics and Biophysics, University of Muenster, Münster, Germany; Department of Structural Biochemistry, Max Planck Institute of Molecular Physiology, Dortmund, Germany*

Samuel Gebremedhn • *Animal Reproduction and Biotechnology Laboratory (ARBL), Department of Biomedical Sciences, College of Veterinary Medicine and Biomedical Sciences, Colorado State University, Fort Collins, CO, USA; J.R. Simplot Company, Boise, ID, USA*

Patrícia A. Gomes • *Life and Health Sciences Research Institute (ICVS), School of Medicine, University of Minho, Braga, Portugal; ICVS/3B's—PT Government Associate Laboratory, Braga, Guimarães, Portugal*

Sowmya Shree Gopal • *Department of Medicine, Cardeza Foundation for Hematologic Research, Sidney Kimmel Medical College, Thomas Jefferson University, Philadelphia, PA, USA*

André Görgens • *Department of Laboratory Medicine, Division of Biomolecular and Cellular Medicine, Karolinska Institutet, Stockholm, Sweden; Department of Cellular Therapy and Allogeneic Stem Cell Transplantation (CAST), Karolinska University Hospital Huddinge and Karolinska Comprehensive Cancer Center, Stockholm, Sweden; Karolinska ATMP Center, ANA Futura, Huddinge, Sweden*

Cristina Grange • *Department of Medical Sciences, University of Torino, Torino, Italy*

David W. Greening • *Molecular Proteomics, Baker Heart and Diabetes Institute, Melbourne, VIC, Australia*

Morgan Greig • *University of Virginia, Charlottesville, VA, USA*

Sriram Gummadi • *Department of Biochemistry, La Trobe Institute for Molecular Science, La Trobe University, Melbourne, VIC, Australia*

Megan Hong • *Department of Neurology, David Geffen School of Medicine, University of California, Los Angeles, CA, USA; Molecular Biology Institute, University of California, Los Angeles, CA, USA*

Iasmin Inocencio • *Baker Department of Cardiometabolic Health, University of Melbourne, Melbourne, VIC, Australia; Molecular Proteomics, Baker Heart and Diabetes Institute, Melbourne, VIC, Australia*

Filippos Katsaitis • *Laboratory of Brain Exosomes and Pathology—ExoBrain, Institute of Biosciences and Applications NCSR "Demokritos", Athens, Greece*

Mandeep Kaur • *Department of Medicine, Cardeza Foundation for Hematologic Research, Sidney Kimmel Medical College, Thomas Jefferson University, Philadelphia, PA, USA*

Marzyeh Kheradmand • *Fischell Department of Bioengineering, University of Maryland, College Park, MD, USA*

Sophie Lanzkron • *Department of Medicine, Cardeza Foundation for Hematologic Research, Sidney Kimmel Medical College, Thomas Jefferson University, Philadelphia, PA, USA*

Alicia Llorente • *University of Oslo, Oslo, Norway*

PATRICIA B. MAGUIRE • *UCD Conway SPHERE Research Group, Conway Institute, University College Dublin, Dublin, Ireland; School of Biomolecular and Biomedical Science, University College Dublin, Dublin, Ireland; AI for Healthcare Hub, Institute for Discovery, O'Brien Centre for Science, University College Dublin, Dublin, Ireland*
MỸ G. MAHONEY • *Departments of Pharmacology, Physiology, and Cancer Biology, Thomas Jefferson University, Philadelphia, PA, USA; Department of Otolaryngology – Head and Neck Surgery, Thomas Jefferson University, Philadelphia, PA, USA*
ELENA MARTENS-UZUNOVA • *Erasmus University Medical Center, Rotterdam, Netherlands*
CHIPPY K. MATHEW • *Stem Cell Facility- DBT Centre of Excellence for Stem Cell Research, All India Institute of Medical Sciences, New Delhi, India*
SURESH MATHIVANAN • *Department of Biochemistry, La Trobe Institute for Molecular Science, La Trobe University, Melbourne, VIC, Australia*
NICO G. MENJIVAR • *Animal Reproduction and Biotechnology Laboratory (ARBL), Department of Biomedical Sciences, College of Veterinary Medicine and Biomedical Sciences, Colorado State University, Fort Collins, CO, USA*
MASSIMO MICARONI • *Centre for Cellular Imaging Core Facility, Sahlgrenska Academy, University of Gothenburg, Gothenburg, Sweden*
SUJATA MOHANTY • *Stem Cell Facility- DBT Centre of Excellence for Stem Cell Research, All India Institute of Medical Sciences, New Delhi, India*
KIMIA MOHSENI • *Department of Neurology, David Geffen School of Medicine, University of California, Los Angeles, CA, USA*
AXL A. NEURAUTER • *Life Technologies AS, Oslo, Norway*
MONICA SUET YING NG • *Kidney Health Service, Royal Brisbane and Women's Hospital, Herston, QLD, Australia; Faculty of Health, Medicine and Behavioural Sciences, University of Queensland, Herston, QLD, Australia; Institute for Molecular Bioscience, university of Queensland, Herston, QLD, Australia; Conjoint Internal Medicine Laboratory, Chemical Pathology, Pathology Queensland, Herston, QLD, Australia; Infection and Inflammation Programm, QIMR Berghofer, Herston, QLD, Australia*
TUULA ANNELI NYMAN • *University of Oslo, Oslo, Norway*
DANIEL O'REILLY • *UCD Conway SPHERE Research Group, Conway Institute, University College Dublin, Dublin, Ireland; School of Biomolecular and Biomedical Science, University College Dublin, Dublin, Ireland; Department of Neonatology, Rotunda Hospital, Dublin, Ireland*
BARON K. OSBORN • *Department of Surgery, Division of GI, Trauma, and Endocrine Surgery, School of Medicine, University of Colorado Denver, Aurora, CO, USA*
KETIL WINTHER PEDERSEN • *Life Technologies AS, Oslo, Norway*
MAIJA PUHKA • *Institute for Molecular Medicine Finland FIMM, University of Helsinki, Helsinki, Finland*
ALIN RAI • *Molecular Proteomics, Baker Heart and Diabetes Institute, Melbourne, VIC, Australia*
GURNEET S. SANGHA • *Fischell Department of Bioengineering, University of Maryland, College Park, MD, USA*
SANJAY SHAHI • *Department of Biochemistry, La Trobe Institute for Molecular Science, La Trobe University, Melbourne, VIC, Australia*
YASHVI SHARMA • *Stem Cell Facility- DBT Centre of Excellence for Stem Cell Research, All India Institute of Medical Sciences, New Delhi, India*
IBRAR SIDDIQUE • *Department of Neurology, David Geffen School of Medicine, University of California, Los Angeles, CA, USA*

KALLIOPI SKOURTI • *Laboratory of Brain Exosomes and Pathology—ExoBrain, Institute of Biosciences and Applications NCSR "Demokritos", Athens, Greece*
LAUREN V. SMITH • *Fischell Department of Bioengineering, University of Maryland, College Park, MD, USA*
IOANNIS SOTIROPOULOS • *Life and Health Sciences Research Institute (ICVS), School of Medicine, University of Minho, Braga, Portugal; ICVS/3B's—PT Government Associate Laboratory, Braga, Guimarães, Portugal; Laboratory of Brain Exosomes and Pathology—ExoBrain, Institute of Biosciences and Applications NCSR "Demokritos", Athens, Greece*
NUNO SOUSA • *Life and Health Sciences Research Institute (ICVS), School of Medicine, University of Minho, Braga, Portugal; ICVS/3B's—PT Government Associate Laboratory, Braga, Guimarães, Portugal*
LUCA A. SOUZA • *Animal Reproduction and Biotechnology Laboratory (ARBL), Department of Biomedical Sciences, College of Veterinary Medicine and Biomedical Sciences, Colorado State University, Fort Collins, CO, USA; Department of Veterinary Medicine, College of Animal Science and Food Engineering, University of São Paulo, Pirassununga, SP, Brazil*
AMIT K. SRIVASTAVA • *Department of Medicine, Cardeza Foundation for Hematologic Research, Sidney Kimmel Medical College, Thomas Jefferson University, Philadelphia, PA, USA*
DAWIT TESFAYE • *Animal Reproduction and Biotechnology Laboratory (ARBL), Department of Biomedical Sciences, College of Veterinary Medicine and Biomedical Sciences, Colorado State University, Fort Collins, CO, USA*
KIMBERLY A. THOMAS • *Vitalant Research Institute, Denver, CO, USA; Department of Pathology, University of Colorado—Anschutz Medical Center, Aurora, CO, USA*
ORNELLA URZÌ • *Sahlgrenska Center for Cancer Research and Wallenberg Centre for Molecular and Translational Medicine, Department of Surgery, Institute of Clinical Sciences, Sahlgrenska Academy, University of Gothenburg, Gothenburg, Sweden; St. Anna Children's Cancer Research Institute (CCRI), Vienna, Austria*
MARTIN E. VAN ROYEN • *Erasmus University Medical Center, Rotterdam, Netherlands*
GERALD W. VERHAEGH • *Radboud University Medical Center, Nijmegen, The Netherland*
CLARISSA L. WAITES • *Department of Pathology and Cell Biology, Taub Institute for Research on Alzheimer's Disease and the Aging Brain, Columbia University Medical Center, New York, NY, USA*
JOSHUA A. WELSH • *Advanced Technology Group, BD Biosciences, San Jose, CA, USA*
PETER J. WERMUTH • *Departments of Pharmacology, Physiology, and Cancer Biology, Thomas Jefferson University, Philadelphia, PA, USA*

Chapter 1

Extracellular Vesicles: Biology, Functions, and Applications

Sowmya Shree Gopal, Mandeep Kaur, Sophie Lanzkron, and Amit K. Srivastava

Abstract

Extracellular vesicles (EVs) are increasingly recognized as active mediators of intercellular communication rather than inert cellular byproducts. This chapter presents a comprehensive overview of EVs, encompassing their biogenesis, classification, molecular composition, and functional roles across physiological and pathological states. By examining the distinct origins, formation pathways, and cargos of exosomes, microvesicles, and apoptotic bodies, it highlights how EVs constitute a complex and regulated system of intercellular signaling. EVs transport diverse molecular payloads, proteins, lipids, mRNAs, miRNAs, and other noncoding RNAs, facilitating functional modulation of recipient cells. These vesicles play crucial roles in immune regulation, tissue homeostasis, neurogenesis, and tumor progression, and serve as key effectors in processes such as inflammation, angiogenesis, and metabolic adaptation. Acting both locally and systemically, EVs enable context-dependent modulation of cell behavior, establishing them as essential components of multicellular communication networks. The discussion extends beyond cellular biology to considering the translational promise of EVs. Increasingly, EVs are being investigated as minimally invasive biomarkers due to their stability in biofluids and ability to reflect the molecular state of their cells of origin. In clinical fields such as oncology, neurology, and autoimmune disease, EVs offer molecular snapshots of disease progression, enabling early detection, real-time monitoring, and potential prognostic evaluation. Concurrently, the intrinsic stability, low immunogenicity, and targeting capabilities of EVs position them as compelling candidates for therapeutic delivery systems. Strategies are being developed to engineer EVs for the transport of small RNAs, CRISPR-Cas components, and immunoregulatory molecules, exploiting their native tropism and biocompatibility. However, the field faces unresolved challenges, including heterogeneity in EV subtypes, lack of standardized isolation methods, and limitations in delivery efficiency and scalability. Technological innovations in microfluidics, high-resolution profiling (including proteomics and transcriptomics), and synthetic biology are rapidly expanding the toolkit available for EV characterization and manipulation. These advances are poised to accelerate the clinical translation of EV-based diagnostics and therapeutics. By positioning EVs as both messengers and modulators, the chapter situates them at the intersection of basic science and applied medicine. As the mechanistic understanding of EV biology deepens, these vesicles are poised to shift from observational biomarkers to active agents of precision therapy, underscoring their potential to reshape the landscape of personalized medicine.

Key words Extracellular vesicles, Intercellular communication, Biomarkers, Therapeutic delivery, Disease modulation

Amit K. Srivastava (ed.), *Extracellular Vesicles*, Neuromethods, vol. 228, https://doi.org/10.1007/978-1-0716-4905-3_1,

1 Introduction

Extracellular vesicles (EVs) are nanoscale lipid-bound particles secreted by almost all cell types into the extracellular environment [1]. Over the past few decades, EVs have garnered significant attention due to their involvement in intercellular communication, their role in physiological processes, and their potential as biomarkers and therapeutic tools [2]. EVs are heterogeneous in terms of size, content, and function, and they carry bioactive molecules such as proteins, lipids, RNA, and metabolites, which facilitate cellular interactions and modulate recipient cells' functions [3, 4] (Fig. 1). Advancements in EV isolation technologies such as microfluidics platforms and affinity-based capture methods have significantly enhanced the yield and purity of EVs uplifting their translational potential in clinical diagnostics [5]. Due to their natural targeting capabilities and low immunogenicity, engineered EVs are now being explored as precision delivery vehicles for RNA therapeutics and CRISPR-Cas9 systems [6]. Advanced research has uplifted the involvement of EVs in immune modulation, viral transmission, and

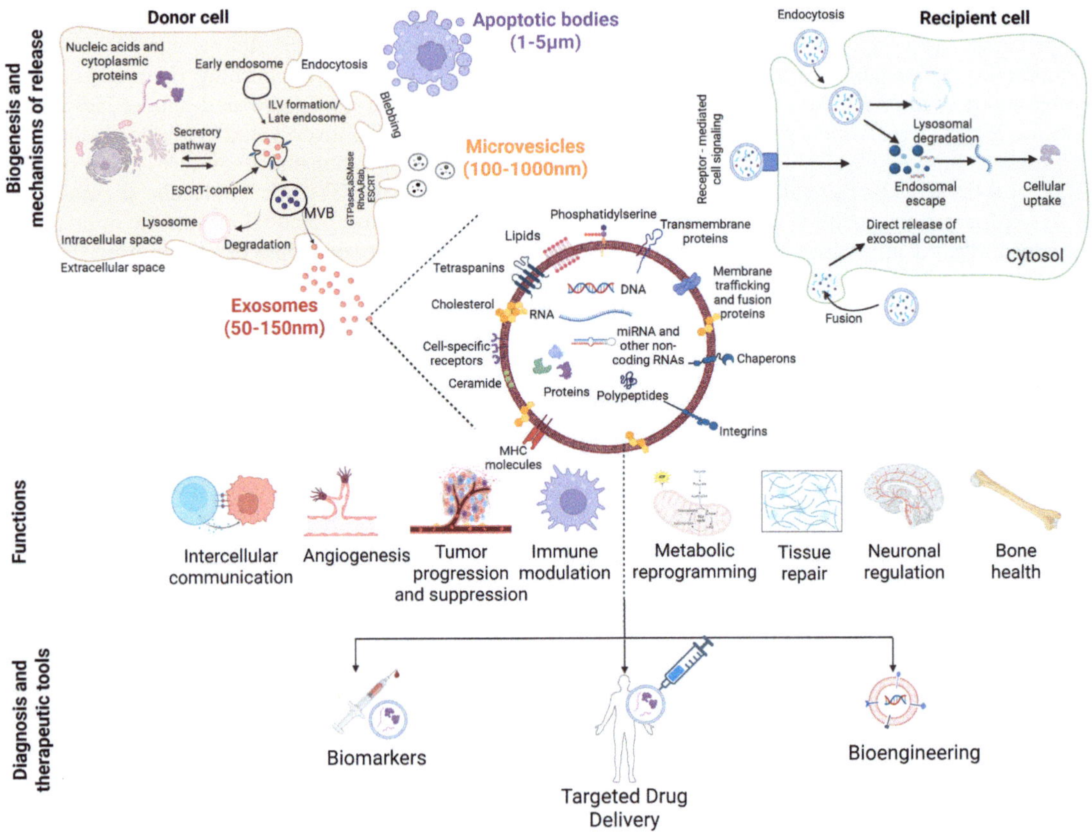

Fig. 1 Overview of extracellular vesicles: biogenesis, classifications, functions, and therapeutic applications

horizontal gene transfer [7]. This chapter provides an overview of the biology, classification, mechanisms of release, functions, and the growing interest in EVs as diagnostic and therapeutic agents.

2 Classification and Types of Extracellular Vesicles

EVs are typically classified based on their size, biogenesis, and cellular origin. The three main types of EVs include exosomes, microvesicles, and apoptotic bodies. Each type of EV differs in its size, cargo, and mechanism of release [8–11].

2.1 Exosomes

Exosomes are small EVs ranging from 30 to 150 nm in diameter. They are formed within the endosomal compartment and are released upon the fusion of multivesicular bodies (MVBs) with the plasma membrane [12]. Exosomes contain a variety of bioactive molecules, including proteins (tetraspanins, heat shock proteins), lipids (sphingolipids, phosphatidylserine), and RNA species (mRNA, miRNA, long non-coding RNAs) [13, 14]. Exosomes have been shown to play a key role in intercellular communication by transferring their cargo to target cells, influencing their behavior and function [15]. Additionally, exosomes are reported to selectively package CRISPR-Cas constituents, favoring gene editing therapies with reduced immunogenicity [16]. Also, exosome profiling integrated with machine learning technologies has been shown to improve the accuracy of cancer diagnostics by employing protein signatures and RNA cargo analysis [17].

2.2 Microvesicles

Microvesicles, also known as ectosomes or shedding vesicles, are larger in size than exosomes, ranging from 100 to 1000 nm. They are formed directly from the outward budding of the plasma membrane [18]. Like exosomes, microvesicles contain proteins, lipids, and nucleic acids, but their cargo profile may differ due to their distinct biogenesis pathway [19, 20]. Microvesicles are involved in various physiological processes, including coagulation, immune responses, and tissue remodeling [21–23]. Furthermore, microvesicles are also known to be implicated in interorgan communication, specifically in metabolic diseases, where the adipose tissue-derived microvesicles have been reported to regulate insulin resistance in liver and muscle in vitro [24]. In addition, engineered therapeutic microvesicles have been demonstrated to deliver anti-inflammatory molecules in autoimmune conditions like rheumatoid arthritis [25].

2.3 Apoptotic Bodies

Apoptotic bodies are the largest EVs, typically ranging from 1 to 5 μm. They are released during the process of apoptosis (programmed cell death) and are composed of fragmented cellular material, including organelles, cytoplasm, and nuclear remnants

[26, 27]. Apoptotic bodies serve as a mechanism for the clearance of dying cells and prevent the release of potentially harmful intracellular components into the extracellular environment [28]. Recent studies have highlighted the active role of apoptotic bodies in immune tolerance and inflammation [29]. In cancer, tumor-derived apoptotic bodies are known to modulate the tumor microenvironment by transferring regulatory RNAs and oncogenic DNA to the neighboring cells [30].

3 Biogenesis and Mechanisms of Release

The formation and release of EVs are complex, involving several cellular processes that are tightly regulated.

3.1 Exosome Biogenesis

Exosomes are generated from endosomal compartments known as multivesicular bodies (MVBs) [31]. During the maturation of MVBs, intraluminal vesicles (ILVs) are formed by the inward budding of the limiting membrane of the MVB. These ILVs eventually become exosomes [32]. The formation of ILVs involves the endosomal sorting complex required for transport (ESCRT) machinery, although ESCRT-independent pathways have also been described [33]. After the MVBs mature, they can either fuse with lysosomes for degradation or with the plasma membrane to release the exosomes into the extracellular space [34]. Recent studies have unveiled YBX1 as a key regulator of exosome biogenesis by facilitating selective miRNA packaging [35]. Furthermore, mitochondrial metabolites such as succinate and fumarate have been found to impact exosome production by modifying endolysosomal pH and vesicle trafficking [36].

3.2 Microvesicle Biogenesis

Microvesicles are generated by direct outward budding of the plasma membrane. This process is regulated by changes in membrane lipid composition and cytoskeletal dynamics [37]. The membrane lipid asymmetry is disrupted, leading to the exposure of phosphatidylserine on the outer leaflet of the membrane, which is a hallmark of microvesicle formation [38]. The release of microvesicles is often facilitated by small GTPases such as RhoA and Rab proteins, which modulate the cytoskeleton and membrane trafficking events [39]. Recent findings highlight calcium signaling and lipid-modifying enzymes like acid sphingomyelinase (aSMase) as critical regulators of microvesicle shedding under stress or inflammation [40]. Moreover, proteomic analyses show that microvesicles from senescent cells carry SASP-related proteins, linking them to aging and chronic inflammation [41].

3.3 Regulation of EV Release

The release of EVs is regulated by multiple signaling pathways and can be influenced by various factors such as cellular stress, inflammation, or stimuli from the microenvironment [42]. For example, stress signals such as oxidative stress or hypoxia can upregulate the release of EVs. Additionally, EV release is often tightly regulated to ensure that the cargo is delivered to the appropriate target cells in a controlled manner [43]. Advance research unveils the role of purinergic signaling on EV secretion via P2X7 receptor activation mediated by extracellular ATP [44]. In addition, EV release is efficiently linked to metabolic cycles and circadian rhythms, speculating biologically programmed mechanism for optimal cellular communication [45, 46].

4 Functions of Extracellular Vesicles

EVs serve a variety of functions in cellular and physiological processes, including intercellular communication [47, 48], immune modulation [23, 49], and disease pathogenesis [50, 51].

4.1 Intercellular Communication

One of the primary roles of EVs is to mediate communication between cells. By carrying a variety of bioactive molecules, including proteins, lipids, and RNA, EVs can transfer molecular signals from one cell to another [47, 48, 52]. This enables cells to exchange information over short and long distances, influencing processes such as immune responses, tissue repair, and metabolic regulation [15]. Exosomes, in particular, are involved in long-range signaling, as they can be taken up by distant target cells, thereby altering their gene expression, protein activity, or cellular behavior [53]. Exercise-mediated skeletal muscle-derived EVs are shown to carry myokines that influence metabolic activity in the liver and brain [54].

4.2 Immune Modulation

EVs play an essential role in regulating immune responses. They can carry antigens, cytokines, and other immune modulators that influence immune cell function [23, 49, 55]. For example, exosomes derived from dendritic cells are capable of presenting antigens to T cells, thus initiating adaptive immune responses [56]. Conversely, tumor-derived EVs may contain immunosuppressive molecules that inhibit immune responses and promote tumor immune evasion [57]. Macrophage-derived EVs enriched in miR155 was shown to promote M1 polarization, intensifying inflammation in rheumatoid arthritis [58]. EVs are also involved in the regulation of inflammation and autoimmunity by modulating the activity of both innate and adaptive immune cells [59, 60].

4.3 Pathological Roles

While EVs play essential roles in maintaining cellular harmony under normal conditions, they also serve as critical mediators of disease. In cancer, tumor-derived EVs act as facilitators of metastasis, transporting oncogenic proteins and RNA species that promote the spread of malignant cells [61]. Glioblastoma-derived EVs carrying mutant IDH1 proteins alters the epigenetic landscape of surrounding astrocytes, promoting tumorigenesis [62]. Pancreatic cancer cell derived EVs have demonstrated the ability to establish a premetastatic niche in the liver via recruiting immunosuppressive myeloid cells and upregulating fibronectin deposition [63]. These vesicles are not mere bystanders in the cancer microenvironment, they actively contribute to reshaping the immune response, fostering an environment conducive to tumor survival, and even "preparing" distant organs for metastatic colonization [50]. This ability to modulate the immune landscape, by carrying immunosuppressive molecules or tumor antigens, highlights EVs as active participants in the tumor's survival strategy, making them invaluable targets for early detection and therapeutic intervention [57]. In neurodegenerative diseases like Alzheimer's and Parkinson's, EVs facilitate the spread of neurotoxic proteins such as amyloid-beta and alpha-synuclein, driving the pathological cascade that leads to neuronal damage [64–66]. Here, EVs act as vectors, propagating toxic aggregates across neuronal networks, which raises the tantalizing possibility of using EVs as biomarkers for these diseases. By detecting specific protein signatures in EVs, we may be able to diagnose neurodegeneration long before symptoms manifest, offering a window for early intervention [67].

Even in autoimmune diseases, EVs take center stage, influencing immune cell behavior and contributing to disease progression. For example, EVs derived from activated T cells or dendritic cells can present antigens to other immune cells, influencing the direction of immune responses [68]. Conversely, in diseases such as systemic lupus erythematosus (SLE), EVs have been implicated in promoting autoimmunity by delivering self-antigens to immune cells, triggering harmful immune activation [69, 70]. In all these cases, EVs are not passive bystanders—they actively shape the trajectory of disease, providing a crucial link between molecular mechanisms and clinical outcomes.

5 Extracellular Vesicles as Diagnostic and Therapeutic Tools

The potential of EVs as diagnostic tools and therapeutic vehicles has generated significant interest in recent years.

5.1 Diagnostic Biomarkers

One of the most compelling applications of EVs lies in their potential as diagnostic tools [71]. As a mirror of the cellular state, EVs carry molecular signatures that can reveal much about the health of the body. The analysis of these signatures, particularly in biofluids such as blood, urine, or saliva, offers a noninvasive method of diagnosing disease, potentially even before the onset of clinical symptoms [72]. This is especially powerful in the context of cancer, where exosomes released by tumor cells carry tumor-specific RNA, proteins, and microRNAs. An unique exosomal signature identified in breast cancer patients enables early-stage detection through specific miRNAs [73]. The presence of these markers in the bloodstream or other bodily fluids could serve as a novel diagnostic approach, offering a "liquid biopsy"that complements or even replaces traditional tissue biopsy methods and significantly improves patient outcomes by allowing earlier intervention [74]. What makes this approach particularly appealing is its potential for monitoring disease progression and therapeutic response in real time. As tumors evolve, the molecular content of their EVs may shift, allowing clinicians to track the changing landscape of cancer biology without invasive procedures [75]. This real-time molecular snapshot could not only improve early detection but also enable the precise tailoring of treatment strategies. Moreover, in neurodegenerative diseases, the ability to detect disease-specific proteins in EVs circulating in the blood could revolutionize diagnostic timelines [76]. Exosomes carrying neurotoxic proteins such as tau and amyloid-beta in cerebrospinal fluid (CSF) of Alzheimer's patients serve as an early indicators of disease progression [77]. Increased levels of alpha-synuclein containing EVs in Parkinson's disease correlate with disease severity and progression [78]. These findings underscore the efficacy of EVs as a promising biomarker, enabling clinicians to identify patients in the earliest stages of disease when therapeutic interventions may still be effective.

5.2 Therapeutic Applications

The promise of EVs extends beyond diagnostics. Their capacity to transport and deliver biologically active molecules, including RNA and small molecules, makes them a compelling platform for targeted drug delivery [79]. Unlike traditional drug delivery systems, which often suffer from poor targeting and low efficiency, EVs offer natural biocompatibility and the ability to cross biological barriers, such as the blood-brain barrier, a major hurdle in treating neurological disorders [80]. This makes them ideal candidates for delivering therapeutics precisely where they are needed most. For instance, in cancer treatment, EVs can be engineered to deliver chemotherapeutic drugs directly to tumor cells, minimizing off-target effects and enhancing treatment efficacy [81, 82]. In addition, EVs offer an exciting opportunity to deliver RNA-based therapies, such as small interfering RNAs (siRNAs) or messenger RNAs (mRNAs), to target cells, enabling the correction of genetic defects or the silencing of harmful genes [6, 83]. This approach is particularly promising for genetic diseases where traditional methods of gene delivery face significant challenges.

Beyond cancer and genetic diseases, EVs could be harnessed to deliver immune-modulating molecules or vaccines, opening new avenues for immunotherapy [84]. Their natural ability to mimic cell membranes and facilitate targeted delivery places EVs at the forefront of a new era in therapeutic design, where treatments are not only personalized but also finely tuned to the molecular signatures of individual patients.

5.3 Challenges and Future Directions

While the scientific advancements in EV research are extraordinary, several challenges remain. The isolation and characterization of EVs are still in their infancy, and the current methods for EV separation are not always precise or reproducible. There is also a need for standardization in the way EVs are analyzed, both in the laboratory and in clinical settings. Furthermore, for EVs to reach their full therapeutic potential, more research is required to understand the optimal methods for loading, targeting, and delivering therapeutic cargo. Despite these challenges, the momentum behind EV research is undeniable. With the rapid advances in nanotechnology, molecular biology, and high-throughput analysis, the ability to harness EVs for diagnostic and therapeutic applications is within reach. By addressing these challenges, we stand at the precipice of a new era in medicine, where EVs can redefine how we detect, treat, and manage diseases.

6 Conclusion

Extracellular vesicles are poised to transform the landscape of medicine. Their critical role in intercellular communication, their potential as diagnostic biomarkers, and their promise as therapeutic delivery systems make them one of the most exciting areas of biomedical research. The scientific community is on the verge of unlocking the full potential of EVs, which could lead to groundbreaking advances in early detection, treatment, and prevention of a wide array of diseases. As we continue to unravel the complexities of EV biology, the clinical applications of these vesicles hold the promise of revolutionizing how we approach patient care, ushering in a new era of precision medicine.

Acknowledgments

This study was supported by the US Department of Defense awards (HT94252310138 and W81XWH2110682) to Amit K. Srivastava.

References

1. Doyle LM, Wang MZ (2019) Overview of extracellular vesicles, their origin, composition, purpose, and methods for exosome isolation and analysis. Cells 8:727
2. Ciferri MC, Quarto R, Tasso R (2021) Extracellular vesicles as biomarkers and therapeutic tools: from pre-clinical to clinical applications. Biology 10:359
3. Di Bella M (2022) Overview and update on extracellular vesicles: considerations on exosomes and their application in modern medicine. Biology 11(6):804
4. Welsh JA, Goberdhan DC, O'Driscoll L, Buzas EI, Blenkiron C, Bussolati B, Cai H, Di Vizio D, Driedonks TA, Erdbrügger U (2024) Minimal information for studies of extracellular vesicles (MISEV2023): from basic to advanced approaches. J Extracell Vesicles 13:e12404
5. Zhang S, Deng J, Li J, Tian F, Liu C, Fang L, Sun J (2022) Advanced microfluidic technologies for isolating extracellular vesicles. TrAC Trends Anal Chem 157:116817
6. Muskan M, Abeysinghe P, Cecchin R, Branscome H, Morris KV, Kashanchi F (2024) Therapeutic potential of RNA-enriched extracellular vesicles: the next generation in RNA delivery via biogenic nanoparticles. Mol Ther 32:2939–2949
7. Kojima R, Bojar D, Rizzi G, Hamri GC-E, El-Baba MD, Saxena P, Ausländer S, Tan KR, Fussenegger M (2018) Designer exosomes produced by implanted cells intracerebrally deliver therapeutic cargo for Parkinson's disease treatment. Nat Commun 9:1305
8. Van Niel G, d'Angelo G, Raposo G (2018) Shedding light on the cell biology of extracellular vesicles. Nat Rev Mol Cell Biol 19:213–228
9. van Niel G, Carter DR, Clayton A, Lambert DW, Raposo G, Vader P (2022) Challenges and directions in studying cell–cell communication by extracellular vesicles. Nat Rev Mol Cell Biol 23:369–382
10. Couch Y, Buzàs EI, Di Vizio D, Gho YS, Harrison P, Hill AF, Lötvall J, Raposo G, Stahl PD, Théry C (2021) A brief history of nearly EV-erything–the rise and rise of extracellular vesicles. J Extracell Vesicles 10:e12144
11. Théry C, Witwer KW, Aikawa E, Alcaraz MJ, Anderson JD, Andriantsitohaina R, Antoniou A, Arab T, Archer F, Atkin-Smith GK (2018) Minimal information for studies of extracellular vesicles 2018 (MISEV2018): a position statement of the International Society for Extracellular Vesicles and update of the MISEV2014 guidelines. J Extracell Vesicles 7: 1535750
12. Crescitelli R, Lässer C, Lötvall J (2021) Isolation and characterization of extracellular vesicle subpopulations from tissues. Nat Protoc 16: 1548–1580
13. Qin J, Xu Q (2014) Functions and application of exosomes. Acta Pol Pharm 71:537–543
14. Chavda VP, Pandya A, Kumar L, Raval N, Vora LK, Pulakkat S, Patravale V, Duo Y, Tang BZ (2023) Exosome nanovesicles: a potential carrier for therapeutic delivery. Nano Today 49: 101771
15. Gurung S, Perocheau D, Touramanidou L, Baruteau J (2021) The exosome journey: from biogenesis to uptake and intracellular signalling. Cell Commun Signal 19:47
16. Balaraman AK, Babu MA, Moglad E, Mandaliya V, Rekha M, Gupta S, Prasad GS, Kumari M, Chauhan AS, Ali H (2024) Exosome-mediated delivery of crispr-Cas9: a revolutionary approach to cancer gene editing. Pathol Res Pract 266:155785
17. Lin X, Zhu J, Shen J, Zhang Y, Zhu J (2024) Advances in exosome plasmonic sensing: device integration strategies and AI-aided diagnosis. Biosens Bioelectron 266:116718
18. Cocucci E, Meldolesi J (2015) Ectosomes and exosomes: shedding the confusion between extracellular vesicles. Trends Cell Biol 25: 364–372
19. Clancy JW, Schmidtmann M, D'Souza-Schorey C (2021) The ins and outs of microvesicles. FASEB BioAdv 3:399–406
20. Dai YD, Dias P (2018) Exosomes or microvesicles, a secreted subcellular organelle contributing to inflammation and diabetes. Diabetes 67:2154–2156
21. Yuana Y, Sturk A, Nieuwland R (2013) Extracellular vesicles in physiological and pathological conditions. Blood Rev 27:31–39
22. Zhang Y, Wu D, Zhou C, Bai M, Wan Y, Zheng Q, Fan Z, Wang X, Yang C (2024) Engineered extracellular vesicles for tissue repair and regeneration. Burns Trauma 12: tkae062
23. Aloi N, Drago G, Ruggieri S, Cibella F, Colombo P, Longo V (2024) Extracellular vesicles and immunity: at the crossroads of cell communication. Int J Mol Sci 25:1205
24. Mladenović D, Vesković M, Šutulović N, Hrnčić D, Stanojlović O, Radić L, Macut JB, Macut D (2024) Adipose-derived extracellular vesicles–a novel cross-talk mechanism in insulin

resistance, non-alcoholic fatty liver disease, and polycystic ovary syndrome. Endocrine 85:18–34

25. Lee ES, Ko H, Kim CH, Kim H-C, Choi S-K, Jeong SW, Lee S-G, Lee S-J, Na H-K, Park JH (2023) Disease-microenvironment modulation by bare-or engineered-exosome for rheumatoid arthritis treatment. Biomater Res 27:81
26. Santavanond JP, Rutter SF, Atkin-Smith GK, Poon IK (2021) Apoptotic bodies: mechanism of formation, isolation and functional relevance. New Front Extracell Vesicles 97:61–88
27. Kakarla R, Hur J, Kim YJ, Kim J, Chwae Y-J (2020) Apoptotic cell-derived exosomes: messages from dying cells. Exp Mol Med 52:1–6
28. Han M, Ryu G, Shin S-A, An J, Kim H, Park D, Lee D-H, Lee CS (2021) Physiological roles of apoptotic cell clearance: beyond immune functions. Life 11:1141
29. Khalilzad MA, Mohammadi J, Amirsaadat S, Najafi S, Zare S, Nilforoushzadeh MA, Khalilzad M, Amirkhani MA, Peyrovan A, Khalili SFS (2025) Therapeutic potential of apoptotic vesicles in modulating inflammation, immune responses, and tissue regeneration. J Nanobiotechnol 23:260
30. Valcz G, Újvári B, Buzás EI, Krenács T, Spisák S, Kittel Á, Tulassay Z, Igaz P, Takács I, Molnár B (2022) Small extracellular vesicle DNA-mediated horizontal gene transfer as a driving force for tumor evolution: facts and riddles. Front Oncol 12:945376
31. Gurunathan S, Kang M-H, Qasim M, Khan K, Kim J-H (2021) Biogenesis, membrane trafficking, functions, and next generation nanotherapeutics medicine of extracellular vesicles. Int J Nanomedicine 16:3357–3383
32. Lee YJ, Shin KJ, Chae YC (2024) Regulation of cargo selection in exosome biogenesis and its biomedical applications in cancer. Exp Mol Med 56:877–889
33. Krylova SV, Feng D (2023) The machinery of exosomes: biogenesis, release, and uptake. Int J Mol Sci 24:1337
34. Xu M, Ji J, Jin D, Wu Y, Wu T, Lin R, Zhu S, Jiang F, Ji Y, Bao B (2023) The biogenesis and secretion of exosomes and multivesicular bodies (MVBs): intercellular shuttles and implications in human diseases. Genes Dis 10:1894–1907
35. Liu X-M, Ma L, Schekman R (2021) Selective sorting of microRNAs into exosomes by phase-separated YBX1 condensates. elife 10:e71982
36. Zhang Y, Tan J, Miao Y, Zhang Q (2021) The effect of extracellular vesicles on the regulation of mitochondria under hypoxia. Cell Death Dis 12:358
37. Akers JC, Gonda D, Kim R, Carter BS, Chen CC (2013) Biogenesis of extracellular vesicles (EV): exosomes, microvesicles, retrovirus-like vesicles, and apoptotic bodies. J Neuro-Oncol 113:1–11
38. Pollet H, Conrard L, Cloos A-S, Tyteca D (2018) Plasma membrane lipid domains as platforms for vesicle biogenesis and shedding? Biomolecules 8:94
39. Rädler J, Gupta D, Zickler A, Andaloussi SE (2023) Exploiting the biogenesis of extracellular vesicles for bioengineering and therapeutic cargo loading. Mol Ther 31:1231–1250
40. Colombo F, Bastoni M, Nigro A, Podini P, Finardi A, Casella G, Ramesh M, Farina C, Verderio C, Furlan R (2018) Cytokines stimulate the release of microvesicles from myeloid cells independently from the P2X7 receptor/acid sphingomyelinase pathway. Front Immunol 9:204
41. Hou J, Chen K-X, He C, Li X-X, Huang M, Jiang Y-Z, Jiao Y-R, Xiao Q-N, He W-Z, Liu L (2024) Aged bone marrow macrophages drive systemic aging and age-related dysfunction via extracellular vesicle-mediated induction of paracrine senescence. Nat Aging 4:1562–1581
42. Crewe C (2023) Energetic stress-induced metabolic regulation by extracellular vesicles. Compr Physiol 13:5051
43. Chiaradia E, Tancini B, Emiliani C, Delo F, Pellegrino RM, Tognoloni A, Urbanelli L, Buratta S (2021) Extracellular vesicles under oxidative stress conditions: biological properties and physiological roles. Cells 10:1763
44. Carotti V, Rigalli JP, van Asbeck-van der Wijst J, Hoenderop JG (2022) Interplay between purinergic signalling and extracellular vesicles in health and disease. Biochem Pharmacol 203:115192
45. Khalyfa A, Gaddameedhi S, Crooks E, Zhang C, Li Y, Qiao Z, Trzepizur W, Kay SA, Andrade J, Satterfield BC (2020) Circulating exosomal miRNAs signal circadian misalignment to peripheral metabolic tissues. Int J Mol Sci 21:6396
46. Yeung C-YC, Dondelinger F, Schoof EM, Georg B, Lu Y, Zheng Z, Zhang J, Hannibal J, Fahrenkrug J, Kjaer M (2022) Circadian regulation of protein cargo in extracellular vesicles. Sci Adv 8:eabc9061
47. Ståhl A-L, Johansson K, Mossberg M, Kahn R, Karpman D (2019) Exosomes and microvesicles in normal physiology, pathophysiology, and renal diseases. Pediatr Nephrol 34:11–30
48. Berumen Sánchez G, Bunn KE, Pua HH, Rafat M (2021) Extracellular vesicles: mediators of

intercellular communication in tissue injury and disease. Cell Commun Signal 19:1–18
49. Buzas EI (2023) The roles of extracellular vesicles in the immune system. Nat Rev Immunol 23:236–250
50. Kumar MA, Baba SK, Sadida HQ, Marzooqi SA, Jerobin J, Altemani FH, Algehainy N, Alanazi MA, Abou-Samra A-B, Kumar R (2024) Extracellular vesicles as tools and targets in therapy for diseases. Signal Transduct Target Ther 9:27
51. Amin S, Massoumi H, Tewari D, Roy A, Chaudhuri M, Jazayerli C, Krishan A, Singh M, Soleimani M, Karaca EE (2024) Cell type-specific extracellular vesicles and their impact on health and disease. Int J Mol Sci 25:2730
52. Salomon C, Das S, Erdbrügger U, Kalluri R, Kiang Lim S, Olefsky JM, Rice GE, Sahoo S, Andy Tao W, Vader P (2022) Extracellular vesicles and their emerging roles as cellular messengers in endocrinology: an endocrine society scientific statement. Endocr Rev 43:441–468
53. Shah N, Ishii M, Brandon C, Ablonczy Z, Cai J, Liu Y, Chou CJ, Rohrer B (2018) Extracellular vesicle-mediated long-range communication in stressed retinal pigment epithelial cell monolayers. Biochimica et Biophysica Acta (BBA)-Mol Basis Dis 1864:2610–2622
54. Whitham M, Parker BL, Friedrichsen M, Hingst JR, Hjorth M, Hughes WE, Egan CL, Cron L, Watt KI, Kuchel RP (2018) Extracellular vesicles provide a means for tissue crosstalk during exercise. Cell Metab 27:237–251. e234
55. Robbins PD, Morelli AE (2014) Regulation of immune responses by extracellular vesicles. Nat Rev Immunol 14:195–208
56. Matsuzaka Y, Yashiro R (2022) Regulation of extracellular vesicle-mediated immune responses against antigen-specific presentation. Vaccine 10:1691
57. Kuang L, Wu L, Li Y (2025) Extracellular vesicles in tumor immunity: mechanisms and novel insights. Mol Cancer 24:45
58. Lin Y, Wang Z, Liu S, Liu J, Zhang Z, Ouyang Y, Su Z, Chen D, Guo L, Luo T (2024) Roles of extracellular vesicles on macrophages in inflammatory bone diseases. Mol Cell Biochem 479:1401–1414
59. Abebaw D, Akelew Y, Adugna A, Teffera ZH, Tegegne BA, Fenta A, Selabat B, Amare GA, Getinet M, Jemal M (2024) Extracellular vesicles: immunomodulation, diagnosis, and promising therapeutic roles for rheumatoid arthritis. Front Immunol 15:1499929
60. Wu W-C, Song S-J, Zhang Y, Li X (2020) Role of extracellular vesicles in autoimmune pathogenesis. Front Immunol 11:579043
61. Bao Q, Huang Q, Chen Y, Wang Q, Sang R, Wang L, Xie Y, Chen W (2022) Tumor-derived extracellular vesicles regulate cancer progression in the tumor microenvironment. Front Mol Biosci 8:796385
62. Zeng A, Wei Z, Rabinovsky R, Jun HJ, El Fatimy R, Deforzh E, Arora R, Yao Y, Yao S, Yan W (2020) Glioblastoma-derived extracellular vesicles facilitate transformation of astrocytes via reprogramming oncogenic metabolism. Iscience 23:101420
63. Li D, Chu X, Ma Y, Zhang F, Tian X, Yang Y, Yang Y (2024) Tumor-derived exosomes: unravelling the pathogenesis of pancreatic cancer with liver metastases and exploring the potential for clinical translation. Cancer Lett 217403:217403
64. Dutta S, Hornung S, Taha HB, Bitan G (2023) Biomarkers for parkinsonian disorders in CNS-originating EVs: promise and challenges. Acta Neuropathol 145:515–540
65. McAllister BB, Lacoursiere SG, Sutherland RJ, Mohajerani MH (2020) Intracerebral seeding of amyloid-β and tau pathology in mice: factors underlying prion-like spreading and comparisons with α-synuclein. Neurosci Biobehav Rev 112:1–27
66. Gomes P, Tzouanou F, Skolariki K, Vamvaka-Iakovou A, Noguera-Ortiz C, Tsirtsaki K, Waites CL, Vlamos P, Sousa N, Costa-Silva B (2022) Extracellular vesicles and Alzheimer's disease in the novel era of precision medicine: implications for disease progression, diagnosis and treatment. Exp Neurol 358:114183
67. Chen J, Tian C, Xiong X, Yang Y, Zhang J (2025) Extracellular vesicles: new horizons in neurodegeneration. EBioMedicine 113: 105605
68. Wang Y, Li C, Wu F, Mao J, Zhu J, Xie H, Zhou X, Wen C, Tian J (2024) The negative effects of extracellular vesicles in the immune system. Front Immunol 15:1410273
69. Pan L, Lu M-P, Wang J-H, Xu M, Yang S-R (2020) Immunological pathogenesis and treatment of systemic lupus erythematosus. World J Pediatr 16:19–30
70. Radhwi ZO, Laswad BMB, AlSulami AA, Bahakeem FY, Bawazer SM, Bajodah AA, Khairo YM, Hafiz WA, Bahakeem FY Sr, Bajodah A (2023) Knowledge and awareness of systemic lupus erythematosus among the population of Makkah, Saudi Arabia: a cross-sectional study. Cureus 15:e41294

71. Carney RP, Mizenko RR, Bozkurt BT, Lowe N, Henson T, Arizzi A, Wang A, Tan C, George SC (2025) Harnessing extracellular vesicle heterogeneity for diagnostic and therapeutic applications. Nat Nanotechnol 20: 14–25
72. Gualerzi A, Picciolini S, Carlomagno C, Roda F, Bedoni M (2021) Biophotonics for diagnostic detection of extracellular vesicles. Adv Drug Deliv Rev 174:229–249
73. Zhao J, Xu L, Yang D, Tang H, Chen Y, Zhang X, Xu Y, Ou R, Li D (2022) Exosome-driven liquid biopsy for breast cancer: recent advances in isolation, biomarker identification and detection. Extracell Vesicle 1:100006
74. Wu X (2022) Advances in clinical liquid biopsy: biofluid extracellular vesicles as promising sources for drug metabolism and efficacy surveillance. Purdue University Graduate School
75. Lawrence SR, Shah KM (2024) Prospects and current challenges of extracellular vesicle-based biomarkers in cancer. Biology 13:694
76. Dehghani S, Ocakcı O, Hatipoglu PT, Özalp VC, Tevlek A (2025) Exosomes as biomarkers and therapeutic agents in neurodegenerative diseases: current insights and future directions. Mol Neurobiol 62:1–26
77. He A, Wang M, Li X, Chen H, Lim K, Lu L, Zhang C (2023) Role of exosomes in the pathogenesis and Theranostic of Alzheimer's disease and Parkinson's disease. Int J Mol Sci 24: 11054
78. Chan L, Chung C-C, Hsieh Y-C, Wu R-M, Hong C-T (2023) Plasma extracellular vesicle tau, β-amyloid, and α-synuclein and the progression of Parkinson's disease: a follow-up study. Ther Adv Neurol Disord 16: 17562864221150329
79. Lehmann TP, Golik M, Olejnik J, Łukaszewska M, Markowska D, Drożdżyńska M, Kotecki A, Głowacki M, Jagodziński PP (2023) Potential applications of using tissue-specific EVs in targeted therapy and vaccinology. Biomed Pharmacother 166: 115308
80. Yadav K, Vijayalakshmi R, Sahu KK, Sure P, Chahal K, Yadav R, Dubey A, Jha M, Pradhan M (2024) Exosome-based macromolecular neurotherapeutic drug delivery approaches in overcoming the blood-brain barrier for treating brain disorders. Eur J Pharm Biopharm 199:114298
81. Song H, Chen X, Hao Y, Wang J, Xie Q, Wang X (2022) Nanoengineering facilitating the target mission: targeted extracellular vesicles delivery systems design. J Nanobiotechnol 20: 431
82. Liu X, Xiao C, Xiao K (2023) Engineered extracellular vesicles-like biomimetic nanoparticles as an emerging platform for targeted cancer therapy. J Nanobiotechnol 21:287
83. Gil-Cabrerizo P, Simon-Yarza T, Garbayo E, Blanco-Prieto MJ (2024) Navigating the landscape of RNA delivery systems in cardiovascular disease therapeutics. Adv Drug Deliv Rev 208: 115302
84. Palma M (2025) Advancing breast cancer treatment: the role of immunotherapy and cancer vaccines in overcoming therapeutic challenges. Vaccine 13:344

Chapter 2

A Protocol to Capture the Extracellular Vesicle Surface (Surfaceome) Combining Biotinylation and Mass Spectrometry

Iasmin Inocencio, Alin Rai, and David W. Greening

Abstract

Extracellular vesicles (EVs) are cell-derived particles carrying cytoplasmic content enclosed by a lipid bilayer and range in diameter between 30 nm and several μm. EVs are broadly classified based on their size, categorized into large EVs (L-EVs, 100–1000 nm) and small EVs (S-EVs, 30–150 nm). Collectively, EVs function in mediating cell-to-cell communication through their cargo, which includes specific proteins (signaling molecules, receptors, integrins, and cytokines), bioactive lipids, and nucleic acids (mRNAs, microRNAs, long noncoding RNAs, and DNA). Proteins on the surface of EVs can help regulate the interaction of EVs with their microenvironment, influencing EV function, distribution in the body, time in circulation, and pharmacokinetics. The EV surface protein network (surfaceome) acts as a fundamental signaling gateway by bridging intra- and extracellular signaling networks and is a key resource of potential circulating biomarkers of health and disease and therapeutic targets for enhanced delivery and retention. Here, we outline a systematic "how to" protocol (along with useful insights/tips) to label and capture the EV surface using membrane-impermeant derivative of biotin and obtain the EV surfaceome using mass spectrometry. Further, we have applied this approach to isolated EV subtypes (small and large EVs) to understand the surface proteome of different EVs.

Key words Extracellular vesicles, Surfaceome, Surface proteins, Proteomics, Vesicle heterogeneity

1 Introduction

The extracellular vesicle (EV) surface protein network (defined surfaceome [1]) regulates the interaction of EVs with their extracellular environment (including recipient cells, cell matrix) [2, 3], cell function (e.g., mobilization [4], uptake/internalization [5]), dictate biodistribution [6], regulate immune escape (e.g., PD-L1 [7]), half-life in circulation [8], and pharmacokinetics [9, 10]. Moreover, given their orientation/localization, EV surface proteins are a source of potential disease biomarkers [11, 12], and suitable as EV-based diagnostic tools [13–15]. EV surface proteins also have functional and therapeutic potential in bone healing [16], cartilage

Amit K. Srivastava (ed.), *Extracellular Vesicles*, Neuromethods, vol. 228, https://doi.org/10.1007/978-1-0716-4905-3_2,

repair and regeneration [17], cardiac protection [18], and neuronal function [19]. Such surface proteins may be exploited to direct enhanced targeting and delivery to target sites (i.e., leading to higher therapeutic effect [20, 21]), facilitate enhanced cell uptake and intracellular cargo release [22], while further to enhance retention of EV-based therapies in specific regions [23, 24] (reviewed [25]). For example, CD47-enriched EVs loaded with chemotherapeutic nanoparticles evade phagocytosis, while a homing peptide promotes their retention by cells in tumor tissue [26], while other strategies can express specific surface ligands to simultaneously target tumor and immunosuppressive stromal cells [27].

Based on their size, EVs can be broadly categorized into large EVs (L-EVs, 100–1000 nm) and small EVs (S-EVs, 30–150 nm) [28, 29]. While several studies have cataloged role of S-EVs in many pathologies (e.g., cancer [30], cardiovascular disease [31], and neurodegenerative diseases [32]), L-EVs are gaining prominence for their signaling role, either complementing or contrasting S-EV function [33]. We have previously shown that L-EVs can be further segregated based on their varying densities, namely, L-EVs displaying buoyant density of 1.07–1.11 g/mL, which are distinct from L-EVs of relatively higher buoyant density (1.22–1.30 g/mL) [1].

The focus of this protocol is to describe the surface protein landscape of different types of EVs (S- and L-EVs, isolated based on differential ultracentrifugation and density-based fractionation) by using membrane-impermeant biotin to label surface proteins and perform proteomic-based profiling using mass spectrometry [1]. The workflow used in this study to define EV surfaceome using quantitative mass spectrometry-based proteomics is outlined in Fig. 1. By definition, the surfaceome encompasses all plasma membrane proteins with at least one amino-acid residue exposed to the extracellular space [34]. We initially isolated small and large EVs based on differential centrifugation combined with density (1.07–1.11 g/mL). For both EV subtypes, we mapped experimentally verified cell-surface proteins (Cell Surface Protein Atlas (CSPA)) [35] and surfaceome proteins based on cell surfaceome predictor SURFY [34] onto global EV protein profiles (*see* **Note 1**). Next, we verified their EV surface localization by enriching surface proteins using amine reactive, membrane-impermeant and thiol-cleavable Sulfo-NHS-SS-Biotin combined with mass spectrometry (*see* **Note 2**). Because EDTA can successfully chelate metal ions such as calcium to release surface-bound proteins without compromising EV integrity [36, 37] (*see* **Note 3**), we incubated EVs with EDTA to define the released proteins as the "EV surface-associated proteome" (*see* **Notes 4** and **5**). This led to the identification of 291 and 456 surface, and 122 and 67 surface-associated proteins in L-EVs and S-EVs, respectively (Fig. 2).

Collectively, this protocol details the EV surfaceome and heterogeneity in surface proteome between EV subtypes (*see* **Note 6**).

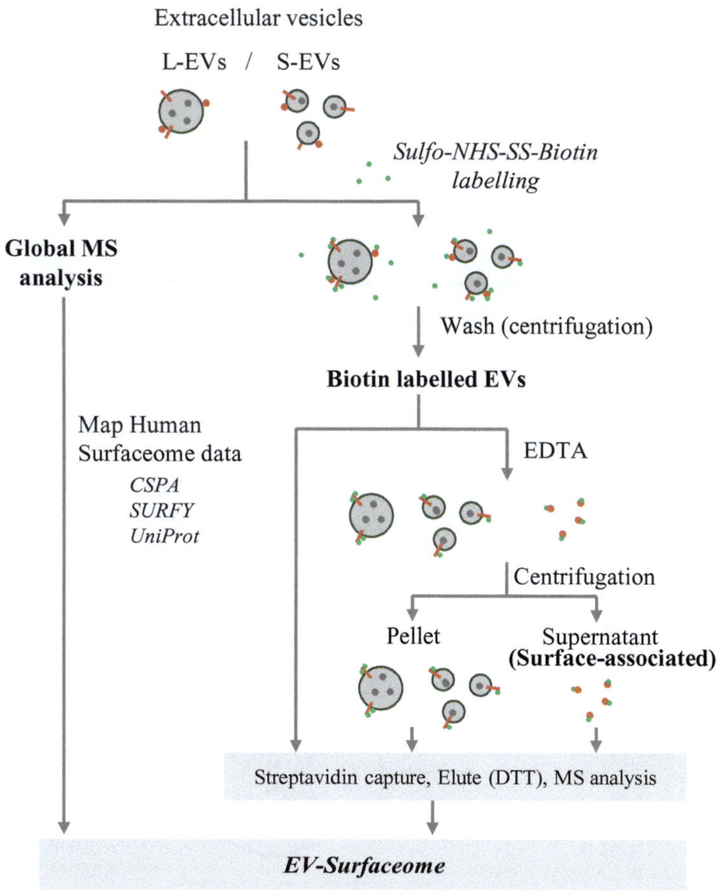

Fig. 1 Workflow for capturing surface proteins of EVs. Small and large EVs isolated using density-based centrifugation. From global mass spectrometry (MS) analysis of EVs, known/predicted cell-surface proteins were mapped. Surface proteins were captured using membrane-impermeant biotin and identified by MS. EV-surface associated proteins were obtained by treating EVs with EDTA and assessing released proteins

2 Materials

2.1 EVs from Cell Culture

1. Cell culture medium (e.g., RPMI-1640 (Life Technologies) supplemented with 5% (v/v) Fetal Bovine Serum (FBS, Life Technologies) and 1% (v/v) Penicillin Streptomycin (Pen/-Strep, Life Technologies).
2. Phosphate-buffered saline (PBS).
3. Trypsin-EDTA (Gibco).
4. Tissue culture plates (e.g., 10–15 cm^2)/flasks (e.g., T75–175).
5. Phosphate-buffered saline (PBS).
6. 50 mL polypropylene centrifuge tubes

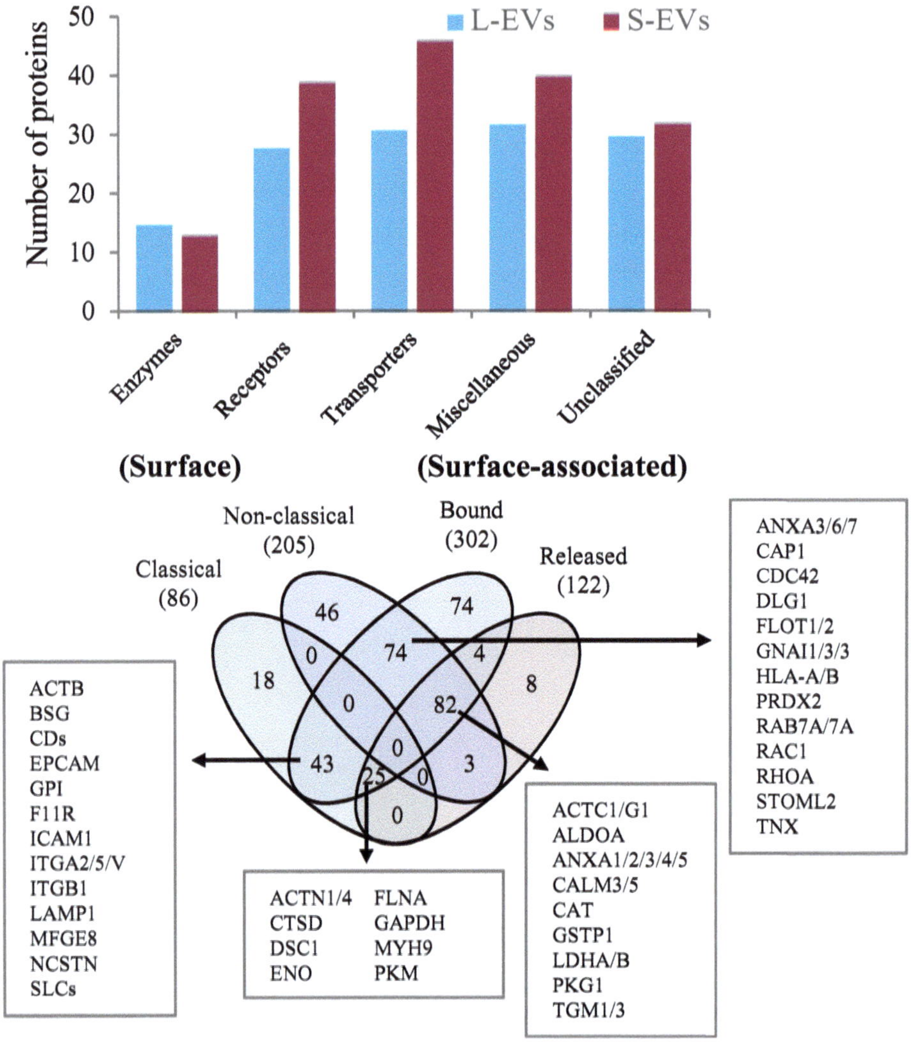

Fig. 2 EV surfaceome *Top panel*: Different classes of surface proteins identified in each EV-subtype (categorised based on SURFY database surface protein annotation). *Bottom panel:* Distribution of large EV surface proteins identified into (i) EV biotin surface proteome previously associated with cell surfaceome (i.e. classical cellular surfaceome) or non-CSPA/SURFY proteins (i.e. non-classical surface proteins) and (ii) EV surface associated following EDTA treatment that remain either bound or are released from EV surface. Gene names for each subset are indicated

7. Refrigerated centrifuge.
8. Incubator (37 °C with 5% CO_2).

2.2 EVs from Cell Bioreactor

1. CELLine adhere bioreactor flask (CELLine AD 1000, Integra Biosciences).

2. Cell culture medium (e.g., RPMI-1640, DMEM) with cell culture supplements, 1% (v/v) Pen/Strep, 5% (v/v) Fetal Bovine Serum (FBS, Life Technologies), 0.5% (v/v) insulin transferrin selenium (ITS, Invitrogen).
3. Incubator (37 °C with 5% CO_2).

2.3 Isolation of Large and Small EVs Using OptiPrep™ Density Gradient

1. Conditioned medium (i.e., 5–13 mL).
2. 50 mL Falcon tubes
3. Sterile/filtered PBS.
4. Optima XPN Ultracentrifuge and matched rotor.
5. OptiPrep™ stock solution: 60% (w/v) aqueous iodixanol.
6. SW28/SW32 Ti swinging-bucket (large scale spins) with 38.5 mL Open-Top Thin wall Ultra-Clear Tube, 25 × 89 mm.
7. TLA-55 fixed angle (small scale/washing) with 1.5 mL Polypropylene Tube with Snap-on Cap, 9.5 × 38 mm.

2.4 Protein Quantitation: Micro BCA Assay

1. MicroBCA™ Protein Assay Kit (23235, Thermo Fisher Scientific.
2. Bovine Serum Albumin (BSA) as standard protein. BSA purity >98%.
3. Working reagent (WR): 100 μL for each standard/sample. Mix 25 parts Reagent A, 24 parts Reagent B, 1 part Reagent C.
4. 1% (w/v) SDS in ultrapure water
5. Protein lysate sample for measurement.
6. 1.5 mL microcentrifuge tubes
7. Flat-bottom, clear 96-well plates.
8. Incubator (37 °C).
9. Microplate reader with 562 nm filter.

2.5 Immunoblotting

1. MicroBCA™ Protein Assay Kit (23235, Thermo Fisher Scientific (*see* Subheading 2.4).
2. iBlot 2 Dry Blotting System, Thermo Fisher Scientific.
3. 96-well Bio-Dot (Bio-Rad Laboratories).
4. EVs, cell lysate preparations (~10–20 μg protein).
5. Lysis buffer (50 mM HEPES (1% SDS)).
6. NuPAGE Bis-Tris Precast gels: 1 mm thickness, 10- or 12-well, 4–12% gradient.
7. 1× MES running buffer: 50 mM MES, 50 mM Tris Base, 0.1% SDS, 1 mM EDTA, pH 7.3. Do not use acid or base to adjust the pH.

8. XCell Surelock™ gel tank, with compatible power supply (e.g., Bio-Rad Laboratories, Hoefer, Thermo Fisher Scientific).
9. BenchMark™ Protein Ladder standard of known protein concentration: 1.7 μg/μL.
10. TTBS solution: 0.05% Tween® 20 in Tris-Buffered Saline (TBS).
11. Blocking buffer: 5% (w/v) skim milk powder in TTBS.
12. Orbital shaker.
13. Odyssey Infrared Imaging System, v3.0.
14. Primary antibodies (1:500 in TTBS); rabbit anti-MET (Santa Cruz Biotechnology), rabbit anti-CD63 (Santa Cruz), rabbit anti-ANXA1 (Abcam), rabbit anti-GAPDH (Cell Signalling), and rabbit anti-GFP (Abcam), mouse anti-ALIX (BD Biosciences) and mouse anti-TSG101 (BD Biosciences).
15. Secondary antibodies (1:15000 in TTBS): IRDye 800 goat anti-mouse IgG or IRDye 700 goat anti-rabbit IgG (LI-COR Biosciences).

2.6 Biophysical Characterization of EVs

2.6.1 Cryo-Electron Microscopy

1. Cryo-electron microscopy (Tecnai G2 F30).
2. EV preparations (~1–2 μg protein).
3. Aurion Protein-G gold 10 nm.
4. Sterile/filtered PBS.
5. Glow-discharged C-flat holey carbon grids.
6. Vitrobot or automated sample preparation device.
7. Liquid ethane.
8. Liquid nitrogen.
9. Gatan cryoholder.

2.6.2 Nanoparticle Tracking Analysis (NTA)

1. EV preparation (~1 μg protein).
2. NanoSight NS300 system or related single particle detection device.
3. Ultrapure water.
4. Disposable 1 mL syringe for sample loading.

2.7 Biotin Labelling and Capture of EVs

1. L-EVs or S-EVs (e.g., ~50 μg protein).
2. EZ-Link Sulfo-NHS-SS-Biotin (Pierce™ Cell Surface Biotinylation and Isolation Kit (A44390, Thermo Fisher Scientific).
3. Phosphate-buffered saline (PBS).
4. Tris-Buffered Saline (TBS).
5. 1.5 mL low protein bind Eppendorf tubes.

2.7.1 Separation of EV Surface Proteins

1. Biotinylated L-EVs or S-EVs (e.g., 100 μg) in PBS.
2. 0.5 M EDTA stock solution
3. Centrifuges (Eppendorf 5430R, TLA-55 rotor; Optima MAX-MP Tabletop Ultracentrifuge.
4. Lysis buffer: 2% SDS (1:100 protease cocktail inhibitor).

2.7.2 NeutrAvidin Purification of Biotinylated EV Surface Proteins

1. EV pellet (L-EVs or S-EVs) post EDTA incubation (see section.
2. Centrifuges (Eppendorf 5430R, TLA-55 rotor; Optima MAX-MP Tabletop Ultracentrifuge.
3. Phosphate-buffered saline (PBS).
4. 50% NeutrAvidin Agarose slurry (Pierce™ Cell Surface Biotinylation and Isolation Kit (A44390, Thermo Fisher Scientific)
5. Wash buffer (A44390, Thermo Fisher Scientific).
6. 1.5 mL safe-lock Eppendorf tubes
7. epTIPS (Eppendorf, 200 μL),
8. 20 μm pore size nylon net (NY2004700, Merck Millipore)
9. 1.5 mL low protein bind Eppendorf tubes
10. 10 mM DTT in 100 mM TEAB
11. Centrifuge (Eppendorf 5430R).

2.8 Proteomics: Sample Preparation

2.8.1 EV Surface Proteomic Sample Preparation

1. EV surface eluates in 10 mM DTT in 100 mM triethylamonium bicarbonate (TEAB).
2. 100 mM TEAB, pH 8.0, in LC-MS grade water. For pH adjustment use 1 M hydrochloride in LC-MS grade water.
3. 50 mM TEAB, pH 8.0, in LC-MS grade water. For pH adjustment use 1 M hydrochloride in LC-MS grade water.
4. 1 M iodoacetamide (IAA) stock solution in 100 mM TEAB, pH 8.0. Prepare fresh before the alkylation step. Prepare fresh before the alkylation step.
5. 500 mM (dithiothreitol (DTT) stock solution in 100 mM triethylamonium bicarbonate (TEAB) pH 8.0. Prepare fresh before the reduction step.
6. Trypsin (Promega, V5111) sequencing grade. Store lyophilized or frozen at −80 °C.
7. Formic acid (FA, LC-MS grade), prepare 2% (v/v) stock.
8. Water (LC-MS grade).
9. 1.5 mL Protein Lo-Bind Eppendorf tubes.
10. ThermoMixer (1.5 mL microfuge tube capacity).
11. Thermostat incubator at 37 °C.

12. pH strip, range 1–14.
13. Water Bath Sonicator.

StageTip Sample Cleanup

1. Empore SDB-RPS (Styrene Divinyl Benzene–Reversed Phase Sulfonate) solid-phase extraction disks for StageTip preparation.
2. 100% (v/v) acetone (LC grade). Stored at −20 °C.
3. 90% (v/v) acetone in water (LC grade). Stored at −20 °C.
4. SDB-RPS StageTip loading buffer: 1% (v/v) trifluoroacetic acid (TFA) in acetonitrile (ACN).
5. SDB-RPS StageTip wash buffer 1: 1% (v/v) TFA in ACN.
6. SDB-RPS StageTip wash buffer 2: 0.2% (v/v) TFA in 5% (v/v) ACN.
7. SDB-RPS StageTip elution buffer: 0.1% TFA in 30–80% (v/v) ACN. Elution buffer must be prepared fresh (within 1 h of use), because the pH will begin to increase due to its high volatility, thereby reducing its elution strength.
8. Water (LC-MS grade).
9. Vacuum centrifuge (lyophilizer).
10. MS loading buffer: 0.07% (v/v) TFA in LC-MS water. This buffer is stable for >6 months at RT.

2.9 Global Proteomic Sample Preparation of EVs

1. EVs (5 μg in 50 μL).
2. 100 mM HEPES buffer, pH 8.
3. Lysis Buffer 1 mL (2% SDS, 50 mM HEPES, pH 8).
4. 500 mM (dithiothreitol (DTT) stock solution in 100 mM HEPES pH 8.0. Prepare fresh before the reduction step.
5. 500 mM iodoacetamide (IAA) stock solution in 100 mM HEPES, pH 8.0. Prepare fresh solution before the alkylation step.
6. Magnetic beads were prepared by mixing SpeedBeads™ magnetic carboxylate-modified particles (65152105050250, 45152105050250, Cytiva) at 1:1 (v:v) ratio and washing twice with 200 μL MS-water. Magnetic beads were reconstituted to a final concentration of 100 μg/μL.
7. 100% ethanol (EA043, ChemSupply).
8. 80% ethanol (EA043, ChemSupply: MS water).
9. Digestion buffer: 50 mM TEAB and digested with trypsin (Promega, V5111) at a 1:50 enzyme-to-substrate ratio.
10. Formic acid (LC-MS grade) 100 %.
11. Water (LC-MS grade).

12. MS loading buffer: 0.07% (v/v) TFA in LC-MS water (This buffer is stable for >6 months at RT.).
13. 1.5 mL Protein Lo-Bind Eppendorf tubes.
14. ThermoMixer (1.5 mL microfuge tube capacity).
15. Thermostat incubator at 37 °C.
16. pH strip, range 1–14.
17. Water Bath Sonicator.

2.10 Fluorometric Peptide Assay

1. Fluorometric Peptide Assay (23290, Thermo Fisher Scientific).
2. Fluorescent-compatible, 96-well microplate (96-well black plate, e.g., 88378, Thermo Fisher Scientific).
3. Fluorescence plate reader compatible with Ex 390 nm/Em 475 nm.

2.11 UHPLC-MS/MS and Proteomic Analysis Parameters

1. MS loading buffer: 0.07% (v/v) TFA in LC-MS water.
2. Transparent MS sample vials, 300 μL: snap ring vial with PP insert.
3. UHPLC System coupled to a Q Exactive HF-X Hybrid Quadrupole-Orbitrap MS System (or a comparable high resolution/sensitivity mass spectrometry system).
4. Load column (Acclaim PepMap100 C18 5 μm beads with 100 Å pore-size).
5. Analytical/separation column (1.9-μm particle size C18, 0.075 × 250 mm, Nikkyo Technos Co. Ltd). (or a comparable high resolution nano-LC).
6. Mobile Phase A: 0.1% formic acid (FA).
7. Mobile Phase B: 0.1% formic acid in acetonitrile.

2.12 Bioinformatics and Statistics

1. MaxQuant *(v2.6.7.0 or similar/latest version) software.*
2. GraphPad Prism (v10.1.2 or latest).
3. Microsoft Excel.
4. Perseus software and statistical significance defined at $p < 0.05$.

3 Methods

3.1 EVs from Cell Culture

1. Culture cells to 60–70% confluency on 15 cm^2 tissue culture plate in suitable culture media with supplements as required: 5% (v/v) FCS, 1% (v/v) Pen/Strep at 37 °C with 5% CO_2 (*see* **Notes 7** and **8**).
2. Gently wash cells three times with 10 mL (prewarmed at 37 °C) of PBS.

3. Culture for 24–48 h in cell culture medium.
4. After 24–48 h incubation, collect conditioned medium (CM) into 50 mL polypropylene tubes and centrifuge: 480× *g* for 5 min followed by 2000× *g* for 10 min) at 4 °C to remove intact cells and cell debris. Carefully transfer the supernatant into a new tube and subject to desired exosome isolation protocol or store at −20 °C until further use. This supernatant contains both soluble and vesicle components.
5. Cell viability should be assessed in cell culture medium (i.e., 24 or 48 h incubation).

3.2 EVs from Cell Bioreactor

1. Bioreactor flasks contain a lower cell-cultivation chamber separated from an upper nutrient supply chamber by a semipermeable membrane with a molecular weight cut-off of 10 kDa. This membrane allows continuous diffusion of nutrients from the upper to the lower chamber, and waste elimination from the lower to the upper chamber. EVs released by cells are retained in the lower chamber.
2. Add 500 mL of prewarmed (37 °C) culture media with 5% (v/v) FCS, 1% (v/v) Pen/Strip) to the nutrient-supply chamber.
3. Prepare cells (>3–4 × 10 [7]) in 15 mL of complete culture media with 5% (v/v) FCS and 1% (v/v) Pen/Strep.
4. Transfer cells to lower cell-cultivation chamber and incubate for 72 h at 37 °C with 10% CO_2.
5. Remove the lower cell-cultivation chamber medium (the cells have now adhered to cultivation chamber) and wash the cultivation chamber three times with serum-free media.
6. Add 15 mL of media supplemented with 0.5% (v/v) ITS and 1% Pen/Strep to the lower cell-cultivation chamber.
7. The CM in the nutrient chamber is replaced every 4 days (cell type dependent).
8. The CM in the cultivation chamber nutrient chamber was collected every 2 days (cell type dependent).
9. Subject the CM from the cultivation chamber to EV isolation protocol.

3.3 Isolation of Large and Small EVs Using OptiPrep™ Density Gradient

1. Centrifuge CM (~30 mL) at 500× *g* for 5 min at 4 °C. Collect supernatant and centrifuge at 2000× *g* for 10 min at 4 °C. Retain the supernatant to be processed for EV isolation or stored at −20 °C until further use.
2. For large EV (L-EV) isolation, centrifuge the supernatant at 10,000× *g* (SW28 rotor; Optima XPN Ultracentrifuge,

Beckman Coulter) for 30 min at 4 °C to pellet crude L-EVs (*see* **Notes 9**).

3. For small EV (S-EV) isolation, Centrifuge the supernatant at 10,000× *g* (SW28 rotor; Optima XPN Ultracentrifuge, Beckman Coulter) for 30 min at 4 °C to pellet crude S-EVs. Centrifuge the supernatant at 100,000× *g* (41 Ti rotor; Optima XPN Ultracentrifuge) for 1 h at 4 °C to pellet crude S-EVs.
4. Resuspend EV pellet (L-EVs or S-EVs) in 200 μL sterile/filtered PBS or immediately store at −80 °C.
5. Prepare Opti Prep™ density gradient solutions by diluting a stock solution of 60% (w/v) OptiPrep™ with 0.25 M sucrose/10 mM Tris–HCl, pH 7.5, to obtain 40% (w/v), 20% (w/v), 10% (w/v), and 5% (w/v) solutions of iodixanol. These solutions are made immediately prior to centrifugation and allow up to 30 min mixing time.
6. To each 1.5 mL polypropylene tube, the gradient is formed by adding 3 mL of 40% iodixanol solution, followed by careful layering of 3 mL each of 20% and 10% solutions, and 2 mL of the 5% solution.
7. Overlay redispersed EVs (200 μL) on top of the discontinuous iodixanol gradient. For the control (blank) tube, overlay 500 μL of 0.25 M sucrose/PBS, pH 7.5, on top of the discontinuous iodixanol gradient.
8. Centrifuge at 100,000× *g* (SW41Ti swinging bucket) for 18 h at 4 °C.
9. After centrifugation, collect twelve individual 1 mL gradient fractions (with increasing density: top-bottom collection) using a 1 mL pipette. Isolation from the meniscus is imperative for this careful procedure.
10. Dilute fractions with 2 mL PBS and centrifuge at either 10,000× *g* (30 min, 4 °C, Eppendorf 5430R) or 100,000× *g* (1 h, 4 °C, TLA-55 rotor; Optima MAX-MP Tabletop Ultracentrifuge) to collect L-EV- and S-EV-containing fractions, respectively.
11. Determine the density of each fraction of the control OptiPrep™ gradient run in parallel. Collect 12 fractions from the control gradient, serially dilute 1:10,000 with distilled water, and determine the iodixanol concentration based on absorbance at 244 nm using a molar extinction coefficient of 320 $L.g^{-1}$ cm^{-1}.
12. Repeat **Step 10** to further wash EV pellet, redisperse in PBS, and quantitate protein amount.
13. Unused material is stored at −80 °C.

3.4 Protein Quantitation: Micro BCA Assay

1. Prepare a set of standard dilutions with BSA, starting with 200 μg/mL, performing dilutions in 1% SDS solution to 0.5 μg/mL and place them at RT (1% SDS solution can crystallize on ice).
2. Thaw protein sample (CM, EVs or protein lysates samples) and place on ice.
3. In a flat-bottom 96-well plate, load 100 μL of each standard BSA dilution/blank. The blank is dependent on what stock/lysis solution exosomes are prepared in.
4. Load 1–4 μL of CM, exosomes, or cell lysate samples into wells, making up the total volume to 100 μL using 1% SDS solution.
5. Prepare working reagent (WR) by mixing reagent A, B and C (as provided in the commercial kit) with the ratio of 25:24:1.
6. Add 100 μL of WR to each sample/standard.
7. Incubate at 37 °C for 1.5–2 h.
8. Cool plate to RT.
9. Measure absorbance at 562 nm.

3.5 Immunoblotting

1. Lyse the samples (~10–15 μg protein, *see* Subheading 2.3) in 50 mM HEPES (1% SDS) with or without DTT (depending on the primary antibody reducing/nonreducing compatibility, respectively) and heat for 5 min at 95 °C (closed cap, although cap can be opened for evaporation to reduce load volume).
2. Perform electrophoresis on lysed samples at 150 V for 1 h.
3. Following electrophoresis, electro-transfer proteins onto nitrocellulose membranes using iBlot™ 2 Dry Blotting System.
4. Incubate membranes with blocking buffer for 1 h at RT. Care should be taken to not touch and disrupt the membrane (only the immediate corners/edges).
5. Probe membranes with primary antibody solution for 1 h in TTBS followed by incubation with secondary antibody solution for 1 h in dark. Carry out antibody incubations using gentle orbital shaking at RT.
6. Wash western blots three times in TTBS for 10 min after each antibody incubation step and subsequently visualize at 800 nm using the Odyssey Infrared Imaging System.

3.6 Biophysical Characterization of EVs

3.6.1 Cryo-Electron Microscopy

1. Optional: Mix EV preparations (~2 μg protein) with Aurion Protein-G gold at a ratio of 1:3. This will allow vesicle diameters to be determined and the possibility for tomographic data collection.
2. Dilute samples as required in PBS.

3. Transfer samples onto glow-discharged C-flat holey carbon grids and blot off excess liquid (Optional, use of Vitrobot allows automated specimen preparation).
4. Immediately plunge-freeze grids in liquid ethane.
5. Mount grids in a cryoholder in liquid nitrogen.
6. Acquire images at 300 kV using a Tecnai G2 F30 in low-dose mode.
7. Typically, 10–20 fields of view are obtained.

3.6.2 Nanoparticle Tracking Analysis (NTA)

1. Dilute sample (exosome or CM) to a final volume of 1 mL (final concentration 1 μg/μL in ultrapure water (1:10,000 dilution)).
2. Load sample onto NS300 flow-cell top plate using syringe pump.
3. Set analysis settings: Detection threshold: 10; Flow rate: 100; Temperature: 25 °C.
4. Adjust camera level and screen gain until particles are visible.
5. For video capture, set conditions as follows: Number of captures: 3 (for triplicate); capture duration: 60 s.
6. For acquisition, change base file name to preferred destination folder.
7. Create and run script to begin capture and analysis with NTA software. Ensure that detection threshold is set to include as many particles as possible with the restriction that 10–100 red crosses were counted while only <10% were not associated with distinct particles. Repeat dilution and analysis as required, ensuring particle counts are within 20–100 particles/frame.
8. Set report details as prompted and view NTA report in destination folder: PDF graphs and batch summaries show size distribution profile data and statistics; CSV files contain raw data for further processing; AVI files contain video capture.

3.7 Biotin Labeling and Capture of EVs

1. Using 100 μg EVs (L-EVs or S-EVs), pellet 50–100 μg μg EVs (L-EVs or S-EVs).
2. Redisperse in 1 mL of 0.25 mg/mL EZ-Link Sulfo-NHS-SS-Biotin for 10 min at room temperature (*see* **Note 2**).
3. Pellet EVs and remove supernatant. Wash twice with ice-cold TBS (10,000× *g* for 30 min for L-EVs or 100,000× *g* for 1 h for S-EVs).

3.7.1 Separation of EV Surface-Associated Proteome

1. Redisperse EVs in 1 mL PBS and add 10 μL of 0.5 M EDTA, incubate for 15 min at room temperature.
2. Centrifuge biotinylated L-EVs at 10,000× *g* (30 min, 4 °C, Eppendorf 5430R) and biotinylated S-EVs at 100,000× *g* (1 h,

4 °C, TLA-55 rotor; Optima MAX-MP Tabletop Ultracentrifuge).

3. Collect the supernatant and put it aside for SP3-based tryptic protein digestion. (Continue to Subheading 3.9).
4. To the resultant EV surface proteome pellet, resuspend in 1 mL PBS, recentrifuge and reconstitute pellet fractions in 50 μL of lysis buffer, and lyse for 30 min on ice. (continue to Subheading 3.7.2).

3.7.2 NeutrAvidin Purification of Biotinylated EV Surface Proteins

1. Assemble protein purification device (e.g., epTIP (Eppendorf, 200 μL) fitted with a 20 μm nylon net (NY2004700, Merck Millipore) and a 1.5 mL Eppendorf collection tube) and add xx μl of NeutrAvidin (NA) Agarose slurry (10–20 μg protein/μL NA). Centrifuge at 1000× *g* for 1 min to remove NA storage solution.
2. Transfer the 50 μL of biotinylated EV surface proteins into the epTIP and incubate with NA for 30 min at room temperature with end-over-end mixing on a rotor.
3. Centrifuge samples at 1000× *g* for 1 min and discard flow-through solution.
4. Add 150 μL of wash buffer per sample, centrifuge samples at 1000× *g* for 1 min, and discard flow-through solution. Repeat two more times.
5. Add 80 μL 10 mM DTT in 100 mM TEAB to each sample and incubate for 45 min at room temperature.
6. To collect EV surface proteins, centrifuge at 1000× *g* for 2 min and collect eluate containing proteins of interest. Continue for proteomic sample preparation (Subheading 3.8.1).

3.8 EV Proteomics: Sample Preparation

1. Prepare fresh 500 mM DTT stock (vortex briefly until dissolved).
2. Prepare fresh 500 mM IAA stock (vortex briefly until dissolved).

3.8.1 EV Surface Protein Proteomic Sample Preparation

1. Add 3.6 μL of 500 mM IAA to each sample to achieve 20 mM IAA in solution and incubate for 30 min at 25 °C in the dark.
2. Add 4 μL of 500 mM DTT to each sample to achieve 20 mM DTT in solution.
3. Trypsin (Promega, V5111) at a 1:50 enzyme-to-substrate ratio for 16 h at 37 °C.
4. 2 μL of FA (100%) was added to each sample and desalted using SDB-RPS Stage-Tips (See Subheading 3.8.1.1).

StageTip Peptide Cleanup

1. SDB-RPS StageTips are prepared as described [38], using Empore solid-phase extraction disks. Prepare one SDB-RPS StageTip for each sample.
2. We use two plugs of SDB-RPS material for the respective StageTips, punched out by using a blunt-end 14-gauge syringe to pierce the ends of 200 μL pipette tips.
3. Add 30 μL of SDB-RPS loading buffer to the top of each sample. Place the StageTips into centrifuge adapters and centrifuge the sample through to dryness (1500× *g*, for ~1 min at RT).
4. Wash the SDB-RPS StageTips using 30 μL of SDB-RPS wash buffer 1 to each sample and centrifuge for ~1 min. Repeat.
5. Transfer each peptide sample to the top of an SDB-RPS StageTip.
6. Place the StageTips into the same centrifuge adapter and centrifuge the sample through to dryness (1500× *g* for ~5–8 min at RT).
7. Wash the StageTips with 100 μL of SDB-RPS wash buffer 1 and centrifuge the sample through to dryness (1500× *g* for ~5–8 min at RT). Repeat with SDB-RPS wash buffer 2.
8. Elute the peptides by adding 60 μL of SDB-RPS elution buffer of the StageTip, and centrifuge through to dryness (1500× *g* for ~5 min at RT). Collect the eluates into clean protein LoBind tubes.
9. Immediately place the tubes into an evaporative concentrator and concentrate under vacuum at 45 °C to dryness. The time to dryness will depend on the number of samples being processed. This typically takes 40–60 min; however, after the first 30 min have elapsed, samples should be checked at intervals of 10 min and concentrated until no liquid remains.
10. Reconstitute peptide by adding 12 μL of MS loading buffer (e.g., 0.07% trifluoroacetic acid) and quantify peptide amount by Fluorometric Peptide Assay.

3.9 Global Proteomic Sample Preparation of EVs

1. Determine 5–10 μg of EV protein (S-EV, L-EV or EV surface associated) and normalize sample volume to 25 μL using MS water.
2. Add 25 μL 2× lysis buffer to each sample and incubate at 95 °C for 5 min.
3. Add 1 μL 500 mM DTT to each sample (cap and spin down) and incubate at room temperature (RT) for 1 h at 350 RPM.
4. Add 1 μL 500 mM IAA to each sample (cap and spin down) and incubate at room temperature (RT) for 30 min at 25 °C in the dark.

5. Add 1 μL 500 mM DTT to each sample (cap and spin down). Record final volume.
6. The bead stock (100 μg/μL) is prepared; add 2 μL to each sample, vortex, and spin down.
7. Add 100% ethanol (volume = sample volume) to each sample and shake for 10 min at 350 RPM).
8. Place samples on a magnetic rack and remove supernatant (discard).
9. Remove samples from magnetic rack and add 200 μL of 80% ethanol, vortex, and spin down (repeat this **step 2** more times).
10. After removing the final 80% ethanol wash, transfer tubes open to the fume hood, lid open to dry.
11. Prepare and add 100 μL of digestion buffer with enzymes (bath sonicate for 10 s).
12. Set ThermoMixer to 37 °C and incubate at 37 °C for 16 h at 1000 rpm.
13. Centrifuge samples at 20,000× *g* for 1 min at RT.
14. Place on a magnetic rack and carefully transfer supernatant to 1.5 mL low protein binding Eppendorf's.
15. Add 2 μL 100% FA to each sample and freeze for 30 min at −80 °C (Peptides can be stored in −80 °C for an extended period before lyophilization.).
16. Transfer samples to Lyophilize in the Speed Vacuum Concentrator (approx. 1 μL per minute).
17. Add 12 μL of MS loading buffer (e.g., 0.07% trifluoroacetic acid) and determine peptide amount by Fluorometric Peptide Assay.
18. Check pH is <2, add 1 μL 100% FA to each sample.

3.10 Fluorometric Peptide Assay

1. Prepare peptide standards from the Assay Kit starting with 1000 μg/mL and perform serial dilutions in MS loading buffer down to 7.8 μg/mL as per the manufacturer's instructions (8 standards plus MS loading buffer as blank).
2. Load the 10 μL of each standard/blank into the black 96 well plate.
3. Load 2 μL of each sample into the plate and make up to 10 μL by adding 8 μL MS loading buffer.
4. Add 70 μL of Fluorometric Peptide Assay Buffer to each well.
5. Add 20 μL of Fluorometric Peptide Assay Reagent to each well.
6. Incubate at RT for 5 min.
7. Measure fluorescence using Ex/Em at 390 nm/475 nm.

8. Use the standard curve to calculate the concentrations of your samples (ensuring to subtract blank (abs) from all standards/ samples (abs).

3.11 UHPLC-MS/MS and Proteomic Analysis Parameters

1. All settings and parameters for nano-LC and mass spectrometer are provided [1]. Briefly, peptides are analyzed on a Dionex UltiMate NCS-3500RS nanoUHPLC coupled to a Q-Exactive HF-X hybrid quadrupole-Orbitrap mass spectrometer equipped with nanospray ion source in positive mode.
2. Peptides were separated with a gradient of 2–28% acetonitrile containing 0.1% formic acid over 95 mins at 300 nL min^{-1} followed by 28–80% from 95 to 98 min at 300 nL min^{-1} and heated at 55 °C (butterfly portfolio heater, Phoenix S&T).
3. The MS-based proteomics data is deposited to the ProteomeXchange Consortium via the PRIDE partner repository and is available via ProteomeXchange with identifier (PXD026658).

3.12 Bioinformatics and Statistics

1. For database searching, RAW files were analyzed with cysteine carbamidomethylation as a fixed modification and N-terminal acetylation and methionine oxidations as variable modifications; for biotin surface proteome analysis, additional Thioacyl (DSP) (C(3)H(4)OS)) was employed (*see* **Note 10**).
2. All other search and analytical parameters, including protein/ peptide false discovery rate, enzyme specificity, and identification requirements, are provided [1].
3. Cellular surfaceome data used include proteins previously experimentally verified as a cell-surface (CSPA) [35] or predicted as surfaceome proteins based on SURFY [34]. (*see* **Note 1**).
4. Online web-based bioinformatics resources such as gProfiler (https://biit.cs.ut.ee/gprofiler/gost), STRING (https:// string-db.org/), Cytoscape [39] can be used to generate EnrichmentMap [40] (plugin v3.7.1). Database for Annotation, Visualization and Integrated Discovery (DAVID) (https://david.ncifcrf.gov/), Sankey diagram of ligand-receptor interactions can be generated using http://www.rna-society.org/cellinker/ [41].

4 Notes

1. Databases that comprise cellular surfaceome data can be employed to refine data analyses based on proteins previously experimentally verified as a cell-surface (CSPA) [35] or predicted as surfaceome proteins based on SURFY [34] and UniProt. Although the CSPA represents the most extensive

experimentally derived surfaceome database so far, the provided surfaceome maps are not complete. Further, challenges in specificity of the surfaceome are likely in such resources, due in part to the presence of intracellular components due to their abundance. Other resources included Human Protein Atlas Plasma Membrane Map [42] and in silico human surfaceome [34] that provide a resource and interactive tool for exploring the complete list of predicted surfaceome proteins and further understanding of surfaceome landscape that are grouped into their functional classes.

2. In this study, we employed membrane-impermeant and thiol-cleavable Sulfo-NHS-SS-Biotin and subsequently identifying using mass spectrometry [43]. This type of biotin provides unique disulfide linkage (-SS-) that is cleavable, and highly soluble in aqueous buffers, making it a popular reagent for studying cell surface proteins [44–46]. We combined this using enrichment of bona fide membrane proteins in affinity-purification using avidin-beads. We are confident that our experimental data reliably provides a comprehensive map of EV surfaceome [34, 35].
3. Shaving the EV: Previously we have shown that surface membrane protein enrichment techniques, using proteases to "shave" proteins from the surface of EVs and identify surface-exposed proteins [47]. Here, sodium carbonate extraction/Triton X-114 phase separation and mild proteolysis (proteinase K, PK) of intact exosomes is used in combination with label-free quantitative mass spectrometry to identify 1025 exosomal proteins of which 208 are predicted to be integral membrane proteins.
4. Surface and the corona: Previous studies have shown that a pool of proteins can associate with EV surface and are functional [36, 48–50]. In this protocol, we highlight that in the EV, surfaceome contains proteins co-identified in CSPA/SURFY (86/291 as classical surface proteins) as well as 205/291 as non-CSPA/SURFY proteins—suggesting that EVs contain a subset of proteins associated on the surface. Recently, the formation of protein corona on EVs has been shown [10], important in influencing function, distribution, and cellular uptake [51]. The protein corona is formed when EVs are exposed biological fluids, and proteins are adsorbed onto the EV surface, including from additives in cell culture conditions. In this study, we found that EDTA-treatment effectively released 122 proteins (now referred to as EV surface-associated) from the surface of EVs, including annexins and 14-3-3 proteins, but not membrane proteins including CDs, receptors, and transporters.

5. To capture the surface-associated proteins of EVs, we employed EDTA. Because EDTA can successfully chelate metal ions such as calcium to release surface-bound proteins (such as ANXA1 [36]) without compromising EV integrity [36, 37], we also incubated EVs with EDTA and defines the released proteins as the "EV-surface-associated proteome" (Fig. 2, *bottom panel*). This led to the identification of 291 and 456 surface, and 122 and 67 surface-associated proteins in L-EVs and S-EVs, respectively. We show that membrane-surface-associated proteins (that are released by divalent ion chelator EDTA) are implicated in actin cytoskeleton regulation, junction organization, glycolysis, and platelet activation.
6. We highlight differential abundance of biotin surface proteins in S-EVs and L-EVs, including various receptors, transporters, claudins, and cadherins selectively identified in EV subtypes. We also identify protein networks similarly expressed in different subtypes including various integrins, cell surface antigens, solute carriers, and metabolite transporters (refer Supp Table 4) [1]. For L-EVs, biotin-captured proteins were enriched for "plasma membrane" proteins, "membrane raft" proteins, "cell projection," "leading edge," and "trailing edge," highlighting their membrane origin during formation.
7. Typically, cells grown to 80% confluency on 15 cm^2 tissue culture plate yield ~1–2 μg (24 h incubation) of EVs. Adherent cells or cells grown in suspension should be cultured until they reach 60–70% confluency. This yield is cell type dependent and should be determined for each cell type. For media obtained from bioreactor, this is again cell type dependent based on yield and growth/media conditions should be optimized to ensure cell viability.
8. Standard growth media for cell culture typically require FCS, which contains EVs that can interfere with studies of cell-derived EVs. To mitigate this, serum-free or EV-depleted FCS supplemented with ITS (0.2–1.5% (w/v)) or BSA (1% (w/v)) is used. Additional methods include ultracentrifugation (100,000× *g*, 16 h) or commercial EV depleted FCS products to ensure optimal conditions during EV production without contamination of serum origin EVs present in FCS.
9. We isolated large EVs (L-EVs) and small EVs (S-EVs) released by different cell lines using differential centrifugation coupled to density gradient separation as previously described [52–54] and characterized them (for their buoyant densities, size, morphology, presence of specific proteins) to meet the experimental requirements as set out by the International Society for Extracellular Vesicles (ISEV) guidelines [55, 56]. Ensuring

you have EVs, purified and characterized prior to this experimental procedure, is critical. Further, maintaining the integrity of the EV membrane [36, 37] is critical to ensure that the oxidation/biotinylation only occurs on extracellular domains, and not ruptured or damaged EV membranes or luminal content. Prior to culture media collection, ensuring cell viability is maintained to ensure that cell lysis is not contributing to the pool of cell-released protein content.

10. Data processing: We employed DDA-based MS acquisition in this study and used MaxQuant for data processing. Other modes of data acquisition and platforms for data processing are encouraged, including DIA-based surfaceome analysis [45]. In this protocol, we employed sequence analysis using variable modification analysis for biotin surface proteome analysis Thioacyl (DSP) (C(3)H(4)OS)). The thioacyl group is a remnant of the labeling reagent and was therefore only used in the biotin condition (i.e., variable). Identifications were filtered to only include biotinylated peptides for further analysis. Carbamidomethylation is an expected modification of cysteines caused by the incubation with IAA prior to labeling. For DIA-based analyses, the use of maximal variable modifications set to 2 and without M(ox) during predicted library generation, with the option to combine predicted library suited for biotin modification, should be employed (e.g., FragPipe, MSFragger).

Conflict of Interest

This work was funded by grants from the National Health and Medical Research Council (NHMRC; MRF1201805, MRF2024039) to D.W.G, Hains Fellowship (Hains Foundation, Australia), Department of Defense (PR23006), and National Heart Foundation Australia (105072). Further, I.I was Supported by Australian Government Research Training Program scholarship.

References

1. Rai A, Fang H, Claridge B, Simpson RJ, Greening DW (2021) Proteomic dissection of large extracellular vesicle surfaceome unravels interactive surface platform. J Extracell Vesicles 10:e12164
2. Hu Q, Su H, Li J, Lyon C, Tang W, Wan M, Hu TY (2020) Clinical applications of exosome membrane proteins. Precis Clin Med 3:54–66
3. Jankovicova J, Secova P, Michalkova K, Antalikova J (2020) Tetraspanins, more than markers of extracellular vesicles in reproduction. Int J Mol Sci 21:7568
4. Peinado H, Aleckovic M, Lavotshkin S, Matei I, Costa-Silva B, Moreno-Bueno G, Hergueta-Redondo M, Williams C, Garcia-Santos G, Ghajar C, Nitadori-Hoshino A, Hoffman C, Badal K, Garcia BA, Callahan MK, Yuan J, Martins VR, Skog J, Kaplan RN, Brady MS, Wolchok JD, Chapman PB, Kang Y, Bromberg J, Lyden D (2012) Melanoma exosomes educate bone marrow progenitor cells

toward a pro-metastatic phenotype through MET. Nat Med 18:883–891

5. Meng Q, Chen C, Yang N, Gololobova O, Shi C, Dunn CA, Rossi M, Martindale JL, Basisty N, Ding J, Delannoy M, Basu S, Mazan-Mamczarz K, Shin CH, Yang JH, Johnson PF, Witwer KW, Biragyn A, Sen P, Abdelmohsen K, De S, Gorospe M (2023) Surfaceome analysis of extracellular vesicles from senescent cells uncovers uptake repressor DPP4. Proc Natl Acad Sci USA 120: e2219801120
6. Hoshino A, Costa-Silva B, Shen TL, Rodrigues G, Hashimoto A, Tesic Mark M, Molina H, Kohsaka S, Di Giannatale A, Ceder S, Singh S, Williams C, Soplop N, Uryu K, Pharmer L, King T, Bojmar L, Davies AE, Ararso Y, Zhang T, Zhang H, Hernandez J, Weiss JM, Dumont-Cole VD, Kramer K, Wexler LH, Narendran A, Schwartz GK, Healey JH, Sandstrom P, Labori KJ, Kure EH, Grandgenett PM, Hollingsworth MA, de Sousa M, Kaur S, Jain M, Mallya K, Batra SK, Jarnagin WR, Brady MS, Fodstad O, Muller V, Pantel K, Minn AJ, Bissell MJ, Garcia BA, Kang Y, Rajasekhar VK, Ghajar CM, Matei I, Peinado H, Bromberg J, Lyden D (2015) Tumour exosome integrins determine organotropic metastasis. Nature 527:329–335
7. Kim DH, Kim H, Choi YJ, Kim SY, Lee JE, Sung KJ, Sung YH, Pack CG, Jung MK, Han B, Kim K, Kim WS, Nam SJ, Choi CM, Yun M, Lee JC, Rho JK (2019) Exosomal PD-L1 promotes tumor growth through immune escape in non-small cell lung cancer. Exp Mol Med 51:1–13
8. Buzas EI, Toth EA, Sodar BW, Szabo-Taylor KE (2018) Molecular interactions at the surface of extracellular vesicles. Semin Immunopathol 40:453–464
9. Charoenviriyakul C, Takahashi Y, Morishita M, Nishikawa M, Takakura Y (2018) Role of extracellular vesicle surface proteins in the pharmacokinetics of extracellular vesicles. Mol Pharm 15:1073–1080
10. Toth EA, Turiak L, Visnovitz T, Cserep C, Mazlo A, Sodar BW, Forsonits AI, Petovari G, Sebestyen A, Komlosi Z, Drahos L, Kittel A, Nagy G, Bacsi A, Denes A, Gho YS, Szabo-Taylor KE, Buzas EI (2021) Formation of a protein corona on the surface of extracellular vesicles in blood plasma. J Extracell Vesicles 10: e12140
11. Al-Nedawi K, Meehan B, Micallef J, Lhotak V, May L, Guha A, Rak J (2008) Intercellular transfer of the oncogenic receptor EGFRvIII by microvesicles derived from tumour cells. Nat Cell Biol 10:619–624
12. Melo SA, Luecke LB, Kahlert C, Fernandez AF, Gammon ST, Kaye J, LeBleu VS, Mittendorf EA, Weitz J, Rahbari N, Reissfelder C, Pilarsky C, Fraga MF, Piwnica-Worms D, Kalluri R (2015) Glypican-1 identifies cancer exosomes and detects early pancreatic cancer. Nature 523:177–182
13. Yoshioka Y, Kosaka N, Konishi Y, Ohta H, Okamoto H, Sonoda H, Nonaka R, Yamamoto H, Ishii H, Mori M, Furuta K, Nakajima T, Hayashi H, Sugisaki H, Higashimoto H, Kato T, Takeshita F, Ochiya T (2014) Ultra-sensitive liquid biopsy of circulating extracellular vesicles using ExoScreen. Nat Commun 5:3591
14. Piibor J, Waldmann A, Dissanayake K, Andronowska A, Ivask M, Prasadani M, Kavak A, Kodithuwakku S, Fazeli A (2023) Uterine fluid extracellular vesicles proteome is altered during the estrous cycle. Mol Cell Proteomics 22:100642
15. Ekstrom K, Crescitelli R, Petursson HI, Johansson J, Lasser C, Olofsson Bagge R (2022) Characterization of surface markers on extracellular vesicles isolated from lymphatic exudate from patients with breast cancer. BMC Cancer 22:50
16. Furuta T, Miyaki S, Ishitobi H, Ogura T, Kato Y, Kamei N, Miyado K, Higashi Y, Ochi M (2016) Mesenchymal stem cell-derived exosomes promote fracture healing in a mouse model. Stem Cells Transl Med 5:1620–1630
17. Feng K, Xie X, Yuan J, Gong L, Zhu Z, Zhang J, Li H, Yang Y, Wang Y (2021) Reversing the surface charge of MSC-derived small extracellular vesicles by epsilonPL-PEG-DSPE for enhanced osteoarthritis treatment. J Extracell Vesicles 10:e12160
18. Vicencio JM, Yellon DM, Sivaraman V, Das D, Boi-Doku C, Arjun S, Zheng Y, Riquelme JA, Kearney J, Sharma V, Multhoff G, Hall AR, Davidson SM (2015) Plasma exosomes protect the myocardium from ischemia-reperfusion injury. J Am Coll Cardiol 65:1525–1536
19. Yang J, Zhang X, Chen X, Wang L, Yang G (2017) Exosome mediated delivery of miR-124 promotes neurogenesis after ischemia. Mol Ther Nucleic Acids 7:278–287
20. Zhu LP, Tian T, Wang JY, He JN, Chen T, Pan M, Xu L, Zhang HX, Qiu XT, Li CC, Wang KK, Shen H, Zhang GG, Bai YP (2018) Hypoxia-elicited mesenchymal stem cell-derived exosomes facilitates cardiac repair through miR-125b-mediated prevention of cell death in myocardial infarction. Theranostics 8:6163–6177
21. Wiklander OP, Nordin JZ, O'Loughlin A, Gustafsson Y, Corso G, Mager I, Vader P,

Lee Y, Sork H, Seow Y, Heldring N, Alvarez-Erviti L, Smith CI, Le Blanc K, Macchiarini P, Jungebluth P, Wood MJ, Andaloussi SE (2015) Extracellular vesicle in vivo biodistribution is determined by cell source, route of administration and targeting. J Extracell Vesicles 4:26316
22. Nakase I, Futaki S (2015) Combined treatment with a pH-sensitive fusogenic peptide and cationic lipids achieves enhanced cytosolic delivery of exosomes. Sci Rep 5:10112
23. Tian T, Zhang HX, He CP, Fan S, Zhu YL, Qi C, Huang NP, Xiao ZD, Lu ZH, Tannous BA, Gao J (2018) Surface functionalized exosomes as targeted drug delivery vehicles for cerebral ischemia therapy. Biomaterials 150: 137–149
24. Antes TJ, Middleton RC, Luther KM, Ijichi T, Peck KA, Liu WJ, Valle J, Echavez AK, Marban E (2018) Targeting extracellular vesicles to injured tissue using membrane cloaking and surface display. J Nanobiotechnol 16:61
25. Rai A, Claridge B, Lozano J, Greening DW (2024) The discovery of extracellular vesicles and their emergence as a next-generation therapy. Circ Res 135:198–221
26. Belhadj Z, He B, Deng H, Song S, Zhang H, Wang X, Dai W, Zhang Q (2020) A combined "eat me/don't eat me" strategy based on extracellular vesicles for anticancer nanomedicine. J Extracell Vesicles 9:1806444
27. Guo Y, Wang H, Liu S, Zhang X, Zhu X, Huang L, Zhong W, Guan L, Chen Y, Xiao M, Ou L, Yang J, Chen X, Huang AC, Mitchell T, Amaravadi R, Karakousis G, Miura J, Schuchter L, Flowers A, Zheng Q, Mou H, Gimotty P, Herlyn M, Guo W, Xu X (2025) Engineered extracellular vesicles with DR5 agonistic scFvs simultaneously target tumor and immunosuppressive stromal cells. Sci Adv 11:eadp9009
28. Xu R, Greening DW, Rai A, Ji H, Simpson RJ (2015) Highly-purified exosomes and shed microvesicles isolated from the human colon cancer cell line LIM1863 by sequential centrifugal ultrafiltration are biochemically and functionally distinct. Methods 87:11–25
29. Kowal J, Arras G, Colombo M, Jouve M, Morath JP, Primdal-Bengtson B, Dingli F, Loew D, Tkach M, Thery C (2016) Proteomic comparison defines novel markers to characterize heterogeneous populations of extracellular vesicle subtypes. Proc Natl Acad Sci USA 113:E968–E977
30. Xu R, Rai A, Chen M, Suwakulsiri W, Greening DW, Simpson RJ (2018) Extracellular vesicles in cancer - implications for future improvements in cancer care. Nat Rev Clin Oncol 15: 617–638
31. Bellin G, Gardin C, Ferroni L, Chachques JC, Rogante M, Mitrecic D, Ferrari R, Zavan B (2019) Exosome in cardiovascular diseases: a complex world full of Hope. Cells 8:8
32. Howitt J, Hill AF (2016) Exosomes in the pathology of neurodegenerative diseases. J Biol Chem 291:26589–26597
33. Tkach M, Kowal J, Thery C (2018) Why the need and how to approach the functional diversity of extracellular vesicles. Philos Trans R Soc Lond Ser B Biol Sci 373:20160479
34. Bausch-Fluck D, Goldmann U, Muller S, van Oostrum M, Muller M, Schubert OT, Wollscheid B (2018) The in silico human surfaceome. Proc Natl Acad Sci USA 115:E10988–E10997
35. Bausch-Fluck D, Hofmann A, Bock T, Frei AP, Cerciello F, Jacobs A, Moest H, Omasits U, Gundry RL, Yoon C, Schiess R, Schmidt A, Mirkowska P, Härtlová A, Van Eyk JE, Bourquin JP, Aebersold R, Boheler KR, Zandstra P, Wollscheid B (2015) A mass spectrometric-derived cell surface protein atlas. PLoS One 10:e0121314
36. Rogers MA, Buffolo F, Schlotter F, Atkins SK, Lee LH, Halu A, Blaser MC, Tsolaki E, Higashi H, Luther K, Daaboul G, Bouten CVC, Body SC, Singh SA, Bertazzo S, Libby P, Aikawa M, Aikawa E (2020) Annexin A1-dependent tethering promotes extracellular vesicle aggregation revealed with single-extracellular vesicle analysis. Sci Adv 6:6
37. Santucci L, Bruschi M, Del Zotto G, Antonini F, Ghiggeri GM, Panfoli I, Candiano G (2019) Biological surface properties in extracellular vesicles and their effect on cargo proteins. Sci Rep 9:13048
38. Rai A, Poh QH, Fatmous M, Fang H, Gurung S, Vollenhoven B, Salamonsen LA, Greening DW (2021) Proteomic profiling of human uterine extracellular vesicles reveal dynamic regulation of key players of embryo implantation and fertility during menstrual cycle. Proteomics 21:e2000211
39. Shannon P, Markiel A, Ozier O, Baliga NS, Wang JT, Ramage D, Amin N, Schwikowski B, Ideker T (2003) Cytoscape: a software environment for integrated models of biomolecular interaction networks. Genome Res 13:2498–2504
40. Merico D, Isserlin R, Stueker O, Emili A, Bader GD (2010) Enrichment map: a network-based method for gene-set enrichment visualization and interpretation. PLoS One 5:e13984
41. Zhang Y, Liu T, Wang J, Zou B, Li L, Yao L, Chen K, Ning L, Wu B, Zhao X, Wang D (2021) Cellinker: a platform of ligand–receptor

interactions for intercellular communication analysis. Bioinformatics 37:2025–2032

42. Thul PJ, Akesson L, Wiking M, Mahdessian D, Geladaki A, Ait Blal H, Alm T, Asplund A, Bjork L, Breckels LM, Backstrom A, Danielsson F, Fagerberg L, Fall J, Gatto L, Gnann C, Hober S, Hjelmare M, Johansson F, Lee S, Lindskog C, Mulder J, Mulvey CM, Nilsson P, Oksvold P, Rockberg J, Schutten R, Schwenk JM, Sivertsson A, Sjostedt E, Skogs M, Stadler C, Sullivan DP, Tegel H, Winsnes C, Zhang C, Zwahlen M, Mardinoglu A, Ponten F, von Feilitzen K, Lilley KS, Uhlen M, Lundberg E (2017) A subcellular map of the human proteome. Science 356:eaal3321
43. Nair JD, Henley JM, Wilkinson KA (2021) Surface biotinylation of primary neurons to monitor changes in AMPA receptor surface expression in response to kainate receptor stimulation. STAR Protoc 2:100992
44. Li Y, Qin H, Ye M (2020) An overview on enrichment methods for cell surface proteome profiling. J Sep Sci 43:292–312
45. Inocencio I, Rai A, Donner D, Greening DW (2025) The proteomic landscape of the coronary accessible heart cell Surfaceome. Proteomics 25:e202400320
46. Lango T, Kuffa K, Toth G, Turiak L, Drahos L, Tusnady GE (2022) Comprehensive discovery of the accessible primary amino group-containing segments from cell surface proteins by fine-tuning a high-throughput Biotinylation method. Int J Mol Sci 24:273
47. Xu R, Greening DW, Chen M, Rai A, Ji H, Takahashi N, Simpson RJ (2019) Surfaceome of exosomes secreted from the colorectal cancer cell line SW480: peripheral and integral membrane proteins analyzed by proteolysis and TX114. Proteomics 19:e1700453
48. Costa-Silva B, Aiello NM, Ocean AJ, Singh S, Zhang H, Thakur BK, Becker A, Hoshino A, Mark MT, Molina H, Xiang J, Zhang T, Theilen TM, Garcia-Santos G, Williams C, Ararso Y, Huang Y, Rodrigues G, Shen TL, Labori KJ, Lothe IM, Kure EH, Hernandez J, Doussot A, Ebbesen SH, Grandgenett PM, Hollingsworth MA, Jain M, Mallya K, Batra SK, Jarnagin WR, Schwartz RE, Matei I, Peinado H, Stanger BZ, Bromberg J, Lyden D (2015) Pancreatic cancer exosomes initiate pre-metastatic niche formation in the liver. Nat Cell Biol 17:816–826
49. Sung BH, Ketova T, Hoshino D, Zijlstra A, Weaver AM (2015) Directional cell movement through tissues is controlled by exosome secretion. Nat Commun 6:7164
50. Ringuette Goulet C, Bernard G, Tremblay S, Chabaud S, Bolduc S, Pouliot F (2018) Exosomes induce fibroblast differentiation into cancer-associated fibroblasts through TGFbeta signaling. Mol Cancer Res 16:1196–1204
51. Liam-Or R, Faruqu FN, Walters A, Han S, Xu L, Wang JT, Oberlaender J, Sanchez-Fueyo A, Lombardi G, Dazzi F, Mailaender V, Al-Jamal KT (2024) Cellular uptake and in vivo distribution of mesenchymal-stem-cell-derived extracellular vesicles are protein corona dependent. Nat Nanotechnol 19:846–855
52. Tauro BJ, Greening DW, Mathias RA, Mathivanan S, Ji H, Simpson RJ (2013) Two distinct populations of exosomes are released from LIM1863 colon carcinoma cell-derived organoids. Mol Cell Proteomics 12:587–598
53. Rai A, Greening DW, Chen M, Xu R, Ji H, Simpson RJ (2019) Exosomes derived from human primary and metastatic colorectal cancer cells contribute to functional heterogeneity of activated fibroblasts by reprogramming their proteome. Proteomics 19:e1800148
54. Suwakulsiri W, Rai A, Xu R, Chen M, Greening DW, Simpson RJ (2019) Proteomic profiling reveals key cancer progression modulators in shed microvesicles released from isogenic human primary and metastatic colorectal cancer cell lines. Biochim Biophys Acta Proteins Proteom 1867:140171
55. Thery C, Witwer KW, Aikawa E, Alcaraz MJ, Anderson JD, Andriantsitohaina R, Antoniou A, Arab T, Archer F, Atkin-Smith GK, Ayre DC, Bach JM, Bachurski D, Baharvand H, Balaj L, Baldacchino S, Bauer NN, Baxter AA, Bebawy M, Beckham C, Bedina Zavec A, Benmoussa A, Berardi AC, Bergese P, Bielska E, Blenkiron C, Bobis-Wozowicz S, Boilard E, Boireau W, Bongiovanni A, Borras FE, Bosch S, Boulanger CM, Breakefield X, Breglio AM, Brennan MA, Brigstock DR, Brisson A, Broekman ML, Bromberg JF, Bryl-Gorecka P, Buch S, Buck AH, Burger D, Busatto S, Buschmann D, Bussolati B, Buzas EI, Byrd JB, Camussi G, Carter DR, Caruso S, Chamley LW, Chang YT, Chen C, Chen S, Cheng L, Chin AR, Clayton A, Clerici SP, Cocks A, Cocucci E, Coffey RJ, Cordeiro-da-Silva A, Couch Y, Coumans FA, Coyle B, Crescitelli R, Criado MF, D'Souza-Schorey C, Das S, Datta Chaudhuri A, de Candia P, De Santana EF, De Wever O, Del Portillo HA, Demaret T, Deville S, Devitt A, Dhondt B, Di Vizio D, Dieterich LC, Dolo V, Dominguez Rubio AP, Dominici M, Dourado MR, Driedonks TA, Duarte FV, Duncan HM, Eichenberger RM, Ekstrom K, El Andaloussi S, Elie-Caille C,

Erdbrugger U, Falcon-Perez JM, Fatima F, Fish JE, Flores-Bellver M, Forsonits A, Frelet-Barrand A, Fricke F, Fuhrmann G, Gabrielsson S, Gamez-Valero A, Gardiner C, Gartner K, Gaudin R, Gho YS, Giebel B, Gilbert C, Gimona M, Giusti I, Goberdhan DC, Gorgens A, Gorski SM, Greening DW, Gross JC, Gualerzi A, Gupta GN, Gustafson D, Handberg A, Haraszti RA, Harrison P, Hegyesi H, Hendrix A, Hill AF, Hochberg FH, Hoffmann KF, Holder B, Holthofer H, Hosseinkhani B, Hu G, Huang Y, Huber V, Hunt S, Ibrahim AG, Ikezu T, Inal JM, Isin M, Ivanova A, Jackson HK, Jacobsen S, Jay SM, Jayachandran M, Jenster G, Jiang L, Johnson SM, Jones JC, Jong A, Jovanovic-Talisman T, Jung S, Kalluri R, Kano SI, Kaur S, Kawamura Y, Keller ET, Khamari D, Khomyakova E, Khvorova A, Kierulf P, Kim KP, Kislinger T, Klingeborn M, Klinke DJ 2nd, Kornek M, Kosanovic MM, Kovacs AF, Kramer-Albers EM, Krasemann S, Krause M, Kurochkin IV, Kusuma GD, Kuypers S, Laitinen S, Langevin SM, Languino LR, Lannigan J, Lasser C, Laurent LC, Lavieu G, Lazaro-Ibanez E, Le Lay S, Lee MS, Lee YXF, Lemos DS, Lenassi M, Leszczynska A, Li IT, Liao K, Libregts SF, Ligeti E, Lim R, Lim SK, Line A, Linnemannstons K, Llorente A, Lombard CA, Lorenowicz MJ, Lorincz AM, Lotvall J, Lovett J, Lowry MC, Loyer X, Lu Q, Lukomska B, Lunavat TR, Maas SL, Malhi H, Marcilla A, Mariani J, Mariscal J, Martens-Uzunova ES, Martin-Jaular L, Martinez MC, Martins VR, Mathieu M, Mathivanan S, Maugeri M, McGinnis LK, McVey MJ, Meckes DG Jr, Meehan KL, Mertens I, Minciacchi VR, Moller A, Moller Jorgensen M, Morales-Kastresana A, Morhayim J, Mullier F, Muraca M, Musante L, Mussack V, Muth DC, Myburgh KH, Najrana T, Nawaz M, Nazarenko I, Nejsum P, Neri C, Neri T, Nieuwland R, Nimrichter L, Nolan JP, Nolte-'t Hoen EN, Noren Hooten N, O'Driscoll L, O'Grady T, O'Loghlen A, Ochiya T, Olivier M, Ortiz A, Ortiz LA, Osteikoetxea X, Ostergaard O, Ostrowski M, Park J, Pegtel DM, Peinado H, Perut F, Pfaffl MW, Phinney DG, Pieters BC, Pink RC, Pisetsky DS, Pogge von Strandmann E, Polakovicova I, Poon IK, Powell BH, Prada I, Pulliam L, Quesenberry P, Radeghieri A, Raffai RL, Raimondo S, Rak J, Ramirez MI, Raposo G, Rayyan MS, Regev-Rudzki N, Ricklefs FL, Robbins PD, Roberts DD, Rodrigues SC, Rohde E, Rome S, Rouschop KM, Rughetti A, Russell AE, Saa P, Sahoo S, Salas-Huenuleo E, Sanchez C, Saugstad JA, Saul MJ, Schiffelers RM, Schneider R, Schoyen TH, Scott A, Shahaj E, Sharma S, Shatnyeva O, Shekari F, Shelke GV, Shetty AK, Shiba K, Siljander PR, Silva AM, Skowronek A, Snyder OL 2nd, Soares RP, Sodar BW, Soekmadji C, Sotillo J, Stahl PD, Stoorvogel W, Stott SL, Strasser EF, Swift S, Tahara H, Tewari M, Timms K, Tiwari S, Tixeira R, Tkach M, Toh WS, Tomasini R, Torrecilhas AC, Tosar JP, Toxavidis V, Urbanelli L, Vader P, van Balkom BW, van der Grein SG, Van Deun J, van Herwijnen MJ, Van Keuren-Jensen K, van Niel G, van Royen ME, van Wijnen AJ, Vasconcelos MH, Vechetti IJ Jr, Veit TD, Vella LJ, Velot E, Verweij FJ, Vestad B, Vinas JL, Visnovitz T, Vukman KV, Wahlgren J, Watson DC, Wauben MH, Weaver A, Webber JP, Weber V, Wehman AM, Weiss DJ, Welsh JA, Wendt S, Wheelock AM, Wiener Z, Witte L, Wolfram J, Xagorari A, Xander P, Xu J, Yan X, Yanez-Mo M, Yin H, Yuana Y, Zappulli V, Zarubova J, Zekas V, Zhang JY, Zhao Z, Zheng L, Zheutlin AR, Zickler AM, Zimmermann P, Zivkovic AM, Zocco D, Zuba-Surma EK (2018) Minimal information for studies of extracellular vesicles 2018 (MISEV2018): a position statement of the International Society for Extracellular Vesicles and update of the MISEV2014 guidelines. J Extracell Vesicles 7:1535750

56. Welsh JA, Goberdhan DCI, O'Driscoll L, Buzas EI, Blenkiron C, Bussolati B, Cai H, Di Vizio D, Driedonks TAP, Erdbrugger U, Falcon-Perez JM, Fu QL, Hill AF, Lenassi M, Lim SK, Mahoney MG, Mohanty S, Moller A, Nieuwland R, Ochiya T, Sahoo S, Torrecilhas AC, Zheng L, Zijlstra A, Abuelreich S, Bagabas R, Bergese P, Bridges EM, Brucale M, Burger D, Carney RP, Cocucci E, Crescitelli R, Hanser E, Harris AL, Haughey NJ, Hendrix A, Ivanov AR, Jovanovic-Talisman T, Kruh-Garcia NA, Ku'ulei-Lyn Faustino V, Kyburz D, Lasser C, Lennon KM, Lotvall J, Maddox AL, Martens-Uzunova ES, Mizenko RR, Newman LA, Ridolfi A, Rohde E, Rojalin T, Rowland A, Saftics A, Sandau US, Saugstad JA, Shekari F, Swift S, Ter-Ovanesyan D, Tosar JP, Useckaite Z, Valle F, Varga Z, van der Pol E, van Herwijnen MJC, Wauben MHM, Wehman AM, Williams S, Zendrini A, Zimmerman AJ, Consortium M, Thery C, Witwer KW (2024) Minimal information for studies of extracellular vesicles (MISEV2023): from basic to advanced approaches. J Extracell Vesicles 13:e12404

Chapter 3

Standardized Protocol for Magnetic Bead-Based Enrichment and Separation of Extracellular Vesicles

Ketil Winther Pedersen and Axl A. Neurauter

Abstract

This chapter covers magnetic bead-based enrichment, separation, and analysis of the smallest members of extracellular vesicles (EVs). Exosomes (30–150 nm) are produced through fusion of multivesicular bodies (MVBs) with the plasma membrane, thereby releasing their internal vesicles into the extracellular environment. Due to their small size, researchers need expensive and high-maintenance ultracentrifuges to enrich exosomes. This often leads to loss of material and the risk of co-isolation of impurities.

Here we utilize the negative surface charge and surface antigen expression to enrich and separate exosomes either by using one of two different preparation protocols or by combining them.

Descriptions of how to enrich and analyze exosomes from biologically relevant and complex starting materials are included.

Key words Extracellular vesicles, Magnetic beads, Strong anion exchange, Automation, Resistive pulse sensing, Exosomes, Enrichment, Isolation, Separation, Western blot, Release, Flow cytometry, Electron microscopy, Mass spectrometry, Phenotype, Serum, Urine, Multiomics, Liquid biopsy

1 Introduction

Cancer is a complex and heterogeneous disease that has a significant public health impact worldwide. Although considerable progress has been made in understanding the molecular mechanisms underlying cancer development and progression, much remains uncertain. Liquid biopsy is a rapidly evolving field that provides a noninvasive approach to diagnosing and monitoring diseases and thus poses a promising approach for further advancement of cancer treatment. In contrast to traditional tissue biopsies, liquid biopsy involves the analysis of bodily fluids to detect circulating biomarkers such as circulating tumor cells, circulating tumor DNA, and exosomes that reflect the presence and progression of disease. Exosomes are 30–150 nm extracellular vesicles (EVs) containing bioactive molecules, such as DNA, RNA, proteins, and metabolites, and are observed in most body fluids [1, 2]. Their destination and

Amit K. Srivastava (ed.), *Extracellular Vesicles*, Neuromethods, vol. 228, https://doi.org/10.1007/978-1-0716-4905-3_3,

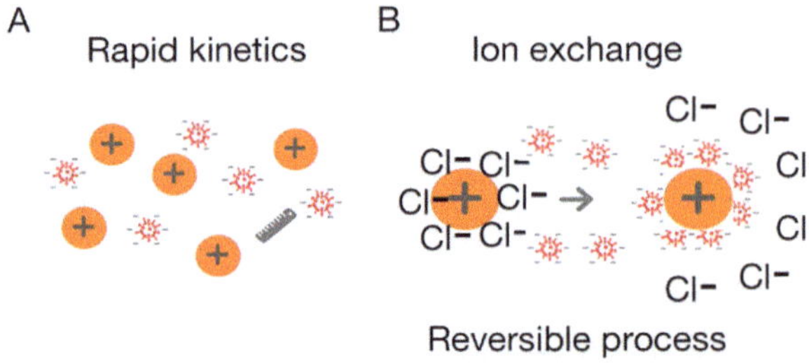

Fig. 1 Enrichment principles. (**A**) The positively charged magnetic beads attract the negative exosomes, enabling rapid binding kinetics and a fast enrichment protocol. (**B**) Cl^- ions are outcompeted with exosomes during the magnetic bead-based enrichment. An anion with greater relative affinity can then be added to replace the exosomes and then release them into the sample

function may be influenced by their biomolecular corona [3–6]. Exosomes are an important factor in intercellular communication by transferring bioactive molecules between cells, indicating the physiological and pathological status of the parent cells that provides their diagnostic potential [7, 8]. In addition, they can act as therapeutic targets or drug delivery vehicles [9]. By analyzing the contents of exosomes isolated from liquid biopsies, researchers can gain insight into cancer progression and drug resistance, as well as the immune response, antigen presentation, programmed cell death, angiogenesis, inflammation, coagulation, and diffusion of oncogenes from tumor cells.

There are currently no standard approaches for exosomes enrichment or separation [10]. Methods typically follow three different primary concepts. The approaches aim to (1) "concentrate" by reducing volume while retaining the target vesicles together with other materials, (2) "enrich" by retaining the vesicles while reducing the number of unwanted materials, or (3) "separate" by utilizing the combination of certain biophysical and/or biochemical properties of the exosoms [10]. The identification and classification of the separated exosomes often relies on a combination of techniques [11, 12]. Methods include measurement of size and concentration, flow cytometry, RT-qPCR, western blot, mass spectrometry, and electron microscopy.

In general, magnetic beads employed in isolation protocols are characterized by the rapid binding kinetics (Fig. 1A). The proximity of the beads to the targets in the solution is directly linked to short incubation times and thus fast protocols. Here we take advantage of the negative surface charge properties of the exosomes (−10 mV to −50 mV) [10, 13] (Fig. 1B) and use positively charged (strong anion-exchange) magnetic beads to enrich exosomes. The enrichment from a variety of biological samples (cell culture medium, serum, plasma, and urine) is conducted within 10 min. The enriched exosomes can be released from the beads by adding anions

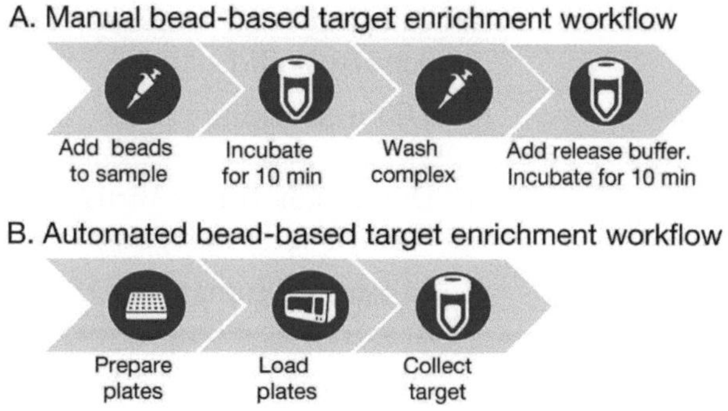

Fig. 2 Exosome enrichment workflow using a manual (**A**) or automated (**B**) method

with stronger relative affinity as a function of charge and hydrated size compared to the bound exosomes. This short and straightforward manual (Fig. 2A) or automated (Fig. 2B) enrichment approach increases the chance of high yield of the exosomes.

To separate out subpopulations of exosomes, different enrichment methods are combined, utilizing different exosome characteristics [10]. Here, surface charge and antigen expression are used to enrich and separate exosomes in a sequential process. Exosomes are enriched using strong anion-exchange magnetic beads followed by release using a high salt buffer. Magnetic beads (Dynabeads™ or DynaGreen™) are used to separate exosomes expressing CD81 on their surface. Extensive characterization of the CD81-expressing exosomes is performed by microfluidic resistive pulse sensing, western blot, electron microscopy, mass spectrometry, and flow cytometry. The enrichment and separation approaches are simplified by using an automated purification system (Fig. 2B).

2 Materials

2.1 Cell Culture

1. SW480 cell line.
2. Jurkat cell line.
3. HCT-116 cell line.
4. HuH-7 cell line.
5. Trophoblast cell line.
6. RPMI 1640 Medium.
7. Heat-inactivated fetal calf serum (FCS).
8. Integra CELLine CL 1000 Culture System bottles.
9. Integra CELLine AD 1000 Culture System bottles.

10. Corning cell culture bottles 225 cm^2 with 0.2 μm ventilation lids.
11. 100 mM sodium pyruvate solution.

2.2 Immunomagnetic Isolation

1. Dynabeads Intact Virus Enrichment (For Research Use Only. Not for use in diagnostic procedure).
2. Exosome-Human CD81 Flow Detection Reagent (For Research Use Only. Not for use in diagnostic procedure).
3. DynaGreen™ CaptureSelect™ Anti-IgG-Fc (Multi-Species) Magnetic Beads (For Research Use Only. Not for use in diagnostic procedure).
4. Wash buffer: Phosphate buffered saline (PBS) (alternative 1).
5. Wash buffer: 10 mM NaCl in 20 mM triethanolamine, pH 6 (alternative 2).
6. 2 mL Sarstedt Tubes.
7. Mixer.
8. Magnet.
9. Buffer for sample dilution: PBS with 0.05% Tween v/v (PBST).
10. Serum, plasma, and urine sample dilution: dilute serum in PBST to a concentration of maximum 50% v/v.
11. Release buffer Alternative 1: 0.25 M KI in 20 mM triethanolamine, pH 6.
12. Release buffer Alternative 2: 0.25 M KI in 20 mM triethanolamine HCl, pH 6.
13. Release buffer Alternative 3: 50 mM citric acid, 50 mM Na-phosphate, pH 4.
14. Magnetic bead concentration: 0.4 mg (10 μL) Dynabeads™ Intact Virus Enrichment per isolation.

2.3 Flow Cytometry

1. Attune NxT Flow Cytometer.
2. Anti-human CD81-PE.

2.4 Electrophoresis

1. Lysis buffer: Radio Immunoprecipitation Assay (RIPA) buffer (3.75 mL 4 M NaCl, 1 mL NP40, 0.5 g SDS, 0.61 g Tris Base, H_2O up to 90 mL, adjust pH to 8 with 5 M HCl, H_2O up to 100 mL).
2. Protease inhibitor: add complete, EDTA-free 25× proteinase inhibitor solution to RIPA lysis buffer.
3. Sample Loading Buffer: 4× Bolt® LDS Sample Buffer (For Research Use Only. Not for use in diagnostic procedure).

4. Precast polyacrylamide gels of various polyacrylamide concentrations and gradient gels are commercially available.
5. Power supply is required for electrophoresis and western blot.
6. Electrophoresis running buffer: 20× Bolt® MES SDS Running Buffer (For Research Use Only. Not for use in diagnostic procedure).

2.5 Western Blot

1. iBlot™ 2 Gel Transfer Device (For Research Use Only. Not for use in diagnostic procedure),
2. iBlot™ 2 Transfer Stacks, PVDF (For Research Use Only. Not for use in diagnostic procedure),
3. iBind™ Western Device (For Research Use Only. Not for use in diagnostic procedure),
4. iBind™ Solution Kit (For Research Use Only. Not for use in diagnostic procedure).

2.6 qPCR

1. MagMAX™ Cell-Free Total Nucleic Acid Isolation Kit (For Research Use Only. Not for use in diagnostic procedure).
2. TaqMan™ MicroRNA Reverse Transcription Kit (miR-16 and miR-24 microRNA) (For Research Use Only. Not for use in diagnostic procedure).
3. TaqMan™ Fast Advanced Master Mix (For Research Use Only. Not for use in diagnostic procedure).
4. TaqMan™ Single Tube Assays (For Research Use Only. Not for use in diagnostic procedure).

2.7 Transmission Electron Microscopy

1. 1% glutaraldehyde v/v in 200 mM cacodylate buffer (pH 7.4), 1.5% v/v Potassium ferricyanide, 1% v/v osmium tetroxide, 1.5% v/v magnesium uranyl acetate
2. 70% ethanol, 90% ethanol, 96% ethanol, absolute ethanol
3. Epon.
4. Ultramicrotome for ultrathin sections.
5. Glow discharged copper grids.
6. Blocking solution: PBS with 0.5% v/v BSA.
7. Protein A gold (10 nm).
8. 0.3% uranyl acetate.

2.8 Resistive Pulse Sensing

1. Spectradyne™ nCS1™ Particle Analyzer.
2. Spectradyne™ ARC™ Particle Analyzer.
3. C-400 cartridge (particle diameters of ~65–400 nm).
4. C-2000 (particle diameters of ~250–2000 nm).

3 Methods

3.1 Exosome Preparation

Exosomes produced by different cells may vary greatly. The Integra CELLine Culture System has demonstrated increased exosome yield and concentration [14] and therefore a good alternative to standard cell culture bottles for is small-scale exosome production. Protocols for exosome production using Integra CELLine Culture System culture flasks are described.

SW480 cells are grown to confluence in RPMI 1640 supplemented with 10% FCS (*see* **Notes 1** and **2**) and 1 mM sodiumpyruvat in Integra CELLine Culture System bottles in a humidified 37 °C incubator with 5% CO_2 atmosphere. The cell-conditioned medium is processed at day 7. After collecting the cell-conditioned medium, two centrifugation steps are performed to remove remaining cells and larger particles (300× *g* for 10 min at 2–8 °C and 2000× *g* for 30 min at 2–8 °C). The precleared conditioned medium is subjected to additional centrifugation (10,000× *g* for 40 min at 2–8 °C, fixed angle) or stored at −80 °C and after thawing, subjected to the 10,000× *g* centrifugation. Exosomes are spiked into serum, plasma, and urine.

3.2 Generic Enrichment of Exosomes Spiked into Serum, Plasma, or Urine

Serum

After centrifugation of whole blood (1,300 rpm for 15 min at 2–8 °C), the serum fraction was collected. The serum was filtered after a second round of centrifugation at 10,000× *g* for 30 min at 4 °C (*see* **Note 3**). The serum was used diluted (10% and 50% in PBST) or undiluted. Exosomes spiked into serum are diluted to 1:20.

Plasma

Human EDTA plasma fraction was collected after centrifugation of EDTA blood (1,300 rpm for 15 min at 2–8 °C) followed by a second round of centrifugation at 10,000× *g* for 30 min at 4 °C and filtered to remove debris (*see* **Note 3**). Plasma fraction was diluted 1:1 in PBST. Exosomes are spiked into plasma at a dilution of 1:20.

Urine

Urine samples are precleared by centrifugation (10,000× *g* for 45 min at 2–8 °C, fixed angle) (*see* **Note 3**). The supernatant is then transferred to a new tube prior to experimentation. Urine samples are used diluted (10% and 50% in PBST) or undiluted. Exosomes spiked into urine are diluted to 1:20.

3.3 Generic Isolation of Exosomes from Serum, Plasma, or Urine

Prepare Magnetic Beads

1. Briefly vortex the beads to resuspend and then place onto a roller for 5 min.
2. Transfer 20 μL of resuspended beads to a new tube.
3. Add 400 μL Washing Buffer 1 to the tube, then mix thoroughly.
4. Apply a magnet for 1 min, then remove the buffer.
5. Magnetic beads are ready for exosome isolation (*see* **Note 6**).

Exosome Enrichment from Serum

1. Dilute serum between 1:1 and 1:10 in Wash Buffer 1 (*see* **Note 4**).
2. Add 1 mL of diluted plasma to prewashed magnetic beads (*see* **Note 5**).
3. Mix an incubate at RT on a roller for 10 min (*see* **Note** 7).
4. Apply a magnet for 1 min, then remove the supernatant.
5. Add 1 mL of Washing Buffer 1, then mix thoroughly by gently pipetting.
6. Apply a magnet for 1 min, then remove the buffer (repeat twice).

Exosome Enrichment from Plasma

1. Dilute plasma 1:1 in Wash Buffer 1 (*see* **Note 4**).
2. Add 1 mL diluted plasma to prewashed magnetic beads (*see* **Note 5**).
3. Mix an incubate at RT on a roller for 10 min (*see* **Note** 7).
4. Apply a magnet for 1 min, then remove the supernatant.
5. Add 1 mL of Washing Buffer 1, then mix thoroughly.
6. Apply a magnet for 1 min, then remove the buffer (repeat step 5–6 twice).

Exosome Enrichment from Urine

1. Add 1 mL of urine to prewashed magnetic beads (*see* **Note 5**).
2. Mix an incubate at RT on a roller for 10 min (*see* **Note** 7).
3. Apply a magnet for 1 min, then remove the supernatant.
4. Add 1 mL of Washing Buffer 1, then mix thoroughly.
5. Apply a magnet for 1 min, then remove the buffer (repeat step 4–5 twice).

The bead-bound exosomes are ready for downstream analysis.

3.3.1 Release of Bead-Bound Exosomes

For certain applications, exosomes must be released from the magnetic beads prior to being used. This is a fast and efficient release protocol to be performed after a generic enrichment of exosomes.

Exosome Release

1. Add 50 μL Release Buffer to the magnetic bead-bound exosomes (*see* **Note 8**).
2. Mix well.
3. Incubate on a roller for 10 min.
4. Mix well by pipetting or vortexing.
5. Apply a magnet for 1 min.
6. Transfer the supernatant containing the enriched exosomes to a new tube (*see* **Note 9**).

The enriched and released exosomes are now ready for downstream analysis.

Size and Concentration Analysis of Released Exosomes

The Spectradyne™ nCS1™ Particle Analyzer is employed to measure the size and concentration of exosomes released from the magnetic beads. Two cartridges are utilized to determine the diameter of released vesicles in the ranges of 65–400 nm (C-400) and 250–2000 nm (C-2000). Released exosomes were diluted in filtered PBS with 1% Tween 20 to 1:10 or 1:1000 and 4 μL of diluted exosomes are applied to the C400 or C2000 cartridges for analysis.

Phenotype Analysis of Released Exosomes

The ARC™ Particle Analyzer is used to determine the identity of the enriched vesicles.

1. Incubate released exosomes with PE-labeled mouse anti-human CD81 antibody for 2 h at RT in the dark (PE-labeled mouse IgG1 antibody is used as an isotype control).
2. Dilute labeled exosomes 1:10 in 1× PBS containing 1% Tween 20 detergent.
3. Insert labeled exosomes into the instrument.

3.3.2 Ultrastructural Analysis of Enriched of Exosomes

Bead-Bound Exosomes for Transmission Electron Microscopy

Exosomes isolated with magnetic beads can be processed for ultrathin sectioning and electron microscopy using a traditional TEM protocol [15]. In brief, for conventional Epon embedding and sectioning, magnetic beads with exosomes can be fixed for 1 h in 1% glutaraldehyde in 200 mM cacodylate buffer (pH 7.4) (*see* **Note 10**), washed repeatedly in aqua distillate, and incubated for 1 h in cacodylate buffer containing 1% OsO_4 and 1.5% $K3Fe(CN)_6$. After two subsequent 30-min incubations in 1% tannic acid and 1.5% magnesium uranyl acetate, the samples are dehydrated by using ethanol and embedded in Epon. Ultrathin sections can be stained with lead citrate.

Immunolabelling and Negative Stain of Released Exosomes for Transmission Electron Microscopy

1. Load exosomes to copper grids at RT for 15 min (*see* **Note 11**).
2. Block exosomes with 0.5% BSA 10 min.
3. Incubate with anti-tetraspanin Abs for 30 min.
4. Wash 5× PBS for 10 min total.
5. Incubate with rabbit & mouse secondary Ab for 30 min.
6. Wash 5× PBS for 10 min total.
7. Incubate with protein A Au (e.g., 10 nm colloidal gold) for 15 min.
8. Wash 5× PBS for 10 min total.
9. Wash 5× H_2O for 10 min total.
10. Embed labelled exosomes in 0.3% Uranyl acetate in methyl cellulose on ice (*see* **Note 12**).
11. TEM analysis.

3.3.3 Separation of Exosome Subpopulations

The separation of subpopulations of exosomes can be done by combining enrichment methods. Two surface-based methods are combined: surface charge enrichment followed by antigen-based separation. The following protocols include generic exosome enrichment followed by specific bead-based capture of exosomes expressing tetraspanin surface markers to distinguish subpopulations of exosomes.

Prepare Magnetic Beads

1. Resuspend the magnetic beads on a roller for 5 min.
2. Transfer 20 μL of resuspended beads to a new tube.
3. Add 400 μL Washing Buffer 1 to the tube, then mix thoroughly.
4. Apply a magnet for 1 min, then remove the buffer.
5. Magnetic beads are ready for exosome separation.

Enrich for Exosomes from Serum, Plasma or Urine

1. Dilute.
   (a) plasma 1:1 in Wash Buffer 1 (*see* **Note 4**).
   (b) serum between 1:1 or 1:10 in Wash Buffer 1 (*see* **Note 4**).

2. Add 1 mL of
   - (a) diluted plasma to prewashed magnetic beads (*see* **Note 5**).
   - (b) diluted serum to prewashed magnetic beads (*see* **Note 5**).
   - (c) urine to prewashed magnetic beads (*see* **Note 5**).
3. Mix and incubate at RT on a roller for 10 min.
4. Apply a magnet for 1 min, then remove the supernatant.
5. Add 1 mL Washing Buffer, then mix thoroughly.
6. Apply a magnet for 1 min, then remove the buffer (repeat step 5–6 twice).

Exosome Release

1. Add 50 μL Release Buffer to the magnetic bead-bound exosomes (*see* **Note 8**).
2. Mix well.
3. Incubate on a roller for 10 min.
4. Mix well by pipetting or vortexing.
5. Apply a magnet for 1 min.
6. Transfer the supernatant containing the enriched exosomes to a new tube.

Separation of Exosome Subpopulations for Flow Cytometry

Magnetic beads also provide a solid support for easy handling during staining and light scatter detection. Enriched exosomes (using strong anion-exchange beads and release buffer) are captured with magnetic beads conjugated with anti-human CD81 antibodies.

Magnetic Bead-Based Separation of Exosome Subpopulation

Preparation of Magnetic Beads

1. Add 300 μL PBS 0.1% BSA in a Sarstedt tube.
2. Add 20 μL of magnetic beads (2×10^5 magnetic beads of a 1×10^7 beads/mL stock solution).
3. Apply a magnet 1 min.
4. Remove the buffer.

Separation of Exosomes

5. Add 100 μL exosome solution to the prewashed magnetic beads (*see* **Note 6**).
6. Incubation overnight (16–24 h) at 2–8 °C with mixing (*see* **Note 13**).
7. Make a quick spin for a few seconds to remove magnetic beads from the lid (on a table top centrifuge).
8. Apply a magnet for 1 min, then remove the supernatant.
9. Add 300 μL PBS 0.1% BSA.

10. Apply a magnet for 1 min, then remove the buffer (repeat once).
11. Resuspend the beads in 150 μL PBS 0.1% BSA.
12. Keep the sample at 2–8 °C until the samples are ready for flow staining.

Staining of Separated Exosomes

13. Transfer 20 μL of staining antibodies to flow tubes (e.g. anti-CD9-PE and isotype control) (*see* **Note 14**).
14. Add 50 μL bead-bound exosomes to the tube. Mix gently by pipetting.
15. Incubate for 45–60 min at RT on a sample shaker (approx. 1000 rpm, protected from light).
16. Add 300 μL of PBS 0.1% BSA to the bead-bound exosomes. Gentle mixing by pipetting (do not vortex).
17. Apply a magnet for 1 min, then remove the supernatant (repeat once).
18. Resuspend the stained, bead-bound exosomes in 300 μL PBS 0.1% BSA.

The stained bead-bound exosomes are ready for flow cytometry analysis.

3.3.4 Rapid Separation of Exosomes from Plasma or Urine

For rapid separation of exosomes with CD81 surface markers, the following protocols are utilized using anti-IgG Fc-conjugated magnetic beads to capture exosomes with CD81 surface markers.

Prepare Magnetic Beads

1. Resuspend anti-IgG Fc-conjugated magnetic beads in the vial (vortex >30 s or tilt and rotate 5 min).
2. Transfer 25 μL magnetic beads to a tube.
3. Apply a magnet for 1 min.
4. Remove the buffer.

Bind Antibody

1. Dilute antibody (2.5 μg) in 200 μL PBST.
2. Add diluted antibody to the magnetic beads.
3. Mix and incubate at RT on a roller for 10 min.
4. Apply a magnet for 1 min, then remove the supernatant, add 200 μL PBST (repeat twice).
5. Apply a magnet for 1 min, then remove the supernatant.
6. The antibody-coated magnetic beads are ready for exosome separation.

Exosome Separation*

1. Add 200 µL of the sample to the antibody-coated magnetic bead (*see* **Note 5**).
2. Mix and incubate at RT on a roller for 10 min (*see* **Note 15**).
3. Apply a magnet for 1 min, then remove the supernatant, add 200 µL PBST mix gently (repeat 3 times).
4. Add 200 µL PBST.
5. Place the tube on the magnet and remove the supernatant.
6. Repeat **step 4–5** 3 times.
7. Place the tube on the magnet and remove the supernatant.

*Alternatively, exosomes can be isolated automatically using the KingFisher instrument.

The magnetic bead-bound exosomes are ready for downstream analysis.

3.4 Analysis of Exosomes

3.4.1 Analysis of Enriched Exosome

Resistive Pulse Sensing Analysis of Exosomes after Generic Exosome Enrichment

Size and Concentration

Exosomes were enriched with strong anion-exchange magnetic beads utilizing the negative charge (10–50 mV) of the exosomes. Magnetic beads were added to cell culture medium with SW480 exosomes, followed by a 10-min incubation period. The enrichment was performed manually or with the KingFisher Instrument. The Spectradyne™ nCS1™ Particle Analyzer is used to measure size and concentration of exosomes released from the magnetic beads. Two cartridges were utilized to measure vesicle diameters in the ranges of 65–400 nm (C-400) and 250–2,000 nm (C-2000). Size and concentration data from particles of diameters 65–400 nm (Fig. 3A) and 250–2000 nm (Fig. 3B) was obtained. Sizes and concentrations of vesicles between 250 nm and 400 nm were measured by both cartridges and demonstrated good alignment (Fig. 3C). Concentrations in replicate samples in the size range observed for exosomes were comparable (C-400).

Phenotype

The Spectradyne™ ARC™ Particle Analyzer was used to determine the identity of the enriched vesicles. The released exosomes were stained with PE-labeled anti-human CD81 antibody. The single-particle fluorescence-based phenotyping showed a >tenfold higher PE-positive signal compared to the isotype control and indicates successful enrichment of exosomes (Fig. 3D).

Flow Analysis of Magnetic Bead-Bound Exosomes

Enrichment Efficiency

The efficiency of the enrichment was evaluated by flow cytometry. Exosomes spiked into PBS were depleted with 10 µL or 25 µL of positively charged magnetic beads and then discarded. The remaining exosomes in the supernatant were captured with magnetic beads coated with anti-human CD81. Exosomes bound to the beads were stained with a PE-labeled anti-CD81 antibody and

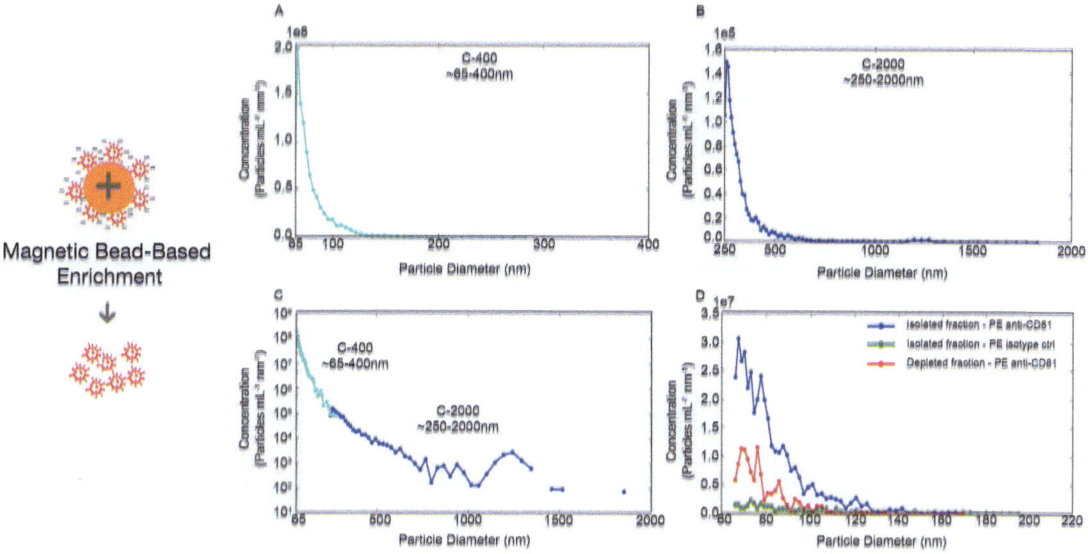

Fig. 3 Size, concentration, and phenotype. Exosomes were enriched with positively charged magnetic beads, followed by release using 0.25 M KI in 20 mM triethanolamine. Size and concentration were measured with the Spectradyne™ nCS1 Particle Analyzer (**A–C**) according to the manufacturer's instructions. Exosome phenotype was determined with the Spectradyne™ ARC™ Particle Analyzer according to the manufacturer's instructions (**D**). (**A**) Size and concentration data using the C-400 cartridge (65–400 nm). (**B**) Size and concentration data using the C-2000 cartridge (250–2,000 nm). (**C**) Comparable size and concentration data across the 65–2,000 nm size range (C-400 and C-2000 combined). (**D**) Size, concentration, and phenotype of exosomes (CD81)

analyzed by flow cytometry after gating the single-bead population (Fig. 4A). The mean fluorescence intensity (MFI) in the PE channel was recorded. Prior to depletion, the MFI was 2,363, and the same sample stained with the isotype control had an MFI of 140, which represents the background noise shown in gray (Fig. 4B). Analysis of exosomes remaining after enrichment using 10 μL or 25 μL of beads decreased the PE signal to 456 and 426, respectively (Fig. 4C, D). The signal decrease indicates that the beads depleted approximately 80% of the exosomes that were present in the sample.

Recovery of Released Exosomes

The exosomes were recaptured using magnetic beads coated with anti-human CD81 antibody and processed for flow staining and analysis (Fig. 5A). Exosomes isolated with 10 μL (Fig. 5C) or 25 μL (Fig. 5D) of magnetic beads, released with 0.25 M KI, and recaptured with magnetic beads coated with anti-human CD81 antibody demonstrated that the exosomes remained structurally undamaged throughout the whole process. For comparison, exosomes were captured directly and analyzed (Fig. 5B). Each sample was compared to the isotype control using an irrelevant antibody of the same isotype (gray peaks).

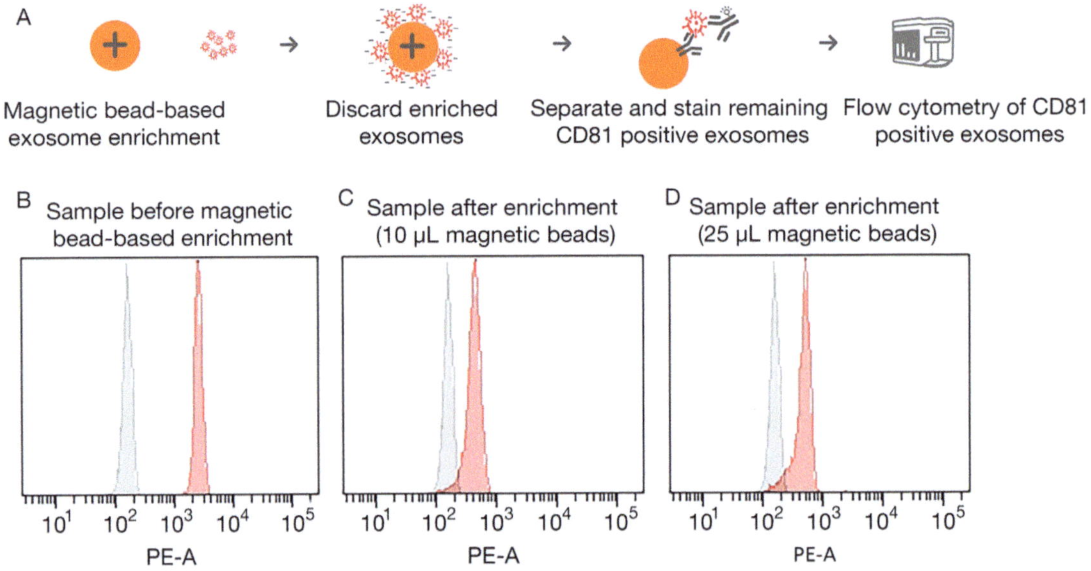

Fig. 4 Flow cytometry analysis of enrichment efficiency. (**A**) Exosomes were enriched from spiked samples (in PBS) with 10 μL or 25 μL of strong anion-exchange magnetic beads. The enriched exosomes were discarded. Residual exosomes in the supernatant were captured with anti-human CD81 magnetic beads and probed with a PE-labeled anti-CD81 detection antibody. The labeled bead-bound exosomes were examined by flow cytometry. (**B**) Exosomes present in the sample prior to enrichment with magnetic beads. (**C**) Exosomes left in the sample after enrichment of exosomes with 10 μL or (D) 25 μL of beads. The gray histograms represent the isotype control (**B**–**D**)

Western Blot Analysis after Generic Exosome Enrichment

Exosomes enriched and released from magnetic beads showed the presence of the exosome marker CD81 (Fig. 6A, B). The CD81 signal in the western blot was increased by increasing the number of beads used for enrichment. Exosomes were efficiently released from the magnetic beads by adding 0.25 M KI in 20 mM triethanolamine to the bead-bound exosomes for 10 min at RT (Fig. 6B).

Electron Microscopy Analysis of Exosomes after Enrichment

Ultrastructural analysis of enriched exosomes was performed by transmission electron microscopy. Intact vesicles were observed to be associated with the positively charged magnetic beads (Arrowheads, Fig. 7A). At greater magnification, the vesicular membranes of the structures were easily observed (Arrowhead, Fig. 7B). For identification, the released exosomes were immunolabeled followed by negative stain contrasting and TEM analysis. The vesicles released were CD81-positive (Arrowheads, Fig. 7C), indicating the exosome nature of the enriched vesicles.

3.5 Complex Starting Materials

3.5.1 Enrichment of Exosomes from Serum, Plasma, and Urine

Exosomes were spiked into serum, plasma, and urine followed by magnetic bead-based enrichment of negatively charged exosomes. Exosomes were released from the magnetic beads with 0.25 M KI. The serum was undiluted or diluted to 10% or 50% in PBS. Plasma was diluted to 50% in PBS while urine was diluted to 10% or

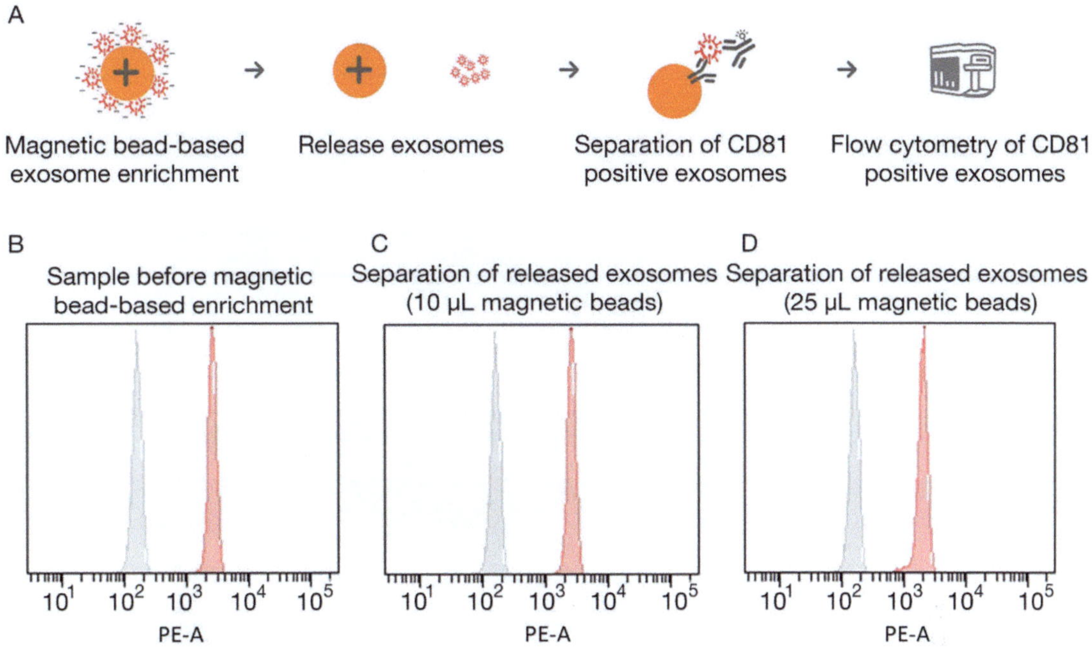

Fig. 5 Flow cytometry analysis of exosomes released from strong anion-exchange magnetic beads. (**A**) Exosomes were enriched from spiked samples (in PBS) with 10 μL or 25 μL of beads. The enriched exosomes were released from the beads with 0.25 M of KI and recaptured with anti-human CD81 magnetic beads. The bead-bound exosomes were probed with PE-labeled anti-CD81 detection antibodies and analyzed by flow cytometry. (**B**) Exosomes present in the sample prior to bead-based enrichment. (**C**) Exosomes released after enrichment with 10 μL or (**D**) 25 μL of magnetic beads. Gray histograms represent the isotype control (**B**–**D**)

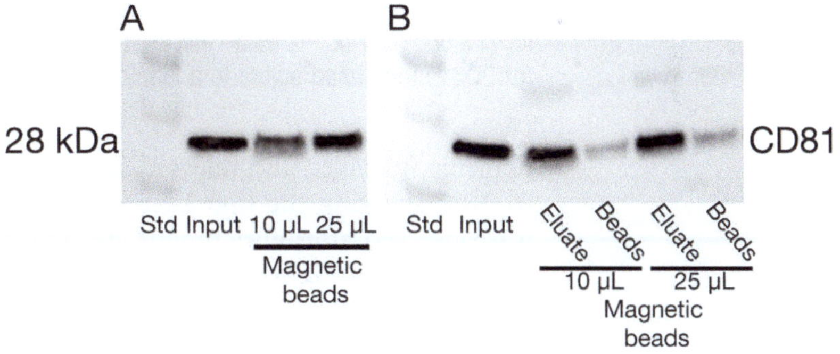

Fig. 6 Enrichment of exosomes with strong anion-exchange magnetic beads. (**A**) Exosomes were enriched with 10 μL or 25 μL of magnetic beads and analyzed for CD81 by western blot. (**B**) Exosomes were isolated with 10 μL or 25 μL of beads, followed by release. The elution and unreleased exosomes were examined for CD81 by western blot

50% in PBS. Released exosomes were analyzed by western blot using an anti-human CD81 antibody (Fig. 8A) or by recapture with magnetic beads conjugated with anti-human CD81 antibodies for downstream flow cytometry analysis (Figs. 8B–I). CD81 was

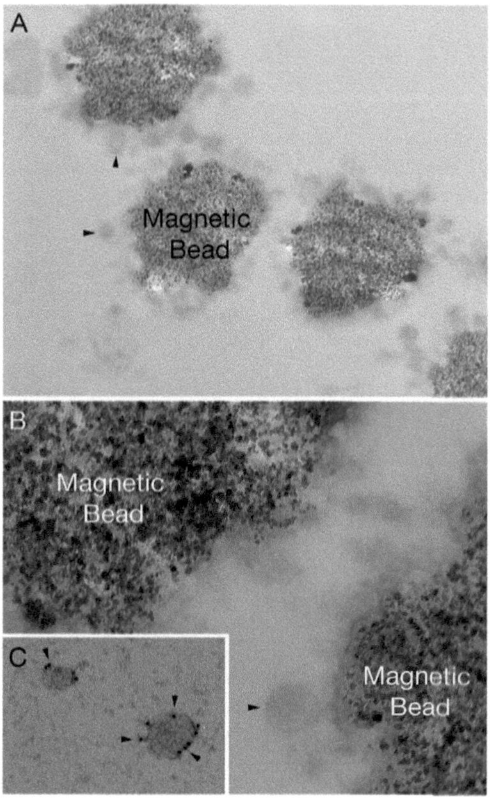

Fig. 7 Ultrastructural analysis of enriched exosomes. Exosomes were enriched with strong anion-exchange magnetic beads and processed for transmission electron microscopy. (**A**, **B**) Exosomes on the surface of magnetic beads (arrowhead in B points to a vesicle structure). (**C**) Released exosomes labeled for CD81 (arrowhead points to a labeled vesicle)

detected in spiked and released serum samples, with serum diluted to 10%, showing the highest band intensity (Fig. 8A). No CD81 was observed after enrichment and release from unspiked PBS or serum samples. CD81 was not detected on the magnetic beads after release, indicating that most of the exosomes were eluted off. The same results were observed by flow cytometry. Exosomes captured and released from magnetic beads were recaptured with magnetic beads conjugated with anti-human CD81 antibodies and analyzed by flow cytometry. Exosomes spiked into PBS, or 10% serum, demonstrated comparable enrichment and release efficiency (Fig. 8B, C). Less dilute serum samples (50% and 100% concentrations) resulted in slightly lower capture efficiency (Fig. 8D, E). Exosomes spiked into urine or plasma demonstrated similar enrichment (Fig. 8F, I). Unspiked samples were included in the experiment. Blue peaks in the histograms represent the signals from unspiked samples and demonstrate that a small fraction of native

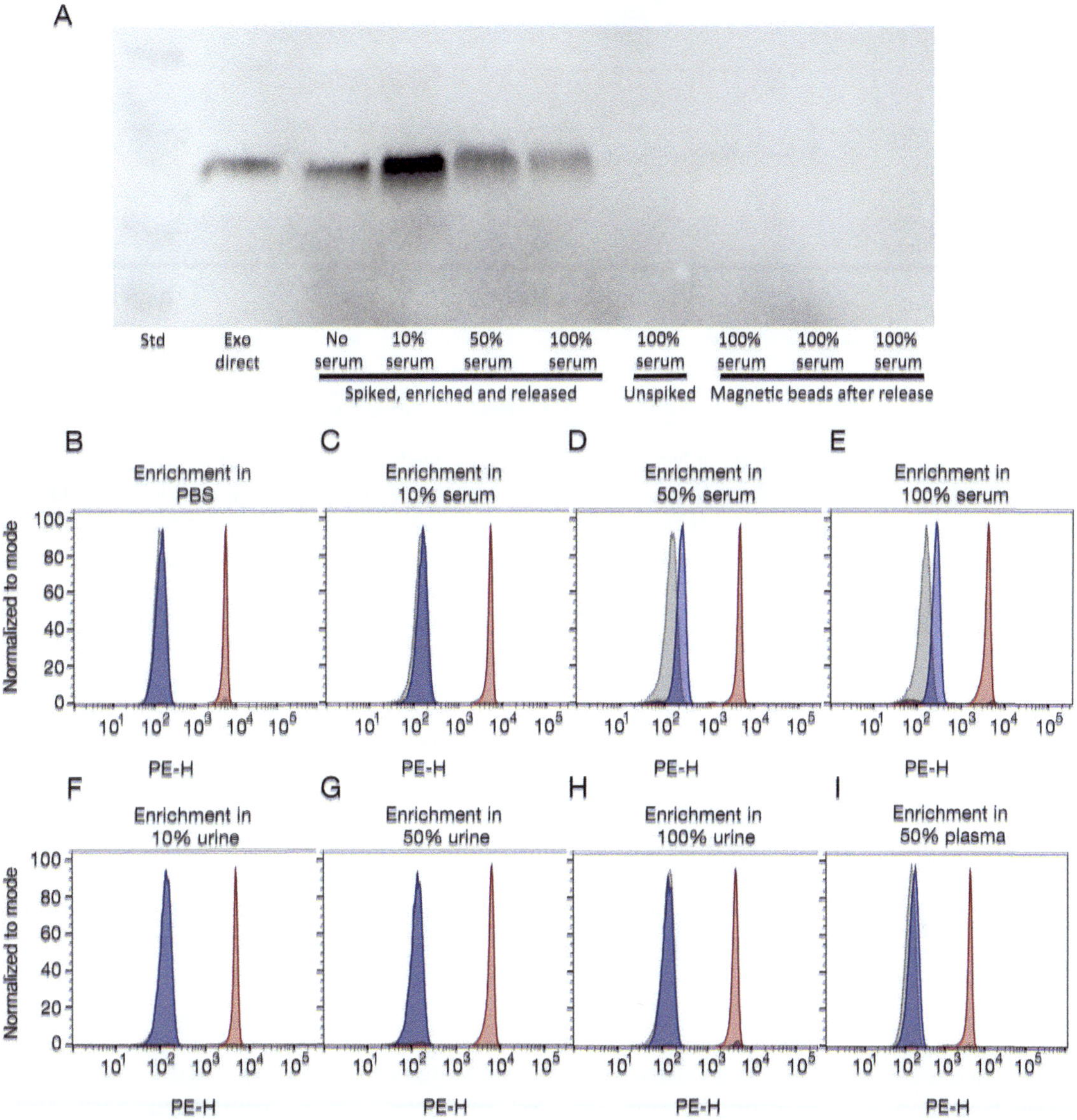

Fig. 8 Enrichment of exosomes from spiked and unspiked serum, plasma, and urine with strong anion-exchange magnetic beads. (**A**) Exosomes were enriched from diluted serum samples (10% or 50% dilutions with PBS) or undiluted (100%) serum samples. The enriched exosomes were released from the beads and analyzed for CD81 by western blot. For comparison, exosomes spiked into PBS, unspiked serum, and beads after release were included. (**B–E**) Flow cytometry analysis of exosomes released from the beads after enrichment from PBS, or 10%, 50%, or 100% serum samples. (**F–I**) Flow cytometry analysis of exosomes released from the beads after enrichment from 10%, 50%, or 100% urine and 50% plasma samples. Released exosomes were captured with anti-human CD81 magnetic beads, stained for CD81 (PE-H), and analyzed by flow cytometry. Gray peaks represent the control with an irrelevant antibody, blue peaks represent unspiked samples, and red peaks represent spiked samples. Peaks were normalized to mode

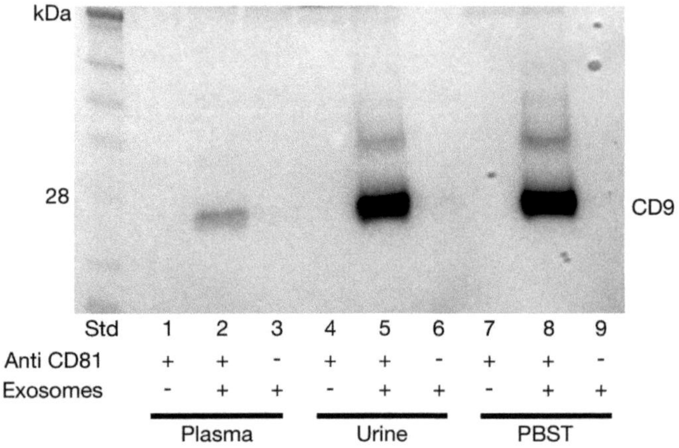

Fig. 9 Purification of exosomes for western blot analysis. Exosomes from spiked and unspiked PBS, plasma, and urine were captured with anti-IgG-Fc magnetic beads conjugated with anti-CD81 antibody (lanes 2, 5, and 8). As negative controls, anti IgG-Fc magnetic beads conjugated with anti-CD81 antibody were incubated with unspiked samples (lanes 1, 4, and 7) and spiked samples with anti IgG-Fc magnetic beads with no antibody bound (lanes 3, 6, and 9). CD9 was detected by western blot

CD81-positive exosomes is enriched. Each sample was compared to the isotype control using an irrelevant antibody of the same isotype (gray peaks).

3.5.2 Analysis After Exosome Separation

Western blot Analysis after Exosome Separation

CD81-positive exosomes were isolated from PBS, plasma, or urine by anti-IgG Fc magnetic beads conjugated with anti-CD81 antibodies and prepared for western blot. Vesicle identity was confirmed by detection of CD9 (Fig. 9).

qPCR Analysis after Specific Exosome Separation

CD81-positive exosomes were separated from PBS, plasma, or urine by anti-IgG Fc magnetic beads conjugated with anti-CD81 antibodies and prepared for qPCR (Fig. 10). cDNA synthesis was performed on miR-16 and miR-24 microRNA. In comparison, the number of exosomes was analyzed by qPCR prior to CD81-dependent capture. The results indicate that most of the exosomes that were present in the sample were captured specifically by the magnetic beads as the Ct values were very similar between bead isolated (blue dots) exosomes and enriched sample (red dots).

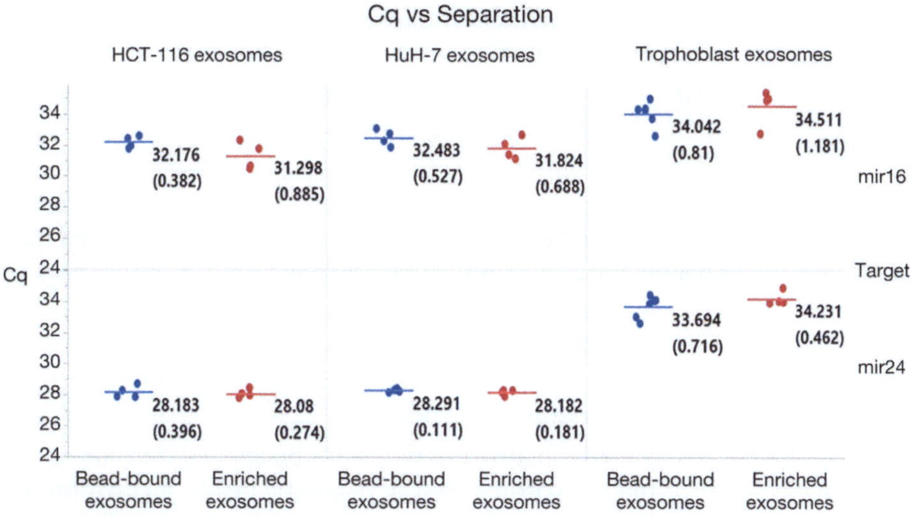

Fig. 10 Purification of exosomes for downstream qPCR. The pre-enriched (precipitated) exosomes samples were divided into two: precipitated sample and CD81 bead isolated sample. Both samples were processed using for qPCR. cDNA synthesis was performed on miR-16 and miR-24 microRNA. Finally, qPCR was conducted on the cDNA for miR-16 and miR-24. These results indicate that all or most of the exosomes that were present in the sample were captured specifically, since the Ct values were very similar between the bead-based purification and the precipitated sample

4 Notes

1. Cells are cultured with exosome-depleted FCS. An alternative is to grow the cells in the absence of FCS if they can handle this.
2. High cell viability is essential to reduce the release of apoptotic bodies.
3. Eppendorf rotor F45-24-11 (85650).
4. General recommendations. Alternative buffers may be used such as 10 mM NaCl in 20 mM triethanolamine, pH 6.
5. As exosome concentration may vary between samples; sample volume must be optimized.
6. More magnetic beads can be added if complete exosome enrichment is required. The protocol can be scaled up by increasing volumes and reagents proportionally.
7. Proper mixing is important. V-shaped tubes are not recommended. The efficiency can be increased by extending the incubation time to 1 h.
8. Alternatives (1) 0.25 M KI in 20 mM triethanolamine, pH 6, (2) 0.25 M KI in 20 mM triethanolamine HCl, pH 6, (3) 50 mM citric acid, 50 mM Na-phosphate, pH 4.
9. The released exosomes can be used for "any" downstream application, including further separation based on specific

surface markers. Release buffer can be replaced by dialysis or spin columns.

10. Magnetic beads with exosomes that are subjected to fixation (GA) require more time on the magnet.
11. Glow discharge carbon-coated copper grids to improve adsorption by making the surface hydrophilic. Use 5–10 L of released exosomes per grid. Cover the grids when incubating.
12. For best possible contrast, different contrast solutions should be considered: Aqueous Uranyl Acetate (for particles stable in low pH, Neutral Phosphotungstic Acid (for particles not stable at low pH, produces less contrast compared to uranyl acetate), Ammonium Molybdate.
13. Shorter incubation times can be applied, but for optimal binding of exosomes to the magnetic beads, incubation up to 21 h is recommended (reason for incubation at 2–8 °C).
14. Titration of the staining antibody is necessary for optimal signal-to-noise ratio (S/N).
15. As exosome concentration may vary between samples; incubation times may be increased optimized.

Acknowledgments

Special thanks to Anette Kullmann, Ruth M. Martin, Juili Kelvekar, and Marc McGowan (Thermo Fisher Scientific) for data; special thanks to Antje Hoenen and Norbert Roos (University of Oslo, Norway) for technical assistance in electron microscopy.

Conflict of Interest The authors have indicated that they have no other conflicts of interest regarding the content of this chapter.

References

1. Villarroya-Beltri C et al (2013) Sumoylated hnRNPA2B1 controls the sorting of miRNAs into exosomes through binding to specific motifs. Nat Commun 4(2980):1–10
2. Ferreira JV et al (2022) LAMP2A regulates the loading of proteins into exosomes. Sci Adv 8 (eabm1140):1–19
3. Tóth EÁ et al (2021) Formation of a protein corona on the surface of extracellular vesicles in blood plasma. J Extracell Vesicles 10(e12140): 1–26
4. Vergauwen G et al (2020) Robust sequential biophysical fractionation of blood plasma to study variations in the biomolecular landscape of systemically circulating extracellular vesicles across clinical conditions. J Extracell Vesicles 10(e12122):1–18
5. Wolf M et al (2021) A functional corona around extracellular vesicles enhances angiogenesis, skin regeneration and immunomodulation. J Extracell Vesicles 11(e12207):1–20
6. Buzas EI (2022) Opportunities and challenges in studying the extracellular vesicle corona. Nat Cell Biol 24:1322–1325
7. Hendrix A (2021) The nature of blood (y) extracellular vesicles. Nat Rev Mol Cell Biol 22:243–243

8. Geeurickx E, Hendrix A (2020) Targets, pitfalls and reference materials for liquid biopsy tests in cancer diagnostics. Mol Asp Med 72(100828):1–17
9. Van Deun J, Consortium E-T (2017) EV-TRACK: transparent reporting and centralizing knowledge in extracellular vesicle research. Nat Methods 14(3):28–32
10. Hendrix A et al (2023) Extracellular vesicle analysis. Nat Rev Method Primers 3:56
11. Théry C et al (2018) Minimal information for studies of extracellular vesicles 2018 (MISEV2018): a position statement of the International Society for Extracellular Vesicles and update of the MISEV2014 guidelines. J Extracell Vesicles 7(1535750):2–43
12. Wever OD, Hendrix A (2019) A supporting ecosystem to mature extracellular vesicles into clinical application. EMBO J 38(e101412):1–5
13. Seo N et al (2022) Distinguishing functional exosomes and other extracellular vesicles as a nucleic acid cargo by the anion-exchange method. J Extracell Vesicles 11:e12205
14. Mitchell JP, Court J, Mason MD, Tabi Z, Clayton A (2008) Increased exosome production from tumour cell cultures using the Integra CELLine culture system. J Immunol Methods 335:98–105
15. Pedersen KW, Meer Yv, Roos N, Snijder EJ (1999) Open reading frame 1a-encoded subunits of the Arterivirus Replicase induce endoplasmic reticulum-derived double-membrane vesicles which carry the viral replication complex. J Virol 73(3):2016–2026

Chapter 4

Isolation of CNS-Originating Extracellular Vesicles from Blood and Cell Culture Media

Ibrar Siddique, Megan Hong, Kimia Mohseni, Seyedshahabadin Bidoki, and Gal Bitan

Abstract

The inaccessibility of the brain and spinal cord makes studying biochemical changes in the central nervous system (CNS) difficult. Traditional methods include brain imaging and biochemical analysis of the cerebrospinal fluid, which are not ideal. CNS-originating extracellular vesicles (EVs) isolated from blood have emerged as a promising source of biomarkers for a variety of CNS studies, because they can cross the blood-brain barrier and protect their cargo, including proteins, nucleic acids, carbohydrates, and lipids, providing a glimpse into the biochemical status of the originating cells. Importantly, immunoprecipitation using antibodies against cell-specific EV surface markers allows isolating CNS-originating EVs from blood and other easily accessible biofluids. High-sensitivity methods can then analyze the EV contents to measure candidate biomarkers or be used for biomarker discovery. However, in some downstream applications, such as mass-spectrometry-based proteomics or surface-enhanced Raman spectroscopy, nonspecific binding of high-abundance biomolecules in the biofluid may prevent using CNS-originating EVs. In these cases, EVs isolated from the culture medium of neurons or glia may provide an alternative, because an immunoprecipitation step is unnecessary. Here, we briefly discuss the strategy of analyzing biomarkers in such EVs and provide detailed protocols for isolating CNS-originating EVs from blood products and neuronal-culture media.

Key words Neurodegenerative disorders, Extracellular vesicles, Blood biomarkers

1 Introduction

Neurodegenerative proteinopathies, such as Alzheimer's disease (AD) and Parkinson's disease (PD), affect millions of people worldwide and are characterized by deposition of protein aggregates in the brain and progressive neurodegeneration [1, 2]. Development of effective therapy for these diseases is highly challenging for multiple reasons, including the difficulty in getting drugs to pass through the blood-brain barrier (BBB) and the intractable nature of the abnormal protein assemblies that injure and kill neurons [3, 4]. Another major challenge is the dearth of reliable biomarkers

Amit K. Srivastava (ed.), *Extracellular Vesicles*, Neuromethods, vol. 228, https://doi.org/10.1007/978-1-0716-4905-3_4,

to assist in determining the diagnosis, severity, progression, prognosis, and treatment effects due to the difficulty in accessing the brain and spinal cord. Although recently, much progress has been made for AD [5, 6] and for PD and other synucleinopathies, [7, 8], sensitive, specific, and easily accessible biomarkers for neurodegenerative diseases are still an urgent public health need. Currently, the diagnosis, prognosis, and the assessment of progression and clinical trial outcome measures rely primarily on a clinical evaluation, which, due to the progressive nature and complex symptoms of neurodegenerative proteinopathies, is prone to errors. Traditional biomarker approaches include measurement of biomolecules in the cerebrospinal fluid (CSF) and various brain-imaging modalities [9, 10]. Early-stage diagnosis is especially difficult because of significant patient-to-patient variation and overlapping symptoms among different diseases, leading to high rates of misdiagnosis [11, 12].

Reliable biomarkers should be sensitive, reproducible, and highly specific. High specificity allows for accurately identifying a particular biological process or condition, distinguishing it from other processes that may manifest via similar symptoms. Thus, a positive amyloid positron-emission tomography (PET) scan may indicate a high likelihood of developing AD, yet some people may have a positive scan and never become demented [13], suggesting that the specificity of this biomarker on its own is limited. Biomarkers ideally also would be noninvasive or minimally invasive. Although CSF has been the medium of choice in many CNS-biomarker studies, because it often reflects biochemical changes in the brain accurately, the invasive nature of the lumbar puncture necessary for obtaining CSF leads many patients to refuse the procedure. CSF is especially problematic as a source of biomarkers in clinical trials requiring analysis at multiple time points. Thus, other bodily fluids that can be obtained using minimally invasive means, such as blood, urine, and saliva, are preferred if the biomarker levels in them reflect reliably the changes in the CNS.

Given these limitations, there is an urgent need to identify and develop reliable, sensitive, specific, and minimally invasive biomarkers for CNS disorders. However, biofluids that can be obtained noninvasively, such as saliva, tears, sweat, or urine, typically do not reflect biochemical changes in the CNS due to the BBB. Thanks to the development of highly sensitive detection methods, such as single-molecule array (Simoa) [14, 15] and electrochemiluminescence immunoassay (ECLIA) [16], substantial progress has been made in our ability to detect certain CNS-originating biomarkers in blood plasma or serum. However, even using high-sensitivity methods, this is often difficult, because the BBB prevents most biomolecules from leaving the CNS, minute amounts that end up in the blood circulation are diluted further, because the volume of the blood is ~40-fold higher than that of the CSF and the blood is richer in catabolic enzymes than the CSF, reducing the concentration of potential biomarkers even more.

An attractive alternative to measuring biomolecules in the CSF is analysis of biomarkers in CNS-originating EVs isolated from the blood [17]. EVs are released by virtually all eukaryotic and prokaryotic cells into the extracellular environment. They act as messengers in cell-to-cell communication, deliver cargo from donor to recipient cells, and modulate the physiological condition of the recipient cells. Importantly, cells use EVs also to help eliminate stressors, such as abnormal oligomers and aggregates of amyloid proteins in proteinopathies, which means that the levels of these stressors in circulating EVs often can be used as a biomarker in distinguishing patients from healthy people [18–20]. CNS-originating EVs can be isolated from various biofluids by immunoprecipitation using surface markers specific for their cell of origin, including neurons and glia [21]. For example, neuronal EVs have been isolated using neuronal cell-adhesion molecule [22], L1 cell-adhesion molecule (L1CAM) [22–25], growth-associated protein 43 (GAP-43) [26], neuroligin 3 (NLGN3) [26], ATPase Na^+/K^+ transporting subunit α3 (ATP1A3) [27], and glutamate ionotropic receptor NMDA-type subunit 2A (NMDAR2A) [28]. Isolation of astroglial EVs has been demonstrated using aspartate-glutamate transporter (GLAST) [29], and oligodendroglial EVs have been immunoprecipitated using myelin oligodendrocyte glycoprotein (MOG) [23] or 2,3-cyclic nucleotide-3-phosphodiesterase [30]. Thus, CNS-originating EVs have emerged in recent years as a promising source of biomarkers for neurological disorders and injuries.

For some downstream applications, such as unbiased, cell-specific, broad-spectrum target discovery by mass spectrometry, immunoprecipitated EVs from biofluids cannot be used, because highly abundant and ubiquitous proteins in these biofluids, for example, albumin, bind nonspecifically to the matrix used in the immunoprecipitation step and obfuscate the signal of less abundant proteins of interest. Instead, cell culture models, such as induced pluripotent stem cell-derived neurons (iPS-Ns), can provide EVs that are biologically relevant, cell-type specific, and easy to isolate for biochemical and biophysical studies without immunoprecipitation.

Here, we provide detailed methods for isolating neuronal, astroglial, or oligodendroglial EVs from serum or plasma using antibodies against NLGN3, GLAST, or MOG, respectively, and neuronal EVs from the culture media of iPS-Ns. It is important to note that these methods are not exclusive. Other isolation approaches and other EV-surface markers may be suitable for similar purposes. Additionally, the isolation of EVs from the media of neurons can be easily adapted for isolating EVs from glial cell cultures.

2 Isolation of CNS-Originating EVs from Blood

The first step in isolating CNS-originating EVs from blood is separating the blood cells and collecting serum or plasma according to standard protocols. Importantly, this step should be performed between 15 and 30 min from the blood collection to avoid erythrocyte hemolysis, which may contaminate the sample. The serum or plasma should be mixed, and any part not used immediately should be aliquoted into 200 μL aliquots (*see* **Note 1**), flash-frozen, and stored at −80 °C.

2.1 Reagents and Lab Equipment Required

1. 1.5 mL microcentrifuge tubes (Eppendorf tubes).
2. Pipettors and pipette tips for up to 10 μL, 20 μL, 100 μL, 200 μL, and 1000 μL volumes.
3. Protease and Phosphatase Inhibitor cocktail (PPI, including serine- cysteine-, and acid-protease inhibitors, and aminopeptidase inhibitors, e.g., from Sigma-Aldrich, St. Louis, MO).
4. Thrombin, (e.g., from Fisher Scientific, Waltham, MA).
5. ExoQuick™ Exosome Precipitating Solution (System Biosciences, Palo Alto, CA); similar reagents are available from different manufacturers. *See* **Note 6**.
6. Temperature-controlled, fixed-angle bench-top centrifuge, for example, from Eppendorf (Enfield, CT).
7. Phosphate-Buffered Saline, 10× Solution (1.37 M NaCl, 0.027 M KCl and 0.119 M phosphate, e.g., from Thermo Fisher Scientific).
8. Dynabeads™ M-270 Epoxy (Thermo Fisher Scientific).
9. Dynabeads™ Antibody Coupling Kit (Thermo Fisher Scientific).
10. For isolation of neuronal EVs—Neuroligin-3 monoclonal antibody (Cat. No sc-137052, Santa Cruz Biotechnology, Santa Cruz, CA). For isolation of astroglial EVs—GLAST (ACSA-1) Antibody (Cat. No 130-095-822, Miltenyi Biotec, Inc., San Diego, CA). For isolation of oligodendroglial EVs—MOG Antibody (D-2) (Cat no. sc-376138, Santa Cruz Biotechnology).
11. Tube rotator, for example, Thermo Scientific Labquake Tube Rotator, Benchmark Roto-Therm Plus™, or Fisherbrand™ Tube Rotator.
12. A 4 °C room or a refrigerator that can accommodate the tube rotator.
13. Bovine serum albumin (BSA), for example, from Sigma-Aldrich.

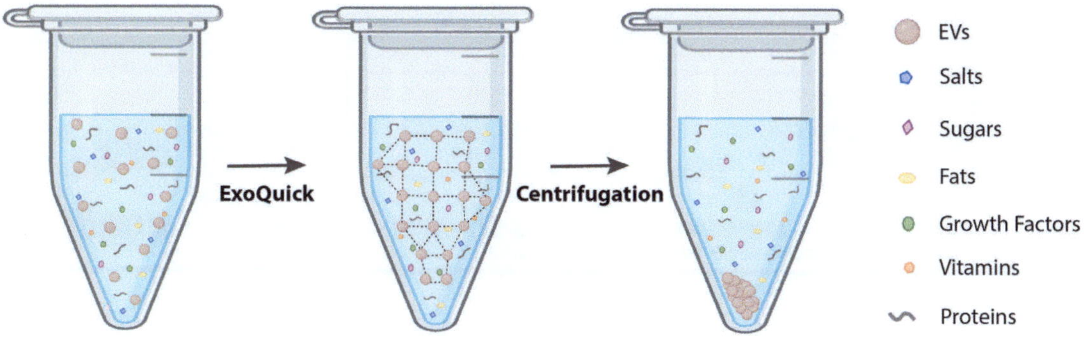

Fig. 1 Schematic representation of polymer-assisted EVs precipitation

14. Magnet for separating the Dynabeads™ from the solution, for example, from Thermo Fisher Scientific.
15. Radioimmunoprecipitation-assay (RIPA) Lysis and Extraction Buffer (25 mM Tris·HCl, pH 7.6, 150 mM NaCl, 1% NP-40, 1% sodium deoxycholate, 0.1% SDS, Thermo Fisher Scientific).
16. Flat-bottom tissue-culture 96-wells plates (CELLTREAT, Ayer, MA).
17. Pierce™ BCA Protein Assay Kit (Thermo Fisher Scientific).
18. Biotek Synergy HTX plate reader (Agilent, Santa Clara, CA).
19. Glycine·HCl buffer (0.1 M glycine adjusted to pH 2.0 using hydrochloric acid).

The protocol below is divided into three parts: (A) Precipitation of EVs (Fig. 1), (B) Preparation of antibody-coupled beads, and (C) Immunoprecipitation of nEVs (Fig. 2).

A. *Precipitation of EVs from Serum or Plasma*

1. Thaw the serum or plasma sample(s) on ice. When thawed, immediately add PPI (*see* **Notes 2** and **3**). If plasma is used, add 1 U/100 μL thrombin (*see* **Notes 4** and **5**). When the sample is completely thawed, incubate for 5 min at room temperature and then centrifuge at 21,000× *g* for 10 min. A small precipitate will be visible. Transfer the supernatant liquid to a fresh tube and add PPI before proceeding. After this point, the protocol uses serum only for clarity and brevity but also refers to thrombin-treated plasma.
2. Add 25.2 μL of ExoQuick™ (*see* **Note 6**) per 100 μL of the serum. Add the ExoQuick™ solution slowly, mixing immediately and continuously by pipetting up and down several times until the solution is clear and homogenous.
3. Incubate on ice for 1 h. Set the benchtop centrifuge temperature to 4 °C.

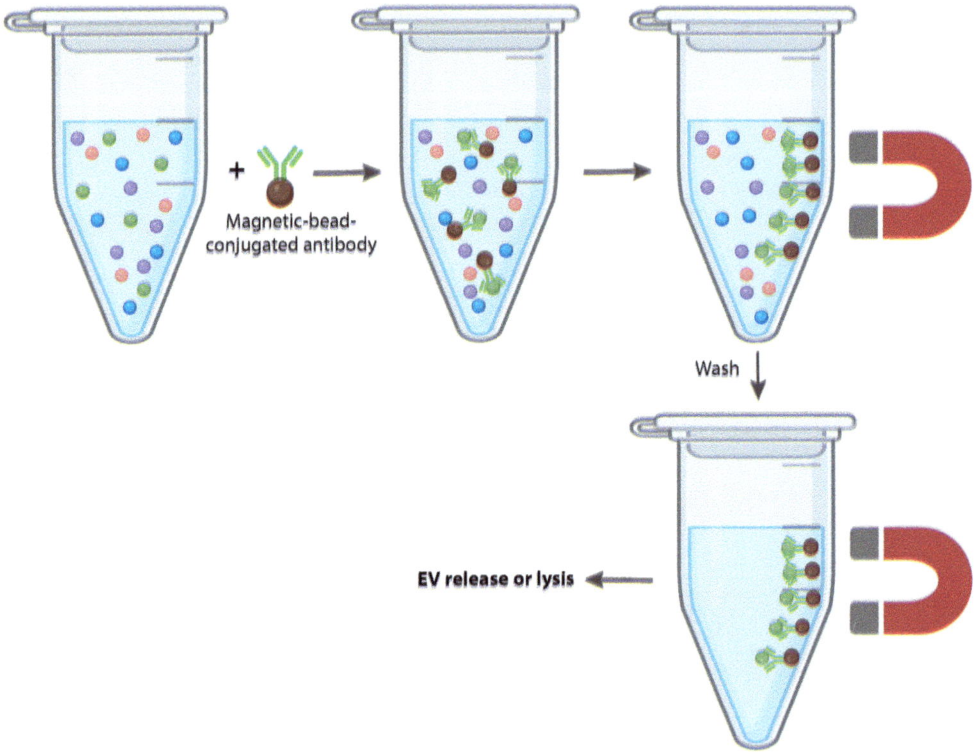

Fig. 2 Schematic representation of immunoprecipitation of EVs using a specific cell-surface marker from a total EV suspension

4. Centrifuge the solution at 4 °C for 30 min at 1500× *g*. The EVs form a yellowish pellet at the bottom of the tube that can be observed by the naked eye (Fig. 3).
5. Remove the supernatant liquid, add 200 μL of PBS containing 1% (w/v) BSA and PPI without disturbing the pellet, and incubate overnight at 4 °C (*see* **Note 7**). The pellet softens and can be resuspended in the next step.

B. *Preparation of Antibody-Coupled Beads*

1. Conjugate 10 mg of Dynabeads™ (*see* **Note 8**) to 20 μg of anti-NLGN3 antibody (*see* **Note 9**) using a Dynabeads™ Antibody-coupling kit from Thermo Fisher Scientific, strictly following the manufacturer's protocol (*see* **Notes 10** and **11**).

C. *Immunoprecipitation of nEVs*

1. Resuspend the pellet generated in step A5 in 200 μL of PBS containing 1% (w/v) BSA and PPI gently to a homogenous suspension and keep it on ice. Transfer 150 μL of the bead-conjugated antibody suspension made in step B1 into a new 1.5-mL vial (*see* **Note 12**). Use the magnet to collect the antibody-conjugated magnetic beads on the tube's wall.

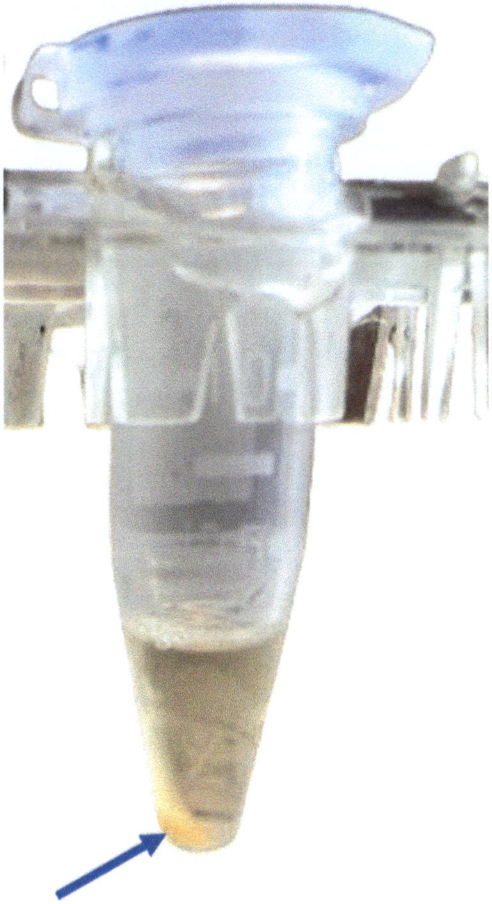

Fig. 3 A yellowish pellet (blue arrow) forms after ExoQuick™ treatment

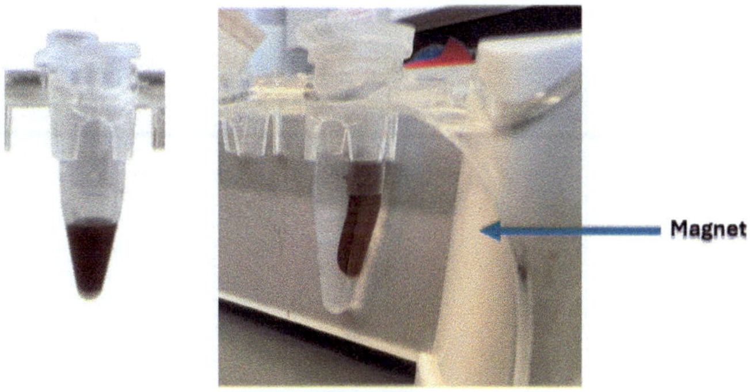

Fig. 4 Dynabeads™ in suspension off the magnet (left) and on the magnet (right)

The beads form a clump against the magnet, and the solution looks clear (Fig. 4). Remove the solution without disturbing the magnetic beads, remove the vial from the magnet, add the EV suspension, and mix thoroughly using a pipette.

2. Transfer the mixture of antibody-conjugated beads and EVs carefully to a new vial. Pipette the mixture to the bottom of the vial, ensuring that the vial walls above the liquid remain dry. Incubate with slow vertical rotation at 4 °C overnight (*see* **Notes 13–17**).
3. If clumps of beads are found at the solution's surface at the end of step C2, push them into the solution using a pipette tip and resuspend gently, without letting the beads adhere to the wall above the liquid. Use the magnet to collect the beads on the side wall and remove the liquid.
4. Remove the magnet, wash the beads by adding 500 μL of ice-chilled 0.1% BSA in PBS and resuspend the beads. Collect the beads again using the magnet and remove as much as possible of the wash buffer. Remove the vial from the magnet. The antibody-conjugated beads are now bound to the desired EVs.

The final step depends on the EVs' downstream processing (e.g., measurement of tetraspanin concentrations). If the content of the EVs is needed following EV lysis, continue to step C5a. If intact EVs are needed, follow step C5b.

5. (a) Lyse the EVs by adding 20 μL of RIPA buffer and incubating on ice for 1 h. Ensure that all the beads are immersed in the buffer, and none are left on the wall outside the solution (*see* **Note 18**).
   (b) Use Glycine·HCl buffer, pH = 2, to denature the antibody and release the intact EVs.

3 Small EV (sEV) Isolation from Culture Media

The protocol below is for isolation of neuronal sEVs from the culture media of iPS-Ns. The same protocol can be used for isolation of sEVs from the media of cell lines, primary cultures, or other iPS cell types.

3.1 Reagents and Lab Equipment

1. Cell culture of choice.
2. Media of choice.
3. SmartSEC™ Single EV Isolation System (System Biosciences).
4. Swinging-bucket rotor ultracentrifuge (*see* **Note 19**).
5. Ultracentrifuge tubes.
6. 15-mL polypropylene conical tubes.
7. 1.5-mL microcentrifuge tubes.
8. Lysis buffer: Neuronal Protein Extraction Reagent (N-PER™, Thermo Fisher Scientific) or similar.

Protocol

1. Collect media from cell culture. Culture cells as per protocol, this method was developed for neural progenitor cells (NeuraCell, Rensselaer, NY) that get differentiated into neurons and plated in 6-well plates with 1.5 mL media per well; 0.5 mL of media was collected for sEV isolation (*see* **Note 20**).
2. Cool the ultracentrifuge to 4 °C.
3. Isolate the sEVs using SmartSEC™ Single columns (*see* **Note 21**). These columns use beads that trap most contaminants in the media, while sEVs flow through the column. Following the manufacturer's protocol, incubate the samples in the column for 30 min at room temperature. At the lowest volume limit, 500 μL, this time can be decreased to 20 min. Elute sEVs following the manufacturer's protocol into ultracentrifuge tubes.
4. Ultracentrifuge the samples at 100,000× *g* at 4 °C for 1–3 h (*see* **Note 22**).
5. After ultracentrifugation, remove as much supernatant liquid as possible without touching the bottom of the ultracentrifuge tube.
6. Resuspend the EV pellet, which may be invisible to the naked eye, in 50 μL of PBS.
7. Add 50 μL of N-PER™ or a similar lysis buffer to the PBS suspension to lyse the EVs. Either process immediately for downstream applications, such as proteomics, or flash-freeze and store at −80 °C for later use.

4 Notes

1. Other volumes may be stored according to the planned downstream use.
2. The serum and plasma are rich in proteins, including proteases that can cleave target proteins. To minimize protein loss, it is crucial to thaw the serum or plasma on ice and add PPI immediately. Avoid repeated freeze-thaw cycles of the samples as this may gradually degrade the molecules of interest. If freeze-thaw cycles are unavoidable, test how they affect the analytes of interest.
3. If the plan does not involve measuring phosphorylated proteins, adding phosphatase inhibitors is unnecessary. If other posttranslational modifications (e.g., acetyl groups) are part of the analysis, the appropriate enzyme inhibitors should be added.

4. The plasma contains fibrinogen. The digestion with thrombin converts the fibrinogen into insoluble fibrin, allowing its precipitation. The treatment with thrombin does not precipitate 100% of the fibrin, but in our experience, it is more than enough for avoiding matrix effects. If both serum and plasma samples are to be analyzed as part of the same experiment, we recommend testing first whether the measurements in serum and thrombin-treated plasma yield comparable results using commercial serum and plasma, for example, from Innovative Research (Los Angeles, CA).
5. Treatment with thrombin is not an absolute must. Biomarkers can be measured in EVs isolated from plasma without this treatment, but the results for certain biomarkers will differ. If this step is skipped, matrix effects for the biomarker of interest must be tested first.
6. ExoQuick™ (System Biosciences, Palo Alto, CA) is a proprietary reagent using the principles of "salting-out" and polymer-assisted precipitation to precipitate EVs, regardless of their size, along with lipid particles, apolipoproteins, and possibly cell debris from various biofluids. Similar products are available from other manufacturers, including Minute™ High-Efficiency Exosome Precipitation Reagent (Invent Biotechnologies, Plymouth, MN) and Immunostep exosome precipitation solution (Immunostep, Salamanca, Spain).
7. The pellet is hard after the centrifugation step and difficult to resuspend. Incubating overnight in the 1% BSA in PBS solution softens the pellet, making it easy to resuspend. Additionally, EVs are sticky and tend to bind to the vial's walls. The BSA competes for the wall surface, reducing nonspecific EVs' adsorption to the plastic wall.
8. Use 12-mm diameter aluminum discs to weigh Dynabeads. Dynabeads can be sticky and may be partially lost if weighing paper is used instead.
9. We use an anti-NLGN3 antibody for the precipitation of nEVs. As mentioned in the Introduction, other options used by different groups include antibodies against L1CAM, GAP-43, ATP1A3, and NMDAR2A.
10. Other magnetic beads also may be suitable, such as MagnaBind (Thermo Fisher Scientific), MagSi Beads (MagnaMedics Diagnostics, Geleen, The Netherlands), SPRI Beads (Beckman Coulter, Indianapolis, Indiana), or MACS MicroBeads (Miltenyi Biotec, San Diego, CA). Some researchers use nonmagnetic beads, such as streptavidin-coated beads. Such beads may be used directly without conjugating the antibody, but in such cases, the antibody should be biotinylated.

11. Other magnetic-particle-conjugation kits may be used, such as Absolute Mag™ Amine Magnetic Particles Conjugation Kit from the CD Bioparticles (Shirley, NY) or SMCC (Succinimidyl 4-(N-maleimidomethyl)cyclohexane-1-carboxylate) as cross-linkers from AAT Bioquest (Pleasanton, CA). Nonmagnetic beads, such as Agarose or Sepharose beads, also may be used. These beads are isolated by centrifugation. Separation of magnetic beads is easier than using centrifugation, but magnetic beads may be more expensive. Magnetic beads also facilitate automation of part C of the process using instruments, such as KingFisher Flex (Thermo Fisher Scientific) or IsoPure™ (Thomas Scientific, Swedesboro, NJ).
12. Autoclave the 1.5-mL vials and the sterile water used to make buffers to avoid bacterial contamination in any solutions.
13. The beads are heavy and tend to precipitate fast. Make sure to mix the beads suspension thoroughly before pipetting.
14. When rotating overnight, the solution must remain at the bottom. The surface tension of the solution keeps it at the bottom (Fig. 5). Keeping the vial walls above the liquid dry is crucial for maintaining the surface tension. If the surface tension breaks, the solution spreads all over the vial, resulting in loss of EVs and low yield.
15. It is important that the rotator motion is very smooth and free of vibration or jerking. Vibration or jerking may break the surface tension and cause sample loss. We found that Roto-Mini™ Plus Rotator, Variable Speed (Benchmark, Sayreville,

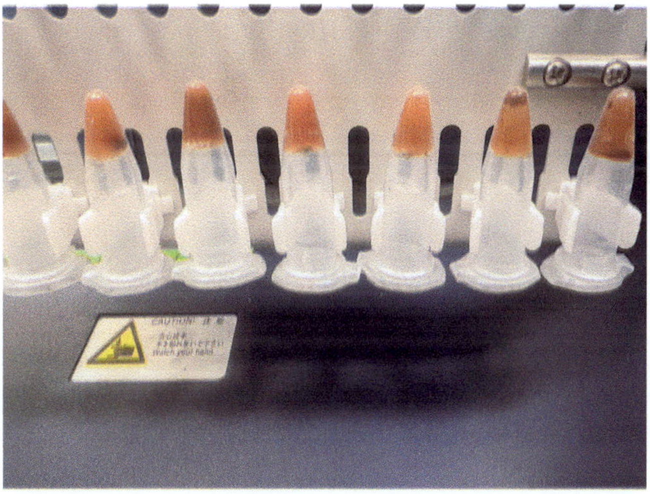

Fig. 5 Antibody-conjugated Dynabeads™ incubated with EV solution at 4 °C on rotation. When the tubes are inverted, the solution's surface tension keeps the beads at the bottom of the tube

NJ), vibrates too strongly and is unsuitable for this task. In contrast, Tube Rotators (Rotisserie, SCILOGEX, Rocky Hill, CT) works well.

16. The maximum volume that maintains the surface tension in a typical 1.5-mL vial is 250 μL. If larger volumes are necessary, they should be divided into 250-μL aliquots and recombined at the end of this step.
17. Incubation at 4 °C can be extended by a few hours, if necessary, without noticeable loss of material. Shortening the incubation time may lead to decreased binding of the EVs to the antibody, potentially reducing the yield.
18. If beads are stuck to the wall, they should be brought down by pouring RIPA buffer on the vial wall several times.
19. A swinging-bucket rotor for ultracentrifugation will pellet sEVs at the bottom of the tube allowing for easier collection, especially when the pellet is not visible. A fixed-angle rotor may be used but leads to lower yields after shorter ultracentrifugation times (*see* **Note 22**).
20. The media can be used immediately, stored at −20 °C for up to one month, or at −80 °C for longer periods. If storage is needed between isolation and final measurements, it is best to perform the EV isolation as close as possible to the measurements, because sEVs are more stable frozen in media than after isolation.
21. Traditional size-exclusion chromatography using a reusable column is cost effective and may be used in this step if sample volume is not an issue. For small volumes, for example, < 1 mL, optimized devices, such as the SmartSEC™ Single EV Isolation System, will minimize sample loss.
22. 1 h will provide enough sample for proteomics; At least 3 h is ideal for ELISA-based analyte quantification.

Acknowledgments

The authors gratefully acknowledge support from NIH/NINDS grants R21NS130326 and RF1NS126406 (GB) and Administrative Supplement to Promote Diversity RF1NS126406-01A1S1 (MH).

References

1. Lamptey RNL et al (2022) A review of the common neurodegenerative disorders: current therapeutic approaches and the potential role of Nanotherapeutics. Int J Mol Sci 23(3):1851
2. Hong M, Bitan G (2024) Recent advances and future therapy development for Alzheimer's disease and related disorders. Neural Regen Res 19(9):1877–1878

3. Rahimi F, Bitan G (2014) Methods for studying and structure–function relationships of non-fibrillar protein assemblies in Alzheimer's disease and related disorders. In: Lahiri DK (ed) Advances in Alzheimer research. Bentham Science Publishers, pp 291–374
4. Rahimi F et al (2016) Modulators of Amyloid β-Protein (Aβ) self-assembly. In: Wolfe MS (ed) Developing therapeutics for Alzheimer's disease. Academic, Boston, pp 97–191
5. Jack CR Jr et al (2024) Revised criteria for diagnosis and staging of Alzheimer's disease: Alzheimer's Association Workgroup. Alzheimers Dement 20(8):5143–5169
6. Jack CR Jr et al (2024) Revised criteria for the diagnosis and staging of Alzheimer's disease. Nat Med 30(8):2121–2124
7. Siderowf A et al (2023) Assessment of heterogeneity among participants in the Parkinson's progression markers initiative cohort using α-synuclein seed amplification: a cross-sectional study. Lancet Neurol 22(5):407–417
8. Gibbons CH et al (2024) Skin biopsy detection of phosphorylated α-Synuclein in patients with Synucleinopathies. JAMA 331(15): 1298–1306
9. Verger A et al (2021) Single photon emission computed tomography/positron emission tomography molecular imaging for parkinsonism: a fast-developing field. Ann Neurol 90(5): 711–719
10. Buchanan SM et al (2021) Mild parkinsonian signs: a systematic review of clinical, imaging, and pathological associations. Mov Disord 36(11):2481–2493
11. Gaugler JE et al (2013) Characteristics of patients misdiagnosed with Alzheimer's disease and their medication use: an analysis of the NACC-UDS database. BMC Geriatr 13:137
12. Beach TG, Adler CH (2018) Importance of low diagnostic accuracy for early Parkinson's disease. Mov Disord 33(10):1551–1554
13. Kozlov M (2024) Should Alzheimer's be diagnosed with a blood test? Proposal sparks controversy. Nature 365:530–531
14. Schubert SM et al (2015) Ultra-sensitive protein detection via single molecule arrays towards early stage cancer monitoring. Sci Rep 5:11034
15. Song L et al (2013) Direct detection of bacterial genomic DNA at sub-femtomolar concentrations using single molecule arrays. Anal Chem 85(3):1932–1939
16. Yoo SM, Jeon YM, Heo SY (2022) Electrochemiluminescence systems for the detection of biomarkers: strategical and technological advances. Biosensors (Basel) 12(9):738
17. Hornung S, Dutta S, Bitan G (2020) CNS-derived blood exosomes as a promising source of biomarkers: opportunities and challenges. Front Mol Neurosci 13:38
18. Perez M, Avila J, Hernandez F (2019) Propagation of Tau via extracellular vesicles. Front Neurosci 13:698
19. Gustafsson G et al (2018) Secretion and uptake of α-Synuclein via extracellular vesicles in cultured cells. Cell Mol Neurobiol 38(8): 1539–1550
20. Liang T et al (2023) The emerging double-edged sword role of exosomes in Alzheimer's disease. Front Aging Neurosci 15:1209115
21. Dutta S et al (2023) Biomarkers for parkinsonian disorders in CNS-originating EVs: promise and challenges. Acta Neuropathol 145(5): 515–540
22. Fiandaca MS et al (2015) Identification of preclinical Alzheimer's disease by a profile of pathogenic proteins in neurally derived blood exosomes: a case-control study. Alzheimers Dement 11(6):600–7 e1
23. Dutta S et al (2021) α-Synuclein in blood exosomes immunoprecipitated using neuronal and oligodendroglial markers distinguishes Parkinson's disease from multiple system atrophy. Acta Neuropathol 142(3):495–511
24. Shi M et al (2014) Plasma exosomal α-synuclein is likely CNS-derived and increased in Parkinson's disease. Acta Neuropathol 128(5):639–650
25. Nogueras-Ortiz CJ et al (2024) Single-extracellular vesicle (EV) analyses validate the use of L1 Cell Adhesion Molecule (L1CAM) as a reliable biomarker of neuron-derived EVs. J Extracell Vesicles 13(6):e12459
26. Eitan E et al (2023) Synaptic proteins in neuron-derived extracellular vesicles as biomarkers for Alzheimer's disease: novel methodology and clinical proof of concept. Extracell Vesicles Circ Nucl Acids 4(1):133–150
27. You Y et al (2023) ATP1A3 as a target for isolating neuron-specific extracellular vesicles from human brain and biofluids. Sci Adv 9(37):eadi3647
28. Tian C et al (2023) Blood extracellular vesicles carrying synaptic function- and brain-related proteins as potential biomarkers for Alzheimer's disease. Alzheimers Dement 19(3): 909–923
29. Goetzl EJ et al (2018) High complement levels in astrocyte-derived exosomes of Alzheimer disease. Ann Neurol 83(3):544–552
30. Yu Z et al (2020) Reduced oligodendrocyte exosome secretion in multiple system atrophy involves SNARE dysfunction. Brain 143(6): 1780–1797

Chapter 5

Isolation of Spontaneously Released Extracellular Vesicles from Human and Mouse Brain

Patrícia A. Gomes, Kalliopi Skourti, Filippos Katsaitis, Cristian Bodo, Minghao Chen, Clarissa L. Waites, Christos Gatsogiannis, Nuno Sousa, Bruno Costa-Silva, and Ioannis Sotiropoulos

Abstract

The isolation of brain extracellular vesicles (EVs), and particularly exosomes, is of the utmost importance for the understanding of their role in physiological and pathological brain processes. Moreover, the detailed characterization of biochemistry and biology of cell-specific small EVs will essentially contribute to the development of novel peripherical biomarkers that allow us to follow the progression of complex and long-lasting brain disorders, such as Alzheimer's disease (AD). Despite growing efforts to understand the 'mirroring relationship' between brain pathology and brain-derived EVs collected peripherally, isolating EVs directly from brain tissue remains challenging. For instance, current methods for the isolation of small EVs (e.g., exosomes) from the brain rely on mechanic and enzymatic digestion of brain tissue, which can disrupt the plasma membrane and contaminate the EV fraction with intraluminal vesicles that would not have been released otherwise. To overcome this drawback, we have developed a novel method for the isolation of brain EVs based on their spontaneous release by the tissue. Thus, the goal of this chapter is to provide methodological guidelines and details of this novel method that produces EV yield enriched in small EVs such as exosomes highlighting the advantages and limitations of this approach.

Key words Extracellular vesicles, Brain, Isolation, Exosomes, Spontaneous, Release

1 Introduction: Brain-Derived Extracellular Vesicles as Novel Biomarker Carriers

Extracellular vesicles (EVs) are recognized as critical mediators of intercellular communication [1] and are secreted by all cell types [2], including the primary cell types of the central nervous system (CNS), such as neurons [3, 4], astrocytes [5], oligodendrocytes [6], and microglial cells [7]. However, our knowledge on the exact role that brain EVs play in both healthy brain function and in pathophysiological conditions remains limited due to the lack of techniques to efficiently isolate pure brain EVs. Recent studies have highlighted the essential role of brain EVs in CNS health and disease [8–11]. In the healthy brain, EVs—including exosomes

Amit K. Srivastava (ed.), *Extracellular Vesicles*, Neuromethods, vol. 228, https://doi.org/10.1007/978-1-0716-4905-3_5,

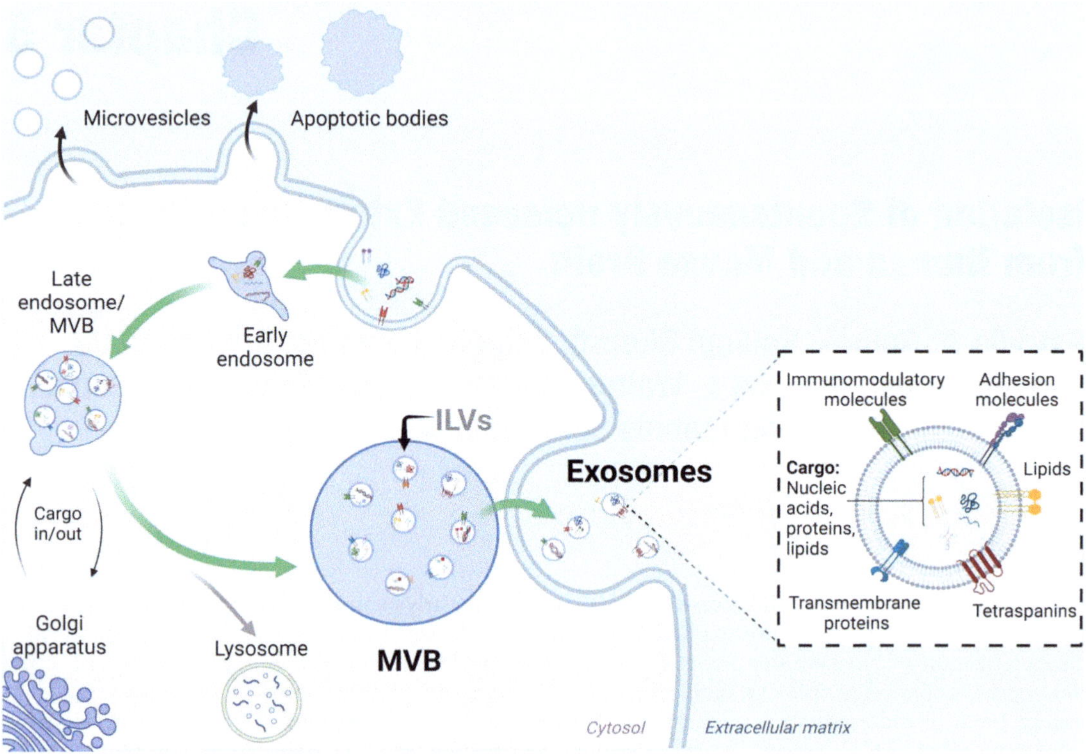

Fig. 1 EV biogenesis and exosome secretion. Exosomes are generated through the endolysosomal pathway. Following endocytosis, plasma membrane gives rise to early endosomes, which further mature into late endosomes or multivesicular bodies (MVBs). Within MVBs, inward budding of the endosomal membrane forms intraluminal vesicles (ILVs), loaded with proteins, lipids, and nucleic acids. MVBs can either fuse with lysosomes for degradation or with the plasma membrane to release ILVs into the extracellular space as exosomes. In light of our inability to distinguish between secreted (extracellular) vesicles and intracellular ILVs, cell membrane damage during mechanical and enzymatic tissue dissociation may lead to the contamination of the EV yield with intracellular components, for example, ILVs, that could otherwise be targeted for degradation instead of exocytosis.

(50–150 nm) and larger microvesicles—play a pivotal role in synaptic plasticity and neuronal homeostasis by regulating key processes such as synaptic pruning [12] and myelination [13]. They are also involved in neuronal development primarily by transferring microRNAs, such as miR-34a, which regulate neuronal differentiation [14], as well as proteins that influence this process [15]—see Fig. 1. Additionally, they orchestrate neuron-glia homeostasis, which is essential for neuronal growth and cell survival [16, 17]. Beyond their physiological roles, brain EVs have also been linked to the progression of neurodegenerative diseases, including Alzheimer's disease (AD) and Parkinson's disease. Specifically, growing evidence indicates that EVs can serve as carriers of pathological proteins such as Aβ and Tau protein [18] as well as α-synuclein [19]. Notably, the inhibition of EVs biogenesis can suppress Tau propagation [20], highlighting the potential of targeting EVs biogenesis and secretion pathways as a therapeutic strategy to mitigate the progression of neurodegenerative brain pathology. Furthermore, due to their ability to cross the blood-brain barrier, brain-derived EVs can be detected in biofluids such as cerebrospinal fluid (CSF) and blood [21, 22]. This accessibility has allowed researchers to isolate brain EVs from plasma using immunoprecipitation techniques, revealing that their cargo reflects the pathological status of the brain [23]. Brain EVs' ability to reflect molecular processes from their originating cells, their cargo-protecting properties, and their long half-lives make them ideal candidates for noninvasive diagnostic and treatment monitoring tools for disease conditions such as brain cancer, multiple sclerosis, AD, Parkinson's disease, major depression, and schizophrenia [24–26]. Therefore, the isolation of brain EVs, which constitute the gold-standard population against which peripheral EV biomarkers ought to be compared to, is of the utmost importance. Conventional methods for brain EVs isolation [27–29] rely on tissue dissociation, which hold the risk of contaminating the EV fraction with intracellular components, for example, immature intraluminal vesicles (ILVs) within late endosomes that would otherwise be targeted for degradation instead of exocytosis [29]. Despite their promising role as biomarkers for CNS disorders, the challenge of isolating brain EVs with high efficiency and low contamination continues to hinder their broader application in diagnostics and therapeutics.

1.1 Conventional Methods to Isolate and Study Brain EVs

As mentioned above, most studies have used in vitro approaches to study the role of EVs in physiological and pathological brain conditions; however, these approaches often fail to depict the multiplex endogenous processes in vivo. Isolating a high-purity yield of the subpopulation of small EVs (sEVs) from biofluids and tissues remains challenging, due to their small diameter and the coexistence with a plethora of "contaminants" of similar size and/or density [30]. Most studies have been able to isolate EVs from cell supernatants [31] or different body fluids (e.g., blood [32], saliva [33], cerebrospinal fluid [34]), with the utilization of several protocol modifications due to their unique and divergent composition [30]). Despite this, EVs' isolation from brain tissue continues to be problematic. Although other research fields may use in vitro approaches that closely mimic in vivo conditions, the complexity of the brain's cellular architecture introduces an additional level of difficulty [35]. Furthermore, another challenge lies in the intricate process of isolating EVs from the extracellular matrix, without disrupting cell membrane integrity [35]. For instance, despite being essential, mechanical disruption of the tissue (e.g., homogenization, vortex, or slicing) is likely to induce at least some degree of membrane damage. In addition, most studies utilize enzymatic digestion, namely papain or collagenase, to disrupt the extracellular matrix and subsequently release trapped EVs. Collagenase does not seem to alter the characteristics of cell or vesicle surfaces and therefore is particularly recommended when the functionality and integrity of cell surface proteins are important [36]. Levy's group and others have proposed a protocol that collects EVs from fresh or frozen brain tissue with minimal damage to cell membranes but with the additional risk of contamination with the indistinguishable, immature intracellular vesicles [35, 37, 38]. In the context of avoiding these issues, studies from the cancer field have demonstrated a methodology for isolating EVs, by placing tissue in culture medium and collecting EVs from the conditioned medium [39–41]. However, the complex architecture of the brain has been preventing the application of this protocol, highlighting the urgent need for an improved methodology that produces a more homogenous EV yield in an effective manner.

In response to this challenge, we propose the "release method" to isolate sEVs upon their release from fresh brain tissue into culture media—the cell's physiological process of sEVs biogenesis. The "released method" is based on conventionally used protocols, with the advantage of incorporating several different steps, aiming to collect EVs released from cells, without mechanical or enzymatic disruption. By allowing EVs to be released spontaneously, this method avoids the need for tissue digestion, resulting in high purity, sEV-enriched yield with EVs being structurally, biochemically, and functionally intact. Additionally, this method could be used for the identification of cell-specific surface markers to distinguish brain EV subpopulations (i.e., neuronal, astrocytic, oligodendrocytic and microglial-derived EVs), which may offer a deeper insight into the pathomechanisms behind complex brain disorders [42]. In conclusion, the "release method" represents a novel tool that lays the foundation for exploring sEVs as potential biomarkers for brain disorders. With minimal intervention, this method preserves the quality of the EVs and offers new opportunities to identify brain-specific biomarkers from EVs collected even from peripheral sources, providing potential for both prognostic and diagnostic applications.

2 Materials

2.1 Brain Tissue

Both human and mouse brain tissue can be used to obtain brain-derived EVs with this method [42]. The most important aspect is to maintain tissue viability, as cell death can influence the release of EVs. Thus, special care should be taken with human brain samples to preserve cell viability by collecting the brain tissue of interest into Hibernate™-A (Gibco™, A1247501) supplemented with serum-free B-27™ (Gibco™, 17504044) immediately after autopsy, and urgently shipped. Consider changing this formulation if using nonadult brain tissue. Based on our work and the expected use of the obtained EVs, the starting tissue amount may vary and may require pooling from different subjects: for instance, 300 mg of tissue should provide $1 - 4 \times 10^9$ particles.

2.2 Dissection Supplies

Lab supplies needed to obtain brain tissue and desired areas include the following:

1. Brain dissection kit of choice
2. Flow hood: mouse brain extraction and microdissection should occur within a flow hood to minimize exposure to contaminants, which could affect EV release [43]
3. Incubator
4. Sterile small petri dishes (same number as of samples to be collected) to weigh the tissue
5. T25 flasks or cell culture dishes (same number as of samples to be collected) to incubate tissue

Reagents necessary to obtain conditioned media:

1. Hibernate™-A (Gibco™, A1247501)
2. EV-release medium (Neurobasal™-A, 1% GlutaMax™, 1% Anti-anti; Gibco™, 10888022, 35050038, 15240062, respectively)
3. Protease and phosphatase inhibitors (cOmplete™, EDTA-free Protease Inhibitor Cocktail, Roche, #04693132001; Phosphatase Inhibitor Cocktail 2 and 3, Sigma-Aldrich, P5726 and P0044, respectively)

2.3 EV Isolation Supplies

Lab supplies needed for EV isolation include the following:

1. 15 mL centrifuge tube (3 times the number of samples) to collect culture media and process it for EV isolation
2. Ultracentrifuge capable of reaching 100,000 g
3. Ultracentrifuge rotor with space for tubes of 25 mL (e.g., 70Ti, Beckman)
4. Ultracentrifuge tubes with a maximum capacity of 25 mL
5. Flow hood: if the isolated EVs are intended to be administered into animals or to treat cells
6. 40 μm cell strainer (e.g., Falcon, 352340)
7. qEV column (IZON©, SP1)
8. Eppendorfs to collect EV fraction
9. 0.22 μm syringe filter

10. 5 mL and 20 mL syringe (same number as number of samples)
11. 18G needle (same number as number of samples)

Reagents necessary for sucrose cushion:

1. Deuterium Oxide (Sigma-Aldrich, 151882-1L)
2. Protease-free sucrose (Sigma-Aldrich, S0389-1Kg)
3. Trizma-base (Sigma-Aldrich, 93352-500g)

Reagents necessary for EV isolation:

1. 0.22 μm filtered-PBS

3 Methods

3.1 Sucrose Cushion

Gradients have been widely used to separate EVs from particles with similar sedimentation coefficients but different densities. These gradients can be continued or discontinued, multilayered or single layered with the former requiring a swinging rotor to not disrupt the gradient. The reagent of choice is either sucrose or optiprep, with the latter being reported to selectively enrich for viral particles within the EV fraction [43]. Herein, we decided to use a single layer of sucrose, that we called "sucrose cushion".

This sucrose cushion can be prepared in advance and stored at 4 °C. Be aware that the heavy water used to make this cushion is not stable for long periods of time.

To prepare 100 mL of sucrose cushion, weigh 30 g protease-free sucrose and 2.4 g Tris-base. Add 50 mL D_2O (heavy water) in the flow hood. Adjust the pH to 7.4 with 10 M HCl (a few drops should be enough), and then add the volume up to 100 mL with D_2O and filter with 0.22 μm filter vacuum inside the laminar.

3.2 qEV Columns

We have decided to use qEV columns to further isolate small EVs. These columns are commercially available or can be manufactured. One should collect all fractions when using these columns for the first time to ascertain which ones contain EVs.

iZon® columns may be used for a maximum of 5 times, after which, they may be regenerated by passing 10 mL of 0.5 M NaOH followed by 30–50 mL of filtered-PBS. Before the isolation, these need to be equilibrated to room temperature; so, remove them from cold storage at least 30 min before the intended use.

3.3 Obtain Brain Tissue: Step 1

Prewarm Hibernate-A™ medium and pour on a small petri dish. Weigh the petri dish with medium for later reference. We suggest that you do not fill all dishes with Hibernate-A™ medium at once; fill it each time when you collect each of your samples, as you are proceeding with brain extraction— see Fig. 2.

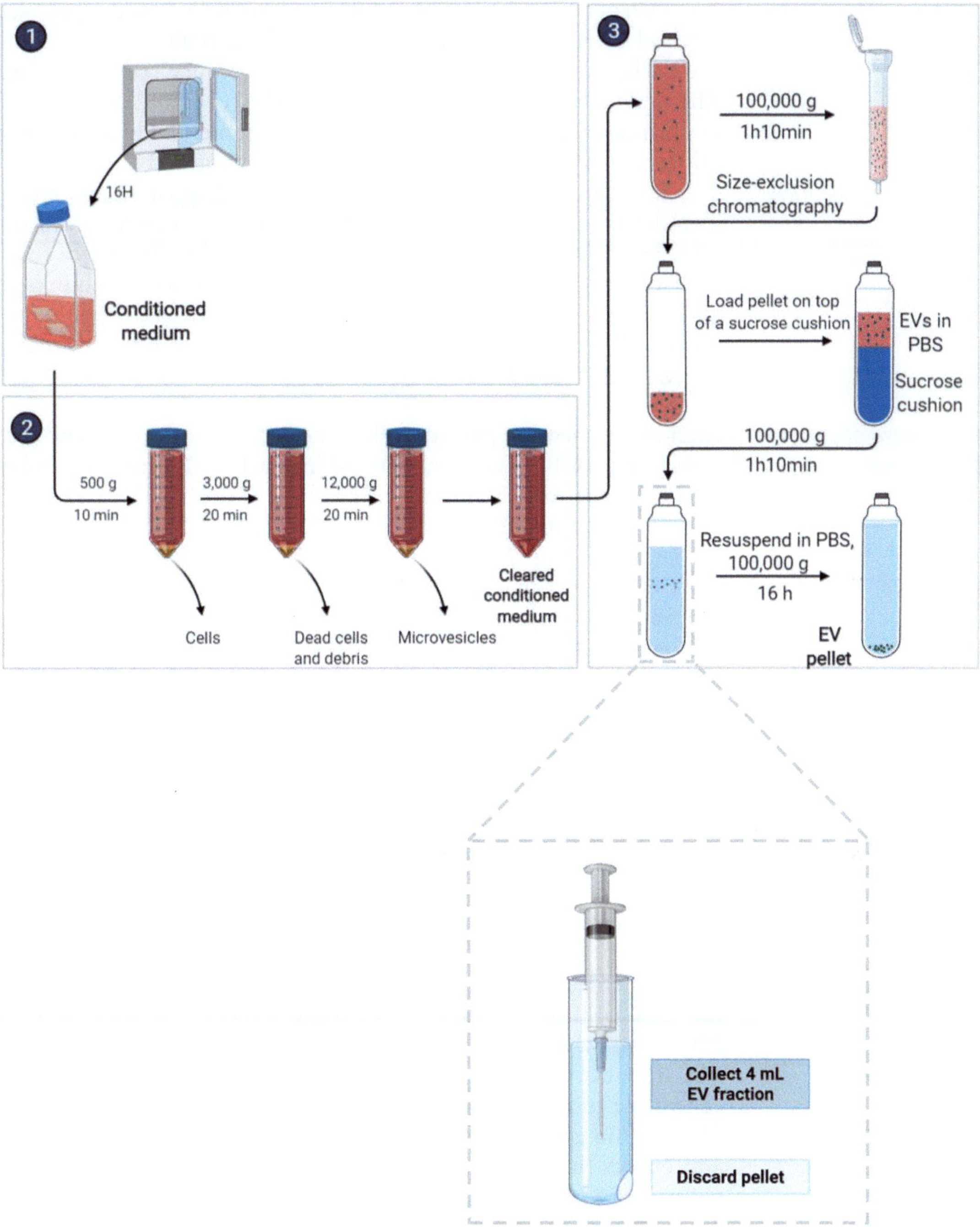

Fig. 2 Schematic representation of the "release method" for the isolation of small EV from brain tissue. After the isolation of brain tissue and its incubation period (Step 1), conditioned media is sequential centrifuged to remove unwanted vesicles and debris (Step 2). A first EV pellet is obtained after 70 min ultracentrifugation, which is resuspended and passed through a size-exclusion chromatography column. The desired fraction is then layered onto a sucrose cushion and ultracentrifuged to obtain a pure EV fraction that is pelleted by an overnight ultracentrifugation (Step 3).

Animals may be sacrificed by the preferred and approved method of euthanasia. The extraction of the brain from the skull should be performed under a flow hood to reduce contamination. The brain should be immediately macrodissected and, if intended, brain areas of interest microdissected and submerged in prewarmed Hibernate-A media in the small petri dish (weighted above).

After the addition of the brain tissue, weigh again the petri dish in order to obtain the weight of the tissue (if necessary, pool the same area from different animals). If you use a big brain area (e.g., whole cortex), it is advisable to cut the tissue into big chunks to maintain tissue viability. Transfer tissue to T25 flasks with pre-warmed EV-release media (usually 15 mL) and put in the incubator for 16 h (37 °C, 5% CO_2).

3.4 EV Isolation: Steps 2 and 3

After the incubation period, collect the conditioned media and pass it through a 40 μm cell strainer and into a 15 mL centrifuge tube. Add the appropriate amount of protease (10 μL/mL) and phosphatase inhibitors (5 μL/mL of each cocktail) and proceed with the EV isolation (Fig. 2).

Centrifuge the sample at 500 g, for 10 min at 10 °C, to remove cell fragments. Transfer the media into a new centrifuge tube and spin it at 3.000 g, for 20 min at 10 °C. Transfer the media into a new centrifuge tube. You can stop at this step and freeze the media, or proceed with the isolation.

Centrifuge at 12.000 g, for 20 min at 10 °C, to remove larger microvesicles.

The supernatant obtained from the 12.000 g centrifugation is filtered through a 0.2 μm syringe filter into an ultracentrifuge tube and centrifuged at 100.000 g for 70 min, at 10 °C.

Discard the supernatant (or, if you prefer, keep it as EV-free conditioned media).

Resuspend the pellet in 500 μL of filtered PBS and collect it into a microcentrifuge tube. Store it at 4 °C while you equilibrate the qEV column with 30 mL of room temperature (RT) filtered PBS. When there are no more PBS drops falling from the column, add the above 500 μL. Collect the flow-through into a 15 mL centrifuge tube. Once the dripping has stopped, add 1 mL of RT filtered PBS and collect to the same tube. Finally, add filtered PBS and collect these drops to the same tube until you collect a total volume of 3 mL. At that moment, start collecting the drops into a microcentrifuge tube until 1.5 mL is collected. Now, the qEV column can be washed with 15 mL of PBS in between samples. After the last usage of the qEV column, wash the column with 30 mL of PBS, add 10% filtered EtOH and store at 4 °C in a vertical position. Be aware that you may need to optimize this process if you are using the newer versions of the qEV column or other size-exclusion chromatography columns.

Mix the desired 1.5 mL fraction with cold filtered PBS to a final volume of 16 mL.

Put 4 mL of the sucrose cushion at the bottom of the ultracentrifuge tube and layer on top of it the 16 mL of sample/PBS mix (one mL at a time). Then, spin it at 100.000 g for 70 min, at 10 °C.

Using a 18G syringe, collect 4 mL from the bottom of the tube and dilute it into 16 mL of cold PBS. Then, centrifuge again at 100.000 g for 8–16 h, at 10 °C. The longer the run, the purer the EV fraction is, but the yield will be smaller.

Finally, resuspend the EV pellet in the desired buffer and store it for further use and/or analysis.

3.5 EV Characterization

Following isolation, the EV fraction should be characterized according to the MISEV guidelines [44]. Accordingly, we have checked the quantity and quality of the EV yield produced by our method (release method) using different techniques—see also [42]. The structure of EVs of this method was checked by cryo-EM imaging, which revealed a homogeneous population of smaller EVs of the mouse brain without major contaminants (Fig. 3a). Quantity and size of isolated EV was also monitored by Nanoparticle Tracking analysis (NTA), confirming that the mean diameter of isolated EVs falls within the classical exosomal size (50–150 nm) with a particular enrichment of these vesicles (Fig. 3b, c). We also monitored the proteome of our EV yield by label-free Mass-Spec proteomic analysis, which revealed classical EV-related proteins as well as cell-specific ones (see Table 1). This data was further validated by WB where we detected EEA-1 and flotillin in our EV preparations, but not Cytochrome C— normally considered a contaminant (Fig. 3d). Moreover, the integrity of EVs isolated by our method was also checked by monitoring their in vivo internalization and movement in the brain (hippocampus) of live animals. Thus, Dil-labelled EVs isolated by our method were injected into the outer molecular layer of the dentate gyrus (DG) and were found, 4 weeks after the injection, in the neuronal soma in the granular cell layer of the DG, indicative of the movement of Dil-labelled EVs within the neurons (Fig. 3e); for more details, see also [42].

4 Conclusion

EVs play important roles in physiological and pathological mechanisms in the brain, mirroring the homeostatic status of their origin cell. While the brain is difficult to access, brain EVs are able to cross the blood-brain barrier and can be obtained from easily accessible biofluids (e.g., plasma), offering a novel tool of brain liquid biopsy. However, we are far from understanding how peripheral EVs relate to brain processes and/or pathology, and there is still a lack of

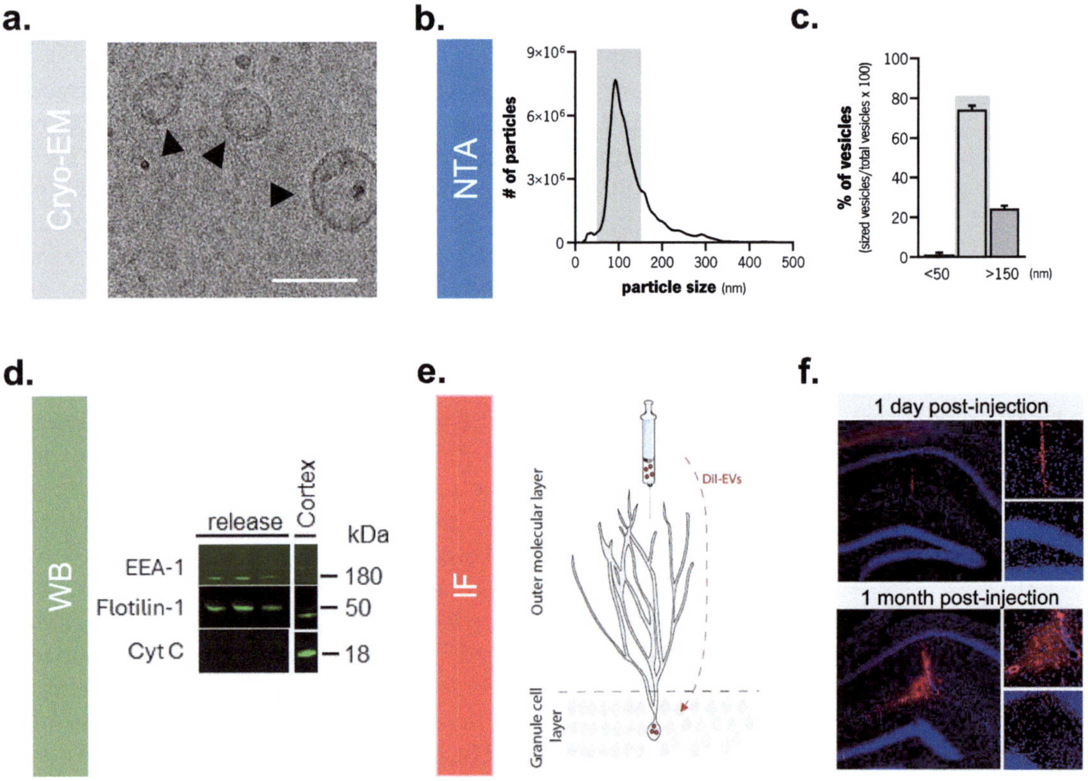

Fig. 3 Quantitative and qualitative characterization of EVs collected by the "release method." (**a**) Cryo-EM micrographs of isolated EVs (black arrowheads; scale bar: 100 nm). (**b**, **c**) Nanoparticle tracking analysis (NTA) of the size distribution (**b**) and the percentage of EVs (**c**) showing that the vast majority of particle size is between 50 nm and150nm. (**d**) Representative immunoblots of mouse brain EVs of the "release method" where we detect EV-associated proteins (Flotillin-1) and endosomal markers (EEA-1) but not the mitochondrial protein Cytochrome C. (**e**) Schematic representation of the DiI-labeled EVs injected into the outer molecular layer (OML) of the dentate gyrus of the mouse hippocampus, showing their subsequent spreading to the granule cell layer (GCL). (**f**) Confocal images of DiI-labelled EVs (red) and nuclei (blue) showing that at 1 day postinjection, DiI-labeled EVs appear only at the injection site (OML), while 4 weeks later DiI-labelled EVs appear also in the GCL. (Adapted from [42])

reliable cell-specific markers for the isolation of cell-specific brain EVs. Therefore, there is a great unmet need for better methods to isolate EVs directly from the brain tissue. While digestion methods have been employed for some time, they face certain challenges with potential contamination of EV preparations with unreleased vesicles, as well as the uncontrolled enzymatic action upon the EV corona that could substantially change EV protein landscape. The method described here (see also ref. 42) offers a great alternative to the classical EV isolation methods from brain tissue and circumvents the major issues related to enzymatic and mechanical digestion. In conclusion, we believe that this method will contribute to

Table 1
Representative proteins identified in the label-free Mass-Spec proteomic analysis of EV yield of the release method

Protein list of EV fraction of the release method	
CNS cell markers	
Neurons	Camk2a, Camk2b, Map1a, Map1b, Map2, Map6, Mecp2, Mif, Ncam1, Ncdn, Pak1, Pik3cb, Pik3r1, Pip4k2a, Plcl1, Ppp1r7, Ppp2r2c, Ppp2r5e, Prdx1, Prkar2b, Prkcb, Prpf19, Prrt2, Pten, Ptgr2, Ptpn5, Pygm, Rab3b, Rab3c, Rab6b, Rala, Rap1a, Rap1b, Rbfox3, Slc16a1, Slc17a7, Slc9a3r1, Snap25, Snx27, Syngr1, Syt1, Thy1
Microglia	C1qa, C1qb, C1qc, Ctsb, P2ry12
Astrocytes	Aldh1l1, Aldoc, Bcan, Dclk1, Gfap, Slc4a4, Slc6a1, Slc6a11
Oligodendrocytes	Cnp, Gpr37, Mbp, Mog, Plp1, Tmem30a, Tspan2, Tubb4a
Common EV proteins	
Tetraspanins	CD9, CD81, CD82
Annexins	Anxa5, Anxa6, Anxa7, Anxa11
Rabs	Rab1A, Rab1b, Rab2a, Rab2b, Rab3a, Rab3b, Rab3d, Rab4a, Rab4b, Rab5a, Rab5b, Rab5c, Rab6a, Rab6b, Rab7a, Rab8a, Rab8b, Rab10, Rab11b, Rab12, Rab14, Rab18, Rab21, Rab23, Rab35
Heat Shock proteins	Hspa1a, Hspa2, Hspa4, Hspa8, Hspa12a, Hspa12b, Hsph1, Hsp90ab1
Endolysosomal pathway	EEA1, Vps4b, Vta1
Absence of non-EV proteins	
Bcl-2, Canx, GM130, Nup, Cyc1	

The complete MS proteomics data are available at Gomes et al. [42] and via PRIDE PXD026712.
In this table, we show the detection of typical EV proteins such as tetraspanins, annexins, heat shock proteins, and the absence of non-EV proteins in the EV yield of our method. Note also the presence of proteins of neuronal, microglial, astrocytic, and oligodendrocytic origin, indicating that our method detects EVs released by different brain cell types.

the (unmet) detailed characterization of brain sEV as it is based on EV spontaneous release from the brain tissue. This will allow the identification of novel cell-specific biomarkers of brain pathology and increase the overall diagnostic potential of brain EVs in complex brain disorders.

Conflicts of Interest

The authors have no conflicts of interest to declare that are relevant to the content of this chapter.

References

1. Petroni D, Fabbri C, Babboni S, Menichetti L, Basta G, Del Turco S (2023) Extracellular vesicles and intercellular communication: challenges for in vivo molecular imaging and tracking. Pharmaceutics 15(6):6. https://doi.org/10.3390/pharmaceutics15061639
2. Zhou X et al (2020) The function and clinical application of extracellular vesicles in innate immune regulation. Cell Mol Immunol 17(4):323–334. https://doi.org/10.1038/s41423-020-0391-1
3. Morel L et al (2013) Neuronal exosomal miRNA-dependent translational regulation of astroglial glutamate transporter GLT1. J Biol Chem 288(10):7105–7116. https://doi.org/10.1074/jbc.M112.410944
4. Fauré J et al (2006) Exosomes are released by cultured cortical neurones. Mol Cell Neurosci 31(4):642–648. https://doi.org/10.1016/j.mcn.2005.12.003
5. Tytell M (2005) Release of heat shock proteins (Hsps) and the effects of extracellular Hsps on neural cells and tissues. Int J Hyperth 21(5):445–455. https://doi.org/10.1080/02656730500041921
6. Krämer-Albers E-M et al (2007) Oligodendrocytes secrete exosomes containing major myelin and stress-protective proteins: trophic support for axons? Proteomics Clin Appl 1(11):1446–1461. https://doi.org/10.1002/prca.200700522
7. Potolicchio I et al (2005) Proteomic analysis of microglia-derived exosomes: metabolic role of the aminopeptidase CD13 in neuropeptide catabolism. J Immunol 175(4):2237–2243. https://doi.org/10.4049/jimmunol.175.4.2237
8. Cano A et al (2023) Extracellular svesicles, the emerging mirrors of brain physiopathology. Int J Biol Sci 19(3):721–743. https://doi.org/10.7150/ijbs.79063
9. Gassama Y, Favereaux A (2021) Emerging roles of extracellular vesicles in the central nervous system: physiology, pathology, and therapeutic perspectives. Front Cell Neurosci 15. https://doi.org/10.3389/fncel.2021.626043
10. Filannino FM, Panaro MA, Benameur T, Pizzolorusso I, Porro C (2024) Extracellular vesicles in the central nervous system: a novel mechanism of neuronal cell communication. Int J Mol Sci 25(3):1629. https://doi.org/10.3390/ijms25031629
11. del Bravo-Miana RC, Arizaga-Echebarria JK, Otaegui D (2024) Central nervous system-derived extracellular vesicles: the next generation of neural circulating biomarkers? Transl Neurodegener 13(1):32. https://doi.org/10.1186/s40035-024-00418-9
12. Lee SH et al (2018) Reciprocal control of excitatory synapse numbers by Wnt and Wnt inhibitor PRR7 secreted on exosomes. Nat Commun 9(1):3434. https://doi.org/10.1038/s41467-018-05858-2
13. Domingues HS, Falcão AM, Mendes-Pinto I, Salgado AJ, Teixeira FG (2020) Exosome circuitry during (De)(Re)myelination of the central nervous system. Front Cell Dev Biol 8:483. https://doi.org/10.3389/fcell.2020.00483
14. Bátiz LF et al (2015) Exosomes as novel regulators of adult neurogenic niches. Front Cell Neurosci 9:501. https://doi.org/10.3389/fncel.2015.00501
15. Mollinari C et al (2015) miR-34a regulates cell proliferation, morphology and function of newborn neurons resulting in improved behavioural outcomes. Cell Death Dis 6(1):e1622–e1622. https://doi.org/10.1038/cddis.2014.589
16. Pascua-Maestro R, González E, Lillo C, Ganfornina MD, Falcón-Pérez JM, Sanchez D (2018) Extracellular vesicles secreted by Astroglial cells transport Apolipoprotein D to neurons and mediate neuronal survival upon oxidative stress. Front Cell Neurosci 12:526. https://doi.org/10.3389/fncel.2018.00526
17. Luarte A et al (2020) Astrocyte-derived small extracellular vesicles regulate dendritic complexity through miR-26a-5p activity. Cells 9(4):930. https://doi.org/10.3390/cells9040930
18. Zhao Y, Gu Y, Zhang Q, Liu H, Liu Y (2023) The potential roles of exosomes carrying APP and tau cleavage products in Alzheimer's disease. J Clin Med 12(5):1883. https://doi.org/10.3390/jcm12051883
19. Emmanouilidou E et al (2010) Cell-produced alpha-synuclein is secreted in a calcium-dependent manner by exosomes and impacts neuronal survival. J Neurosci 30(20):6838–6851. https://doi.org/10.1523/JNEUROSCI.5699-09.2010
20. Asai H et al (2015) Depletion of microglia and inhibition of exosome synthesis halt tau propagation. Nat Neurosci 18(11):1584–1593. https://doi.org/10.1038/nn.4132
21. Saint-Pol J, Gosselet F, Duban-Deweer S, Pottiez G, Karamanos Y (2020) Targeting and crossing the blood-brain barrier with

extracellular vesicles. Cells 9(4):851. https://doi.org/10.3390/cells9040851

22. Chen CC et al (2016) Elucidation of exosome migration across the blood-brain barrier model in vitro. Cell Mol Bioeng 9(4):509–529. https://doi.org/10.1007/s12195-016-0458-3
23. Delgado-Peraza F et al (2021) Neuronal and astrocytic extracellular vesicle biomarkers in blood reflect brain pathology in mouse models of Alzheimer's disease. Cells 10(5):993. https://doi.org/10.3390/cells10050993
24. Thompson AG et al (2016) Extracellular vesicles in neurodegenerative disease - pathogenesis to biomarkers. Nat Rev Neurol 12(6): 346–357. https://doi.org/10.1038/nrneurol.2016.68
25. Saeedi S, Israel S, Nagy C, Turecki G (2019) The emerging role of exosomes in mental disorders. Transl Psychiatry 9(1):1–11. https://doi.org/10.1038/s41398-019-0459-9
26. Gomes P et al (2022) Extracellular vesicles and Alzheimer's disease in the novel era of Precision Medicine: implications for disease progression, diagnosis and treatment. Exp Neurol 358:114183. https://doi.org/10.1016/j.expneurol.2022.114183
27. Perez-Gonzalez R, Gauthier SA, Kumar A, Levy E (2012) The exosome secretory pathway transports amyloid precursor protein carboxyl-terminal fragments from the cell into the brain extracellular space. J Biol Chem 287(51): 43108–43115. https://doi.org/10.1074/jbc.M112.404467
28. Pérez-González R, Gauthier SA, Kumar A, Saito M, Saito M, Levy E (2017) A method for isolation of extracellular vesicles and characterization of exosomes from brain extracellular space. Methods Mol Biol 1545:139–151. https://doi.org/10.1007/978-1-4939-6728-5_10
29. Vella LJ et al (2017) A rigorous method to enrich for exosomes from brain tissue. J Extracell Vesicles 6(1):1348885. https://doi.org/10.1080/20013078.2017.1348885
30. Ramirez MI et al (2018) Technical challenges of working with extracellular vesicles. Nanoscale 10(3):881–906. https://doi.org/10.1039/C7NR08360B
31. Théry C, Amigorena S, Raposo G, Clayton A (2006) Isolation and characterization of exosomes from cell culture supernatants and biological fluids. Curr Protoc Cell Biol Chapter 3:Unit 3.22. https://doi.org/10.1002/0471143030.cb0322s30
32. Caby M-P, Lankar D, Vincendeau-Scherrer C, Raposo G, Bonnerot C (2005) Exosomal-like vesicles are present in human blood plasma. Int Immunol 17(7):879–887. https://doi.org/10.1093/intimm/dxh267
33. Ogawa Y, Kanai-Azuma M, Akimoto Y, Kawakami H, Yanoshita R (2008) Exosome-like vesicles with dipeptidyl peptidase IV in human saliva. Biol Pharm Bull 31(6): 1059–1062. https://doi.org/10.1248/bpb.31.1059
34. Akers JC et al (2015) miRNA contents of cerebrospinal fluid extracellular vesicles in glioblastoma patients. J Neuro-Oncol 123(2): 205–216. https://doi.org/10.1007/s11060-015-1784-3
35. Crescitelli R, Lässer C, Lötvall J (2021) Isolation and characterization of extracellular vesicle subpopulations from tissues. Nat Protoc 16(3): 1548–1580. https://doi.org/10.1038/s41596-020-00466-1
36. Autengruber A, Gereke M, Hansen G, Hennig C, Bruder D (2012) Impact of enzymatic tissue disintegration on the level of surface molecule expression and immune cell function. Eur J Microbiol Immunol (Bp) 2 (2):112–120. https://doi.org/10.1556/EuJMI.2.2012.2.3
37. Yu Z-L et al (2022) Untouched isolation enables targeted functional analysis of tumour-cell-derived extracellular vesicles from tumour tissues. J Extracell Vesicles 11(4):e12214. https://doi.org/10.1002/jev2.12214
38. D'Acunzo P, Kim Y, Ungania JM, Pérez-González R, Goulbourne CN, Levy E (2022) Isolation of mitochondria-derived mitovesicles and subpopulations of microvesicles and exosomes from brain tissues. Nat Protoc 17(11): 2517–2549. https://doi.org/10.1038/s41596-022-00719-1
39. Jeurissen S et al (2017) The isolation of morphologically intact and biologically active extracellular vesicles from the secretome of cancer-associated adipose tissue. Cell Adhes Migr 11(2):196–204. https://doi.org/10.1080/19336918.2017.1279784
40. Jingushi K et al (2018) Extracellular vesicles isolated from human renal cell carcinoma tissues disrupt vascular endothelial cell morphology via azurocidin. Int J Cancer 142(3): 607–617. https://doi.org/10.1002/ijc.31080
41. Lunavat TR et al (2017) BRAFV600 inhibition alters the microRNA cargo in the vesicular secretome of malignant melanoma cells. Proc

Natl Acad Sci USA 114(29):E5930–E5939. https://doi.org/10.1073/pnas.1705206114

42. Gomes PA et al (2023) A novel isolation method for spontaneously released extracellular vesicles from brain tissue and its implications for stress-driven brain pathology. Cell Commun Signal 21(1):35. https://doi.org/10.1186/s12964-023-01045-z

43. Sanwlani R, Gangoda L (2021) Role of extracellular vesicles in cell death and inflammation. Cells 10(10):2663. https://doi.org/10.3390/cells10102663

44. Welsh JA et al (2024) Minimal information for studies of extracellular vesicles (MISEV2023): from basic to advanced approaches. J Extracell Vesicles 13(2):e12404. https://doi.org/10.1002/jev2.12404

Chapter 6

Methods for Isolating and Characterizing Red Blood Cell Extracellular Vesicles

Gurneet S. Sangha, Marzyeh Kheradmand, Lauren V. Smith, and Alisa Morss Clyne

Abstract

Extracellular vesicles (EVs) derived from red blood cells (RBCs) have emerged as a promising disease biomarker and drug delivery tool due to their intercellular signaling and cargo transporting properties. RBC-EVs have primarily been induced via calcium ionophore treatment. We recently showed that activation of piezo1, a mechanosensitive stretch-activated Ca^{2+} ion channel, stimulates RBCs to release EVs. These RBC-EVs may be more physiologically relevant than calcium ionophore-induced RBC-EVs, since RBCs in vivo may be stimulated to release EVs via piezo1 activation as they squeeze through capillaries and experience mechanical forces such as membrane tension and shear stress. This chapter provides a comprehensive overview of how to generate piezo1-stimulated RBC-EVs in vitro. We describe RBC isolation and treatment, along with subsequent RBC-EV isolation and characterization. Through this chapter, readers will learn how to generate piezo1 RBC-EVs for diverse biomedical applications.

Key words Piezo1, Red blood cells, Erythrocytes, Extracellular vesicles, Exosomes, Microvesicles

1 Introduction

Extracellular vesicles (EVs) have garnered significant interest in biomedical research due to their potential as a drug delivery platform and their role in cell-to-cell communication [1–3]. Among the various EV subtypes, red blood cell extracellular vesicles (RBC-EVs) are a promising choice to transport therapeutic molecules, because they lack genetic material (DNA, RNA, miRNA), can pass through biological barriers, and can be readily generated in large quantities from the patient's RBCs [2, 4]. RBC-EVs are also released in pathological conditions and likely play a role in the progression of various diseases and disorders, such as sickle cell disease, obstructive sleep apnea, diabetes, and inflammation [5–

Amit K. Srivastava (ed.), *Extracellular Vesicles*, Neuromethods, vol. 228, https://doi.org/10.1007/978-1-0716-4905-3_6,

9]. Thus, the ability to produce, isolate, and study RBC-EVs could open new avenues in drug delivery, diagnostics, and personalized healthcare.

There are various methods to induce RBC-EV formation. RBCs can be stimulated to take up calcium (Ca^{2+}) with calcium ionophore, which produces high EV yields (~10^{12} particles/mL); however, this method does not model a cellular pathway that normally occurs in the human body. RBC-EV formation can also be induced via biochemical stresses, for example, by treating RBCs with *tert*-butyl hydroperoxide to induce oxidative stress, storing RBC to deplete ATP, or oxygen cycling to mimic RBC hypoxia [8, 10]. However, RBC-EVs generated through storage and other pathological conditions may promote cardiovascular dysfunction, limiting their use in drug delivery applications [6, 8, 9]. Others have generated RBC-EVs by exposing RBCs to shear stress through needle shearing or high-pressure extrusion [7, 11, 12]. Mechanically stimulating RBC-EV release may model vesiculation as RBCs (8–10 μm diameter) squeeze through capillaries (2–4 μm diameter) during normal circulatory transit. But mechanical stimulation yields fewer RBC-EVs than other methods [12] and may require research equipment not found in standard biomedical laboratories.

The ideal method to generate RBC-EVs for biomedical applications would produce a high RBC-EV yield while mimicking a physiologically relevant process. We recently demonstrated that activating the RBC mechanosensitive ion channel piezo1 via its agonist yoda1 produces equivalent numbers of RBC-EVs as calcium ionophore treatment [12–15]. Piezo1-induced RBC-EV release also likely occurs in vivo as RBCs squeeze through capillaries, experiencing a shear stress that opens piezo1, increases Ca^{2+} influx, and triggers RBC-EV release [12, 16–19]. Yoda1 provides an accessible, rapid, and reproducible means of stimulating piezo1 to trigger RBC-EV release [12].

This chapter provides an overview of methods to generate piezo1-stimulated RBC-EVs (Fig. 1), with a focus on yoda1. We describe how to isolate RBCs, stimulate RBC-EV production, and purify RBC-EVs via ultracentrifugation. We then describe how to characterize RBC-EV markers by Western blot, size and concentration using tunable resistive pulse sensing (TRPS), and size and shape using transmission electron microscopy (TEM). This chapter provides readers with clear protocols to produce and characterize physiologically relevant RBC-EVs, enabling standardized research among laboratories toward developing RBC-EV-based diagnostic and drug delivery tools.

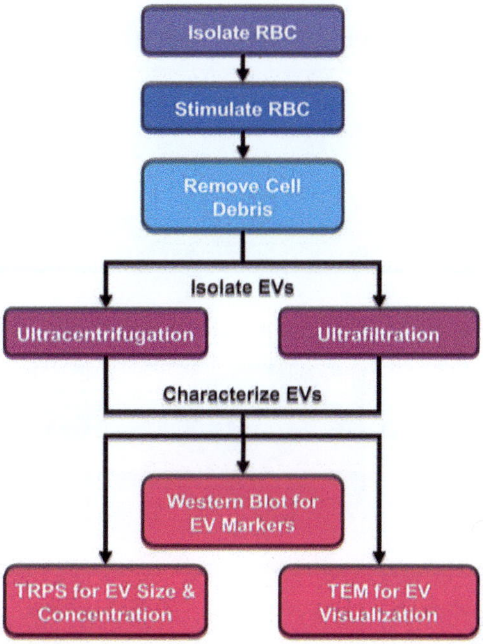

Fig. 1 Overview of the methods to produce, isolate, and characterize piezo1 RBC-EVs

2 Materials and Methods

2.1 RBC Isolation (Table 1)

1. Prepare PBS to wash RBCs.
   (a) 1× PBS: Add 50 mL of 10× PBS to 450 mL deionized (DI) water to obtain a 1× PBS solution.
2. Draw 20–30 mL of blood from a human subject's antecubital vein into vacutainer EDTA tubes and invert the tubes 8–10 times to mix well (Fig. 2a).
3. Store blood on ice until ready to isolate RBCs.
4. Wash RBCs four times with 1× PBS to remove the buffy coat and residual plasma (*see* **Notes 1** and **2**).
   (a) Centrifuge the tube of whole blood at room temperature at 600 × g for 5 min to separate the blood into three phases: RBCs, buffy coat, and plasma (Fig. 2b).
   (b) Carefully aspirate the plasma and buffy coat with a Pasteur pipette connected to a vacuum line, while being careful not to aspirate the RBC layer.
   (c) Add 5 mL 1× PBS to the remaining RBCs in the tube and invert 2–3 times to mix. Ensure RBCs are well mixed in the PBS.

Table 1
Materials and reagents to isolate RBCs from whole blood

Material/reagents	Vendor	Catalog number
Vacutainer EDTA tubes	Becton Dickinson	367856
Phosphate buffered saline (PBS) (10×), pH 7.4	Gibco	70011069
Disposable Pasteur pipettes	VWR	14673-043

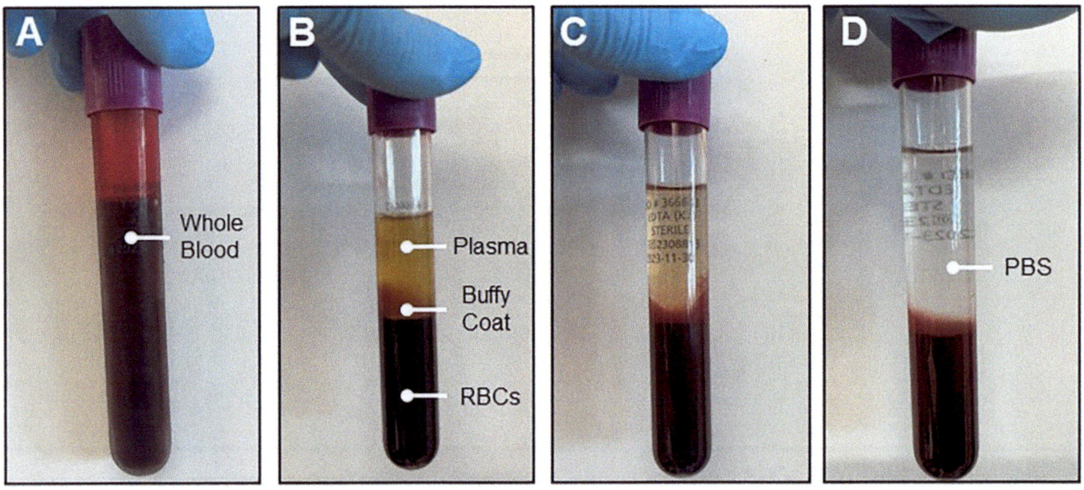

Fig. 2 RBC isolation from whole blood. (**a**) Whole blood mixed with anticoagulant after collection. (**b**) Plasma, buffy coat, and RBCs after first centrifugation at 600 × g for 5 min. (**c**) Plasma/PBS mixture and RBCs after second centrifugation at 600 × g for 5 min. (**d**) Two-phase separated RBCs and PBS after the final centrifugation at 600 × g for 5 min

(d) Centrifuge the tube again at 600 × g for 5 min. The supernatant should be more clear, and there should be a smaller buffy coat layer (Fig. 2c).

(e) Aspirate the PBS and repeat steps b-d until there are only RBCs and PBS (no buffy coat or plasma) after centrifugation (Fig. 2d).

2.2 Stimulate RBCs to Generate RBC-EVs (Table 2)

1. Prepare reagents to stimulate RBC-EVs with yodal

(a) 1× PBS: Add 5 mL of 10× PBS to 45 mL DI water to obtain a 1× PBS solution.

(b) Glucose supplemented PBS: Add 49.5 mg glucose to 50 mL 1× PBS for a final glucose concentration of 99 mg/dL. pH balance glucose supplemented PBS to 7.4.

(c) Yodal: Resuspend 5 mg yodal in 400 μL DMSO to obtain a 35.2 mM yodal stock solution. Aliquot the

Table 2
Materials and reagents to stimulate RBCs to generate RBC-EVs

Material/reagents	Vendor	Catalog number
D-Glucose	Millipore Sigma	G7021
Yodal	Millipore Sigma	SML1558
Dimethyl sulfoxide (DMSO)	Millipore Sigma	D2650
Calcium chloride dihydrate	Thermo Fisher Scientific	C79-500
Phosphate buffered saline (PBS; 10×), pH 7.4	Gibco	70011069
50 mL centrifuge tubes	CELLTREAT	229420
0.5 mL microcentrifuge tubes	CELLTREAT	230400

yodal stock solution into 0.5 mL centrifuge tubes to reduce freeze/thaw cycles. Store the yodal stock solution at −80 °C.

(d) 2 M calcium chloride ($CaCl_2$): Resuspend 4.41 g calcium chloride dihydrate in 15 mL DI water to obtain a 2 M $CaCl_2$ stock solution.

2. Add 47 mL glucose supplemented PBS to a clean 50 mL centrifuge tube.
3. Add 14.2 μL of 35.2 mM yodal to the glucose supplemented PBS in the 50 mL tube to obtain a 10 μM yodal solution. Invert the tube 2–3 times to mix the PBS and yodal.
4. Add 25 μL of 2 M $CaCl_2$ solution to the supplemented PBS + yodal in the 50 mL tube to obtain a 1 mM $CaCl_2$ solution. Calcium will enter the RBCs through the piezo1 channel to trigger RBC-EV production. Invert the centrifuge tube 2–3 times to mix the PBS, yodal, and $CaCl_2$.
5. Add 3 mL RBCs to the PBS supplemented with glucose, yodal, and $CaCl_2$ in the 50 mL tube to obtain 6% hematocrit. Invert the centrifuge tube 2–3 times to mix the PBS, yodal, $CaCl_2$, and RBCs (*see* **Note 3**).
6. Incubate RBCs in the yodal solution for 30 minutes at 37 °C.

2.3 Remove Cell Debris Before RBC-EV Isolation (Table 3)

1. Centrifuge the 6% hematocrit solution at 600 × g for 20 min to separate the RBC-EV-rich solution from the RBCs (Fig. 3a).
2. Using a Pipet-Aid, transfer RBC-EV-rich solution into a new 50 mL centrifuge tube and discard the remaining RBC pellet in a biohazard container.
3. Centrifuge the RBC-EV-rich PBS at 1600 × g for 15 min at room temperature (Fig. 3b) and transfer supernatant to a new 50 mL centrifuge tube, followed by 3250 × g for 15 min at

Table 3
Materials and reagents to remove cell debris before RBC-EV isolation

Material/reagents	Vendor	Catalog number
Optima XE Ultracentrifuge	Beckman coulter	A99842
JA-30.50 Ti Fixed-angle Titanium rotor	Beckman coulter	363421
50 mL Polycarbonate ultracentrifuge tube	Beckman coulter	357000
0.22 μM polyethersulfone syringe filter	Foxx life sciences	371-2115-OEM
50 mL syringe	BD	309654
50 mL centrifuge tubes	CELLTREAT	229420
50 mL serological pipette	CELLTREAT	229030B

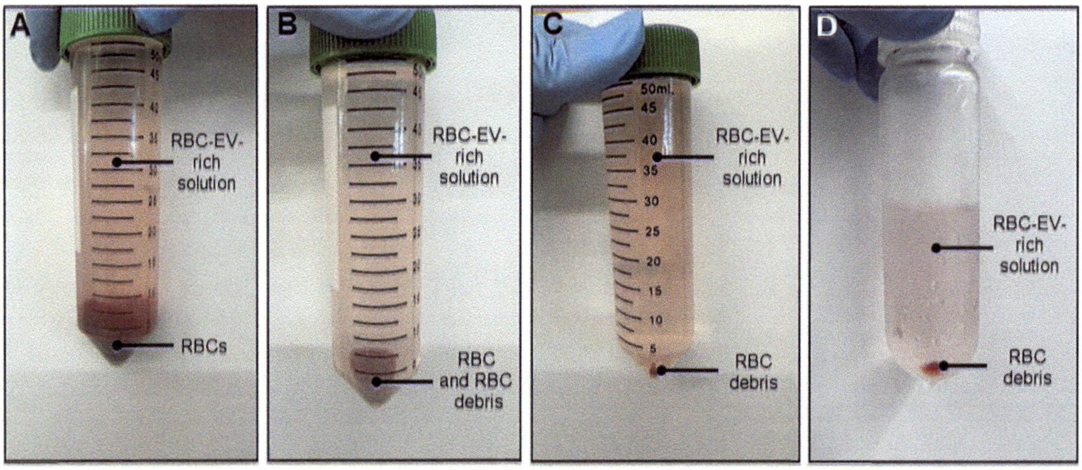

Fig. 3 RBC debris removal before RBC-EV isolation. (**a**) RBC-EV-rich solution separated from RBCs after the first centrifugation step at 600 × g for 20 min. (**b**, **c**) Sequential centrifugation for 15 min at 1600 × g and 3250 × g, respectively, separates the RBC debris from the RBC-EV-rich solution. (**d**) RBC debris separated from the RBC-EV-rich solution after the final centrifugation step at 10,000 × g for 30 min

room temperature (Fig. 3c). Transfer contents to a 50 mL polycarbonate ultracentrifuge tube and centrifuge at 10,000 × g for 30 min at 4 °C (Fig. 3d).

4. After each centrifugation step, pipette the RBC-EV-rich solution to a new 50 mL centrifuge tube. Avoid disturbing the RBC debris pellet.
5. Attach a 0.22 μm polyethersulfone syringe filter to a 50 mL syringe that has the syringe plunger removed.
6. Pour the RBC-EV-rich solution into the 50 mL syringe. Reattach the plunger and pass the RBC-EV-rich solution through the syringe filter to remove remaining cell debris.

Table 4
Materials and reagents to isolate RBC-EVs using ultracentrifugation

Material/reagents	Vendor	Catalog number
Optima XE Ultracentrifuge	Beckman coulter	A99842
JA-30.50 Ti Fixed-angle Titanium rotor	Beckman coulter	363421
50 mL Polycarbonate ultracentrifuge tube	Beckman coulter	357000
Phosphate buffered saline (PBS; 10×), pH 7.4	Gibco	70011069
25 mL serological pipette	CELLTREAT	229245
0.5 mL microcentrifuge tubes	CELLTREAT	230400

7. Store the RBC-EV-rich solution on ice until use. The RBC-EV-rich solution can also be stored long term at -80 °C.

2.4 Isolate RBC-EVs Using Ultracentrifugation (Table 4)

1. If the RBC-EV-rich solution was frozen, thaw it at room temperature before proceeding to the next step. Keep the RBC-EV-rich solution on ice after it has thawed.
2. Cool the ultracentrifuge equipped with a fixed angle titanium rotor to 4 °C.
3. Pipette 35 mL of RBC-EV-rich solution into a 50 mL polycarbonate ultracentrifuge tube. Add 35 mL of water into a second 50 mL polycarbonate ultracentrifuge tube, which will serve as the ultracentrifuge balance. Ensure the tubes weigh within a hundredth of a gram of each other.
4. Centrifuge the RBC-EV-rich solution at 100,000 × g for 70 min at 4 °C.
5. Aspirate the RBC-EV-depleted solution carefully without disturbing the RBC-EV pellet (Fig. 4).
6. Optional: Wash the RBC-EV pellet in 1× PBS to enhance sample purity.
   (a) Add 5 mL of 10× PBS to 45 mL DI water to obtain a 1× PBS solution.
   (b) Add 1 mL 1× PBS to the ultracentrifuge tube containing the RBC-EV pellet. Using a 1 mL pipette, pipette up and down until the RBC-EV pellet is resuspended in the PBS.
   (c) Add an additional 34 mL 1× PBS to the ultracentrifuge tube. Centrifuge the RBC-EV-rich solution again at 100,000 × g for 70 min at 4 °C.
   (d) Aspirate the RBC-EV-depleted PBS carefully without disturbing the RBC-EV pellet.
7. Resuspend the RBC-EV pellet in 400 μL 1× PBS to obtain an RBC-EV concentration of ~10^{12} particles/mL.

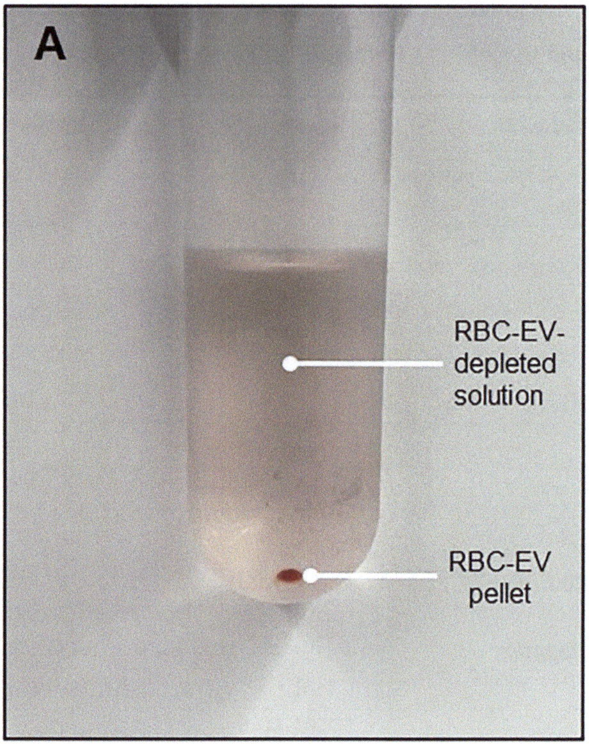

Fig. 4 Piezo1 RBC-EV isolation via ultracentrifugation. RBC-EV pellet after 70 min of ultracentrifugation at 100,000 × g

Table 5
Materials and reagents to isolate RBC-EVs using ultrafiltration

Material/reagents	Vendor	Catalog number
15 mL Amicon Ultra-15 centrifugal filter 100 kDa MWCO	Milipore Sigma	UFC910008
Phosphate-buffered saline (PBS) (10), pH 7.4	Gibco	70011069
25 mL serological pipette	CELLTREAT	229245

8. Aliquot 50–100 μL RBC-EVs into 0.5 mL microcentrifuge tubes and store at −80 °C (*see* **Note 4**).

2.5 Isolate RBC-EVs Using Ultrafiltration (Table 5)

1. Add 12 mL of RBC-EV-rich solution to the top of an Amicon Ultra-15 filter and close the filter with a cap (Fig. 5a).
2. Centrifuge the Amicon filter with the RBC-EV-rich supernatant at 3820 × g for 30 min. There should be ~200–250 μL of concentrated RBC-EVs remaining in the filter (Fig. 5b).
3. Optional: Wash the RBC-EVs in 1× PBS to enhance sample purity.

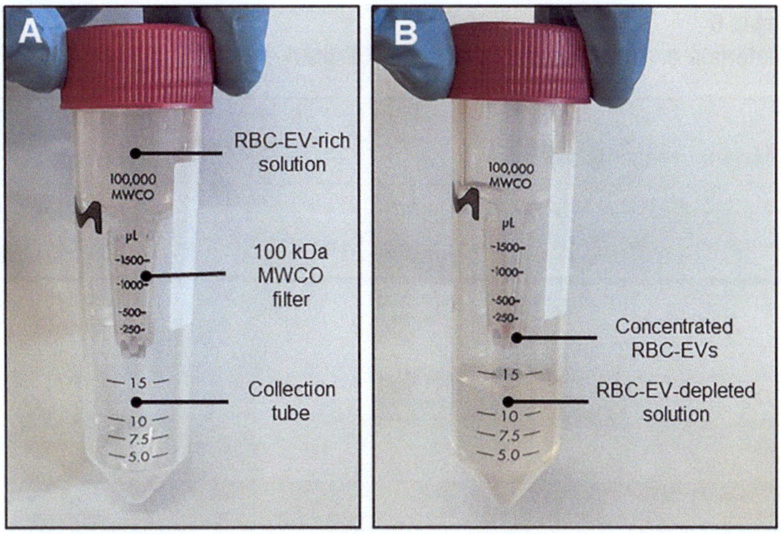

Fig. 5 EV isolation from EV-rich solution via ultrafiltration. RBC-EV-rich solution (**a**) before and (**b**) after centrifugation to concentrate RBC-EVs in the 100 kDa molecular weight cut-off (MWCO) filter

(a) Add 5 mL of 10× PBS to 45 mL DI water to obtain a 1× PBS solution.

(b) Pour the RBC-EV-depleted solution out of the collection tube.

(c) Add 11 mL of 1× PBS to the top of the Amicon filter to wash the RBC-EVs.

(d) Gently pipette the solution up and down several times to mix RBC-EVs and PBS.

(e) Centrifuge the Amicon filter at 3820 × g for an additional 30 min.

4. Wash the sides of the membrane in the Amicon filter with the ~200–250 μL of concentrated RBC-EVs in the filter to ensure complete EV recovery.

5. Aliquot 50–100 μL of RBC-EVs into 0.5 mL microcentrifuge tubes and store at −80 °C (*see* **Note 4**).

2.6 Detect RBC-EV Markers by Western Blot (Table 6)

1. Prepare Western blot reagents.

(a) 1× MOPS Running Buffer: Dilute 50 mL 20× MOPS in 950 mL DI water.

(b) TBS-T: Mix 500 mL 10× TBS and 25 mL Tween in 4.5 L DI water.

(c) 5% bovine serum albumin (BSA) in TBS-T: Dissolve 50 g BSA in 1 L TBS-T.

(d) 1% BSA in TBS-T: Dissolve 10 g BSA in 1 L TBS-T.

Table 6
Materials and reagents to detect RBC-EV markers using Western blot

Material/reagents	Vendor	Catalog number
NuPAGE MOPS SDS running buffer (20×)	Fisher Scientific	NP0001
TBS, Tris buffered saline, 10× solution, pH 7.4	Fisher Scientific	BP24711
Tween™ 20 Surfact-Amps™ Detergent Solution	Thermo Fisher Scientific	85115
Bovine serum Albumin, heat shock fraction	Millipore Sigma	A7906
Alix Antibody (3A9)	Santa Cruz Biotechnology	sc-53538
TSG101 Antibody (C-2)	Santa Cruz Biotechnology	sc-7964
CD9 Antibody (C-4)	Santa Cruz Biotechnology	sc-13118
CD63 Antibody (MX-49.129.5)	Santa Cruz Biotechnology	sc-5275
Anti-Mouse IgG (H + L), HRP Conjugate	Promega	W4021
SuperSignal West Pico PLUS	Thermo Fisher Scientific	34578
Mini Gel Tank with Spatula	Thermo Fisher Scientific	A25977
Owl EC300XL2 Compact Power Supply	Thermo Fisher Scientific	EC300XL
Bolt Bis-Tris Plus Mini Protein Gels, 4–12%, 1.0 mm, WedgeWell format	Thermo Fisher Scientific	NW04120BOX
NuPAGE™ LDS Sample Buffer (4×)	Thermo Fisher Scientific	NP0008
NuPAGE™ Sample Reducing Agent (10×)	Thermo Fisher Scientific	NP0009
Spectra™ Multicolor Broad Range Protein Ladder	Thermo Fisher Scientific	26634
MagicMark™ XP Western Protein Standard	Thermo Fisher Scientific	LC5602
NuPAGE™ Antioxidant	Thermo Fisher Scientific	NP0005
Restore PLUS Western Blot Stripping Buffer	Thermo Fisher Scientific	46430
iBlot 3 Western Blot Transfer Device	Thermo Fisher Scientific	IB31001

(continued)

Table 6
(continued)

Material/reagents	Vendor	Catalog number
iBlot Transfer Stack, PVDF, regular size (contains filter paper and absorbent pad)	Thermo Fisher Scientific	IB401001
Western Blot Roller	Thermo Fisher Scientific	84747
VWR® Advanced Mini Dry Block Heater	VWR	10153-348

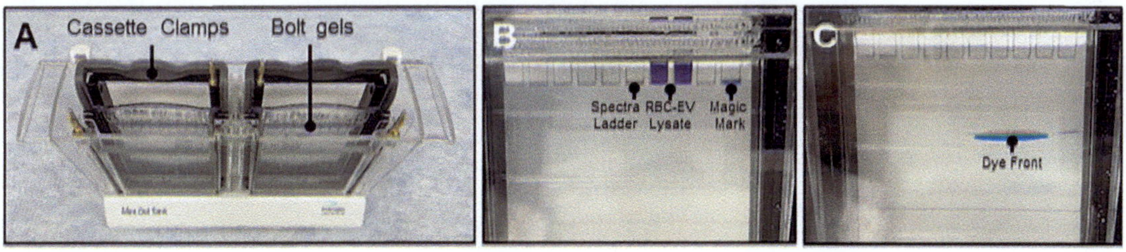

Fig. 6 Preparation of SDS-PAGE for verification of RBC-EV protein markers. (**a**) Bolt gels are secured with cassette clamps. (**b**) Lysate is loaded into the wells to prepare for separating EV protein markers. (**c**) Evenly migrating Spectra Ladder, dye front, and MagicMark as the EV protein markers separate according to their molecular weight

(e) Primary antibody solutions: Add 20 μL antibody to 10 mL of 1% BSA in TBS-T to obtain a 1:500 dilution. Prepare antibody solutions for ALIX, TSG101, CD9, and CD63. Each antibody solution can be used up to 3 times.

(f) Secondary antibody solutions: Add 5 μL mouse horseradish peroxidase-conjugated secondary antibody to 10 mL 1% BSA in TBS-T to obtain a 1:2000 dilution. Make fresh secondary antibodies for each experiment.

(g) Chemiluminescence detection solution: Mix 1 part SuperSignal™ West Pico PLUS Luminol/Enhancer Solution and 1 part SuperSignal™ West Pico PLUS Stable Peroxide Solution. Make fresh chemiluminescence reagent before imaging Western blot membranes.

2. Prepare the Mini Gel Tank.

   (a) Place the left and right cassette clamps into the mini gel tank.

   (b) Remove the Bolt 4–12% Bis-Tris 10 Wedge Well gels from the wrapping and secure the gel into the cassette clamps with the wells facing toward you. Lock the gels in place (Fig. 6a). There should be two gels in the mini gel tank to detect multiple RBC-EV markers.

(c) Add 1× MOPS to the front side of the chamber to the fill line. Make sure MOPS does not leak to the back chamber of the tank (*see* **Note 5**).

(d) Fill in the back chamber of the tank with MOPS.

(e) Carefully remove the comb by placing a finger at both ends of the comb and slowly pushing the comb straight up.

(f) Wash each well of the gel by pipetting up and down with 100 μL MOPS. This will remove the storage buffer and any gel debris in the well.

3. In a 0.5 mL tube, mix 39 μL RBC-EVs, 15 μL sample buffer, and 6 μL reducing agent to make the RBC-EV lysate. This RBC-EV lysate will be used to detect ALIX, TSG101, and CD9.

4. In a second 0.5 mL tube, mix 39 μL RBC-EVs, 15 μL sample buffer, and 6 μL DI water to make the RBC-EV lysate (*see* **Note 6**). This RBC-EV lysate will be used to detect CD63.

5. Vortex the RBC-EV lysates for 5 s to mix the samples. Briefly spin down the lysate using a mini centrifuge to collect the liquid in the bottom of the tube.

6. Place the RBC-EV lysates in a heating block or water bath at 70 °C for 5 min to denature proteins.

7. Load the gel with the molecular weight ladders and RBC-EV lysate starting from the right most well.

(a) In the first well of both gels, pipette 5 μL of Magic Mark protein ladder (Fig. 6b). MagicMark protein ladder is used to determine the molecular weight of each Western blot band (*see* **Note 7**).

(b) Leave the second well of both gels empty, as MagicMark can bleed into adjacent wells during imaging and disrupt protein band visualization.

(c) Pipette 40 μL RBC-EV lysate into wells three through nine. In one gel, add RBC-EV lysate containing reducing agent. This gel will be used to detect ALIX, TSG101, and CD9 (Fig. 6b).

(d) In the other gel, add RBC-EV lysate containing DI water instead of reducing agent. This gel will be used to detect CD63.

(e) In the last well of both gels, pipette 5 μL of Spectra Ladder (Fig. 6b). Spectra Ladder will be used to cut the PVDF membrane at specific molecular weight markers to detect multiple RBC-EV markers (*see* **Note 7**).

8. Add 250 μL antioxidant to the front and back sides of the chamber.

9. Run the gel electrophoresis.
   (a) Put the lid on the tank and place it in a 4 °C cold room to maintain protein sample integrity. You can also perform the Western blot on ice.
   (b) Connect the electrodes from the lid to the power source with its corresponding black and red color.
   (c) Turn on the power source and set it to 175 V and 80 mAmps for 60 min.
   (d) Start the gel electrophoresis. Check that bubbles are evenly forming at the top of the chamber to ensure the machine is properly running (*see* **Note 8**).
   (e) Monitor the mini-gel tank throughout the 60 min to make sure the Spectra Ladder, dye front, and MagicMark are evenly migrating (Fig. 6c) (*see* **Note 8**).
   (f) After 60 min or when the Spectra Ladder and dye front have reached the bottom of the gel, turn off the power source and detach the tank from the power source. The gel is now ready for transfer.
10. Transfer proteins from the gel to a membrane.
   (a) Remove the cathode stack from the polyvinylidene difluoride (PVDF) transfer stack.
   (b) With a spatula, remove the top plastic covering around the gel and cut off the wells (Fig. 7a).
   (c) Invert the gel into a container containing MOPS buffer. Using a spatula, push the gel off the bottom plastic covering through the slit in the bottom of the plastic covering (Fig. 7b).
   (d) Place the gel on top of the PVDF membrane in the transfer stack (Fig. 7c). Cut and remove the foot of the gel with a spatula (Fig. 7d, e).
   (e) Repeat **steps a** through **d** for the second gel.
11. Assemble PVDF transfer stack for protein transfer.
   (a) Dip one piece of filter paper in DI water and put it on top of the gel.
   (b) With a roller, roll from the center of the filter paper to the edges to remove bubbles (Fig. 7f).
   (c) Place the cathode stack onto the filter paper and roll bubbles out of the stack again (Fig. 7g).
   (d) Place the stack in the iBlot 2 semidry transfer system. Then place an absorbent pad on top of the stack and close the iBlot (Fig. 7h).
   (e) Set the iBlot to 20 V for 7 min (P3 setting) for even and complete transfer.

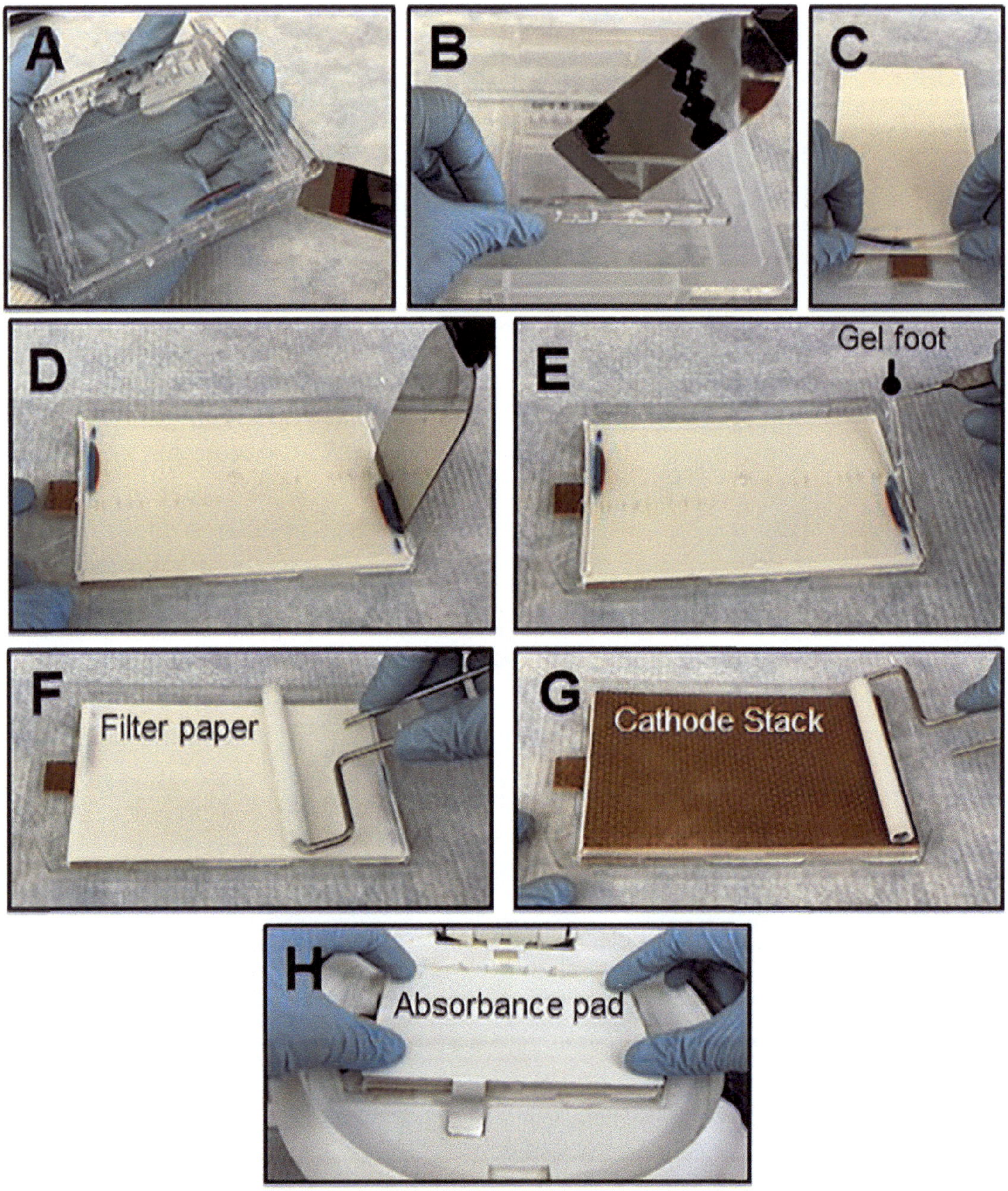

Fig. 7 Western blot protein transfer. (**a**, **b**) Removing the top plastic cover and pushing the gel off the bottom plastic cover. (**c**) Placing the gel on a PVDF membrane in the transfer stack. (**d**, **e**) Cutting and removing the gel feet. (**f**) Removing bubbles with a roller from the paper filter. (**g**) Placing the cathode stack and removing bubbles with a roller. (**h**) Placing the transfer stack into the iBlot 2 semidry transfer system with the absorbance pad on the top

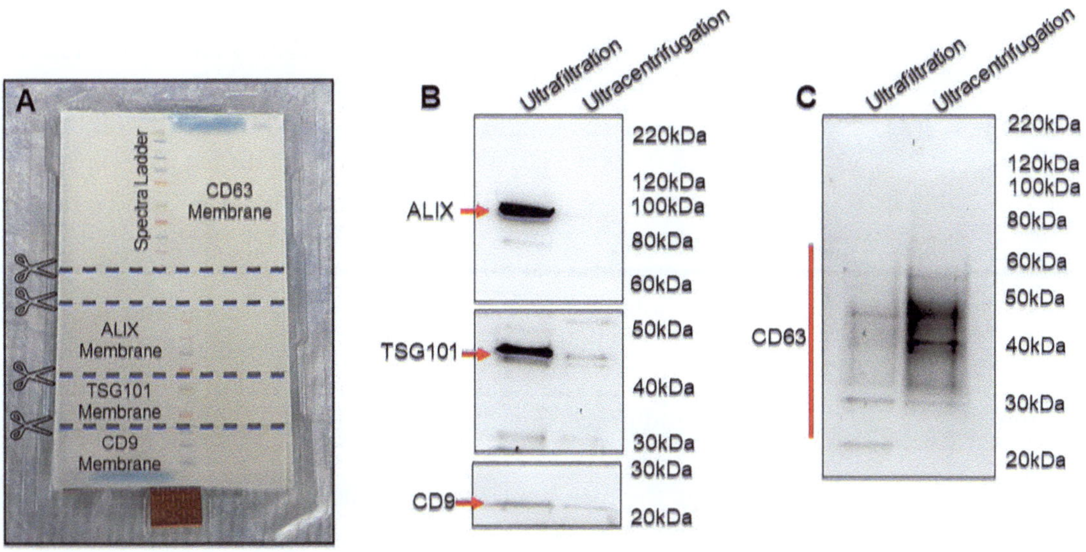

Fig. 8 Western blot to detect cytosolic and membrane EV markers. (**a**) PVDF membrane after RBC-EV protein transfer from Bolt gel. Scissors indicate where to cut membranes using a razor blade to blot for cytosolic EV markers ALIX and TSG101, and membrane EV markers CD9 and CD63. Representative RBC-EV Western blots for (**b**) ALIX, TSG101, CD9, and (**c**) CD63. RBC-EVs were isolated by ultrafiltration and ultracentrifugation using RBCs from two different people. The red arrow and line highlight the proteins of interest

12. Prepare the PVDF membranes for antibody labeling.
    (a) Remove the stack from the iBlot. Remove the top portion of the stack to reveal the membrane.
    (b) Cut the membrane in half where the two gels had previously met.
    (c) For the PVDF membrane containing proteins prepared with reducing agent, cut the membrane between the 35 and the 70 kDa markers. These PVDF membranes will be used to detect ALIX, TSG101, and CD9 (Fig. 8a).
    (d) For the PVDF membrane containing proteins prepared without reducing agent, place the entire membrane in a blot box. This PVDF membrane will be used to detect CD63 (Fig. 8a).
13. Label the membranes with primary antibodies.
    (a) Add 10 mL of 5% BSA in TBS-T to each PVDF membrane in a blot box. Incubate for 1 h at room temperature on a shaker plate at 50 rpm to block non-specific binding.
    (b) Aspirate the 5% BSA in TBS-T and wash the PVDF membrane with TBS-T for 5 min on a shaker at 50 rpm.
    (c) Aspirate the TBS-T and incubate the PVDF membranes with 10 mL primary antibodies for ALIX, TSG101, CD9 or CD63 overnight at 4 °C on a shaker plate at 50 rpm.

14. Label the membranes with secondary antibodies.
    (a) The next day, pipette out the primary antibody solution. Primary antibodies can be reused up to 3 times.
    (b) Wash the PVDF membrane two times with TBS-T for 5 min on a shaker at 50 rpm. Aspirate the TBS-T.
    (c) Incubate the PVDF membranes with 10 mL of the mouse secondary antibody solution for 2 h at room temperature on a shaker at 50 rpm.
    (d) Aspirate the secondary antibody solution and wash the PVDF membrane with TBS-T for 15 min on a shaker.
    (e) Aspirate the TBS-T and wash the PVDF membrane three additional times with TBS-T for 5 min on a shaker at 50 rpm.
15. Image the membranes using chemiluminescence:
    (a) Place the membrane on the imager platform.
    (b) Pipette 1 mL of chemiluminescence substrate onto the membrane.
    (c) Image the membranes so that the protein bands are gray. You may need to adjust the sensitivity and exposure time.
    (d) ALIX, TSG101, and CD9 can be identified by a single band at 96 kDa, 44 kDa, and 22 kDa, respectively (Fig. 8b). CD63 can be identified by a smear ranging from 30–60 kDa (Fig. 8c) (*see* **Note 9**).
16. After imaging, wash the PVDF membrane with TBS-T for 5 min on a shaker.
17. If you want to reprobe the membrane for a different protein, incubate the membrane in a 10 mL stripping buffer for 10 min at room temperature on a shaker. Then wash the PVDF membrane two times with TBS-T for 5 min on a shaker.
18. Membranes can be stored at 4 °C. Wrap the membrane in Parafilm and then in aluminum foil, and clearly label it with the experiment title and antibodies used.

2.7 Quantify RBC-EV Size and Concentration Using Tunable Resistive Pulse Sensing (TRPS) (Table 7)

1. Prepare TRPS reagents.
    (a) 2× PBS: Dilute 10 mL of 10× PBS in 40 mL DI water. 2× PBS is used to establish a current across the nanopore. Filter PBS through 0.22 μm syringe filter provided in the TRPS reagent kit to remove particles that may block nanopore.
    (b) Measuring Electrolyte (ME) Solution: Add 30 μL of wetting solution concentrate from the TRPS reagent kit to 9.97 mL of 2× PBS. Filter PBS through 0.22 μm syringe filter provided in the TRPS reagent kit to remove particles that may block nanopore.

Table 7
Materials for RBC-EV quantification using TRPS

Material/reagents	Vendor	Catalog number
TRPS Measurement System	Izon	qNano Gold or Exoid
Gibco™ PBS (10×), pH 7.4	Thermo Fisher Scientific	70011044
TRPS Reagent Kit (Includes wetting solution and syringe filters)	Izon	RK3
Nanopore Size NP100 (50–330 nm)	Izon	NP100
Pressure Application Device	Izon	PAD
Ethyl alcohol	Millipore Sigma	E7023
Kimwipes Delicate Task Wipes	Kimtech Science	34120
Ultrafree-MC Centrifugal Filter, 0.22 μm pore size, hydrophilic PVDF, 0.5 mL volume, sterile	Millipore Sigma	UFC30GV0S
Digital Caliper	Uline	H-7352
Calibration Particles - TRPS Measurement Size CPC100 (100 nm)	Izon	CPC100

(c) 70% Ethanol: Add 7 mL of ethanol to 3 mL of DI water.

2. Thaw RBC-EV aliquot on ice. Spin down aliquot in a mini-centrifuge for 1 s so that the liquid is collected in the bottom of the tube.
3. Filter RBC-EV samples to remove large particles that could clog the nanopore by pipetting the RBC-EV aliquot into a 0.22 μm pore size ultrafree-MC centrifugal filter. Centrifuge for 1 min at 12,000 × g. The liquid in the bottom of the tube is the filtered RBC-EVs to be used for size and concentration measurements.
4. Turn on the computer connected to the qNano gold TRPS system and plug the qNano into an outlet. Open the IZON app and select Classic Capture (Fig. 9a).
5. Wash the fluid cell to remove contaminants that could block the nanopore.
   (a) Add 75 μL 70% ethanol to the lower fluid cell of the qNano using reverse pipetting (Fig. 9b) (*see* **Note 10**). Aspirate the ethanol and wipe the bottom chamber with a Kimwipe to dry.

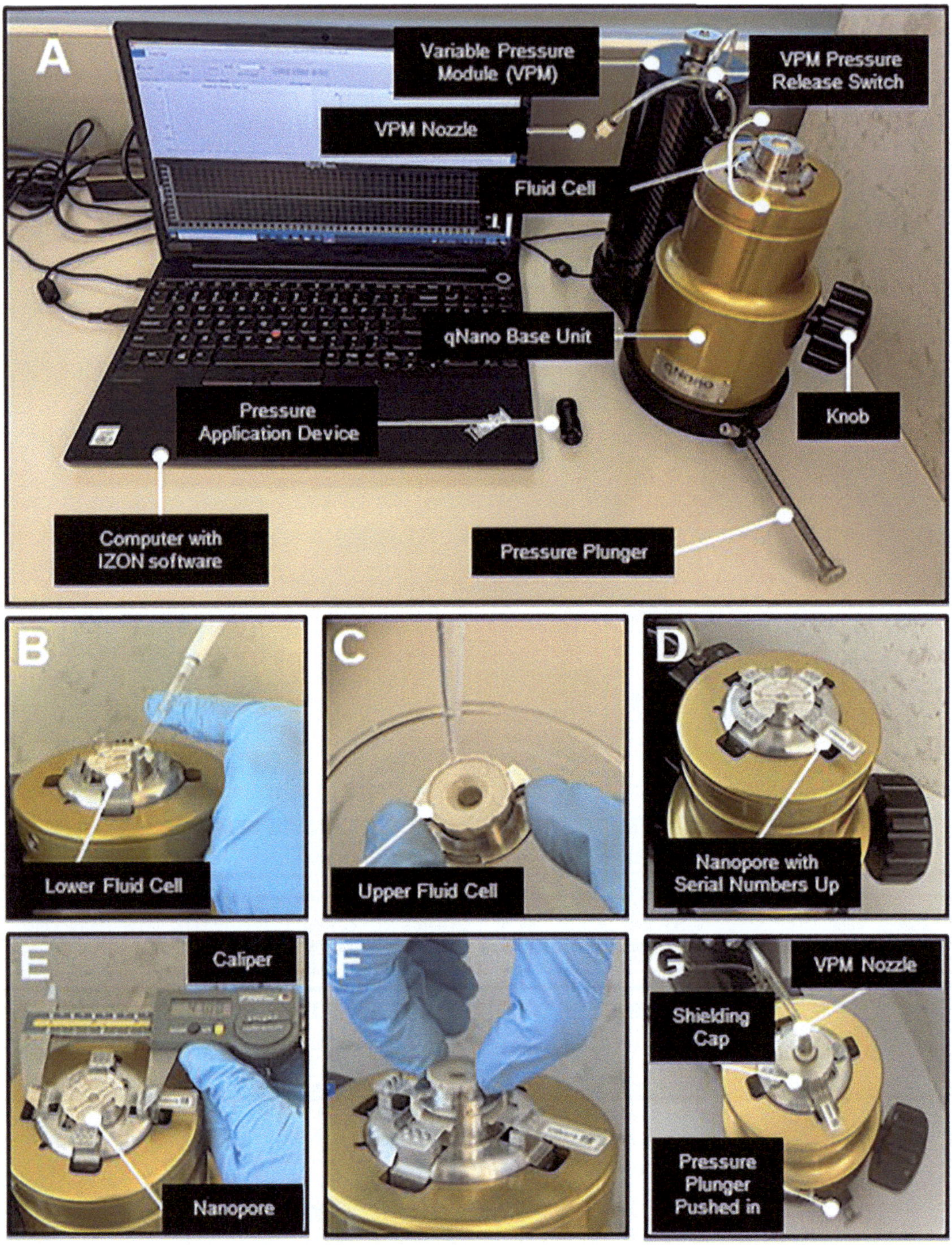

Fig. 9 TRPS methods to quantify the RBC-EV size and concentration. (**a**) TRPS qNano setup with software opened. (**b**) The lower fluid cell is washed using ethanol and ME solution. (**c**) The upper fluid cell is washed with ethanol and 2× PBS. (**d**) Nanopore equipped to the lower fluid cell with serial numbers facing up. (**e**) Nanopore stretched to 47 mm using a caliper. (**f**) The upper fluid cell is equipped above the nanopore. (**g**) VPM nozzle is connected to the shielding cap with the pressure plunger pushed to apply pressure to the nanopore

   (b) Repeat **step 5(a)** with 75 μL of ME solution.
   (c) Rinse the upper fluid cell with ethanol, followed by two rinses with 2× PBS (Fig. 9c). Gently tap the upper fluid cell onto a Kimwipe to dry after each rinse.
6. Prepare the nanopore.
   (a) Rinse the NP100 nanopore with 2× PBS, focusing on the center. Pat dry with a Kimwipe. Do not rub the center of the NP100 nanopore as this can cause the nanopore to rupture. Each nanopore can be used 1–3 times (*see* **Note 11**).
   (b) Put the nanopore over the lower fluid cell with the serial number on the nanopore facing up (Fig. 9d).
   (c) Stretch the nanopore to 47 mm by turning the black knob on the side of the qNano clockwise (Fig. 9e). To measure 47 mm, place a caliper outside the metal teeth holding the nanopore while stretching the nanopore. This measures the distance between teeth that hold on to the polyurethane membrane containing the nanopore in its center.
   (d) In the IZON software, set the stretch to 47 mm and press the "calibrate stretch" button.
   (e) Add 75 μL ME into the lower fluid cell using reverse pipetting (*see* **Note 10**).
   (f) Put the upper fluid cell on the top of the nanopore and twist to secure it in place (Fig. 9f).
   (g) Add 35 μL of ME into the upper fluid cell using reverse pipetting. Be careful not to touch the nanopore with the pipette tip.
7. Open the nanopore.
   (a) Add the fluid cell cap and variable pressure module (VPM) nozzle to the top of the upper fluid cell.
   (b) Close the VPM pressure release switch and apply maximum pressure by pushing in the pressure plunger to 20 mbar (Fig. 9g).
   (c) Set the current to 0.10 V and wait 2 min for the nanopore to open. The nanopore is open when the current increases to 13–47 nA.
   (d) If the nanopore is not open after 2 min, follow the steps below one at a time until the current increases to 13–47 nA:
      (i) Use the pressure application device to apply pressure to the upper fluid cell 3–5 times.
      (ii) Click the shielding cap by twisting and shutting the cap onto the upper fluid cell 3 times.

(iii) Stretch the nanopore to 48 mm.

(iv) Use the pressure application device on the upper fluid cell 3–5 times.

(v) Click the shielding cap on the upper fluid cell 2–3 more times.

8. Once the nanopore is open, increase the voltage until the current is stable at 120 nA (*see* **Note 12**).

9. Begin counting particles when the root mean square (RMS) noise stabilizes below 10 pA. The particle rate should be linear throughout the measurement (*see* **Note 13**).

10. Calibrate the TRPS system using 100 nm carboxylated polystyrene (CPC100) particles.

    (a) Dilute the CPC100 calibration particles by adding 1 μL CPC100 in 2 mL ME solution. Adjust the dilution as needed to ensure that particle rate is 200–1500 particles/minute.

    (b) Remove the ME solution from the upper chamber and reverse pipette 35 μL diluted CPC into the upper chamber.

    (c) Reattach the VPM nozzle. Push the pressure plunger to apply 5 m bar pressure.

    (d) Press the record button in the IZON software to record particle detection. Record at least 500 particles at a particle rate of 200–1500 particles/minute (*see* **Note 14**).

    (e) Repeat recording with the pressure set to 10 mbar (*see* **Note 15**).

    (f) Recalibrate the system every 20–30 min while measuring RBC-EVs, as the nanopore will enlarge over time.

11. Measure RBC-EV size and concentration.

    (a) All RBC-EV measurements should be performed at the same nanopore stretch and pressure used to calibrate the system.

    (b) Dilute the RBC-EVs 1:100 by adding 5 μL RBC-EVs to 495 μL ME solution. Adjust the RBC-EV dilution if the particle rate is not 200–1500 particles/minute [15].

    (c) Remove the solution from the upper chamber and reverse pipette 35 μL diluted RBC-EV into the upper chamber.

    (d) Reattach the shielding cap and the VPM nozzle. Apply 5 mbar pressure.

    (e) Press the record button in the IZON software to capture the data.

    (f) Repeat recording with the pressure set to 10 mbar [14].

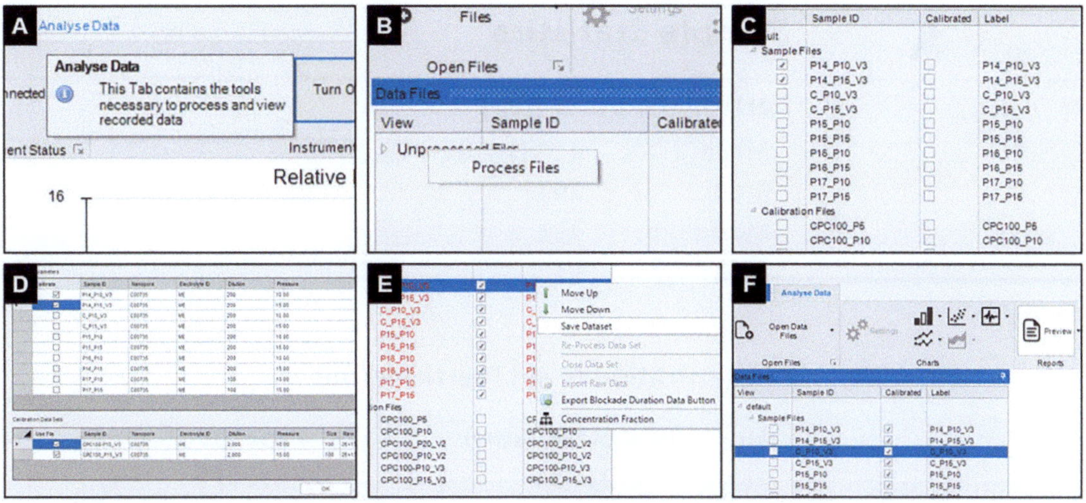

Fig. 10 TRPS analysis to visualize RBC-EV size and concentration. (**a**) Select Analyze Data to access tools needed to process and view data. (**b**) Right-click to process files to access sample and calibration data. (**c**) Select two sample files and (**d**) two calibration files acquired at the same pressure. (**e**) Right-click to save data set and (**f**) select preview to visualize data

(g) Remove the measured sample. Repeat **step 11** for all of your samples [16].

12. Data analysis and visualization.

(a) Click the "Analyze Data" tab, right click on "Unprocessed Files," and select "Process Files" (Fig. 10a, b).

(b) Under "Sample Files," select the RBC-EV measurements taken at two different pressures (Fig. 10c).

(c) A new window will open with the "Sample Files" selected. Then select two calibration measurements taken at the same pressures as the RBC-EV measurements (Fig. 10d). The calibration files will automatically appear at the bottom of the screen. Select OK at the bottom of the window when all the sample and calibration files have been selected.

(d) Right click over red sample files and select "Save dataset" to save the calibrated data (Fig. 10e).

(e) Repeat **steps b** through **d** for each sample.

(f) Select "Preview" to visualize the size and concentration data (Fig. 10f). Preview will provide the RBC-EV sample statistics (Fig. 11a) and the RBC-EV size and concentration histogram (Fig. 11b). The raw data will be automatically saved in the IZON software folder.

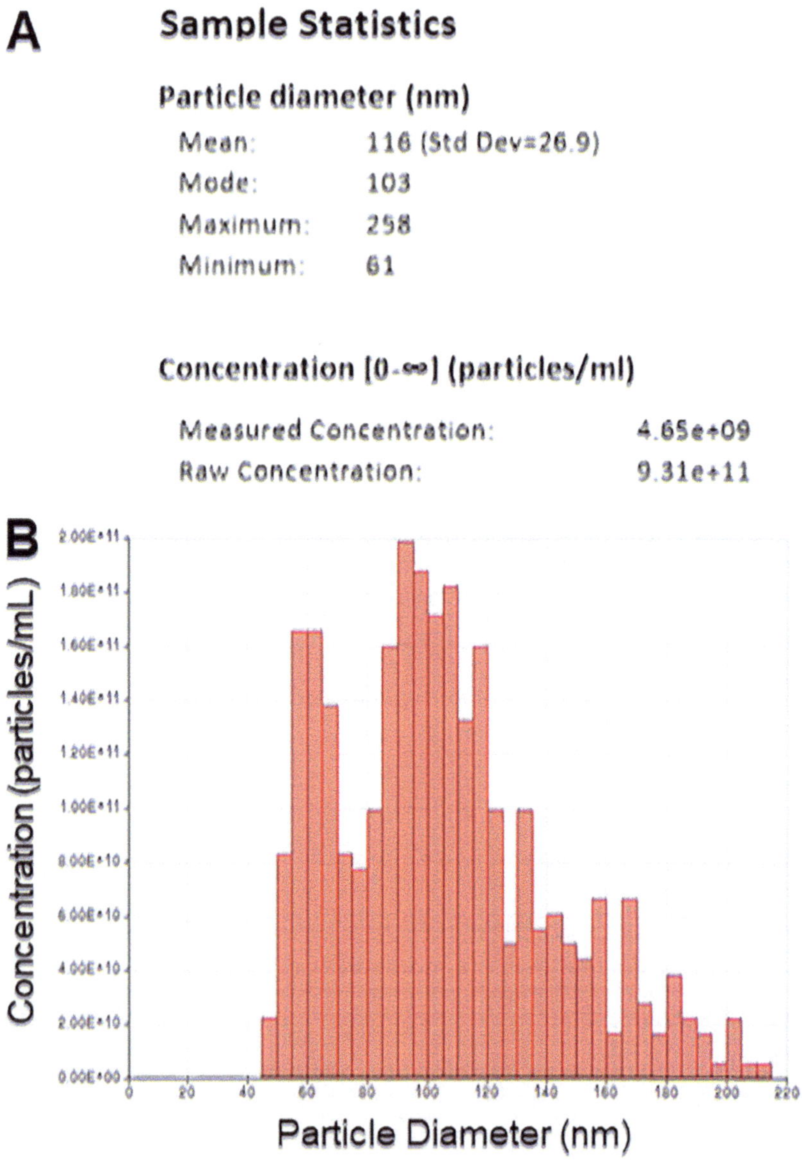

Fig. 11 Representative RBC-EV size and concentration data. (**a**) Sample statistics of RBC-EV diameter and concentration. (**b**) RBC-EV size histogram describing particle concentration with increasing particle diameter

2.8 Visualize RBC-EVs Using Transmission Electron Microscopy (TEM) (Table 8)

1. Prepare TEM reagents.
   (a) 2% uranyl acetate: Add 2 g uranyl acetate into 98 mL DI water.
2. Grip the 200 mesh copper grids with tweezers using a rubber band to hold the tweezers shut (Fig. 12a).
3. Add 5 μL RBC-EV samples (~10^{11} to 10^{12} RBC-EVs/mL) to the copper grid for 30 seconds (Fig. 12b).
4. Wick the RBC-EV sample off using a Kimwipe and add 5 μL DI water to the copper grid (Fig. 12c).

Table 8
Materials and reagents needed to prepare RBC-EVs for visualization via TEM

Material/reagents	Vendor	Catalog number
Dumostar super thin tip tweezers	Electron Microscopy Sciences	72705-01
Copper Grids	Electron Microscopy Sciences	FCF200-Cu
Uranyl Acetate	Electron Microscopy Sciences	22400

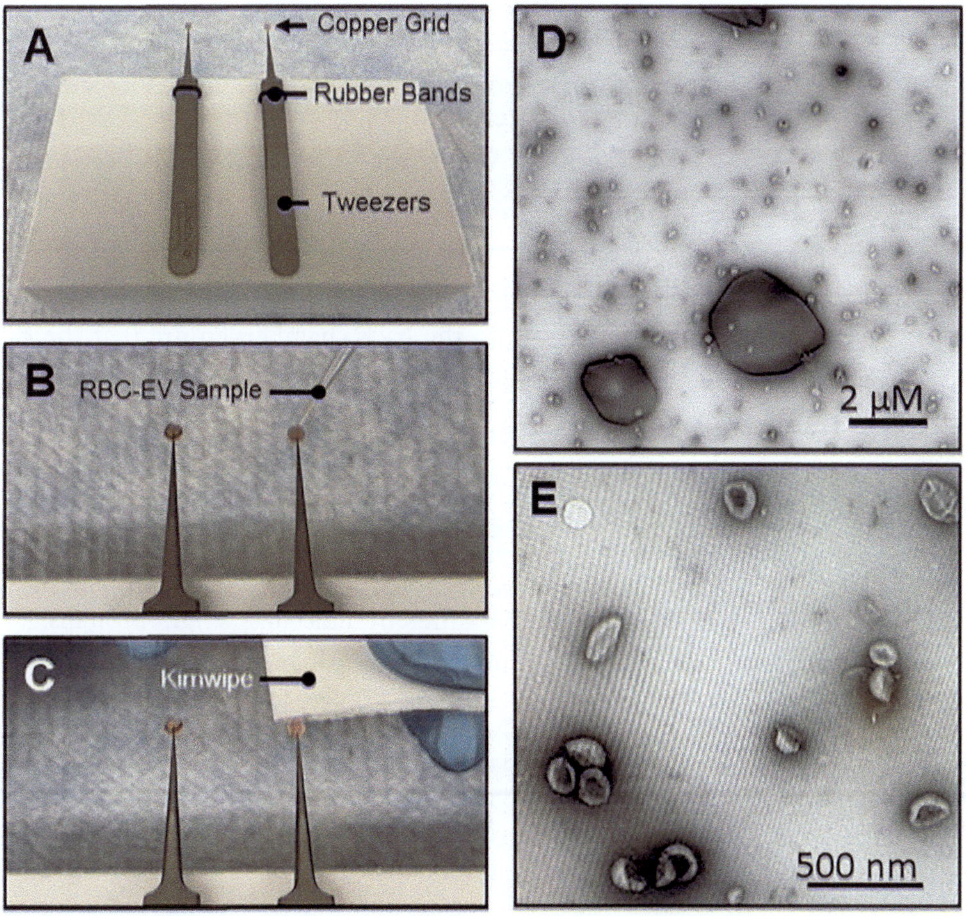

Fig. 12 TEM visualization of RBC-EV size and morphology. (**a**) Copper grids secured to tweezers using rubber bands. (**b**) RBC-EV samples pipetted onto the copper grids, with (**c**) excess sample removed by wicking. Representative TEM images of RBC-EVs at (**d**) 6,000× magnification and (**e**) 30,000× magnification

5. Wick the DI water off immediately using a Kimwipe.
6. Apply 5 µL 2% uranyl acetate to the copper grid for 15 seconds. Wick the uranyl acetate off the copper grid with a Kimwipe.
7. Acquire images at 6000× magnification (Fig. 12d) and 30,000× magnification (Fig. 12e) to visualize the RBC-EVs.

3 Notes

1. The buffy coat is the fraction of anticoagulated blood that contains the white blood cells and platelets following centrifugation. The buffy coat is found between the RBC pellet and plasma and generally is a white/yellow color. Removing this layer will minimize white blood cell and platelet-derived EV contamination.
2. You can also use other isotonic salt solutions like Hanks Buffered Saline Solution (HBSS) or cell culture media, depending on your application.
3. Adding yoda1 to the isotonic salt solution before adding RBCs will minimize cell lysis.
4. We found no significant differences in EV concentration, purity, or size between RBC-EVs isolated using fresh or frozen RBC-EV-rich PBS. Storage at 4 °C for more than a day or multiple freeze-thaw cycles will decrease RBC-EV size and concentration.
5. If MOPS leaks to the back chamber, ensure that the cassette clamps are completely locked in place around the Bolt 4–12% Bis-Tris Wedge Well protein gels. If this does not fix the problem, consider purchasing new cassette clamps.
6. Add water instead of reducing agent if you are detecting CD63.
7. MagicMark is visible under chemiluminescence and is used to identify the correct protein band during imaging. Spectra Ladder is visible under regular light and is used to cut the membrane to label for different molecular weight proteins.
8. If bubbles are not evenly forming at the top of the chamber or proteins do not migrate evenly, ensure that the mini-gel tank cover is securely connected to the electrodes on the cassette clamps and that the negative and positive electrodes are connected correctly to the power supply.
9. The band intensity for ALIX, TSG101, CD9, and CD63 will vary depending on the RBC donor, disease state, and RBC-EV concentration.
10. Use reverse pipetting when loading and removing liquid from the bottom and top chamber to minimize bubbles, which will clog the nanopore. To reverse pipette, first set the pipette to the desired volume. Depress the plunger past the first and second stop. Immerse the tip into the liquid and slowly release the plunger to full extension. Dispense the liquid by pressing the plunger to the first stop. A small amount of liquid will remain in the tip.

11. NP100 will measure particles between 50 and 330 nm. Use the appropriate nanopore for the size range you are interested in quantifying. The number of times you can use a nanopore will depend on the number of measurements performed and the experience of the user.
12. A voltage of 0.5–0.7 V should be sufficient to obtain 120 nA current. The maximum voltage that should be applied is 1.6 V.
13. The RMS noise may increase and particle rate may become nonlinear during measurement of calibration particles or RBC-EVs due to particle aggregates or bubbles. To minimize particle aggregates, vortex the CP100 before and after dilution and ensure that the RBC-EVs are filtered through a 0.22 μm ultrafree-MC tube to remove RBC-EVs or RBC-EV aggregates larger than 220 nm. Bubbles can also increase RMS noise and decrease particle rate by impairing particle flow through the nanopore. Ensure there are no bubbles in the upper fluid cell by removing and repipetting your calibration particle or RBC-EV sample. Finally, check if bubbles have formed in the lower fluid cell, which will happen over time. If you see bubbles in the lower fluid cell, remove the nanopore and restart the protocol from the nanopore washing, stretching, and wetting steps.
14. Dilute the CPC100 or RBC-EVs if the particle rate exceeds 1500 particles/minute, because the analysis software will begin to count multiple particles passing through the nanopore as one particle. If the particle rate is below 200 particles/minute, the CPC100 or RBC-EVs are too dilute or the nanopore may not be fully open. In this case, use more concentrated CPC100 or RBC-EV solutions. If this does not fix the problem, repeat the "open the nanopore" steps in the TRPS protocol and the troubleshooting steps in **Note 13**.
15. You can also apply higher pressures to speed up the particle count measurements: for example, 15 and 20 mbar. Higher pressures may put more stress on the nanopore and decrease its longevity.
16. The qNano system will stop recording data if 500 particles are not measured within 10 min.
17. When saving calibration or RBC-EV measurement, it is suggested to save the data as "Sample"_"pressure"_"calibration attempt." For example, "piezo1 RBC-EVs_P10_v2" for piezo1 RBC-EVs measured at 10 mbar using the second CPC100 calibration recording. You may have to acquire multiple calibration measurements if your nanopore becomes clogged or as the nanopore enlarges past ~30 min. This information is important when selecting the correct calibration measurements with the corresponding RBC-EV measurements.

Acknowledgement

This study was funded by the University of Maryland Presidential Postdoctoral Fellowship, American Heart Association Postdoctoral Fellowship [916512], and the National Institutes of Health [R01HL165193).

References

1. Biagiotti S, Abbas F, Montanari M, Barattini C, Rossi L, Magnani M, Papa S, Canonico B (2023) Extracellular vesicles as new players in drug delivery: a focus on red blood cells-derived EVs. Pharmaceutics 15(2). https://doi.org/10.3390/pharmaceutics15020365
2. Chiangjong W, Netsirisawan P, Hongeng S, Chutipongtanate S (2021) Red blood cell extracellular vesicle-based drug delivery: challenges and opportunities. Front Med (Lausanne) 8:761362. https://doi.org/10.3389/fmed.2021.761362
3. Thangaraju K, Neerukonda SN, Katneni U, Buehler PW (2020) Extracellular vesicles from red blood cells and their evolving roles in health, coagulopathy and therapy. Int J Mol Sci 22(1). https://doi.org/10.3390/ijms22010153
4. Usman WM, Pham TC, Kwok YY, Vu LT, Ma V, Peng B, Chan YS, Wei L, Chin SM, Azad A, He AB, Leung AYH, Yang M, Shyh-Chang N, Cho WC, Shi J, Le MTN (2018) Efficient RNA drug delivery using red blood cell extracellular vesicles. Nat Commun 9(1): 2359. https://doi.org/10.1038/s41467-018-04791-8
5. Almizraq RJ, Holovati JL, Acker JP (2018) Characteristics of extracellular vesicles in red blood concentrates change with storage time and blood manufacturing method. Transfus Med Hemother 45(3):185–193. https://doi.org/10.1159/000486137
6. Berezin AA, Obradovic Z, Kopp K, Berezina TA, Lichtenauer M, Wernly B, Berezin AE (2022) The association of glucose control with circulating levels of red blood cell-derived vesicles in type 2 diabetes mellitus patients with atrial fibrillation. Int J Mol Sci 24(1). https://doi.org/10.3390/ijms24010729
7. Jana S, Strader MB, Meng F, Hicks W, Kassa T, Tarandovskiy I, De Paoli S, Simak J, Heaven MR, Belcher JD, Vercellotti GM, Alayash AI (2018) Hemoglobin oxidation-dependent reactions promote interactions with band 3 and oxidative changes in sickle cell-derived microparticles. JCI Insight 3(21). https://doi.org/10.1172/jci.insight.120451
8. Peng L, Li Y, Li X, Du Y, Li L, Hu C, Zhang J, Qin Y, Wei Y, Zhang H (2021) Extracellular vesicles derived from intermittent hypoxia-treated red blood cells impair endothelial function through regulating eNOS phosphorylation and ET-1 expression. Cardiovasc Drugs Ther 35(5):901–913. https://doi.org/10.1007/s10557-020-07117-3
9. Straat M, Boing AN, Tuip-De Boer A, Nieuwland R, Juffermans NP (2016) Extracellular vesicles from red blood cell products induce a strong pro-inflammatory host response, dependent on both numbers and storage duration. Transfus Med Hemother 43(4):302–305. https://doi.org/10.1159/000442681
10. Sudnitsyna J, Skverchinskaya E, Dobrylko I, Nikitina E, Gambaryan S, Mindukshev I (2020) Microvesicle formation induced by oxidative stress in human erythrocytes. Antioxidants (Basel) 9(10). https://doi.org/10.3390/antiox9100929
11. Jy W, Johansen ME, Bidot C Jr, Horstman LL, Ahn YS (2013) Red cell-derived microparticles (RMP) as haemostatic agent. Thromb Haemost 110(4):751–760. https://doi.org/10.1160/TH12-12-0941
12. Sangha GS, Weber CM, Sapp RM, Setua S, Thangaraju K, Pettebone M, Rogers SC, Doctor A, Buehler PW, Clyne AM (2023) Mechanical stimuli such as shear stress and piezo1 stimulation generate red blood cell extracellular vesicles. Front Physiol 14: 1246910. https://doi.org/10.3389/fphys.2023.1246910
13. Cahalan SM, Lukacs V, Ranade SS, Chien S, Bandell M, Patapoutian A (2015) Piezo1 links mechanical forces to red blood cell volume. elife 4. https://doi.org/10.7554/eLife.07370
14. Gottlieb PA (2017) A tour de force: the discovery, properties, and function of piezo channels. Curr Top Membr 79:1–36. https://doi.org/10.1016/bs.ctm.2016.11.007

15. Parpaite T, Coste B (2017) Piezo channels. Curr Biol 27(7):R250–R252. https://doi.org/10.1016/j.cub.2017.01.048
16. Danielczok JG, Terriac E, Hertz L, Petkova-Kirova P, Lautenschlager F, Laschke MW, Kaestner L (2017) Red blood cell passage of small capillaries is associated with transient Ca (2+)-mediated adaptations. Front Physiol 8: 979. https://doi.org/10.3389/fphys.2017.00979
17. Dupire J, Socol M, Viallat A (2012) Full dynamics of a red blood cell in shear flow. Proc Natl Acad Sci USA 109(51): 20808–20813. https://doi.org/10.1073/pnas.1210236109
18. Larsen FL, Katz S, Roufogalis BD, Brooks DE (1981) Physiological shear stresses enhance the Ca2+ permeability of human erythrocytes. Nature 294(5842):667–668. https://doi.org/10.1038/294667a0
19. Leverett LB, Hellums JD, Alfrey CP, Lynch EC (1972) Red blood cell damage by shear stress. Biophys J 12(3):257–273. https://doi.org/10.1016/S0006-3495(72)86085-5

Chapter 7

Analysis of Urinary Extracellular Vesicles

Uta Erdbrügger, Martin E. van Royen, Gerald W. Verhaegh, Cristina Grange, Dylan Burger, Morgan Greig, Joshua A. Welsh, Tuula Anneli Nyman, Alicia Llorente, Elena Martens-Uzunova, Maija Puhka, Karina A. Barreiro, and Monica Suet Ying Ng

Abstract

After blood, urine is the second most studied biofluid, because it is easy to collect and available in large quantities. However, urine is also the most dynamic biofluid with large variations in concentration of its diverse content. Separation, storage, and characterization of urine extracellular vesicles have special considerations to tackle these large variations of contents by addressing preanalytical (e.g., standardized collection) and analytical factors (e.g., normalization strategies). This chapter will provide real-world, step-by-step protocols for urine collection and storage, along with urine extracellular vesicle cargo isolation and characterization. Single urine extracellular vesicle characterization methods include nanoparticle-tracking analysis (NTA) and flow cytometry. Bulk urine extracellular vesicle characterization methods include protein analysis using immunoblotting and mass spectrometry, and, ribonucleic acid (RNA) analysis with untargeted (e.g., next-generation sequencing) and targeted (e.g., quantitative real-time polymerase chain reaction [PCR]) approaches. This chapter will also discuss urine-specific strategies to address variations in urine concentration, ultralow cargo inputs for downstream analysis, and uromodulin contamination.

Key words Urine, Nanoparticle-tracking analysis, Protein analysis, Flow cytometry, Immunoblot, Mass spectrometry, RNA analysis, Next-generation sequencing, Quantitative PCR, Uromodulin

1 Introduction

Urinary Extracellular Vesicles (uEVs) are released from living cells and pathophysiological processes of organs in the urogenital tract and possibly other, more distant organs. Their cargo is seen as a molecular mirror of their cells of origin. Interest in leveraging uEVs now for clinical diagnostics and therapeutics has grown exponentially with research focusing on urogenital cancers, acute and chronic kidney injury, kidney transplant rejection, urinary tract infection, pregnancy, and prenatal diagnostics. After blood, urine is the second most studied biofluid as it is easy to collect and

Amit K. Srivastava (ed.), *Extracellular Vesicles*, Neuromethods, vol. 228, https://doi.org/10.1007/978-1-0716-4905-3_7,

available in large quantities. However, urine is also the most dynamic biofluid with large variations in concentration of its diverse content. Therefore, uEV research deserves special considerations to tackle these large variations of content by addressing preanalytical (e.g., standardized collection) and analytical factors (e.g., normalization strategies).

In this chapter, we are providing protocols for characterization of uEV size, concentrations and uEV protein, and RNA cargo starting with a discussion of their collection and storage. For single EV analysis, most uEV researchers are using particle analyzers such as nanoparticle-tracking analysis and tunable resistive pulse sensing tools, which provide information on uEV size, charge, and concentration. Some NTA instruments have also lasers to fluorescently label uEV cargo. A second major single EV analysis tool is EV nano flow cytometry, which allows high-throughput detection and characterization of uEV cargo and origin. Special algorithms are used to also provide information on uEV size. Many bulk analysis tools are also utilized in addition to single EV analysis tools. Immunoblotting is often employed for validation of uEV origin and uEV biomarkers as a targeted approach. uEV mass spectrometry allows unbiased uEV proteome analysis. Finally, we describe special consideration for uEV RNA cargo analysis with targeted (e.g., PCR) and unbiased analysis approaches (Transcriptomic analysis).

2 Collection and Storage of Urine Specimens for uEV Analysis

2.1 Introduction

Urine collection and storage procedures influence the outcome of uEV analysis due to variable or insufficient representation of organ-specific EVs in the urine sample, or suboptimal uEV preservation during storage. To address the need for a consensus on the best methodological practices tailored for uEVs, the International Society of Extracellular Vesicles' (ISEV) Urine Task Force (UTF) recently published a community position paper with recommendations to improve the rigor and reproducibility in uEV analysis, later accompanied by a Quick Reference Card with recommendations on storage of uEVs [1, 2].

Optimization of urine collection protocols is partly dictated by the analytical parameter of interest (e.g., absolute uEV concentration and size, bulk uEV RNA or protein analyses, uEV biomarkers on individual EVs or in the bulk uEVs population) and possibly the EV-secreting organ of interest (e.g., kidney, bladder or prostate). Nevertheless, practical considerations like timing of medical consultations or home sampling partly determine the choice for random or timed single void collection (spot urine), or timed urine collections over longer time courses (e.g., 24-h urine), either collecting full or first void or midstream urine. The composition of the uEV pool in these collected urines is strongly influenced by the overall fitness and circadian rhythm of the donor [3], variations in recent physical activity, food intake, and hydration status [4]. Moreover, medical interventions can significantly influence the contribution of the organ of interest to EVs in the collected urine. Digital rectal examination (DRE) shortly before urine collection strongly increases the contribution of prostate-derived EVs in the urine [5]. Therefore, strict collection protocols and detailed instructions for the clinical staff and the donors are essential and should be customized toward the EV subpopulation of interest and must be reported in publications.

In addition to these biological and methodological factors of influence, technical variables in preanalytical processing and storage of the urine samples will introduce variation in the sample quality and the potential outcome of uEV analysis. Such variables include the following:

1. Time and temperature during sample handling and storage before freezing
2. Steps to remove (sub)cellular and protein complex contaminants before freezing
3. Sample volumes collected
4. Aliquoting to speed up freezing and thawing
5. Avoidance of multiple freeze/thaw cycles of the urine samples
6. Freezing temperature
7. Storage time

To minimize the effects of these variables, protocols should be standardized as much as possible, and variables should be recorded and reported in detail for the individual collected urine samples [1, 2]. If done properly, these detailed sample records will enable researchers to recognize influencing parameters on subsequent uEV analysis and identify sample outliers that might contaminate the analytical data.

Here we describe example protocols for urine collection and storage for EV analysis based on protocols approved and adopted by the European Association of Urology Research Foundation (EAU-RF) and deposited at the Biospecimen Research Database of the National Institutes of Health/National Cancer Institute (NIH/NCI) [6]. We complement these protocols with recommendations from the ISEV UTF [1, 2].

2.2 Recommendations for Urine Collection for uEV Analysis

1. Dependent on the local or national regulation on the use of clinical materials, an ethical approval by the regulating body of the institute and informed consent of the donors may be required.
2. Patients should be instructed how to collect the appropriate type of urine (24-h urine, first void urine, or spot urine) and

volume of urine without contamination. The use of a urine collection device (e.g., Colli-Pee, Novosanis) may be considered.

3. Urine should be collected in a certified urine collection cup free of any preservatives and containing a lid. The cup should be appropriately coded for patient identification and possible link to clinical information, preferably with use of a clinical database to ensure data protection and adherence to patient privacy regulations.
4. Upon collection, the urine in the urine collection cup must be maintained at 2–8 °C in a fridge or cool box and processed within 4 h after collection to avoid microbial growth, as well as sedimentation and degradation of more labile biomolecules. Samples kept in a cool box should never come into direct contact with the ice pack to avoid damaging the EVs.
5. Collected urine should be protected from direct exposure to light to prevent degradation of potentially light-sensitive urine constituents (e.g., bilirubin and porphyrins).
6. Urine quality should be verified using standard urine quality control measurements (e.g., dipsticks). pH and total protein content of the urine can be measured to identify sample outliers caused by microbial outgrowth or high level of urinary protein.
7. The use of protease inhibitors may be considered when appropriate and if it does not interfere with subsequent uEV analysis.

2.3 Urine Storage for uEV Analysis

Generally, uEV isolation or direct uEV analysis is performed using the urine supernatant. However, collection and storage of whole urine and urinary sediment aliquots is recommended whenever logistics and availability of storage space allows. The availability of these additional aliquots makes it possible to include additional controls in the uEV purification and analysis process. The minimal volume required for the uEV analyses and additional sample types (e.g., whole urine, urine sediment) or analyte-specific storage containers should be determined in advance and implemented in an appropriate aliquoting scheme before the start of collection. Aliquoting in multiple smaller volumes is advised to avoid freeze/thaw cycles of samples. When less than the required volume is collected, the urine sample should be aliquoted according to a priority scheme that indicates

1. The priority of collecting different sample types (e.g., urine supernatant, whole urine, urine sediment)
2. The analyte-specific storage containers

   The number of vials and the maximum sample volumes per vial (*see* Table 1).

Table 1
Example priority scheme indicating the number of aliquots with indicated sample types (e.g., whole urine or urine supernatant) and volumes. In this example, the preferred collection is 50 ml (1 × 8 ml: sample 1, 2 × 5 ml: sample 2, and 6 × 5 ml: sample 3). When a smaller urine volume is collected, priorities are indicated for sample types and aliquot volumes

Collected Urine	Priority	Aliquot type 1	Aliquot type 2	Aliquot type 3
(ml)		(e.g. 8ml)	(e.g. 5 ml)	(e.g. 1 ml)
50	**Preferred collection volume**	**1**	**8**	**2**
45	**1**	1	7	2
40	**2**	1	6	2
35	**3**	1	5	2
30	**4**	1	4	2
25	**5**	1	3	2
20	**6**	1	2	2

2.3.1 Storage of Whole Urine Samples

1. Invert the urine collection cup five times to resuspend cells and potential protein aggregates in the urine.
2. Transfer the appropriate volumes to prelabelled urine preservation vials or cryovials until it is filled to 75% at maximum to accommodate sample expansion. This should be done immediately after collecting the urine.
3. Close the vials tightly and store them immediately at −70 °C or lower. Freeze samples rapidly in the −80 °C freezer or alternatively snap freeze in liquid nitrogen.
4. Store the urine collection cup back at 2–8 °C in a fridge or cool box for further aliquoting.
5. Record time of transfer of urine, time of freezing, and the location in the storage system. The use of worksheets as provided by the EAU-RF should be considered [6].

2.4 Preparation and Storage of Urine Supernatant

1. Transfer the total required volumes for urine for the urine supernatant from the urine collection cup in a prelabelled 50 ml conical centrifuge tubes. If less than the preferred urine volume is collected, the volumes and number of cryovials should be adjusted according to the preestablished priority scheme.
2. Centrifuge the tube(s) at 4 °C for 20 min at 500 × G without applying the centrifuge brake.
3. Slowly pipette the appropriate volumes (up to a maximum 80–90% of the total centrifuged volume) of the urine supernatant from the centrifuge tube into an appropriate number of prelabelled certified cryovials (compatible with the available biobank storage system) until it is filled up to a maximum of

75% of the vial volume to accommodate for sample expansion after freezing. The sediment at the bottom of the centrifuge tubes should be left undisturbed (and can be further processed, see below). Aliquoting in multiple smaller volumes is advised to avoid freeze/thaw cycles of the samples for individual experiments.

4. Close the vials with urine supernatant tightly and store them immediately at −70 °C or lower. Freeze samples rapidly in the −70 °C freezer or alternatively snap freeze in liquid nitrogen.
5. Record time of transfer of urine, time of freezing, and the location in the storage system. The use of worksheets as provided by the EAU-RF should be considered [6].

2.5 Storage of Urine Sediment

1. Centrifugal tubes used for the preparation of urine sediment should be appropriately coded with the ID according to own routine procedures.
2. Remove residual volume supernatant without disturbing the sediment.
3. Mix the sediment in the prelabelled conical centrifuge tubes used for the preparation of the urine supernatant with 2 ml ice-cold buffered sodium chloride solution pH 7.4 (PBS) and transfer the resuspended sediment to cryovial(s) labelled with the appropriate ID.
4. Centrifuge the cryovials at 4 °C for 10 min at 3000 × G with the centrifuge brake activated.
5. Discard the supernatant from each cryovial.
6. Immediately freeze the two cryovials with urine sediment in liquid nitrogen and subsequently, place it into a freezer unit of −70 °C or lower. If freezing in liquid nitrogen is not feasible, the cryovials should be immediately frozen at −70 °C or lower. The time and method of freezing and the location in the storage system should be recorded. The use of worksheets as provided by the EAU-RF should be considered [6].

3 Nanoparticle-Tracking Analysis (NTA) for uEV Characterization

3.1 Introduction

Nanoparticle-tracking analysis (NTA) enables the determination of both the hydrodynamic size distribution and concentration of small particles, including extracellular vesicles (EVs). NTA is a microscope-based technique that allows the characterization of nanoparticles in suspension within the approximate range of 50/200–2000 nm, enabling real-time tracking of individual nanoparticles, in scatter or fluorescence modes. By tracking the particles over time based on their fluorescence of light scatter signals, the particle diameter can be approximated using mean-squared displacement and using assumptions about the particle's composition. This technique is based upon the principles of Brownian motion. There are many devices available, from different companies, with variations in hardware and software that can lead to discrepancies in measurement results. The majority of nanoparticle analyzers are capable of measuring particles in the approximate size range of 50 nm to 1 μm with precision and customizable configurations such as interchangeable laser modules. The software's programmable temperature control ensures accurate measurements under optimal conditions.

NTA is commonly used to quantify the size, concentration, and, in certain cases, zeta potential of inorganic particles (e.g., polymers, silicon dioxide, minerals) and biological nanoparticles (e.g., EVs, liposomes, protein aggregates, viruses). Focusing on the EV field, NTA is widely used to rapidly quantify the concentration and size of EV samples, helping with the characterization of EV heterogenicity and understanding of EV involvement in pathophysiological conditions. Moreover, it can be utilized to monitor sample characterization and stability over time among different storage conditions (Fig. 1).

3.1.1 The Principle of Nanoparticle-Tracking Analysis

A laser beam is directed through a sample chamber containing particles. A microscope camera records the random thermal motion of each particle, known as Brownian motion (Fig. 2). This information enables particle counting and analysis of particle size

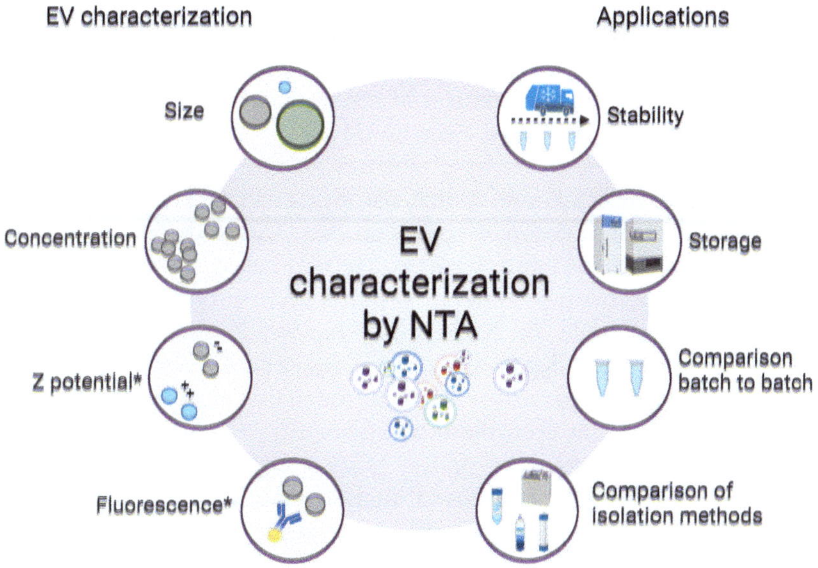

Fig. 1 Graphical representation of NTA applications. Availability depending on the type of instrument. Moreover, fluorescence testing cannot be done to best practice as we do not have calibrated units yet. Created in BioRender. Grange, C. (2025) https:/BioRender.com/vgedqcv

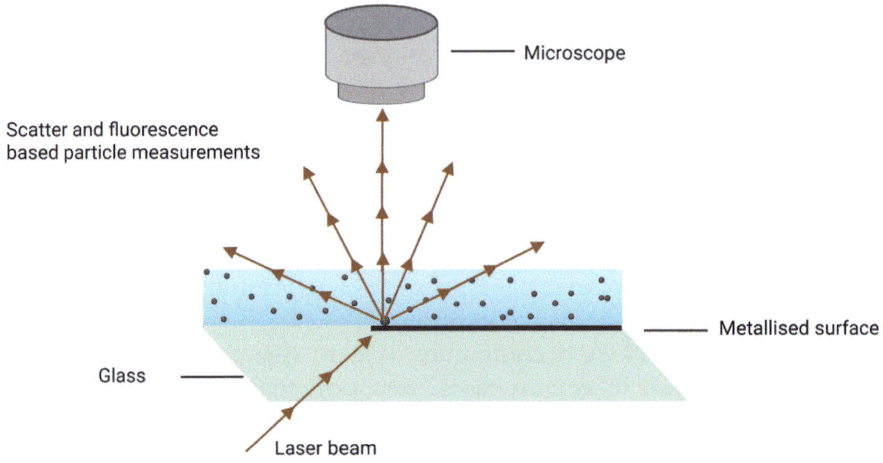

Fig. 2 Graphical representation of NTA principle. Created in BioRender. Grange, C. (2025) https:/BioRender.com/a5y7252

distribution. The number of particles in relation to the analyzed volume allows for the quantification of particle concentration. Additionally, the Stokes-Einstein equation can be used to calculate the hydrodynamic radius of each particle from the Brownian motion observed in video analysis. The smallest detectable particle size depends on different factors such as the particle's scattering properties, fluorophore brightness and photobleaching properties, the microscope's magnification power, and the camera's sensitivity. Metallic nanoparticles, with their high refractive index, scatter light strongly, enabling detection down to approximately 10 nm. In contrast, biological nanoparticles like EVs have lower refractive indices (around 1.37–1.45), limiting detection to 50–100 nm. A detailed explanation of the principle of size estimation by NTA has been published elsewhere [7].

3.2 Methodology

3.2.1 Equipment and General Considerations

There are two major companies that produce NTA devices: Malvern, a company based in the UK produces "NanoSight" systems (Fig. 3), and Particle Metrix, based in Germany, produces the "ZetaView" line of systems. While both instruments utilize the principle of NTA, their hardware and software configurations exhibit significant differences. Factors such as camera magnification, sample movement mechanisms, and standardized settings vary between these two devices.

Regardless of the NTA system used, an important consideration in NTA-based analysis of urinary EVs is the potential for uromodulin, a highly abundant urinary protein, to interfere with NTA analysis [4]. Approaches to isolation of urinary EVs, including those that can remove uromodulin from urine, have been discussed elsewhere [8]. Accordingly, this methodology will focus on the analysis of processed urinary EVs beginning at the stage of sample

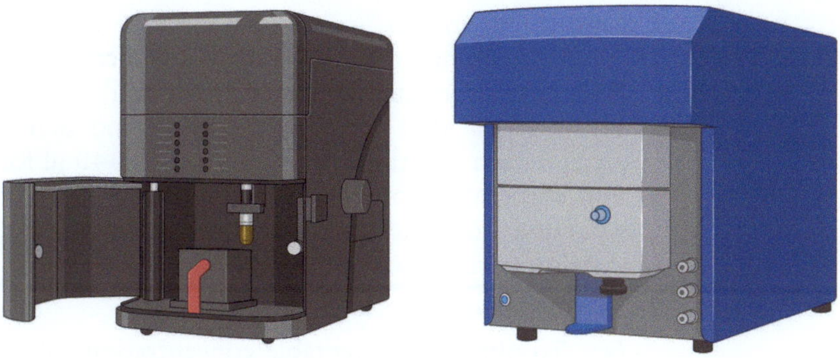

Fig. 3 Graphical representation of the two major instruments used. Created in BioRender. Grange, C. (2025) https:/BioRender.com/mxqlmfh

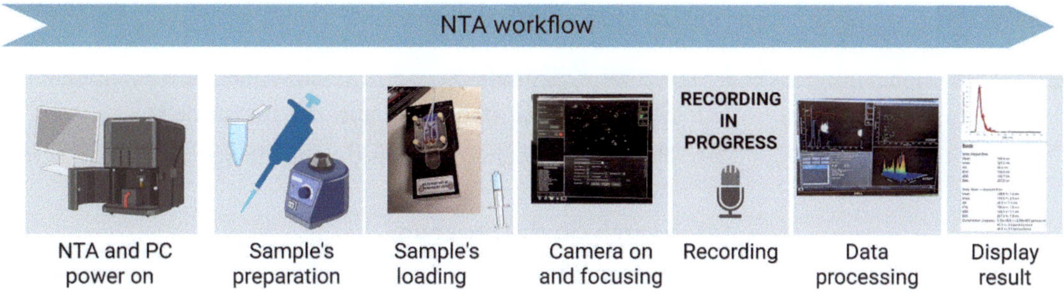

Fig. 4 Nanoparticle-tracking analysis work flow [9]. Created in BioRender. Grange, C. (2025) https:/BioRender.com/ejgmfhg

dilution and loading. As the precise approaches can differ depending on the instrument, protocols are divided by instrument. Most instruments also detect Zeta potential.

3.2.2 Protocols for NanoSight

For a better performance of the instruments (Fig. 4), accurate EV isolation protocols should be applied based on the downstream analysis and on the type of fluid taking into account removal of uromodulin. The NanoSight is equipped with different lasers, but the most common one is the 488 laser.

1. *Required Materials*
   (a) For sample dilution: particle-free PBS or saline solution; in alternative filtering solutions with a 0.1 μm filter
   (b) For rinsing: particle-free water
   (c) Pipettes and tips: 100–1000 μl; 2–20 μl
   (d) Syringe: 1 ml without a needle
   (e) Vortex
   (f) Biological safety cabinet (for handling and preparation of biohazardous samples)

2. *Sample Preparation*

   It is recommended to prepare a dedicated aliquot for the NTA analysis, containing an adequate amount of sample (usually 5–10 μl after ultracentrifugation and density gradient, precipitation isolation techniques and 20–40 μl for size exclusion chromatography and tangential flow filtration methods) separated from the rest of the samples to avoid unnecessary thawing of the whole samples [7].

3. *Sample Dilution:*
   (a) To obtain an acceptable concentration, all samples must be diluted in particle-free PBS according to the manufacturer's recommendations, with a final volume of 1 ml.
   (b) Vortex the sample for about 30 s.

       Note: The ideal particle per frame values are around 20–100 particles/frame.

4. *Instrument Preparation*
   (a) Switch on the computer and the instrument.
   (b) Position the chamber inside the device using the screws provided.
   (c) Switch on the program.

5. *Instrument Setting*

   Note: Instrument settings are highly context and model dependent. The precise settings should be determined during instrument installation and refined with experience. The settings below are provided only as a guide.

   (a) Camera level: Increase the number of camera level until all particles are distinctly visible not exceeding a particle signal saturation over 20%, usually for biological particles, the camera levels are around 13–16.
   (b) Flow condition: Constant flow rate is 30–100.
   (c) Video: 3 to 6 successive videos of 60 s.

6. *Loading of Sample*
   (a) Aspirate 1 ml of EV-containing solution, avoiding to create bubble at the top and bottom of the syringe.
   (b) Carefully load the sample into the injector tube.

7. *Acquisition*
   (a) Autofocus is adjusted so that indistinct particles are avoided.
   (b) Start recording.

8. *Data Analysis*

   The NTA software analyzes this video, by tracking individual particles frame by frame and measuring their Brownian motion. Data are analyzed using the most updated available NTA software with a detection threshold of about 5. The ideal detection threshold is set to capture the largest number of particles, while adhering to these limitations: a count of 10 to 100 red crosses, where less than 10% remain unmatched to specific particles, and a blue cross count restricted to a maximum of 5.

9. *Cleaning of Instrument*

   Cleaning of the instrument should be completed as per manufacturer's instructions.

10. *Suggestions to Improve the Accuracy of NTA Measurement*

    For samples with a low particle number, the NTA's precision may be low due to variations in the particle count between repeated measurements of the same sample. To improve the accuracy for low-particle-count samples, it's recommended to change the samples' dilution or increase the number of video replicates [9, 10].

3.2.3 Protocols for ZetaView

Similar to NanoSight systems, the ZetaView line of systems are capable of being equipped with one or more different lasers for analysis with 405 and 488 being the most common. The primary distinguishing features of ZetaView systems are a lower magnification and the use of up to 11 different measuring positions. The system allows for the removal of individual measuring positions from analysis if they are impacted by large particles or air bubbles.

1. *Materials*
   - (a) For sample dilution: Use a particle-free PBS or saline solution; in alternative filtering solutions with a 0.1 μm filter (Amicon).
   - (b) For rinsing: Use particle-free water.
   - (c) Pipettes and tips: 100–1000 μl; 2–20 μl.
   - (d) Syringe: 1 ml and 10 ml without needle or luer lock.
   - (e) Alignment solution: 100 nm alignment beads (provided by the manufacturer). The final stock will be ~1:250,000 dilution.
   - (f) Vortex.
   - (g) Biological safety cabinet (for handling and preparation of biohazardous samples).
   - (h) Cleaning detergent (1 ml of liquid soap in 50 ml of water).
   - (i) Lint-free pipe cleaners (i.e., BJ Long or equivalent).
2. *Start-Up and Calibration*
   - (a) Turn on the ZetaView system.
   - (b) Turn on the computer and start the ZetaView.

(c) The ZetaView instrument will initialize. Progress will be indicated in the bar at the lower right of the window.

(d) After initialization, follow the instrument prompts first for a Cell Quality Check. Using a 10 ml syringe, rinse the instrument with 1–10 ml of particle-free water and click OK. Results should be reported as "Very Good." If the results are not "Very Good," then the system may require cleaning or camera realignment (see below).

(e) When prompted for "AutoAlignment/Optimize Focus, load the system with ~1 ml of alignment solution and click OK. The system will start autoalignment and focusing process. If successful, you will receive a message stating "The ZetaView instrument is now ready for experiments." If this does not appear, then the system may require cleaning or camera realignment (see below).

(f) Optional: Run an analysis on the 100 nm beads to confirm that the system is operating properly (see "Sample Analysis" below for instructions).

(g) Click on the "Pump and Temp" tab and set the instrument to the desired temperature for analysis (Recommended 25 °C).

3. *Instrument Settings*

Note: Instrument settings are highly context and model dependent. The precise settings should be determined during instrument installation and refined with experience. The settings below are provided only as a guide.

(a) Shutter: In general, set to maximize particle visibility while limiting overexposure. Seventy is common for general analysis, while 32 (the most sensitive) may be appropriate for fluorescent-labelled samples.

(b) Sensitivity: Set to ensure that all particles are distinctly visible but not overexposed. For urine samples, this is generally in the 75–90 range.

(c) Frame rate: In general, 30 for urine samples. If focusing explicitly on larger EVs, then 15 may be appropriate.

(d) Positions: 11, number of cycles: 2, High.

(e) Post Acquisition Parameters: Min Brightness: 30, Max Area: 1000, Min Area: 10, Tracelength: 5, nm/class: 15, Classes/Decade: 64.

**It is advisable to save the settings used for urinary EV analysis as an SOP within the software.*

4. *Sample Analysis*
   (a) Click on the "Measurement" Tab.
   (b) Rinse out the system prior to sample loading. A "clean" cell should read less than 5 under "No. of Detected Particles with low scattering intensity."
   (c) Slowly load 1 ml of the sample into the cell.
   (d) Ensure that your sample is stable and not moving within the cell by clicking on "Check Particle Drift at 0 V" and ensuring limited drift (indicated by a green line within the crosshairs).
   (e) Ensure that your sample is diluted to within the working range of the system (indicated by a green line within the "Number of Detected Particles": ideally 80–120).
   (f) Click on "Run Video Acquisition." Select your folder and name your file. Confirm your parameters for analysis (see Instrument settings above).
5. *Cleaning of Instrument*

   Cleaning of the instrument should be done as per manufacturer's instructions. A video tutorial can be found at https://support.particle-metrix.com/video-tutorials/zetaview.

4 uEV Flow Cytometry

4.1 Introduction

Flow cytometry (FCM) was first applied to detect EVs released by platelets in 1988 [11]. At this time, EVs were crudely characterized as any particle with lower forward and side scatter compared to platelets in arbitrary units. EVs are orders of magnitude smaller and dimmer than the cells that FCM equipment was originally intended to characterize. However, the *major strength of EV flow cytometry is the capacity for single EV analysis in a high-throughput manner.* Advancements in flow cytometry optics, fluidics, and electronics over the last several decades in conjunction with deepened understanding of biofluid composition (e.g., urine) and EV characterization optimization protocols and controls have moved the needle toward greater confidence in single EV characterization and reproducibility between laboratories [12]. Yet, as a newcomer to the field of single uEV FCM, we advise the user to follow the recommendation of ISEV regarding urinary EV analysis (see position statement on uEVs [1] and the quick reference card on storage of uEVs [2]). In addition, the user needs to *characterize the sensitivity of their own FCM instruments and tools to allow cross-platform comparisons* [13]. The principles underlying current standards in single EV FCM are explored in depth by Welsh et al. in "A compendium of single EV FCM." [12] In general, there is no gold standard for EV collection, handling, isolation, and characterization using FCM

[14]. For FCM, the ISEV-ISAC-ISTH EV Flow Cytometry Working Group developed "The Minimum Information about a Flow Cytometry experiment for Extracellular Vesicles (MIFlowCyt-EV)" as a framework for standardized reporting of EV FCM to improve reproducibility and transparency of results [15]. *Urine is a dynamic biofluid* with the urine concentration varying by fluid intake, time of collection, diet, exercise, age, biological sex, medication, and health and disease status. *For uEV data normalization, data for uEVs and non-uEV urine parameters need to be collected* to estimate absolute (normalization, e.g., to total protein, uEV numbers, uEV biomarker) or relative excretion rates (normalization to time collection, urinary creatinine/osmolality, or organ-specific markers such as prostate-specific antigen) [1].

This chapter will exemplify how EV flow cytometry principles have been applied to real-world protocols for characterizing uEVs. However, for experimental design, it is crucial that several considerations and parameters are considered:

1. Choice of uEV separation.
2. Panel development (choice of antibodies and fluorochromes/dyes/careful titration of reagents) and sample preparation.
3. Controls.
4. Optimization of instrument settings.
5. Data calibration to standard units.

4.2 Materials

1. Centrifuge (is needed (a) to generate cell-free urine and (b) to pre-enrich, e.g., studying the low centrifugation pellet [16] vs. ultracentrifugation pellet [17]).
2. Size exclusion chromatography column (Izon, qEV original columns) [18].
3. Triton X, 0.5%.
4. Antibodies of choice.
5. 0.2 μm filtered PBS.
6. Single labeled uEVs and serially diluted EVs.
7. Flow cytometer (many FCM companies are offering now special FCM with small particle analysis capabilities, conventional FCM's sensitivity needs to be tested).
8. NIST-traceable size beads, polystyrene (ThermoFisher, Polysciences).
9. MESF calibration beads (BD biosciences, Bangs Laboratories, Spherotech, e.g., APC, FITC, PE).
10. Multipeak multispectral ("Rainbow") beads (Spherotech).
11. FCMPASS.

4.3 Methods

4.3.1 uEV Separation

uEV separation methodology impacts significantly on results on EV population(s) available for downstream analysis. For uEV FCM, ideally cell-free urine and/or cell-free urine biobanks should be utilized when samples are frozen to avoid artificial generation of uEVs from cells [2]. Cellular debris, lipoproteins and uromodulin can be co-isolated with uEVs. Most laboratories do not remove uromodulin for FCM; however, this needs to be further investigated [8]. Urinary EVs can be detected in cell-free urine or in preparations which pre-enrich uEVs. Figure 5 reports two methods to isolate two subpopulations of uEVs, utilizing sequential centrifugation with low and high speed and additional use of size exclusion chromatography (SEC) [18]. Figure 6 demonstrates that the type of isolation method (here), namely, (a) low centrifugation and (b) ultracentrifugation, determines the range of uEVs (e.g., by size range) characterized by FCM. If centrifugation is used, it needs to be known which K factor is used so that rotor efficiencies are known for different sample volumes utilized. Imaging flow cytometry has also been used for EV characterization directly from body fluids, but this technology is not discussed in this chapter [19].

4.3.2 Addressing Dynamic Nature of Urine as a Body Fluid

Urine samples vary in concentration from subject to subject, and thus, it is important to consider concentration when characterizing uEVs using FCM. Using a more basic protocol, concentration is not considered, and each sample is diluted to the same degree. The more advanced protocol, and likely more representative, uses a counting technique to standardize uEV concentration throughout a cohort. This can be estimated using Nanoparticle-Tracking Analysis (NTA), which calculates the concentration of uEVs isolated from a sample. Diluting cell-free urine (SN4.6) in a buffer at a 1: 100 ratio is sufficient to determine this variable using an NTA instrument.

4.3.3 Panel Development

Panel development for uEV FCM is similar to panel development for cellular flow cytometry (e.g., avoid overlapping spectra, brighter fluorophores for low abundance antigens) with three extra considerations: (1) unbound fluorophore background, (2) ability for standardization using Molecules of Equivalent Soluble Fluorochrome (MESF) or Antibody Binding Capacity (ABC) beads, and (3) Steric hinderance when using 2 or more antibodies. Fluorophores available in corresponding MESF beads include phycoerythrin (PE), brilliant violet 421 (BV421), allophycocyanin (APC), alexa fluor 647 (AF647), alexa fluor 488 (AF488), fluorescein (FITC), enhanced green fluorescent protein (eGFP). Steric hinderance limits the number of antigens that can bind simultaneously to a single EV . Due to the large range in uEV concentration (10^6–10^{13} EVs/ml urine), protocols should also adjust antibody concentration for EV concentration or urine volume [20]. Also of note, some investigators utilize a low dose of detergents to detect

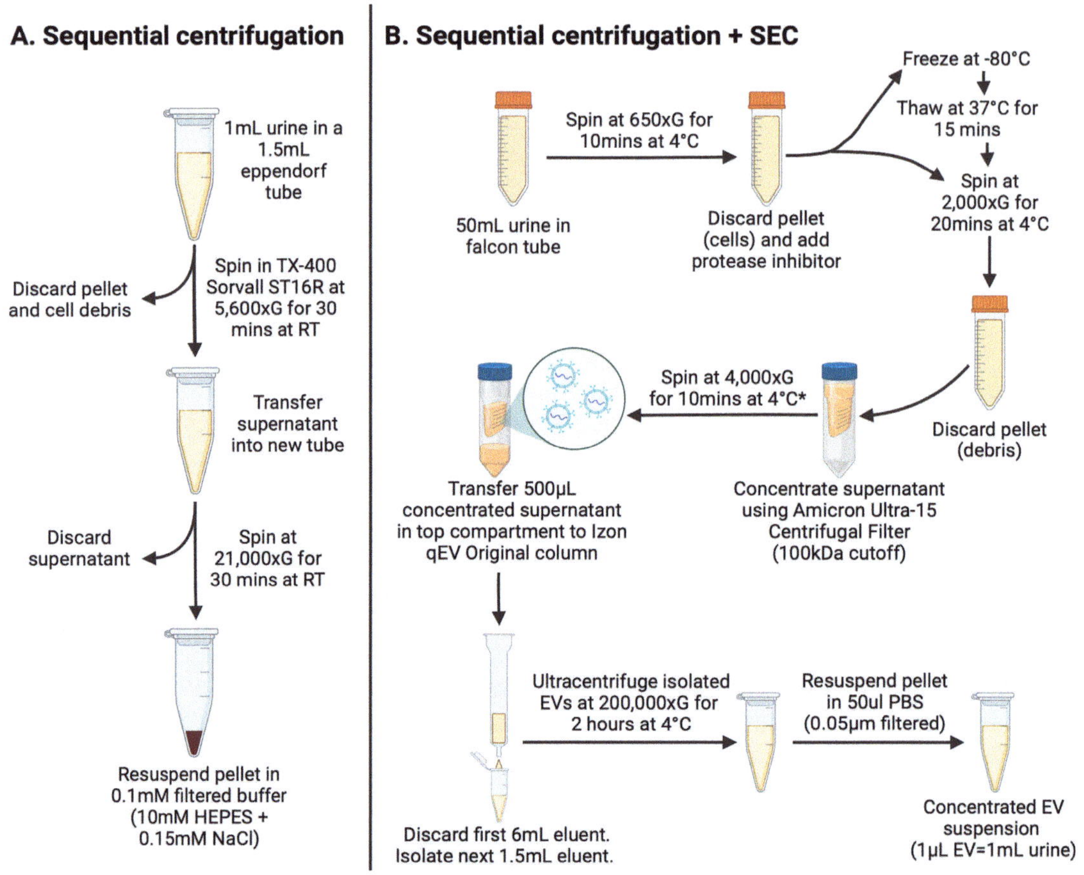

Fig. 5 Two protocol options for urine EV isolation. Protocol A involves sequential centrifugation 1 mL of raw urine sample is loaded into a 1.5 mL low-binding microcentrifuge tube and spun at 5,600 g for 30 min, e.g., in a TX-400 Sorvall ST16R) or similar CF. Without disrupting the pellet, the supernatant (SN4.6) is placed into a new 1.5 mL low-binding microcentrifuge tube and spun for a second time at 21,130 g in the same rotor. The supernatant (SN20) is removed and the pellet (P20) is resuspended in buffer (0 mM HEPES +0.15 mM NaCl 0.1 mm filtered) to obtain approximately a concentration of 2×10^9 particles mL^{-1} (measured by NTA). Protocol B involves a combination of sequential centrifugation, size exclusion chromatography (SEC), and ultracentrifugation [18]. A 50 ml urine is first centrifuged at 650 × G and 2000 × G to remove cells and cellular debris, respectively. The supernatant is then concentrated in a Amicron Ultra-15 centrifugal filter. The concentrated supernatant is then added to a size exclusion chromatography column to separate uEVs (Izon, qEV original). uEVs are further concentrated using ultracentrifugation for 2 h at 200,000 × G. Abbreviations: C Celsius, EV extracellular vesicles, HEPES 4-(2-Hydroxythyl)piperazine-1-ethane-sulfonic acid, kDa kilodaltons, mL milliliters, L microliters, NaCl sodium chloride, PBS phosphate-buffered saline, RT room temperature. Created in BioRender. Ng, M. (2025) https:/BioRender.com/p49f994

intracellular/intravesicular epitopes [4] (protocol discussed here [4]). A list of uEV markers of kidney origin is provided in the position paper of the Urine Task Force of ISEV [1]. Table 2 provides some examples of EV surface markers detected using flow cytometry that can indicate nephron segment source.

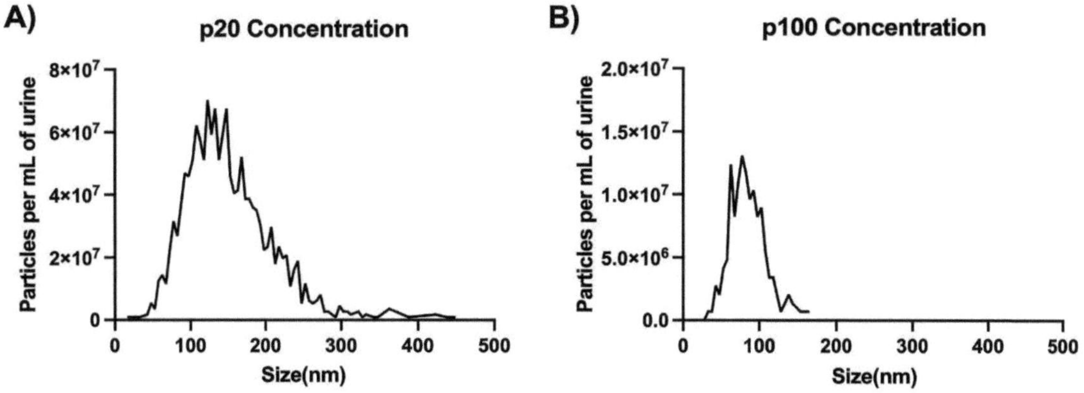

Fig. 6 This figure demonstrates examples of particle counting and sizing of urine by NTA after enrichment with the two protocols: (**a**) For the unwashed p20 pellet from 15 mL of raw urine, the median size was 138.9 nm and the concentration was 1.5×10^9 particles/mL of urine. (**b**) For the unwashed p100 pellet from 15 mL of raw urine, the median size was 89.7 nm and the concentration was 1.2×10^8 particles/mL of urine

Table 2
uEV surface markers indicating nephron segment source

Nephron segment	Surface markers
Podocyte	Anti-podoplanin (Biolegend, 337021) [21] Anti-podocalyxin (R&D, MAB1658) [22]
Proximal tubule	Anti-URAT1 (Bioss, bs-10357RFITC) [23] Anti-megalin (Abcam, AB317333)
Loop of Henle	Anti-SLC12A1 (Invitrogen, PA5-41995)
Distal tubule	Anti-NCC (Bioss, bs-7694R-FITC) [24]
Collecting duct	Anti-AQP2 (Biorbyt, orb13806) [24]

4.3.4 Controls for (u) EV Flow

As single uEV FCM works at the detection limit of current flow cytometers, multiple controls are required to ensure that detected events are EVs. These include buffer only controls, detergent controls, serial dilution controls, single stained controls, Isotype controls and antibody cocktail controls (Fig. 7). Events are recorded using the same settings as for EVs (with exception of single stained controls with compensation beads). We will refer to the MiFlowCyt-EV guidelines for experimental design for EVs in general and more detailed discussions of need for several controls [15].

Example Controls

According to the MIFlowCyt-EV guidelines, various controls are necessary to ensure accuracy and reproducibility when using EVs in flow cytometry [15].

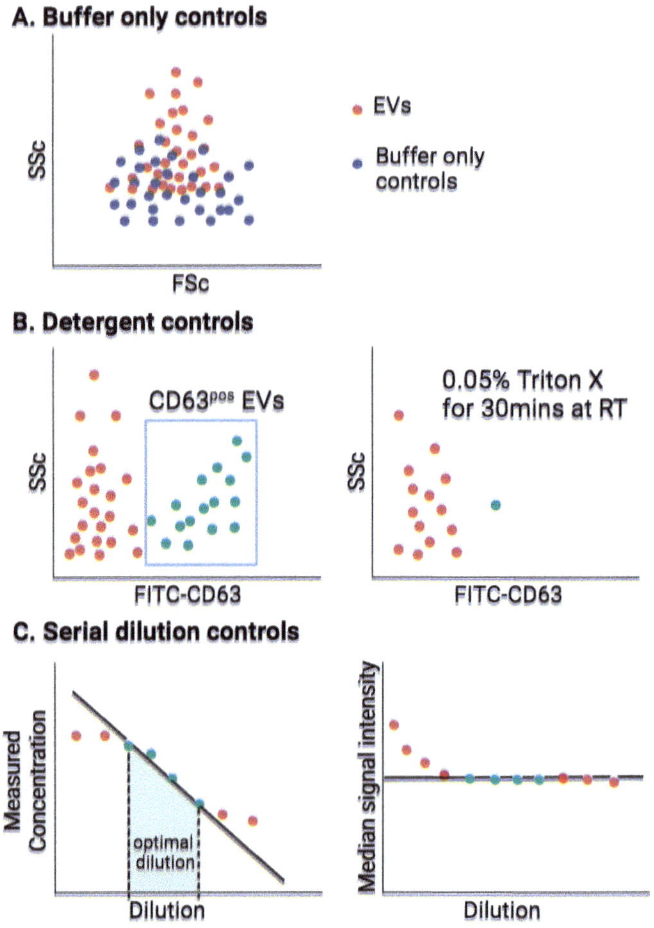

Fig. 7 Buffer only, detergent, and serial dilution controls. (**a**) Buffer only controls show the background events from the buffer. (**b**) Detergent controls can assist in determining if measured events are from lipid-encased particles. (**c**) Serial dilution controls are used to determine the optimal dilution for assessing extracellular vesicles. The optimal dilution is when measured concentration is proportional to the dilution. Created in BioRender. Ng, M. (2025) https:/ BioRender.com/o69b706

1. 0.2 μm filtered PBS buffer control is loaded into the plate.
2. Ab cocktail is diluted 1:2000 in buffer to generate the Ab cocktail control.
3. Unstained uEVs are diluted 1:2000 in buffer to generate the unstained control.
4. Isotype controls are added to uEVs at the same concentration as the single stain.
5. Proportionally diluted antibody is diluted 1:2000 in buffer to generate a single stain control for each antibody used in the panel independently.

6. Dilutions (D250, D500, D1000, and D2000) are run to find a concentration at which the abort rate is less than 10% of the count rate.

4.3.5 Sample Acquisition

It is important to note that each instrument varies in sensitivity, so the running settings will vary from instrument to instrument. We will refer to published protocols to assess the suitability of a flow cytometer for small particle analysis [13]. The background number of events should fall to 1000–1500 events/second at optimized instrument settings [13]. See Fig. 9 for examples of acquired plots.

1. We recommend the use of FCMPASS to optimize and calibrate both fluorescence and light scatter detectors.
2. One should consider using positive controls such as recombinant EVs expressing GFP (Sigma), 6 peak rainbow beads, MESF beads, and NIST traceable polystyrene beads for calibration.
3. The 6-peak rainbow beads can be cross-calibrated with MESF beads. This allows a cost-effective and ergonomic approach to optimize and calibrate assays longitudinally.
   (a) *Step 1: Flow cytometer set up protocol (Fig. 8): Run daily QC and clean the instrument*: Before starting any small particle flow cytometry experiments, we must first check that the instrument is within its operational range. Once QC has passed, ensuring the instrument is clean and free from background signals. This can be achieved using bleach and water. It may be necessary to perform a flow cell clean using Contrad or other liquid detergent to clean equipment.
   (b) *Step 2: Cross-calibrate multispectral beads.*
   (c) *Step 3: Optimization of detector settings.* This step is achieved by analyzing fluorescent beads.
   (d) *Step 4: Calibrate instrument at optimal settings.* This step can be performed with bead standards and optionally positive control EVs to validate the derived limits of detection.
   (e) *Step 5: All fully labeled samples and additional controls for 2 min at the established settings.* This can be done using either single tubes or a thoroughly cleaned, a low-binding 96 well sample plate is loaded into the instrument and run consecutively until completion, starting with single stain and buffer controls and ending with stained samples at established settings.

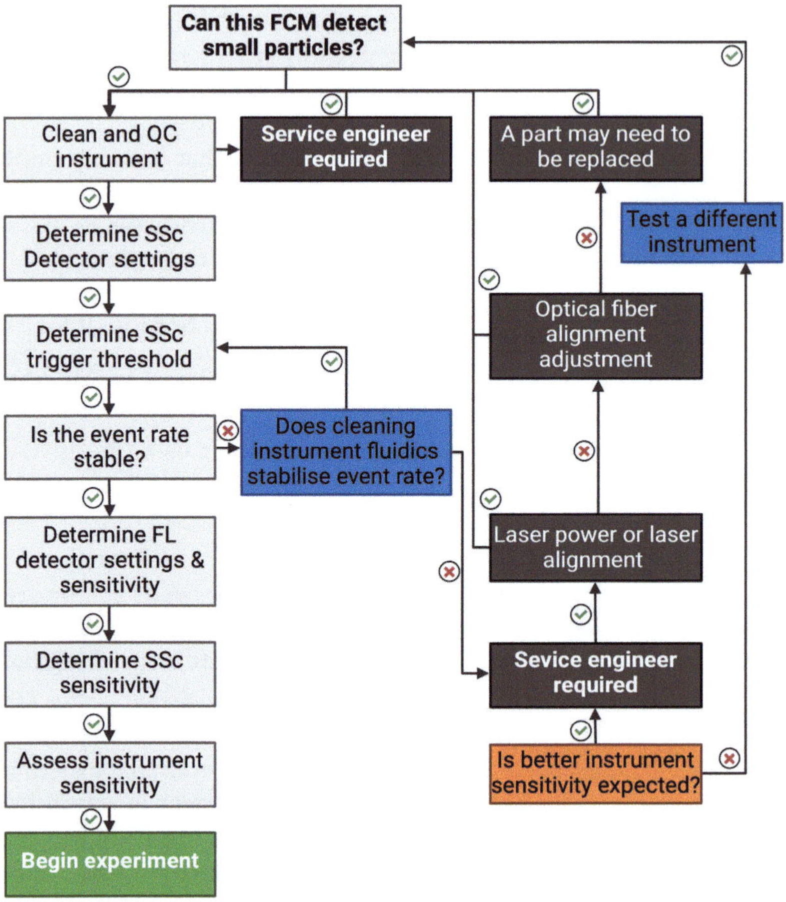

Fig. 8 Sample acquisition workflow (grey boxes) with troubleshooting tips. (Adapted from [13]). The blue boxes represent troubleshooting activities actionable by investigator. The black boxes represent actions to be completed by service engineer. Detailed protocols are provided in the publication by Cook et al. [13]. Created in BioRender. https:/BioRender.com/whyemv2

4.3.6 Flow Data Analysis

It is of outmost importance to identify the limits of the flow cytometer which each individual laboratory is using and to optimize the detector settings. We will refer for more information to the literature [13]. The utility of the FCMPASS program was developed to calibrate signals, also for cross-platform studies and reproducibility. This software performs a particle diameter approximation using Mie Theory and NIST traceable polystyrene particles of known size and refraction index (RI) FCMPASS software is specifically utilized to interpret and reproduce single extracellular vesicle flow cytometry data [25]. This software can be used to generate .fcs files, which convert the raw flow data into standard units, and allow analysis to extract data on uEV sizing and molecules of soluble fluorescence (MESF beads, Bangs Laboratories, Inc). This is done by running various beads, for example,

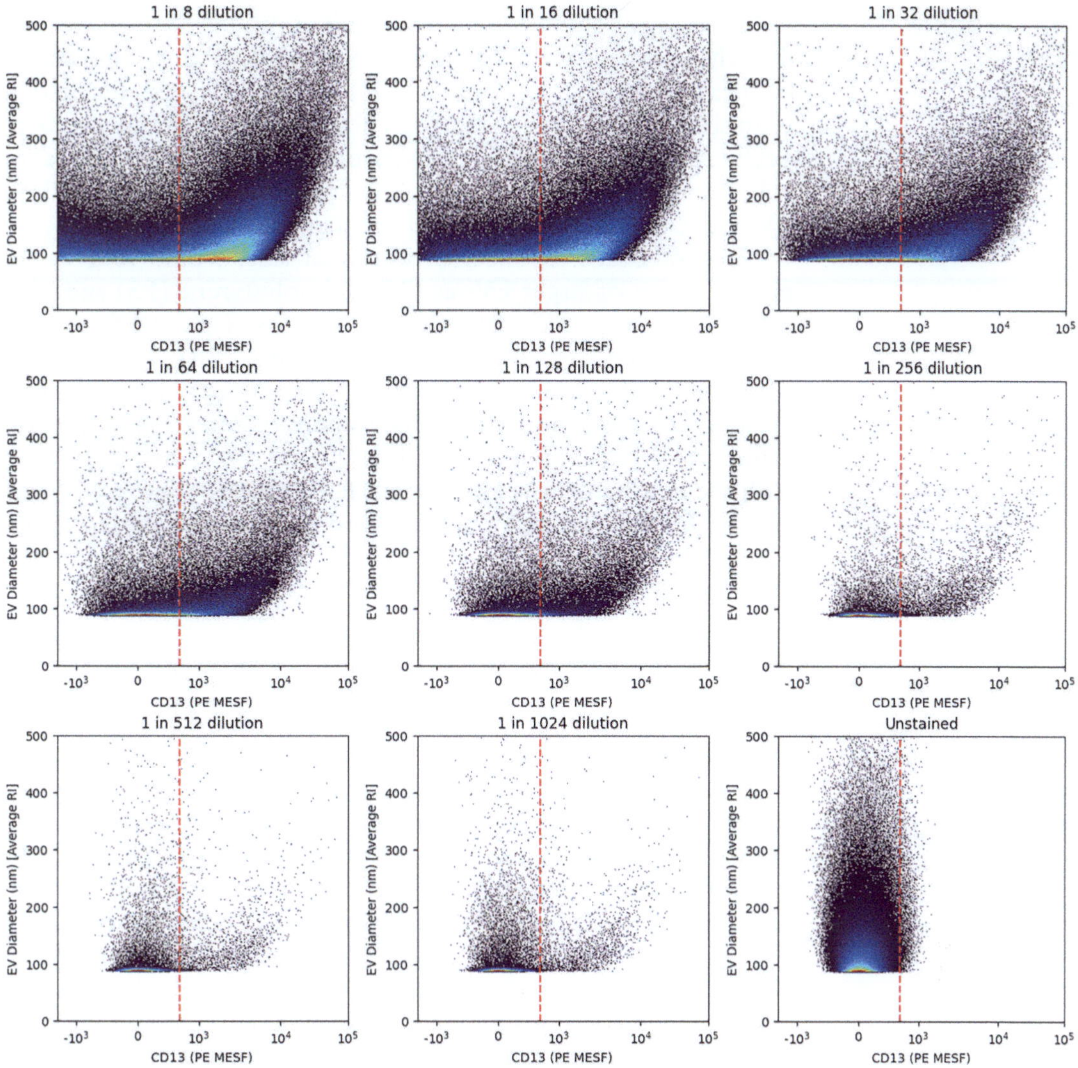

Fig. 9 Serial dilution of uEvs stained with PE-CD13 antibody. EV diameter was calculated using the average RI setting on FCMPass using 100 nm (polystyrene), 400 nm (polystyrene) and 700 nm (silica) beads. The red line is the threshold set at 99% of the unstained uEV population. At 1:256 dilution, the reference noise begins spreading over the threshold line, indicating antibody swarming

100, 122, 152, 203, 240, 303, 401, 496, and 600 nm, on the flow cytometer prior to data collection and uploaded into FCMPASS to use for standardization.

4.3.7 Gating Strategies

Raw or calibrated data files are imported into analysis software, for example, FCS Express, FlowJo, Floreada software, for analysis. An unstained sample is used to establish the range of positively stained uEVs by setting a gate at the boundary of the negative population. This gating strategy is applied consistently across all fluorescent markers, allowing for reproducible plots for each sample in the cohort. See also Fig. 9.

4.3.8 Methods to Count EVs

EV counting should be done by volumetric counting. This method directly calculates the absolute number of EVs (events) in a sample by measuring the exact volume of the sample analyzed during data acquisition, eliminating the need for adding counting beads to determine cell concentration; in particular, the FCM measures the volume of fluid passing through the flow cell, allowing for a direct EV count based on the detected events with that volume.

Use of beads as direct measures of size, would lead to the measurement of larger EVs as EVs and polystyrene (PS) and silica beads have very different refractive indices, e.g., a PS bead of 400 nm would measure an EV of 1000 nm [26]. It is also not recommended to describe the results in percentages, as the denominator of all EVs is not known due to limit of detection of the flow cytometers (depending on the instrument down to 70–100 nm with the new nano flow cytometer). *EV data reporting* (e.g., filling out the MIFlow-Cyt-EV checklist) and *EV data sharing* (to a public repository) should also be performed to support transparency and reproducibility in this field of uEV research.

5 uEV Proteomics

5.1 Introduction and Key Considerations for Immunoblotting of EV Protein Cargo

Immunoblotting of urinary EV protein cargo uses gel electrophoresis and immunoaffinity to visualize and compare levels of select proteins in a mixture. It uses the same principles as used for WB of non-EV proteins. Immunoblotting is a common technique for the semiquantitative measurement of specific proteins found in extracellular vesicles. It is widely used to assess the efficiency of various EV separation techniques as well as to compare levels of EV-associated proteins between samples. EVs are highly heterogeneous in their composition, size, and origin. Since immunoblotting is a bulk analysis method, it cannot be used to resolve this heterogeneity or to examine specific subpopulations of EVs by itself. However, it is useful in analyzing the relative abundance of marker proteins found in subpopulations differentially separated during sample acquisition. While immunoblotting and immunodetection of EV proteins is accomplished by typical, standardized methods, there are some unique aspects of EV samples that should be considered when performing the technique:

1. Conditions of sample lysis
2. Normalization of sample loading on gels
3. Normalization of immunoblot results

5.2 Example of uEV Characterization

As an example, we will provide the protocol to detect two tubular markers in the kidney: Sodium chloride cotransporter and

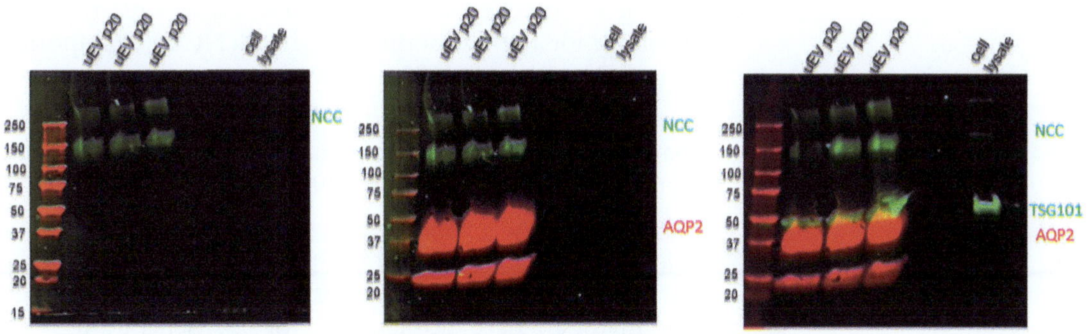

Fig. 10 Example of immunoblot of uEVs showing NCC dimer and Aquaporin and TSG101. Assembled on one blot

aquaporin 2, using TSG101 as a control protein marker for EV origin and cell lysate from HEK293 cells as negative control (Fig. 10).

1. Materials:
   (a) Bio-Rad Mini-Protean Precast Gels
   (b) Bio-Rad Precision Plus Protein Standard
   (c) Blotting Filter Paper
   (d) Fiber Foam Pads
   (e) Running Module, Electrodes
   (f) Transfer Cell
   (g) Blocking Buffer
   (h) Abcam SLC12A3
   (i) Sigma-Aldrich TSG101
   (j) Santa Cruz AQP2 Alexa Fluor 680
2. Protocol:
   (a) Urinary extracellular vesicle pellets are solubilized in Electrophoresis Solubilization Buffer including 50 mM dithiothreitol, protease, and phosphatase inhibitor cocktail.
   (b) Samples are run on a polyacrylamide gel.
   (c) A molecular weight ladder, three uEV p20 samples, and 25 μg of HEK293T cell lysate (a negative control for cellular proteins) are loaded into the gel.
   (d) The gel is run at 100 V for 90 min.
   (e) Protein on the gel is transferred to a PVDF membrane and the transfer cell is run at 0.2 amps for 2 h at 4 ° C.

(f) The gel is blocked overnight in a blocking buffer before probing for the antibodies SLC12A3(NCC), aquaporin-2 (AQP2), and TSG101.

(g) Images are obtained using an Odyssey CLx imager, and bands imaged in green are in the 800 nm channel and the one in red is imaged in the 700 nm.

5.3 Normalization of Sample Loading on Gels

1. *Load samples from equivalent starting volumes*

   One option is to load samples from equivalent volumes of starting material, here urine, used for EV separation. The results can then be interpreted as the total amount of the target protein carried by EVs within that volume. This approach will minimize discrepancies due to contaminating proteins left in the EV preparation but will not be informative for the amount of target protein found per EV.

2. *Normalize by total protein content of uEVs*

   Another option is to normalize the amount of EV lysate loaded per lane by total protein content. In this scenario, equal mass quantities of proteins are loaded per lane on gels prior to electrophoresis. Again, the results from this type of normalization will not be informative for the number of EVs represented on the blot, or the amount of target protein per EV.

   Normalizing by the amount of protein loaded per lane will also introduce discrepancies due to possible variation in the presence of contaminating proteins retained from the starting biofluid sample. The amount of contaminating protein may not be trivial since the protein content of EVs is typically low compared to the total protein content found in biofluids like blood or urine.

3. *Run Lysates of equal number of EVs*

   A third option to normalize gel loading is to run lysates of equal number of EVs per lane, as determined by NTA or FCM. This is the only method that allows comparison of the amount of target protein per EV between samples. However, errors may be introduced due to the inability of techniques like NTA to discriminate between EVs and particles consisting of protein aggregates etc. Regardless of the technique used for normalization of gel loading, it is always desirable to stain for total protein with a reversible stain after transfer to a membrane, or to stain a replica gel with the same samples loaded on it. This allows assessment of the separation of the protein samples by gel electrophoresis and the amount of abundant protein contaminants, like uromodulin in urinary EV preps or albumin from plasma or serum EV samples.

5.4 Normalization of Results

Normalization of the results from immunoblotting can also be extended to analysis of the quantified bands detected. Immunoblot results for target proteins from cell lysates are often normalized to the expression of "housekeeping" proteins such as GAPDH or b-actin detected by blotting the same membrane. While no analogous housekeeping EV proteins have been identified, proteins like syntenin-1 or TSG101 are frequently detected in EVs from a wide range of sources, and there may be some utility in using quantification of their expression for normalization. Normalization of immunoblotting of proteins from urinary EVs poses a particular problem, due to the highly dynamic nature of urine and large fluctuations in the concentration of the starting material. The levels of the metabolite creatinine is used in the clinic as a surrogate for urine concentration, and urine creatinine concentration has been shown to correlate with uEV concentration. Therefore, it may prove useful to normalize the signal from immunoblot of uEV proteins to the creatinine level of the urine samples.

6 Proteomic Analysis of Extracellular Vesicles Isolated from Urine

6.1 Introduction

Proteomics technologies enable large-scale characterization of EVs isolated from urine. Mass spectrometry (MS)-based proteomics has been extensively used to characterize uEVs (Fig. 11). Here, we will summarize the most important aspects of uEV analysis using MS-based proteomics including uEV isolation, uromodulin removal, proteomics sample preparation, LC-MS data acquisition, and data analysis [17, 27–29].

6.2 uEV Separation for MS-Based Proteomics

The most common protocol to isolate uEVs for proteomics analysis is sequential centrifugation. In general, this protocol starts with a low-speed centrifugation at 500–4600 × G for 10–30 min at 4 °C to remove cells and large debris. Then, a population of uEVs is pelleted by centrifugation at 10,000–20,000 × G for 30 min (also called low centrifugation pellet), and other uEV population by ultracentrifugation of the remaining supernatant at 100,000–200,000 × G for 1–2 h to at 4 °C (e.g., use of polycarbonate tubes such as Beckman catalogue number 35563 and use of ultracentrifuge (e.g., 70TiFix angle rotor (Beckman) (also called high centrifugation pellet). Both EV populations (low and high centrifugation pellet) carry valuable uEV biomarkers [16]. For this reason, it is relevant to perform comparative or combined analysis of both fractions for a comprehensive analysis [30, 31] in order to have a comprehensive unbiased analysis of the MS-based uEV proteome analysis. It is discussed if proteinase inhibitors (PI) need to be added. This question has not been fully answered. Most

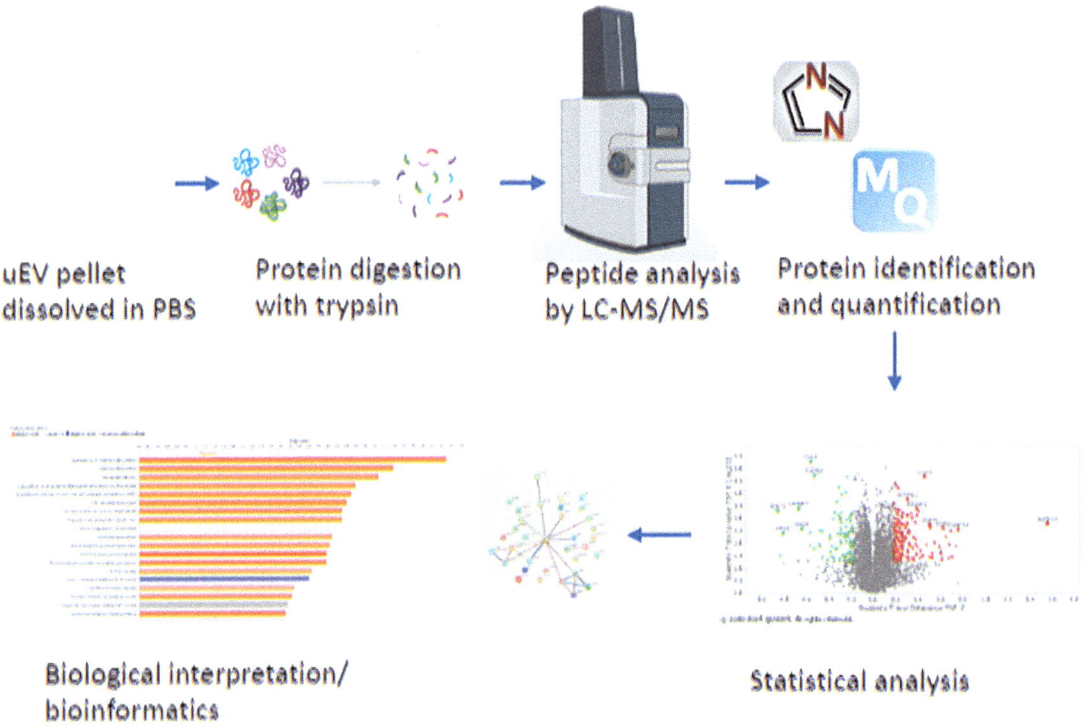

Fig. 11 Mass-spectrometry-based proteomics workflow to analyze uEVs

investigators add PI but base their decision on their experimental questions (*see* also Chapter 2 about "Collection and Storage of urine specimens for uEV analysis").

6.3 Removal of Uromodulin

A second consideration when isolating uEVs is to avoid the coisolation of large amounts of uromodulin, the most abundant protein in urine, in the EV sample. Coisolation of uromodulin with uEVs could mask the identification of low abundant uEV proteins, and also affect total protein measurement of EVs. Simple protocols have been developed to remove uromodulin (see below protocols and publications [8]). See Fig. 12 demonstrating a SDS page/Silverstain showing the removal of uromodulin by washing with PBS and use of filter (protocol 2).

Of note, with advancement of mass spectrometry instruments and increased sensitivity, it needs to be reassessed if uromodulin removal will be necessary to obtain good MS results. It should also be mentioned that the use of different EV isolation methods may lead to different levels of uromodulin coisolation with the EV pellet. It this therefore advised, as mentioned above, to perform an SDS-page/silver staining of each cleaned uEV preparation/pellet.

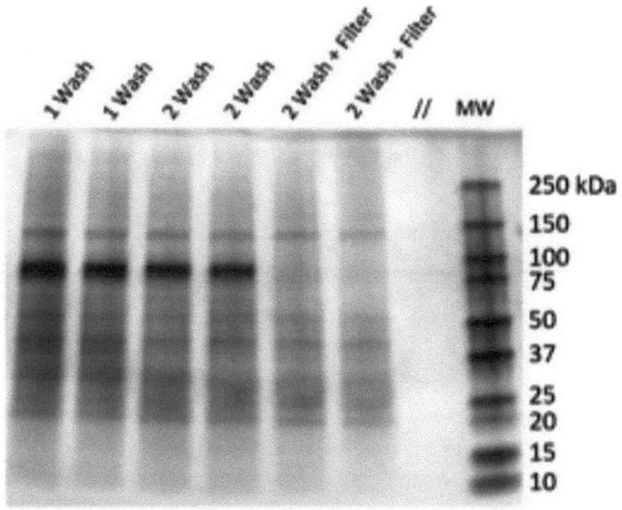

Fig. 12 Silver staining of different PBS washes of the 100,000 pellet. Samples 1 and 2 were washed twice with phosphate-buffered saline solution (PBS) at 4C and then a 200 nm filtration step was performed, which removed the uromodulin (MW around 90kDA) as seen in last 2 lanes. Ideally, silver staining of SDS Page should be done for each sample to understand the protein patterns and uromodulin contamination of each uEV samples. (Special thanks to Krizia Sagini, Håkon Flaten, and Alicia Llorente for providing the figure)

6.3.1 Example Protocols of Uromodulin Removal

There are different options to remove uromodulin; we will refer to different published protocols.

Protocol 1

1. *Using low ionic strength buffer for removal of uromodulin in low centrifugation pellet*: The low centrifugation pellets can be depleted of uromodulin by using low ionic strength buffer (10 mM HEPES pH 7.4) supplemented with EDTA (2.5 mM pH 8.0) and centrifuged at 15,000 rpm; RCF 21,130 g for 30 min at 4 °C in an Eppendorf microcentrifuge 5424 fix angle rotor (FA-45-24-11) (Eppendorf, Hamburg, Germany). This step is repeated twice for a total of 3 solubilization-centrifugation cycles. Pellets are then stored at −80 °C degrees and solubilized after thawing in 0.1 μm filtered (Minisart® PES syringe filter code 16553-K, Sartorious, Göttingen, Germany) phosphate-buffered saline solution (PBS) pH 7.4 no Ca^{2+} and Mg^{2+} (Gibco®, life technology™, Thermo Fisher Scientific) (PBS 0.1 μm) or electrophoresis solubilization buffer (ESB). P100 pellet was cleaned by size exclusion chromatography (SEC) [32].
2. *Using size exclusion chromatography for removal of uromodulin in high centrifugation pellet*: The high-speed centrifugation

pellet is resolubilized in guanidine 6 M in acidic condition (PH = 3.5) and mixed for 20 min in end-over-end wheel before being applied on top of SEC 70 nm original column (Izon). Columns are washed with guanidine buffer before and after samples loading. The collection of the fraction is carried out with the aid of automatic fraction collector. One mL fraction is collected and precipitated by Trichoroacetic Acid and Sodium Deoxycholate (TCA-DOC) Presence of uEV's was validated by Immunoblotting for TSG101 [32].

3. Both pellets can be analyzed individually or combined for MS based proteomics.

Protocol 2: Alternative to Protocol 1b

1. *Using washing with PBS and 200 nm filter for removal of uromodulin in high centrifugation uEV pellets*: The high centrifugation pellet is resuspended in 10 ml PBS, vortexed, filtered through a 200-nm pore Supor syringe filter (Pall, Port Washington, NY, USA) and finally the solution is centrifuged at 100,000 × *g* for 70 min at 4 °C in a swinging bucket SW40 rotor (24,000 RPM, *k* factor = 379) in 10-ml polypropylene tubes (Ref. 358210, Beckman Coulter, Indianapolis, IN, USA). Most of the supernatant is then removed, leaving 50–100 μL PBS at the bottom, which is used to resuspend the pellet. The amount of sample was estimated by measuring the concentration of total protein in the EV pellet and the EV counts with a Zetasizer Nano ZS (Malvern Panalytical, Malvern, UK). Then, the samples were stored at −80 °C until further use [33].

6.3.2 Sample Preparation for uEV Analysis

The isolated uEVs can be analyzed with many common proteomics sample preparation and liquid chromatography-tandem mass spectrometry (LC-MS/MS) analysis strategies. Previously proteomics experiments required considerable amounts of starting material (tens to hundreds of ug protein/sample), which could be a challenge for the analysis of clinical samples where the amount of samples available can be limited. However, with recent improvements in sample preparation, mass spectrometry instrumentation, data acquisition, and data processing tools, few ug (or even less) protein/sample is enough for an in-depth EV characterization. This amount can be generally obtained from 10 to 15 ml of urine.

Sample preparation protocols used for uEV proteomics include *filter-aided sample preparation (FASP)* [27, 34] described already in 2009 as a universal sample preparation method for proteomics. Since then, several other "universal" proteomics sample preparation strategies suitable for uEV analysis have been developed. We routinely use *protein aggregation capture on microparticles (PAC)*, which enables highly reproducible proteomic sample preparation from small amounts of starting material, including uEVs [35]. PAC protocol can easily be adapted to a 96-well format and be done either using multichannel pipetting or fully automated. This is an important consideration for large-scale proteomics projects of clinical sample cohorts, because it minimizes batch-to-batch variation due to sample preparation.

1. *Example protocol for proteomic sample preparation of uEVs:*
   (a) *PAC*:
      (i) Dissolve isolated uEVs into PBS.
      (ii) Measure protein concentration, for example, using BCA protocol and take 1–2 ug/protein/sample for analysis.
      (iii) Add 100 μl 100% acetonitrile to each sample.
      (iv) Add 10 μl (20 mg/ml) magnetic hydroxyl beads (ReSyn Biosciences) to each sample.
      (v) Vortex samples, leave standing for 15 min at RT, and vortex again and keep at RT for 15 min.
      (vi) Transfer to magnetic stand and discard the supernatant.
      (vii) Wash the beads with 200 μl ACN and with 200 μl 70% ethanol and let them air dry.
   (b) *Protein reduction, alkylation, and tryptic digestion*:
      (i) Dissolve the beads in 50 μl of 50 mM ammonium bicarbonate, 10 mM DTT, and incubate at 37 °C for 45 min.
      (ii) Add 50 μl of 50 mM ammonium carbonate with 25 mM IAA and incubate at RT in the dark for 30 min.
      (iii) Add 0.5 μg of Trypsin (Promega) and digest at 37 degrees overnight.

6.3.3 LC-MS/MS and Data Analysis

1. MS-based proteomic analysis requires expensive LC-MS/MS instrumentation and dedicated proteomics personnel to operate them and process the data. Therefore, the EV researcher usually is dependent on having access to a Proteomics core facility/research group to perform the analysis using the instruments available in that facility. For most sensitive, high-coverage proteome analysis, data-independent acquisition (DIA) combined with latest generation high-resolution mass spectrometry instrumentation is the method of choice [36].
   (a) To enable *relative quantification of uEV proteins* (quantification between samples) either label-free quantification (LFQ) or different labelling strategies like tandem mass tags (TMT) can be used. For large clinical sample batches,

LFQ is the preferred option: there each sample is analyzed individually in LC-MS/MS and quantification is based on MS signal intensity. The main advantage of LFQ is that no expensive labelling is needed, and unlimited number of samples can be compared to each other.

(b) For *quantitative proteomics of uEVs*, normalization is an important point to consider already before sample preparation and LC-MS/MS analysis. Usually, in proteomics, the same total protein amount from all of the samples is used to enable relative quantification between the samples. However, with EVs, depending on the scientific question, the total EV secretion can increase markedly. Then normalization can be done in different ways; either the "standard" way of using the same protein amount for all the samples or by starting, for example, with same urine volume for the EV isolation and then loading the same amount based on volume to LC-MS/MS. Zhang et al. recently discussed normalization methods for the proteomic analysis of uEVs: normalization factors are needed [37]. Urine creatinine concentration can be used to minimize the inter-sample variability for EV-based studies. However, creatinine can only be used in a steady state and it is not studied in all kidney disease processes. Research is ongoing to identify normalization factors for uEV research (refer to uEV position paper [1]). Considering that urine contains EVs from different organs, it is possible that different EV population may require specific normalization approaches.

2. *Proteomics data analysis workflow* (protein identification and quantification followed by data quality evaluation, filtering and statistical comparison analysis) for uEVs is the same as for any proteomics experiment. Common ways to visualize proteomics data include volcano plot to highlight the differentially abundant proteins in two groups, and principal component analysis to visualize how the different sample groups separate. The proteomic data can also be compared to existing EV databases like Vesiclepedia and ExoCarta to determine how much of the identified proteins are known EV-proteins [38, 39]. An estimate of the purity of the EVs can also be obtained from proteomic data by comparing the identification results to, for example, ExoCarta list of top 100 proteins that are most often identified in exosomes.

6.3.4 Alternative/ Complementary Techniques to MS-Based Methods

1. Olink-antibody based assay [40]
2. Somascan

In addition to MS-based proteomics, complementary techniques, like Olink and Somascan [41], are also available. Both these are protein-affinity-based technologies; Olink uses specific antibodies and Somalogic protein-specific aptamers to capture proteins from biological samples. These techniques are well suited to biofluids, especially for plasma and serum, which challenging samples for in-depth proteome characterization using MS-based techniques. A very recent article used Olink to characterize uEVs from urinary bladder cancer (UBC) and their results strengthen the rationale of using uEVs as noninvasive biomarkers and prognostic tools for UBC [41].

7 EV RNA Cargo Analysis

7.1 Introduction to uEV RNA Characterization

uEVs and their RNA cargo are investigated for their functional importance as carriers of cellular messages and cellular turnover products, and for their clinical application as biomarkers and delivery vehicles of therapeutic agents. The rich RNA content of uEV includes not only the most studied miRNAs and mRNAs but also a plethora of small and long noncoding RNAs (piRNAs, tRNAs, snoRNAs, snRNAs, circRNAs, etc.), which play various role in epigenetic regulation, transcription, and translation. Adding to uEV's attractiveness, the uEV RNA cargo has been found to reflect the pathophysiological status of the genitourinary tissues: kidney, bladder, or prostate [42–46].

All together this underpins the importance of accurate and sensitive analytical techniques to identify and quantify the total uEV transcriptome as well as to monitor the dynamics of individual uEV RNA species. For over 10 years, uEV RNA have been characterized by many researchers utilizing, for example, quantitative real-time PCR (qPCR) [47, 48], probe-based technologies [49, 50], and next-generation sequencing (NGS) [51, 52]. While these technologies are still in use, digital PCR (dPCR) [53, 54] has become a common tool as well and ventures of native long RNA sequencing are being made [55]. Thus, with the development of improved detection technologies, RNA profiling has become a popular topic among uEV researchers.

Despite the technical advances in existing techniques and the introduction of new ones, the absence of standard workflows, consensus uEV-RNA quality parameters, and the lack of uEV reference genes have led to the use of widely varying protocols and large differences in the quality and quantity of input RNA. Consequently, reported results can be highly variable. For example, the amount of ribosomal RNA in uEV samples has shown great variation between studies, likely owing to the choices made in each step preceding the actual RNA analysis.

The most important preanalytical steps that should be considered when intending to explore the uEVs transcriptome or individual uEV associated transcripts include the removal of urinary cells and cell debris, the selection of uEV target size and subsequent uEV isolation protocol, the size of the target RNA(s), the choice of RNA isolation protocol, the inclusion of a ribodepletion step, and selection of a ribodepletion protocol as well as the choice of reverse transcription and cDNA priming strategies.

The ISEV UTF and Extracelullar RNA Communication Consortium (ERCC) have acknowledged the complexity of uEV RNA analysis and published recommendations [1, 2], explorative studies, and tools [56, 57] striving to steer the uEV field toward improved results reproducibility and more standardized protocols. Here we provide some thoroughly tested workflows and extra tips for extraction of uEV total RNA, RNA quality control by fragment length analysis and analysis by NGS and qPCR (Fig. 13). The protocols have been used in several urinary EV sequencing and qPCR studies [43, 44, 58–61].

7.2 Materials

7.2.1 DNAse Treatment of EV Suspension

1. DNA lobind DNAse- and RNAse-free Eppendorf Tubes (Eppendorf), Cat. No. 10051232
2. DNAse I set (RNAse-free) including DNA digestion buffer (Zymo research), Cat. No. E1010, E1010-1-4
3. Qiazol Lysis Reagent (Qiagen), Cat. No. 79306

7.2.2 RNA Isolation

1. miRNEasy micro kit (Qiagen), Cat. No. 217084 (recommended for NGS and long RNA qPCR)
2. Urine Exosome RNA Isolation Kit (Norgen Biotek), Cat. No. SKU 47200 (recommended for small RNA)
3. Optional, RNAse-free DNAse set (Qiagen), Cat. No. 79254
4. RNAse AWAY™ Surface Decontaminant (Thermo Fischer Scientific), Cat. No. 7003

7.2.3 RNA Sample Quality Control by Fragment Length Analysis

1. RNA 6000 Pico Kit for 2100 Bioanalyzer Systems (Agilent Technologies, Waldbronn, Germany), Cat. No. 5067-1513
2. Small RNA Kit for 2100 Bioanalyzer Systems (Agilent Technologies), Cat. No. 5067-1548
3. High Sensitivity RNA ScreenTape assay (Agilent Technologies), ScreenTape Cat. No. 5067-5579, Ladder Cat. No. 5067-5581, Buffer Cat. No. 5067-5580

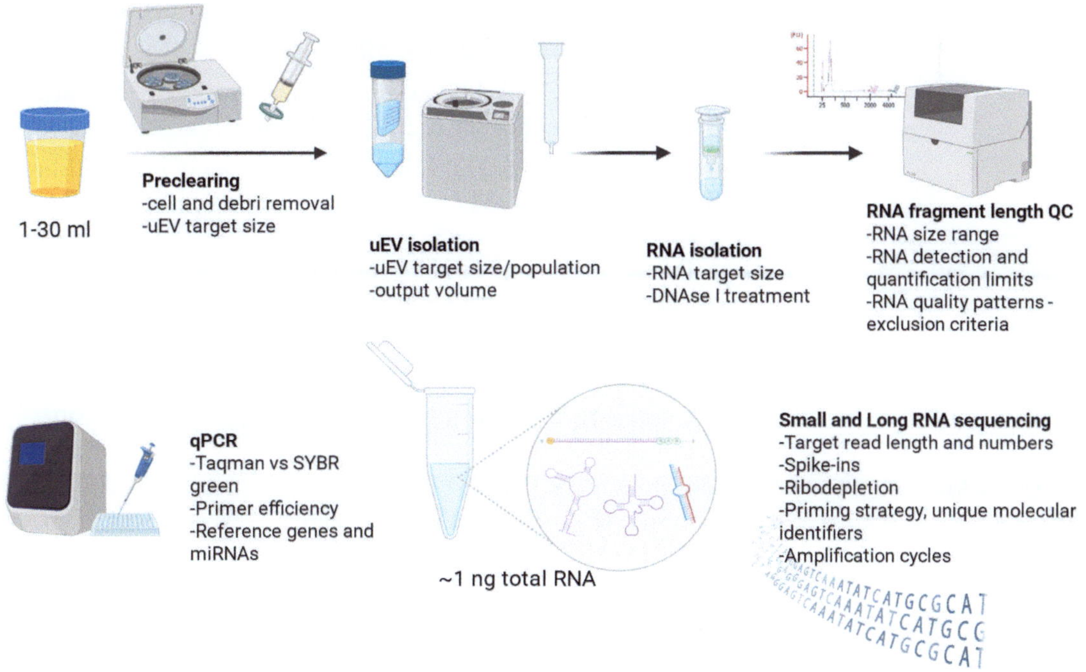

Fig. 13 A workflow of uEV RNA analysis with key considerations. The workflow starts with urine collection and preclearing to remove cells, and cell debris by low to medium speed centrifugation or filtration steps. The desired uEV (sub)population is next enriched by ultracentrifugation, size exclusion hromatography, and/or ultrafiltration. RNA isolation method can be chosen to target, e.g., total or small RNA and performed with different DNAse I treatment options. The uEV RNA quality control through high sensitivity RNA fragment length analysis helps to exclude low quality samples and determine the RNA quantity needed for next steps. Approximately 1 ng of total RNA is sufficient for many qPCR and sequencing applications. For qPCR, selection of the protocol chemistry and reference genes is dependent on the size of the investigated RNAs. Primer efficiency and stability of reference genes has to be evaluated in each study and system; the use of spike-in controls in qPCR is recommended. For sequencing, targeted reads, use of spike-ins, ribodepletion, priming strategy, and optimization of amplification cycle numbers need to be considered. Created in BioRender. Puhka, M. (2025) https:/BioRender.com/q57c562

7.2.4 RNA Analysis

Next-Generation Sequencing

cDNA and Library Preparation and Quality Control

1. SMART-Seq® v4 Ultra Low Input RNA Kit for Sequencing (Takara BIo Inc., Mountain View, CA), for example, for 24 or 96 samples, Cat. Nos. 634889, 634891
2. Nextera XT Index Kit, e.g., 24 unique indexes for sequencing library preparation of up to 96 samples (Illumina, San Diego, CA, USA), Cat No. FC-131-1001
3. Nextera® XT DNA Library Preparation Kit (Illumina), for example, for 24 or 96 samples, Cat. Nos. FC-131-1024, FC-131-1096
4. Agencourt AMPure XP Kit (Beckmann Coulter Inc., Miami, FL, USA), Cat. No. A63881

5. 2100 Bioanalyzer System and High Sensitivity DNA Kit (Agilent Technologies), Cat. No. 5067-4626
6. Qubit™ 1X dsDNA HS Assay Kit (ThermoFisher Scientific), Q33231, assay tubes set of 500, Q32856
7. LabChip GX Touch HT High Sensitivity assay (PerkinElmer, USA), Cat. No. 760517

Quantitative Real-Time PCR for Small RNAs

Reverse Transcription and cDNA Synthesis

1. TaqMan Advanced miRNA cDNA Synthesis Kit (ThermoFisher Scientific), Cat. No: A28007, or
2. miRCURY LNA RT Kit (QIAGEN), Cat. No. 339340
3. RNA Spike-In Kit, For RT (QIAGEN), Cat. No. 339390

Quantitative Real-Time PCR

High Sensitivity PCR Plates

1. AmpliStar-II 96 PCR plate (Westburg), Cat. Nr. WB1900WS, or
2. MicroAmpTM 384-well reaction plate (ThermoFisher, Scientific), Cat. Nr. 4343370
3. MicroAmp™ Optical Adhesive Film (ThermoFisher, Scientific), Cat. Nr. 4311971

Reagents for Probe-Based Chemistry

1. TaqMan™ Fast Advanced Master Mix for qPCR (ThermoFisher Scientific), Cat. No. 4444557
2. TaqMan™ Advanced miRNA Assay (ThermoFisher Scientific), Cat. Nr. A25576, or
3. miRCURY LNA Probe PCR Kit (QIAGEN), Cat. No. 33937
4. miRCURY LNA miRNA Probe PCR Assays (QIAGEN), Cat. No. 339350
5. miRCURY LNA miRNA Custom Probe PCR Assays (QIAGEN), Cat. No. 339351

Reagents for SYBR Green Chemistry

1. miRCURY LNA SYBR Green PCR Kit (QIAGEN), Cat. No. 339345 or equivalent (e.g., SensiMix™ SYBR® Low-ROX Kit, Meridian Bioscience, Cat. NQT625-05)
2. miRCURY LNA miRNA PCR Assay (QIAGEN), Cat. No. 339306
3. miRCURY LNA miRNA Custom PCR Assay (QIAGEN), Cat. No. 339317

7.3 Methods

We recommend to work in an RNA-work dedicated pre-PCR laboratory using an RNA-work-only coat and ice box, single use nitrile gloves, nuclease-free filter tips, tubes, water, and reagents. Work benches, hood, pipette set, and equipment should be wiped clean with the RNAse AWAY Surface Decontaminant before start of the experimental procedures.

The concentration of recovered total RNA is highly variable and depends on multiple factors, including the time of urine collection, average water intake of the donor, medical handlings performed prior to urine collection, etc. If the study protocol allows, first void, morning urine will typically be most concentrated and enriched in uEVs. Nevertheless, individual studies may opt for other collection moment, for example, a post DRE, spot urine sample may be more enriched with prostate-derived vesicles.

Starting volumes of 5–10 ml urine supernatant usually work well and deliver sufficient amount of RNA for subsequent reverse transcription and qPCR. Nevertheless, RNA concentrations are generally low and should ideally be examined using Bioanalyzer Pico or TapeStation High Sensitivity assays or equivalent (see below). It should be noted that uEV total RNA contains many short RNA species and is also usually naturally degraded with a median size peak of around 200–300 nt.

7.3.1 RNA Extraction from Urinary Extracellular Vesicles, DNAse Treatment, and Quality Control for Transcriptomics Studies and qPCR of mRNA and lncRNAs

The total EV RNA including the miRNA fraction can be isolated using miRNEasy micro kit (Qiagen) and protocol: "Purification of Total RNA, Including miRNA, from Animal Cells" (*see* **Note 1**). The produced total RNA is suitable both for long and small RNA sequencing. The protocol can be performed with or without an in-column DNAse treatment (*see* **Note 2**). Total RNA yields of approximately 0.5–10 ng can be expected from 5 to 30 ml of urine supernatant. For sequencing applications, do not add poly-A or other carrier RNA as they may interfere with the sample preparation or sequencing of the sample RNAs. For evaluating technical reproducibility and sequencing quality, spike-in RNAs can be added if a careful optimization of the in-put quantities has been carried out. We have experience with the following spike-ins: UniSp or cel-miR-39-3p RNA Spike-in Templates (Qiagen) and the Ambion ERCC RNA Spike-In Mix 1 (ThermoFisher). The ERCC RNA Spike-in Mix 1 contains a set of 92 polyadenylated transcripts from the ERCC plasmid reference library. The transcripts span approximately a 1,000,000-fold concentration range that is suitable for sequencing platforms. Figure 14 presents a scatter plot from a test sequencing using the SMART-Seq® protocol (below) and an uEV control sample with spike-ins. It shows a good linearity and quality of sequencing and sample preparation. The same plot shows a good dynamic range of spike-ins expression estimated using RNA-seq data.

RNA Extraction with DNAse I Treatment for NGS and mRNA/lncRNA qPCR

Samples: The protocol has been tested with urinary EV isolated by e.g. ultracentrifugation from approximately 5–30 ml minimally 1 ml of urine (*see* **Note 1**). EVs are suspended in 40 μl of filtered

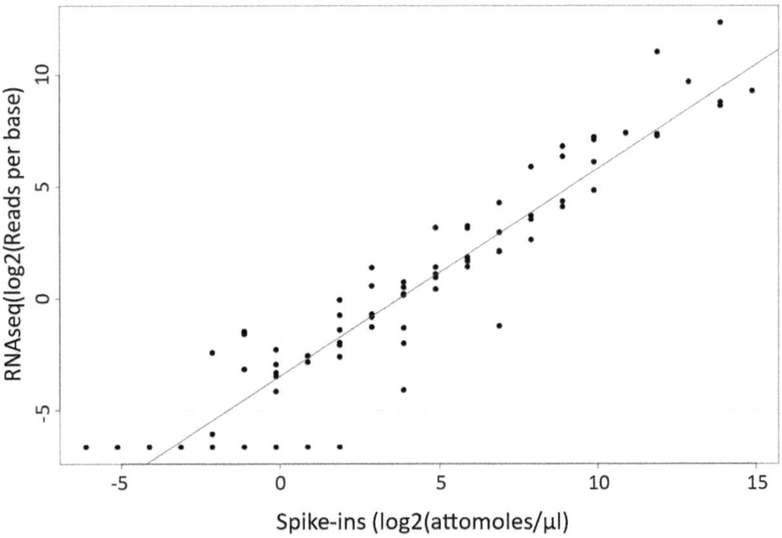

Fig. 14 A scatter plot of read counts mapping to spike-ins. The Ambion ERCC RNA Spike-In Mix 1 was spiked into a test uEV-RNA sample and sequenced using the SMART-Seq® protocol. Spike-ins (attomols/ul) are plotted against the estimated expression of the same spike-ins in RNA-seq data (counts of reads mapping on spike-in per base of spike-in)

(0.1–0.2 μm PES) PBS (*see* **Note 3**). Samples may be fresh or frozen at −80 °C in DNA Lobind Eppendorf tubes (1.5 ml).

DNAse I treatment of EV suspension (*see* **Note 2**)

1. Add 275 μl DNase-/RNase-free Water to reconstitute the lyophilized DNase I at 1 U/μl.
2. Mix by gentle inversion. Aliquot for storage according to need.
3. If needed, thaw the uEV samples at RT or on ice and spin down. Avoid prolonged incubation at RT.
   (a) If the uEV samples are not in RNAse-free DNA low binding tubes, transfer them into RNAse-free DNA low binding tubes.
4. Add 5 μl (5 U) of DNAse I and 5 μl of DNA Digestion buffer to 40 μl of uEV sample (*see* **Note 3**).
5. Mix by gentle inversion and spin down.
6. Incubate 15 min at RT.
7. Inactivate the DNAse I and lyse the EVs by adding 700 μl of Qiazol and vortexing with high speed immediately for 30–60 s.
8. Spin down and freeze at −80 °C or proceed directly to RNA isolation.

RNA Isolation

1. If needed, thaw the uEV samples at RT and spin down.
2. Work according to the miRNEasy micro kit protocol from **step 4** onward.
3. At **step 15**, use nuclease-free water for elution and choose the elution volume according to the volume and concentration needs of your targeted quality control and downstream protocols. Using the default elution volume of 14 μl of RNAse-free water leads to the recovery of 12 μl, because there is a 2 μl dead volume in the column. (*see* **Note 4**)
4. Transfer the eluate tubes immediately on ice after the last centrifugation step.
5. Aliquot and freeze the uEV RNA samples in RNAse-free DNA LoBind tubes at −80 °C. For example, reserve minimally 1.5 μl for evaluation of RNA using Agilent Bioanalyzer Pico chip (or 2.5 μl if using High Sensitivity RNA ScreenTape assay for tapestation) and the rest for downstream RNA work.

Direct Extracellular Vesicle RNA Extraction from Urine Supernatant for qPCR Analysis of Small RNAs

The reverse transcription of the total small RNA transcriptome is a multistep process that, depending on the choice of RT kit, can include some or all of the following enzymatic steps: poly-A synthesis, 5′ adapter ligation, reverse transcription/cDNA synthesis, and cDNA amplification. Enzymes that are used in these steps are sensitive to the presence of carryover inhibitors, for example, organic and/or phenolic compounds used for RNA isolation. Therefore, we opt for the use of nonorganic extraction protocols that avoid the use phenol-chloroform extraction, for example, Urine Exosome RNA Isolation Kit (Norgen Biotek). RNA extraction can be performed on a fraction of isolated uEVs or simultaneously with the uEVs capture, where uEVs are directly lysed after a capture on a resin or membrane. If direct lysis is chosen, an aliquot may be kept to quantify the number and/or size of captured vesicles.

1. If needed, quickly thaw the urine supernatant aliquots at 37 °C.
2. Transfer 11 ml cell-free urine (or combine the thawed urine supernatant aliquots) in a 15 ml conical tube by pipetting.
3. Centrifuge at 2000 g for 10 min to remove debris, bacterial cells, or cryoprecipitates.
4. Transfer 10 mL by pipetting into a new 15 mL conical tube without disturbing the last 1 ml at the bottom of the tube.
5. Work according to the Urine Exosome RNA Isolation Kit protocol (Norgen Biotek) from **step 3** onward.
6. At **step 3**, make sure to add 50 μL Binding buffer for each ml of urine supernatant exceeding the volume of 5 ml. For example,

if 10 ml sample is processed, add 300 mL of Slurry B1 and 250 μL of Binding Buffer A.

7. Mix well by vortexing for 10 s.
8. If a part of the sample is going to be used for EV quantitation or electron microscopy, transfer a homogenous aliquot to a new tube.
   (a) Centrifuge at 2500 rpm for 2 min to pellet the resin and carefully pipet away the supernatant in order not to dislodge the slurry pellet.
   (b) Resuspend the pellet in 100–200 μL of 1× PBS pH 7.4 by vortexing for 10 s.
   (c) Centrifuge at 2500 rpm for 2 min to pellet the resin and collect the 1× PBS containing intact exosomes for quantitative analysis.
9. Continue RNA isolation with the rest of the sample following the Urine Exosome RNA Isolation Kit (Norgen) protocol from **step 5** onward.
10. At step 12, add 100 μL of Elution Solution A to the column and centrifuge for 2 min at 2000 RPM, followed by 2 min at 14,000 RPM.
11. Transfer the eluate tubes immediately on ice after the last centrifugation step.
12. Aliquot and freeze the uEV small RNA samples in RNAse-free DNA low-bind tubes at −80 °C. For example, reserve 3 μl for evaluation of RNA using Agilent Bioanalyzer Pico chip and the rest for downstream RNA work.

RNA Sample Quality Control

RNA profiles and concentration can be assessed by, for example, using 2100 Bioanalyzer systems or Tapestation manufacturers have been given (*see* **Note 5**). For EV total RNA, either RNA 6000 pico kit for Bioanalyzer or High Sensitivity RNA ScreenTape assay for TapeStation can be used, but attention must be paid to the detection limits of the assays (Fig. 15). Small RNA content can be assessed with the small RNA Kit for Bioanalyzer. Typical electropherograms associated with good sequencing outcomes are presented in Fig. 15. A typical total RNA profile includes a peak below 200 nt and in some samples, it is possible to observe two ribosomal peaks, deviations from the main peak profile presented in Fig. 15 have been associated with DNA contamination or degraded RNA possibly from cellular sources [43]. Of note, sometimes, the main RNA peak merges with the marker partially or fully and this could be a sign of RNA degradation (Fig. 15). EV Small RNA electropherograms are characterized by a miRNA peak below 40 nt and a second peak between 40 and 80 nt corresponding to the tRNA region. Importantly, to obtain reliable RNA profiles, the

a

	Pico chip	Small RNA chip	High Sensitivity RNA ScreenTape
Sample volume (µl)	1	1	2
Qualitative range (pg/µL)	50 – 5000 in water	50-2000 of purified miRNA in water	NA
Quantitative range (pg/µL)	NA	50-2000 of purified miRNA in water	500-10000
Sensitivity (S/N>3) (pg/µL)	50 in water or 200 in TE	50 in water	100
Quantitative Reproducibility (within a chip) (%CV)	20	25	15
Quantitative accuracy (%)	±30	NA	±30
RNA Sizing Range	NA	6-150 nt	NA
Representative uEV RNA electropherograms	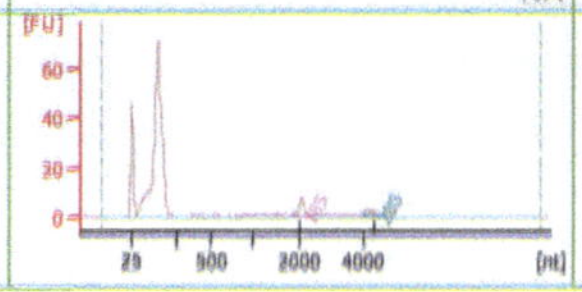	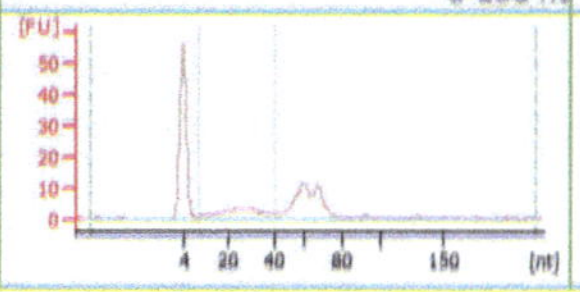	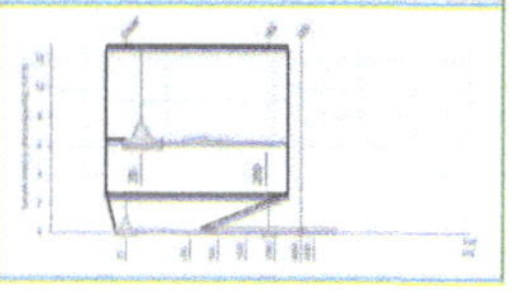

b

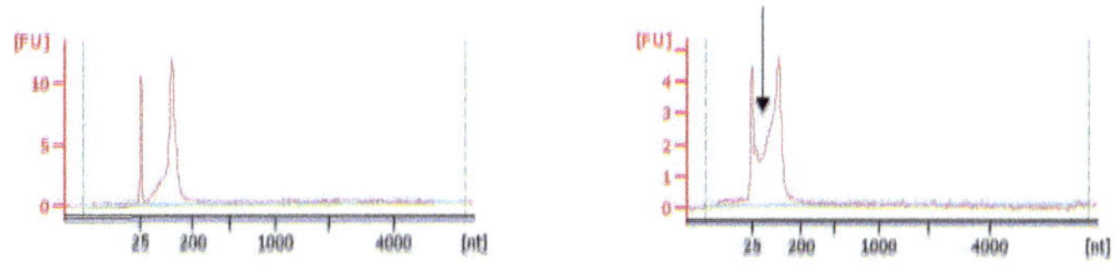

Fig. 15 uEV RNA sample quality control using fragment length analysis. (**a**) Comparative parameters between RNA quality control assays and representative electropherograms of good quality samples. (**b**) Electropherogram of a sample isolated without (left) and with (right) in-column DNAse I treatment. In this case, the DNAse treatment resulted in RNA degradation, which is evidenced by the overlapping of the marker peak (25 nt) with the RNA peak (25–200 nt) (black arrow). Not available (NA)

instrument should be installed on a stable surface and far from other instruments that vibrate like vortexes. Instrument vibration could cause spikes or a noisy baseline in the chip run. The chip should be pipetted on a clean surface without dust and using particle-free plastics and solutions. Dust or particles on the solutions or the chip, dirty chip bottom, and improperly mixed dye are the most likely causes of artifactual peaks and noisy baselines. Please refer to the Agilent 2100 Bioanalyzer System maintenance and troubleshooting guide for possible solutions when unexpected peaks or noise are detected. (https://www.agilent.com/cs/library/usermanuals/public/2100-Bioanalyzer_Maintenance-Troubleshooting_USR_ENG.pdf). Electrodes should be cleaned often and sonicated whenever signs of buildup appear (aborted run due to voltage errors in particular wells or artifacts like late migration).

7.3.2 cDNA Synthesis, Amplification, and Library Preparation for NGS

Before starting the sample preparation to sequencing, a number of decisions need to be made. First, the targeted RNA species need to be defined as separate protocols and kits exist for small RNA, miRNA, mRNA, long RNA, and total RNA. They differ in the utilized technologies, fragmentation, priming strategies, the need for rRNA depletion, and indexing systems. Because the uEV RNA, like any EV RNA, is mainly short, protocols with and without fragmentation can be applied [62]. Ribosomal RNA depletion helps to reserve the sequencing capacity for the other (more interesting) RNA species. Different uEV types might carry different amounts of rRNAs; however, the quantity of rRNA is typically low in the small-EV dominated uEV populations. Thus, the depletion is not always needed, for example, when using a poly-A-targeting priming strategy for mRNAs like the SMARTSeq protocol presented here. Polyadenylated rRNAs are rare in human cells [63]. If there is a large quantity of rRNA, it suggests the presence of contaminating cellular debris in the uEV preparation. In that case, the ribodepletion does not rescue from detecting the contaminating cell-derived non-rRNA transcripts. Unlike poly-A targeted priming, random priming results in sequencing of a wider presentation of RNA species even from degraded samples. However, also higher amounts of rRNA will be sequenced and thus, solutions to reduce the rRNA sequencing are needed [64]. Use of unique molecular identifiers (UMIs) provides a means for more reliable quantitation of the transcripts and variants—tagging each transcript with this unique barcoding system helps in filtering out duplicate reads and PCR errors and allows identification of unique reads. High duplication rate is a typical problem in low in-put RNA sequencing.

Similar to long RNA sequencing, diverse approaches are available for small RNA sequencing. The most common approach is the two-adaptor ligation protocol (e.g., Lexogen). This protocol has shortcomings such as ligation bias and preferences to certain small RNA classes (e.g., miRNAs—attention must be paid to whether this approach captures the small RNA of interest). Advanced two-adaptor ligation protocols overcome some of these limitations and some include the use of UMIs (e.g., QIAseq) to tackle the qPCR bias. Other approaches not involving two-adaptor ligation include polyadenylation (SMARTer) and capture probes (EdgeSeq).

The cDNA libraries can be prepared from approximately 1 ng of total uEV RNA in up to 10 μl volume. Make sure no poly-A or other carrier RNA has been added to the sample. This in-put RNA amount applies both for mRNA sequencing using the SMART-Seq® v4 Ultra Low Input RNA Kit for Sequencing and Nextera XT kit and for small RNA or miRNA sequencing with Lexogen Small RNA-Seq Library Prep Kit (Lexogen GmbH, Vienna, Austria) and QIAseq miRNA Library Kit (Qiagen).

Until the amplification step, work in a clean, pre-PCR RNA laboratory using a dedicated coat and ice box, single use nitrile gloves, nuclease-free filter tips, tubes, water, and reagents. Work benches, hood, pipette set, and equipment should be wiped before starting with the RNAse AWAY Surface Decontaminant.

cDNA Synthesis and Amplification for mRNA Sequencing

1. Thaw the total RNA samples on ice and spin down.
2. Work according to the SMART-Seq Protocol section V from A Protocol: First-Strand cDNA Synthesis onward.
3. Continue through the B Protocol: cDNA Amplification by LD PCR **steps 1–3**.

Before amplification in B Protocol: cDNA Amplification by LD PCR step 4 onward: Shift to work in a PCR laboratory.

4. At step B. Protocol: cDNA Amplification by LD PCR, a good starting point for a 1 ng total EV RNA input is 15 cycles of amplification (*see* **Note 6**). However, it is best to refer to Appendix B for optimizing the number of amplification cycles according to the in-put RNA quantity and quality.
5. After the B Protocol, the samples may be stored at 4 °C overnight.
6. Continue with C Protocol: Purification of Amplified cDNA using the Agencourt AMPure XP Kit. After the protocol, make a small aliquot, for example, 2 μl for quality check. All the samples may be now stored at −20 °C.
7. Perform cDNA validation according to D Protocol: Validation Using the Agilent 2100 Bioanalyzer and High Sensitivity DNA Kit or Agilent TapeStation 4200 and High Sensitivity D5000 ScreenTape system.
   (a) The recommended quantity for library preparation is 100–150 pg of cDNA.

Library Preparation for Sequencing

1. Follow the Protocol section VI. Library Preparation for Sequencing on Next-Generation Sequencing A. Protocol: Illumina Library Preparation Using Nextera DNA Library Preparation Kits and work according to Illumina Nextera XT Reference Guide.
2. Prepare the cDNA libraries with Nextera XT DNA Library Preparation and a nonredundant indexing system, for example, unique dual index kits.
3. Perform library quality and concentration check using, for example, LabChip GX Touch HT High Sensitivity assay and Qubit™ 1X dsDNA HS Assay Kit.

4. Library pooling, quantification, and sequencing can be performed according to the available systems at the sequencing laboratory (*see* **Note 6**). It may be necessary to subject the pooled libraries to size selection in case there are some unwanted peaks visible in the quality control by fragment length analysis, for example, linker-linker artifacts that may arise due to too low EV sample nucleic acid in-put. With size selection, it should be considered that the targeted size range covers the wanted RNA species, for example, when investigating miRNAs, our samples have been size selected using BluePippin (Sage Science, Beverly, Massachusetts, USA) targeting 130–180 bp: the size of adapters plus the sequences of interest (a peak at 120 bp corresponds to linker-linker artifact and peaks above 180 bp could correspond, e.g., to tRNAs, mRNAs, snRNA).

Sequencing Output

Sequencing output will depend on the targeted read numbers, that is, the selection of the flow cell, cycle numbers (read lengths) as well as the number and quality of the samples. For the SMARTSeq protocol presented here, we have typically used a paired-end approach and targeted to get 20–100 M paired end reads per sample. This can be achieved by, for example, applying 20 uEV samples to a flow cell (NextSeq 500/550 High-Output v2 Kit, 150 cycles, FC-404-2002, Illumina), with 800 M reads (400 M per each end) and read length of 2 × 150 bp. Another example is a larger set of 72 samples, sequenced with NovaSeq6000 and S1 flow cell (200 cycles, Illumina) offering 3200 M reads (1600 M per each end) with 2 × 101 bp read length. These set-ups have produced 10–50 M raw reads per sample [43]. For small RNA sequencing with QIASeq or Lexogen, we obtained an average number of 6.5 M reads (for both) using Illumina HiSeq 4000 and NovaSeq6000, respectively.

The SMARTSeq protocol takes advantage of the reverse transcriptase template switching activity to enrich for full-length cDNAs, from the 3′ poly-A to the 5′ end [65, 66]. In practice, this can be noticed from the relatively even gene-body coverage of the reads on the transcripts (*see* example in Fig. 16). In addition, the percentage of exonic reads is typically ~80% and that of rRNA, intronic, and intergenic reads small, < 1–5% each (Fig. 16). This result indicates the quality of the sample preparation is good and the SMARTSeq protocol helps to preserve the sequencing capacity for sequencing of the mRNAs of interest.

7.3.3 cDNA Synthesis and Quantitative Real-Time PCR for the Analysis of Small RNAs

If multiple small RNAs are to be examined, we recommend the use of universal cDNA synthesis kits that generate cDNA simultaneously for all small RNAs present in the sample. This approach avoids technical bias in RT conditions used for the reverse

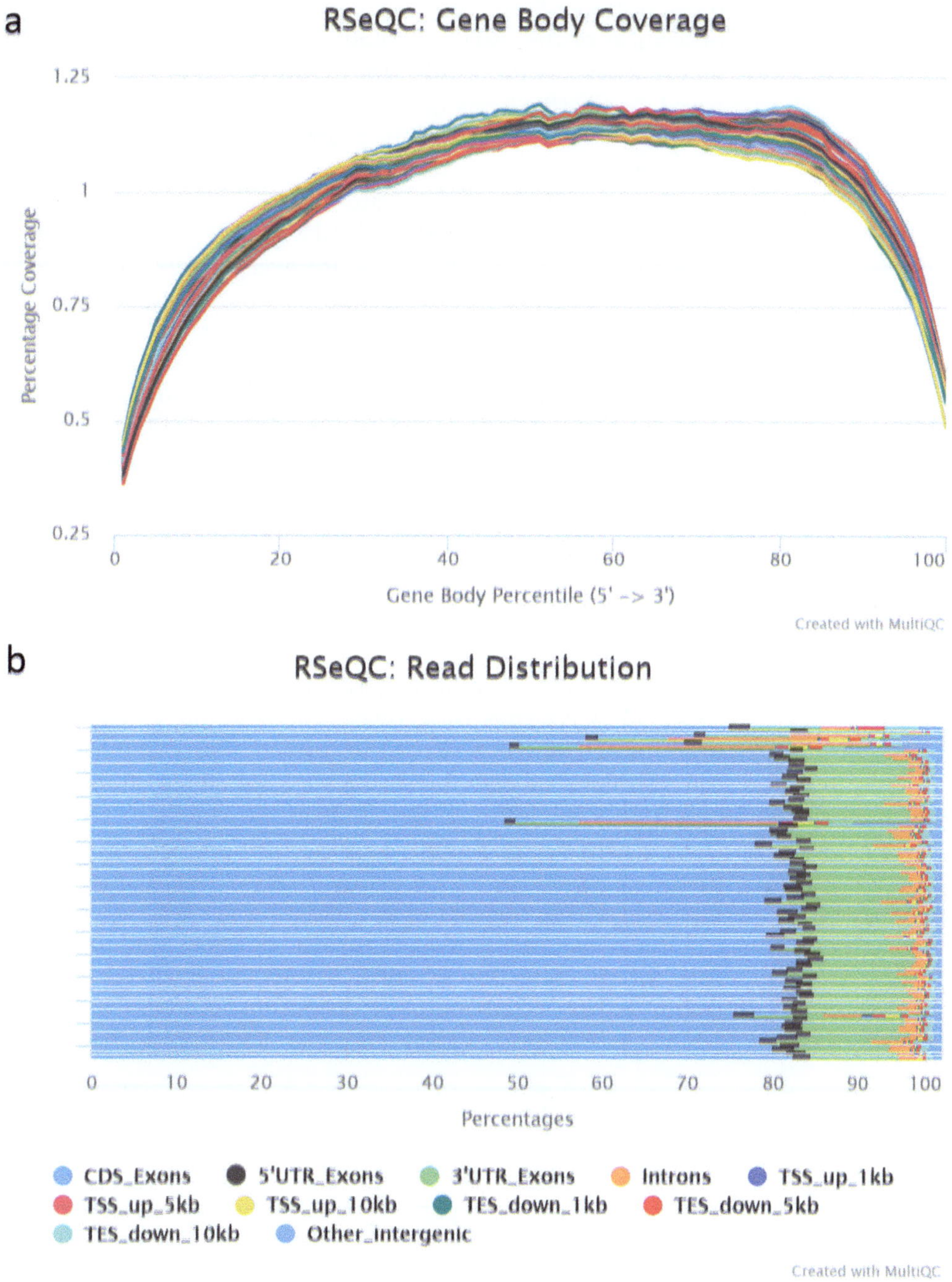

Fig. 16 Sequencing output quality control examples from the SMARTSeq protocol. (**a**) RSeQC gene body coverage analysis (**b**) RSeQC read distribution analysis showing proportions of tags mapping to exons (CDS, 3'UTR and 5'UTR), intergenic regions (TSS or TES), other intergenic regions, and introns. Transcriptional end site (TES), transcriptional start site (TSS), untranslated region (UTR)

transcription of different small RNAs and allows most efficient utilization of the usually limited amount of RNA sample. Several different kits for universal cDNA synthesis compatible with probe-based or SYBR-Green quantitative real-time PCR chemistry are available in the market and the most commonly used ones are listed in methods section above.

Selection of the appropriate qPCR chemistry depends on the type and number of tested small RNAs. Off-the-shelf SYBR Green and probe-based qPCR assays are available for the majority of miRNAs and some snoRNAs that are used as normalization controls. However, if other small RNA species are investigated, for example, tRNAs or small RNA fragments, a custom RNA design will be required. In such cases, we recommend to perform pilot studies using LNA-based SYBR green assays and move to probe-based chemistry at later stage. Probe-based chemistry provides more accurate quantification results, less false positives, and allows the development of duplex assays for the simultaneous detection of 2 different small RNA species in the same reaction. Nevertheless, duplexing requires additional optimization in order to balance the concentration of reaction mix components during the real-time step.

We strongly recommend the inclusion of synthetic spike-in small RNA controls of known concentration to the total RNA used for cDNA synthesis, as they provide valuable information for the efficiency of cDNA synthesis and allow to monitor for the presence of inhibitory compounds in the sample or technical artifacts during the PCR step. In addition, spike-ins allow for cross-plate and cross-experiment comparison across larger cohorts when relatively a large number of samples are analyzed.

Below we provide two protocols routinely used in our laboratory for the detection of small RNAs and small RNA fragments using either SYBR green chemistry in combination with LNA-modified primer assays or probe-based chemistry.

SYBR™ Green Chemistry Using miRCURY SYBR Green Chemistry and LNA PCR Assays

cDNA Synthesis

1. Set up the RT reactions in PCR tubes or strips and use PCR cycler for the incubation steps.
2. If available, use repetitive pipets to aliquot master mixes and low volume, multichannel pipets (1–2, 1–5, or 1–10 μl) for better accuracy.
3. Thaw the RNA samples on ice and spin down.
4. If using the kit for the first time, prepare the UniSp6 RNA spike-in control according to the kit instructions. Aliquot and store frozen at –20 °C
5. Recommended starting amount of total RNA is 10 ng (2μl RNA with concentration of 5 ng/μl) per RT reaction of 10 μl.

However, if sample concentration is lower (e.g., due to low EV concentration in the urine sample), sample volume can be increased to up of 6.5 μl while reducing the volume of water added to the RT reaction.

6. Add spike-in control directly to the RT master mix to ensure equal distribution across all samples.
7. Include water control sample to monitor for reagent purity and/or cross-contaminations.
8. We recommend to prepare RT control for each individual RNA sample to monitor for possible endogenous DNA contamination.
9. Follow the steps in the miRCURY® LNA® RT Kit protocol (QIAGEN).
10. RT reactions can be stored overnight at 4 °C or longer term at –20 °C.

Quantitative Real-Time PCR

1. If proceeding directly to the qPCR step, dilute an aliquot of the RT reaction 30–80 times (e.g., add 79 μl ultrapure water, to 1 μl RT reaction) depending on the abundance of the target small RNA of interest. Diluted RT reactions cannot be stored and should be used immediately.
2. Do not dilute the entire RT-mix if not intending to use it immediately.
3. If using the 2× miRCURY SYBR® Green PCR Master Mix and an ABI qPCR instrument, add ROX passive reference dye to the Master Mix according to the manufacturer's recommendations.
4. Follow the protocol "Quantitative, Real-Time PCR Using Individual miRCURY LNA miRNA PCR Assays" to set up the real-time reactions and run the PCR. At least 3 replicates per cDNA sample and a no cDNA template water control should be included in the experiment design.
5. After the run is completed, analyze data according to instructions provided with the qPCR instrument software.

Probe-Based Detection Using TaqMan Advanced Chemistry

cDNA Synthesis

1. Set up the RT reactions in PCR tubes, strips, or plates depending on the sample number and use PCR cycler for the incubation steps.
2. If available, use repetitive pipets to aliquot master mixes and low volume, multichannel pipets (1–5 μl, or, 1–10 μl). Prepare the real-time PCR reactions in an area free of artificial templates and siRNA transfections. High-copy-number templates can easily contaminate the real-time PCR reactions. Keep the assays

protected from light and stored as indicated until ready for use. Excessive exposure to light may affect the fluorescent probe.

3. Thaw the RNA samples on ice and spin down.
4. Recommended starting volume of total RNA is 2μl RNA per polyA tailing reaction of 5 μl. Sample concentration before adding to reactions should be ≤5 ng/μl. (*see* **Note** 7 about using spike-in controls with TaqMan Advanced miRNA assays).

   IMPORTANT! Do not denature the total RNA.

   If sample concentration is in the low pg range, sample volume can be increased to up to 3.5 μl while reducing the volume of water added to the RT reaction.
5. Follow instructions as described in the TaqMan™ Advanced miRNA Assays USER GUIDE Single-tube assays, Chapter 2 [68].
6. Proceed to real-time PCR or store the undiluted miR-Amp reaction products at −20 °C for up to 2 months.

Quantitative Real-Time PCR

1. Prepare the real-time PCR reactions in an area free of artificial templates and siRNA transfections. High-copy-number templates can easily contaminate the real-time PCR reactions.
2. Keep the PCR assays protected from light and stored as indicated until ready for use. Excessive exposure to light may affect the fluorescent probe.
3. A minimum of 3 replicates per reaction is recommended for optimal analysis.
4. The minimum real-time reaction volume should be not lower than 10 μl if 96 well plates are used. Scale reagents accordingly.
5. Follow instructions as described in the TaqMan™ Advanced miRNA Assays USER GUIDE Single-tube assays, Chapter TaqMan™ Fast Advanced Master Mix for qPCR, Chapter 3.
6. After the run is completed, analyze data according to instructions provided with the qPCR instrument software.

7.4 Notes

1. When urine sample volumes are above 20 ml, urinary EV RNA can be extracted with miRNEasy mini kit using the minimum 30 μl elution volume. The mini kit has a slightly more straightforward protocol than the miRNeasy micro kit.
2. When the urine samples have good quality—rapid processing to carefully remove cells and freezing at −70 °C or colder—RNA isolation may succeed without a DNAse I treatment. However, if urine samples have less good quality, there are two options to remove contaminating DNA:

(a) One can perform the DNAse treatment in-column during RNA isolation. This can be risky—every now and then, the RNA sample may become partly degraded or give reduced yield. (Or)

(b) Perform DNAse I treatment of the EV before RNA isolation as described here—in most cases, this is sufficient to remove the contaminating DNA.

3. If the uEV samples are resuspended in a higher volume than 40 μl, the DNAse I set reagent volumes can be scaled up. However, the Qiazol volume does not have to be scaled up if the whole reaction volume does not exceed 100 μl. For >100 μl reaction volumes, one can either use Trizol LS (Thermo Fischer Scientific) or follow the miRNeasy Serum/Plasma kit (Qiagen) protocol starting with addition of 5 volumes of Qiazol.

4. For maximum possible yield, the eluate can be passed through the column a second time using the same centrifugation conditions as in **step 15**. For highest possible concentrations using miRNEasy micro kit, the minimum elution volume is 10 μl, but decreased total yields (−20%) can be expected.

5. For pico and small RNA Bioanalyzer chips, we recommend using inverse pipetting for the gel dye mix to avoid the formation of air bubbles that could interfere with the electrophoresis. In our experience, the run can be performed with either heated or native samples, bu we recommend applying the chosen protocol systematically for samples to be compared. Since December 31, 2023, the 2100 Bioanalyzer instrument is no longer available for purchase. For those laboratories/core facilities that own an instrument, it is still possible to use the bioanalyzer, because until 2030, Agilent will continue offering Bioanalyzer kits and reagents, support for instruments, consumables, and software. However, in the long term, alternatives need to be considered.

6. The step B. Protocol: cDNA Amplification by LD PCR gives suggestions about the in-put cell numbers, RNA quantity, and number of amplification cycles to be used. However, the recommendation cannot be applied to EV RNA as such as it does not take into account the different composition of the EV RNA: relatively high short to long RNA ratio. Our experience is that the number of PCR cycles can be estimated by dividing the total EV RNA quantity by 10 and then checking the corresponding amplification cycles from the table. It is recommended to use the lowest number of PCR cycles that gives enough material for library construction.

7. When using exogenous spike-in controls (e.g., ath-miR-159a) with TaqMan Advanced miRNA assays, the assay chemistry

requires that exogenous controls are 5′-phosphorylated. The final concentration of the spike-in control in the sample should be 1–10 pM.

Acknowledgments

This work was supported by the IMMPROVE Alpe d'HuZes grant of the Dutch Cancer Society (EMCR2015-8022). MSYN acknowledges funding from Metro North Clinician Research Fellowship, Queensland Health Clinical Research Fellowship, and RACP Jacquot Research Establishment Fellowship. This study was funded by the IMMPROVE Alpe d'HuZes grant of the Dutch Cancer Society (EMCR2015-8022), Metro North Clinician Research Fellowship (CRF-606-2024), Queensland Health Clinical Research Fellowship (Round 4), Royal Australasian College of Physicians Jacquot Research Establishment Fellowship (2025REF008), SalWe Research Program Personalized Diagnostics and Care (GET IT DONE), Business Finland Grant (3986/31/2013), and Innovative Medicines Initiative 2 Joint Undertaking (115974). The Innovative Medicines Initiative 2 Joint Undertaking receives support from the European Union's Horizon 2020 research and innovation program and European Federation of Pharmaceutical Industries and Associations (EFPIA) and Juvenile Diabetes Research Foundation (JDRF).

This study was also funded through 5K23HL126101 (National Institute of Health/NHLBI) and ADA 7-22-ICTSPM-19 (American Diabetes Association).

Conflict of Interest

The authors have no conflicts of interest to declare that are relevant to the content of this chapter.

References

1. Erdbrügger U, Blijdorp CJ, Bijnsdorp IV, Borràs FE, Burger D, Bussolati B et al (2021) Urinary extracellular vesicles: a position paper by the urine task force of the International Society for Extracellular Vesicles. J Extracell Vesicles 10:e12093. https://doi.org/10.1002/jev2.12093
2. van Royen ME, Soekmadji C, Grange C, Webber JP, Tertel T, Droste M et al (2023) The quick reference card "Storage of urinary EVs" - a practical guideline tool for research and clinical laboratories. J Extracell Vesicles 12:e12286. https://doi.org/10.1002/jev2.12286
3. Koritzinsky EH, Street JM, Chari RR, Glispie DM, Bellomo TR, Aponte AM et al (2019) Circadian variation in the release of small extracellular vesicles can be normalized by vesicle number or TSG101. Am J Physiol Renal Physiol 317:F1098–f1110. https://doi.org/10.1152/ajprenal.00568.2017
4. Blijdorp CJ, Tutakhel OAZ, Hartjes TA, van den Bosch TPP, van Heugten MH, Rigalli JP et al (2021) Comparing approaches to normalize, quantify, and characterize urinary extracellular vesicles. J Am Soc Nephrol 32:1210–1226. https://doi.org/10.1681/asn.2020081142

5. Pellegrini KL, Patil D, Douglas KJS, Lee G, Wehrmeyer K, Torlak M et al (2017) Detection of prostate cancer-specific transcripts in extracellular vesicles isolated from post-DRE urine. Prostate 77:990–999. https://doi.org/10.1002/pros.23355
6. European Association of Urology Research Foundation: European Association of Urology Research Foundation (EAU RF) Available at: https://brd.nci.nih.gov/brd/sop-compendium/show/1583. Accessed Jan 20
7. Dragovic RA, Gardiner C, Brooks AS, Tannetta DS, Ferguson DJ, Hole P et al (2011) Sizing and phenotyping of cellular vesicles using nanoparticle tracking analysis. Nanomedicine 7:780–788. https://doi.org/10.1016/j.nano.2011.04.003
8. Harding MA, Yavuz H, Gathmann A, Upson S, Swiatecka-Urban A, Erdbrügger U (2024) Uromodulin and the study of urinary extracellular vesicles. J Extracell Biol 3:e70022. https://doi.org/10.1002/jex2.70022
9. Parsons MEM, McParland D, Szklanna PB, Guang MHZ, O'Connell K, O'Connor HD et al (2017) A protocol for improved precision and increased confidence in nanoparticle tracking analysis concentration measurements between 50 and 120 nm in biological fluids. Front Cardiovasc Med 4:68. https://doi.org/10.3389/fcvm.2017.00068
10. Auger C, Brunel A, Darbas T, Akil H, Perraud A, Bégaud G et al (2022) Extracellular vesicle measurements with nanoparticle tracking analysis: a different appreciation of up and down secretion. Int J Mol Sci 23. https://doi.org/10.3390/ijms23042310
11. Sims PJ, Faioni EM, Wiedmer T, Shattil SJ (1988) Complement proteins C5b-9 cause release of membrane vesicles from the platelet surface that are enriched in the membrane receptor for coagulation factor Va and express prothrombinase activity. J Biol Chem 263:18205–18212
12. Welsh JA, Arkesteijn GJA, Bremer M, Cimorelli M, Dignat-George F, Giebel B et al (2023) A compendium of single extracellular vesicle flow cytometry. J Extracell Vesicles 12:e12299. https://doi.org/10.1002/jev2.12299
13. Cook S, Tang VA, Lannigan J, Jones JC, Welsh JA (2023) Quantitative flow cytometry enables end-to-end optimization of cross-platform extracellular vesicle studies. Cell Rep Methods 3:100664. https://doi.org/10.1016/j.crmeth.2023.100664
14. Welsh JA, Goberdhan DCI, O'Driscoll L, Buzas EI, Blenkiron C, Bussolati B et al (2024) Minimal information for studies of extracellular vesicles (MISEV2023): from basic to advanced approaches. J Extracell Vesicles 13:e12404. https://doi.org/10.1002/jev2.12404
15. Welsh JA, Van Der Pol E, Arkesteijn GJA, Bremer M, Brisson A, Coumans F et al (2020) MIFlowCyt-EV: a framework for standardized reporting of extracellular vesicle flow cytometry experiments. J Extracell Vesicles 9:1713526. https://doi.org/10.1080/20013078.2020.1713526
16. Musante L, Bontha SV, La Salvia S, Fernandez-Piñeros A, Lannigan J, Le TH et al (2020) Rigorous characterization of urinary extracellular vesicles (uEVs) in the low centrifugation pellet - a neglected source for uEVs. Sci Rep 10:3701. https://doi.org/10.1038/s41598-020-60619-w
17. Correll VL, Otto JJ, Risi CM, Main BP, Boutros PC, Kislinger T et al (2022) Optimization of small extracellular vesicle isolation from expressed prostatic secretions in urine for in-depth proteomic analysis. J Extracell Vesicles 11:e12184. https://doi.org/10.1002/jev2.12184
18. Wang X, Kim CS, Adams BC, Wilkinson R, Hill MM, Shah AK et al (2024) Human proximal tubular epithelial cell-derived small extracellular vesicles mediate synchronized tubular ferroptosis in hypoxic kidney injury. Redox Biol 70:103042. https://doi.org/10.1016/j.redox.2024.103042
19. Wu L, Woud WW, Baan CC, Hesselink DA, van der Pol E, Jenster G et al (2023) Isolation-free measurement of single urinary extracellular vesicles by imaging flow cytometry. Nanomedicine 48:102638. https://doi.org/10.1016/j.nano.2022.102638
20. Pink D, Basu A, Wong M, Pham D, Valencia J, Triana V et al (2023) Antibody titrations are critical for microflow cytometric analysis of extracellular vesicles. Cytometry A 103:670–683. https://doi.org/10.1002/cyto.a.24733
21. Myette RL, Xiao F, Geier P, Feber J, Burger D, Kennedy CRJ (2023) Urinary podocyte-derived large extracellular vesicles are increased in paediatric idiopathic nephrotic syndrome. Nephrol Dial Transplant 38:2089–2091. https://doi.org/10.1093/ndt/gfad086
22. Ng MSY, Wang X, Kassianos A, Healy HG (2023) Diagnostic application of urinary extracellular vesicles in chronic kidney disease. J Extracell Vesicles 12(e12329):12197–12198
23. Turco AE, Lam W, Rule AD, Denic A, Lieske JC, Miller VM et al (2016) Specific renal parenchymal-derived urinary extracellular vesicles identify age-associated structural changes in living donor kidneys. J Extracell Vesicles 5:

29642. https://doi.org/10.3402/jev.v5.29642

24. Chirackal RS, Jayachandran M, Wang X, Edeh S, Haskic Z, Perinpam M et al (2019) Urinary extracellular vesicle-associated MCP-1 and NGAL derived from specific nephron segments differ between calcium oxalate stone formers and controls. Am J Physiol Renal Physiol 317:F1475–f1482. https://doi.org/10.1152/ajprenal.00515.2018
25. Welsh JA, Jones JC, Tang VA (2020) Fluorescence and light scatter calibration allow comparisons of small particle data in standard units across different flow cytometry platforms and detector settings. Cytometry A 97:592–601. https://doi.org/10.1002/cyto.a.24029
26. Welsh JA, Horak P, Wilkinson JS, Ford VJ, Jones JC, Smith D et al (2020) FCMPASS software aids extracellular vesicle light scatter standardization. Cytometry A 97:569–581. https://doi.org/10.1002/cyto.a.23782
27. Bernardino R, Carvalho AS, Hall MJ, Alves L, Leão R, Sayyid R et al (2024) Profiling of urinary extracellular vesicle protein signatures from patients with cribriform and intraductal prostate carcinoma in a cross-sectional study. Sci Rep 14:25065. https://doi.org/10.1038/s41598-024-75272-w
28. Anfaiha-Sanchez M, Santiago-Hernandez A, Lopez JA, Lago-Baameiro N, Pardo M, Martin-Blazquez A et al (2024) Urinary extracellular vesicles as a monitoring tool for renal damage in patients not meeting criteria for chronic kidney disease. J Extracell Biol 3: e170. https://doi.org/10.1002/jex2.170
29. Khoo A, Govindarajan M, Qiu Z, Liu LY, Ignatchenko V, Waas M et al (2024) Prostate cancer reshapes the secreted and extracellular vesicle urinary proteomes. Nat Commun 15: 5069. https://doi.org/10.1038/s41467-024-49424-5
30. Hallal SM, Sida LA, Tűzesi C, Shivalingam B, Sim HW, Buckland ME et al (2024) Size matters: biomolecular compositions of small and large extracellular vesicles in the urine of glioblastoma patients. J Extracell Biol 3:e70021. https://doi.org/10.1002/jex2.70021
31. Lischnig A, Bergqvist M, Ochiya T, Lässer C (2022) Quantitative proteomics identifies proteins enriched in large and small extracellular vesicles. Mol Cell Proteomics 21:100273. https://doi.org/10.1016/j.mcpro.2022.100273
32. Bielopolski D, Musante L, Hoorn EJ, Molina H, Barrows D, Carrol TS et al (2024) Effect of the DASH diet on the sodium-chloride cotransporter and aquaporin-2 in urinary extracellular vesicles. Am J Physiol Renal Physiol 326:F971–f980. https://doi.org/10.1152/ajprenal.00274.2023
33. Ramirez-Garrastacho M, Berge V, Linē A, Llorente A (2022) Potential of miRNAs in urinary extracellular vesicles for management of active surveillance in prostate cancer patients. Br J Cancer 126:492–501. https://doi.org/10.1038/s41416-021-01598-1
34. Wiśniewski JR, Zougman A, Nagaraj N, Mann M (2009) Universal sample preparation method for proteome analysis. Nat Methods 6:359–362. https://doi.org/10.1038/nmeth.1322
35. Batth TS, Tollenaere MX, Rüther P, Gonzalez-Franquesa A, Prabhakar BS, Bekker-Jensen S et al (2019) Protein aggregation capture on microparticles enables multipurpose proteomics sample preparation. Mol Cell Proteomics 18:1027–1035. https://doi.org/10.1074/mcp.TIR118.001270
36. Fröhlich K, Fahrner M, Brombacher E, Seredynska A, Maldacker M, Kreutz C et al (2024) Data-independent acquisition: a milestone and prospect in clinical mass spectrometry-based proteomics. Mol Cell Proteomics 23:100800. https://doi.org/10.1016/j.mcpro.2024.100800
37. Zhang G, Ding Y, Zhang H, Wei D, Liu Y, Sun J et al (2024) Assessment of urine sample collection and processing variables for extracellular vesicle-based proteomics. Analyst 149: 3416–3424. https://doi.org/10.1039/d4an00296b
38. Chitti SV, Gummadi S, Kang T, Shahi S, Marzan AL, Nedeva C et al (2024) Vesiclepedia 2024: an extracellular vesicles and extracellular particles repository. Nucleic Acids Res 52: D1694–D1698. https://doi.org/10.1093/nar/gkad1007
39. Keerthikumar S, Chisanga D, Ariyaratne D, Al Saffar H, Anand S, Zhao K et al (2016) ExoCarta: a web-based compendium of exosomal cargo. J Mol Biol 428:688–692. https://doi.org/10.1016/j.jmb.2015.09.019
40. Steiner L, Eldh M, Offens A, Veerman RE, Johansson M, Hemdan T et al (2025) Protein profile in urinary extracellular vesicles is a marker of malignancy and correlates with muscle invasiveness in urinary bladder cancer. Cancer Lett 609:217352. https://doi.org/10.1016/j.canlet.2024.217352
41. Bader JM, Albrecht V, Mann M (2023) MS-based proteomics of body fluids: the end of the beginning. Mol Cell Proteomics 22: 100577. https://doi.org/10.1016/j.mcpro.2023.100577

42. Bazzell BG, Rainey WE, Auchus RJ, Zocco D, Bruttini M, Hummel SL et al (2018) Human urinary mRNA as a biomarker of cardiovascular disease. Circ Genom Precis Med 11:e002213. https://doi.org/10.1161/circgen.118.002213
43. Dwivedi OP, Barreiro K, Käräjämäki A, Valo E, Giri AK, Prasad RB et al (2023) Genome-wide mRNA profiling in urinary extracellular vesicles reveals stress gene signature for diabetic kidney disease. iScience 26:106686. https://doi.org/10.1016/j.isci.2023.106686
44. Barreiro K, Dwivedi OP, Rannikko A, Holthöfer H, Tuomi T, Groop PH et al (2023) Capturing the kidney transcriptome by urinary extracellular vesicles-from pre-analytical obstacles to biomarker research. Genes (Basel) 14. https://doi.org/10.3390/genes14071415
45. Bajo-Santos C, Brokāne A, Zayakin P, Endzeliņš E, Soboļevska K, Belovs A et al (2023) Plasma and urinary extracellular vesicles as a source of RNA biomarkers for prostate cancer in liquid biopsies. Front Mol Biosci 10. https://doi.org/10.3389/fmolb.2023.980433
46. Larsen JH, Jensen IS, Svenningsen P (2024) Benchmarking transcriptome deconvolution methods for estimating tissue- and cell-type-specific extracellular vesicle abundances. J Extracell Vesicles 13:e12511. https://doi.org/10.1002/jev2.12511
47. Nilsson J, Skog J, Nordstrand A, Baranov V, Mincheva-Nilsson L, Breakefield XO et al (2009) Prostate cancer-derived urine exosomes: a novel approach to biomarkers for prostate cancer. Br J Cancer 100:1603–1607. https://doi.org/10.1038/sj.bjc.6605058
48. Miranda KC, Bond DT, McKee M, Skog J, Păunescu TG, Da Silva N et al (2010) Nucleic acids within urinary exosomes/microvesicles are potential biomarkers for renal disease. Kidney Int 78:191–199. https://doi.org/10.1038/ki.2010.106
49. Ramezani A, Devaney JM, Cohen S, Wing MR, Scott R, Knoblach S et al (2015) Circulating and urinary microRNA profile in focal segmental glomerulosclerosis: a pilot study. Eur J Clin Investig 45:394–404. https://doi.org/10.1111/eci.12420
50. Armstrong DA, Green BB, Seigne JD, Schned AR, Marsit CJ (2015) MicroRNA molecular profiling from matched tumor and bio-fluids in bladder cancer. Mol Cancer 14:194. https://doi.org/10.1186/s12943-015-0466-2
51. Cheng L, Sun X, Scicluna BJ, Coleman BM, Hill AF (2014) Characterization and deep sequencing analysis of exosomal and non-exosomal miRNA in human urine. Kidney Int 86:433–444. https://doi.org/10.1038/ki.2013.502
52. Miranda KC, Bond DT, Levin JZ, Adiconis X, Sivachenko A, Russ C et al (2014) Massively parallel sequencing of human urinary exosome/microvesicle RNA reveals a predominance of non-coding RNA. PLoS One 9: e96094. https://doi.org/10.1371/journal.pone.0096094
53. Brokāne A, Bajo-Santos C, Zayakin P, Belovs A, Jansons J, Lietuvietis V et al (2023) Validation of potential RNA biomarkers for prostate cancer diagnosis and monitoring in plasma and urinary extracellular vesicles. Front Mol Biosci 10. https://doi.org/10.3389/fmolb.2023.1279854
54. Karttunen J, Kalmar L, Grant A, Ying J, Stewart SE, Wang X et al (2024) miR-182, miR-221 and miR-222 are potential urinary extracellular vesicle biomarkers for canine urothelial carcinoma. Sci Rep 14:17967. https://doi.org/10.1038/s41598-024-69070-7
55. Verwilt J, Verniers K, De Geyter S, Roelandt S, Pinheiro C, Hendrix A et al (2024) Intact messenger RNA exists in human blood plasma and urine, and their purified macromolecular compartments. bioRxiv. https://doi.org/10.1101/2024.11.30.626091
56. Murillo OD, Thistlethwaite W, Rozowsky J, Subramanian SL, Lucero R, Shah N et al (2019) exRNA atlas analysis reveals distinct extracellular RNA cargo types and their carriers present across human biofluids. Cell 177:463–477.e415. https://doi.org/10.1016/j.cell.2019.02.018
57. Mateescu B, Jones JC, Alexander RP, Alsop E, An JY, Asghari M et al (2022) Phase 2 of extracellular RNA communication consortium charts next-generation approaches for extracellular RNA research. iScience 25:104653. https://doi.org/10.1016/j.isci.2022.104653
58. Barreiro K, Dwivedi OP, Leparc G, Rolser M, Delic D, Forsblom C et al (2020) Comparison of urinary extracellular vesicle isolation methods for transcriptomic biomarker research in diabetic kidney disease. J Extracell Vesicles 10: e12038. https://doi.org/10.1002/jev2.12038
59. Barreiro K, Dwivedi OP, Valkonen S, Groop PH, Tuomi T, Holthofer H et al (2021) Urinary extracellular vesicles: assessment of pre-analytical variables and development of a quality control with focus on transcriptomic biomarker research. J Extracell Vesicles 10:

e12158. https://doi.org/10.1002/jev2.12158

60. Barreiro K, Karttunen J, Valo E, Viippola E, Quintero I, Käräjämäki A et al (2024) Selected miRNAs in urinary extracellular vesicles show promise for non-invasive diagnostics of diabetic kidney disease. medRxiv. https://doi.org/10.1101/2024.09.12.24312889
61. Puhka M, Thierens L, Nicorici D, Forsman T, Mirtti T, Af Hällström T et al (2022) Exploration of extracellular vesicle miRNAs, targeted mRNAs and pathways in prostate cancer: relation to disease status and progression. Cancers (Basel) 14. https://doi.org/10.3390/cancers14030532
62. Rodosthenous RS, Hutchins E, Reiman R, Yeri AS, Srinivasan S, Whitsett TG et al (2020) Profiling extracellular long RNA transcriptome in human plasma and extracellular vesicles for biomarker discovery. iScience 23:101182. https://doi.org/10.1016/j.isci.2020.101182
63. Slomovic S, Laufer D, Geiger D, Schuster G (2006) Polyadenylation of ribosomal RNA in human cells. Nucleic Acids Res 34:2966–2975. https://doi.org/10.1093/nar/gkl357
64. Gustafsson C, Hauenstein J, Frengen N, Krstic A, Luc S, Månsson R (2023) T-RHEX-RNAseq—a tagmentation-based, rRNA blocked, random hexamer primed RNAseq method for generating stranded RNAseq libraries directly from very low numbers of lysed cells. BMC Genomics 24:205. https://doi.org/10.1186/s12864-023-09279-4
65. Chenchik A, Zhu Y, Diatchenko L, Li R, Hill J, Siebert P (1998) Generation and use of high-quality cDNA from small amounts of total RNA by SMART PCR. In: Siebert P, Larrick J (eds) RT-PCR methods for gene cloning and analysis. BioTechniques Books, Massachusetts, pp 305–319
66. Takara Bio USA: SMART-Seq® v4 Ultra® Low Input RNA Kit for Sequencing User Manual. Available at: https://www.takarabio.com/documents/User%20Manual/SMART/SMART-Seq_V4%20Ultra%20Low%20Input%20RNA%20Kit%20User%20Manual_111017.pdf?srsltid=AfmBOoom8pinXfsNp4HzS5xm0EaDahZdqSpussUdOqm6JePAjh299tbt. Accessed Jan 20
67. Comfort N, Cai K, Bloomquist TR, Strait MD, Ferrante AW Jr, Baccarelli AA (2021) Nanoparticle tracking analysis for the quantification and size determination of extracellular vesicles. J Vis Exp. https://doi.org/10.3791/62447
68. ThermoFisher Scientific: TaqMan™ Advanced miRNA Assays USER GUIDE Single-tube assays for use with: TaqMan™ Advanced miRNA cDNA Synthesis Kit. Available at: https://assets.thermofisher.com/TFS-Assets/LSG/manuals/100027897_TaqManAdv_miRNA_Assays_UG.pdf. Accessed Jan 2

Chapter 8

From Conditioned Media to Therapeutics: Isolation and Characterization of Mesenchymal Stem Cell–Derived Extracellular Vesicles

Chippy K. Mathew, Yashvi Sharma, and Sujata Mohanty

Abstract

Mesenchymal Stem Cells (MSCs) are widely recognized for their immunomodulatory and regenerative properties. In recent years, extracellular vesicles (EVs) derived from MSCs have garnered significant attention for replicating many of the regenerative functions of their parent cells. EVs can be classified based on their origin and biogenesis into exosomes (endosomally derived), ectosomes (budding from the plasma membrane), and apoptotic bodies. They can also be categorized by size into small EVs (<200 nm), large EVs (>200 nm), and apoptotic bodies (≤1 μm). Among these, small EVs, particularly exosomes, are most prominently associated with regenerative and immunomodulatory roles. Despite their potential, the translation of MSC-derived EVs into clinical applications remains limited. A major challenge lies in the lack of standardized protocols for isolating and characterizing EVs from MSC-conditioned media. This chapter explores the hurdles in standardizing these techniques and proposes key factors to optimize the isolation and characterization processes, bridging the gap between their therapeutic potential and clinical implementation.

Key words Mesenchymal Stem Cells, Extracellular Vesicles, Conditioned Media, Ultracentrifugation, MISEV

1 Introduction and Background

In the field of regenerative medicine, mesenchymal stem cells (MSCs) have emerged as prominent candidates for cellular therapy [1]. As adult-origin cells, MSCs pose minimal ethical concerns and are relatively straightforward to isolate [2]. They can be derived from diverse sources, including bone marrow, adipose tissue, umbilical cord, dental pulp, and menstrual blood [3]. Numerous preclinical and clinical studies have demonstrated the safety and efficacy of MSCs in treating a range of conditions, including cardiac diseases, skin wounds, neurological disorders, and more [4].

During the COVID-19 pandemic, when conventional therapies fell short in addressing the complete pathophysiology of the

Amit K. Srivastava (ed.), *Extracellular Vesicles*, Neuromethods, vol. 228, https://doi.org/10.1007/978-1-0716-4905-3_8,

disease, MSCs demonstrated safety and efficacy [5]. A Phase I clinical trial (CTRI/2020/08/027043) in India reported the safety and efficacy of umbilical cord–derived MSCs in managing the cytokine storm and aiding patient recovery, with shortening of hospital durations.

Despite their therapeutic potential, MSCs remain in clinical trial phases and have yet to receive regulatory approval as standard treatments [6]. Challenges such as embolism risk, immunogenicity, tumorigenicity, donor variability, high cost for maintenance of their culture and storage, and issues related to viability and shelf life have hindered their broader adoption [7].

To address these limitations, there has been a growing focus on extracellular vesicles (EVs) derived from MSCs [1]. These nano-sized vesicles have gained attention as potential cell-free therapeutic agents with regenerative and immunomodulatory properties [3]. EVs are broadly classified based on their biogenesis: exosomes (originating from endosomal pathways), ectosomes (budding from the plasma membrane), and apoptotic bodies (larger vesicles formed during apoptosis) [8]. They may also be categorized by size into small EVs (<200 nm) and large EVs (200 nm–1 μm), as defined by the MISEV (Minimal Information for Studies of Extracellular Vesicles) guidelines, which provide a standardized framework for their categorization and characterization [8, 9]. MSCs EVs are primarily considered as mediators of intercellular communication and carriers of bioactive molecules like nucleic acids, proteins, and lipids [1]. The cargo of these EVs may also differ based on the physiological state of the cells [1].

Studies have shown that MSC-derived EVs can replicate the therapeutic mechanisms of their parent cells, promoting regenerative and immunomodulatory effects [2]. Unlike live cells, EVs mitigate concerns regarding viability and tumorigenicity [4]. Their nanoscale size and double-membrane structure facilitate formulation into stable, transportable products [10]. Additionally, their liposome-like structure positions them as natural drug carriers, broadening their therapeutic potential [11].

However, certain barriers remain in translating EVs into clinical practice, including their low yield from MSC-conditioned media and the limited studies addressing dose standardization [12]. However, the most significant hurdle is the absence of standardized protocols for their isolation and characterization, which are essential to ensure consistency and prevent adverse reactions [13]. Addressing these limitations through protocol optimization is a vital step toward harnessing the full clinical potential of MSC-derived EVs and advancing the field of cell-free therapeutics [1].

This evolving paradigm highlights the promise of MSC-derived EVs as a next-generation therapeutic platform, addressing the limitations of traditional cell-based therapies while opening new frontiers in regenerative medicine.

2 Techniques for Isolation of Extracellular Vesicles

EV isolation from cell culture media can be performed using various methods. Differential ultracentrifugation is the most common method, where successive centrifugation steps separate EVs based on size and density [14]. Density gradient centrifugation further purifies EVs using a gradient, typically sucrose or iodixanol [15]. Size-exclusion chromatography (SEC) separates EVs by size, providing high purity [16]. Immunoaffinity capture uses antibodies targeting EVs markers (e.g., CD9, CD63) for selective isolation [17]. Commercial kits combine techniques like precipitation, filtration, or magnetic bead-based methods, offering easier, faster protocols [14]. Each method varies in yield, purity, and efficiency, depending on the specific research requirements (Fig. 1, Table 1).

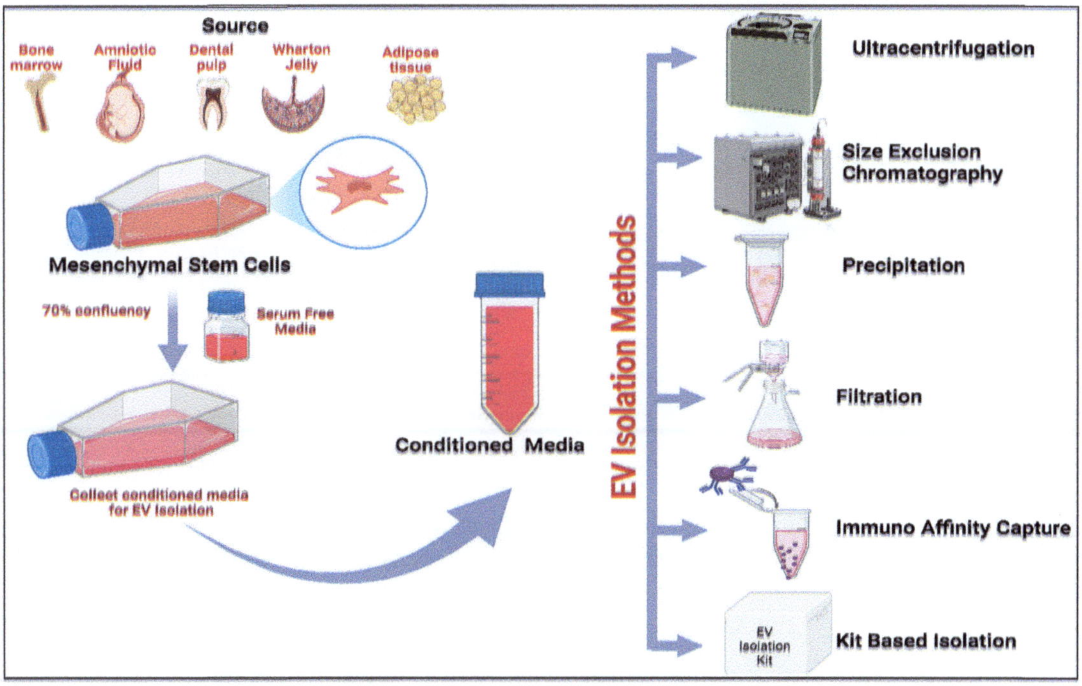

Fig. 1 Methods of isolation of MSC EV from conditioned media

Table 1
Comparison of EV isolation methods

S. No.	Method of EV isolation	Summary of applications	Reference
1	Ultrafiltration, ultracentrifugation, and precipitation	The ultrafiltration method isolates vesicles with a broad size and shape range, lower homogeneity, and reduced functionality compared to precipitation and ultracentrifugation methods. In contrast, precipitation yielded smaller, more homogeneous vesicles, while ultracentrifugation isolated highly homogeneous EVs with a narrow size distribution and preserved functionality.	[14]
2	Combination of PEG precipitation and ultracentrifugation	EVs were efficiently isolated from 10 L of MSC-conditioned media within 24 h using a scalable and high-throughput isolation method.	[18]
3	Ultracentrifugation, Kit Methods (TEIR, ExoQuick-TC and Exo-PREP), and size-exclusion chromatography	EVs with similar morphology, but different sizes and concentrations. The highest EV-yields were obtained with UC, ExoQuick, and Exo-PREP. Similar mean EV size (ExoQuick, UC, and TEIR).	[19]
4	Ultracentrifugation (UC) and sucrose density gradient ultracentrifugation (SUC)	Size of EVs isolated by both the methods was within less than 200 nm. Furthermore, concentration of EVs was higher in the SUC in comparison with the UC method. EVs isolated by SUC demonstrated cup-shaped morphology and were higher in number compared to EVs isolated from UC method.	[20]
5	Tangential flow filtration and ultracentrifugation	Tangential flow filtration (TFF) in combination with 3D MSC cultures improves the yield of EVs (3D-TFF-EVs) seven-fold over 3D-UC-EVs.	[13]

2.1 Ultra-centrifugation

Ultracentrifugation (UC) is a commonly used method for isolating EVs, leveraging high-speed centrifugation to separate them based on size and density [15]. The process typically involves multiple centrifugation steps at different speeds. First, cellular debris is removed by centrifuging at 300 g for 10 min (low speed), followed by centrifugation at 10,000 g for 30 min to eliminate microvesicles (medium speed). The supernatant is then subjected to ultracentrifugation at 100,000 g and 4 °C for 90 min (high speed). After this, the supernatant is discarded and the EV pellet is resuspended in 1× PBS and subjected to a second round of ultracentrifugation at 100,000 g and 4 °C for 90 min to further purify the EVs (Fig. 2) [20–23]. In some protocols, the medium-speed centrifugation is

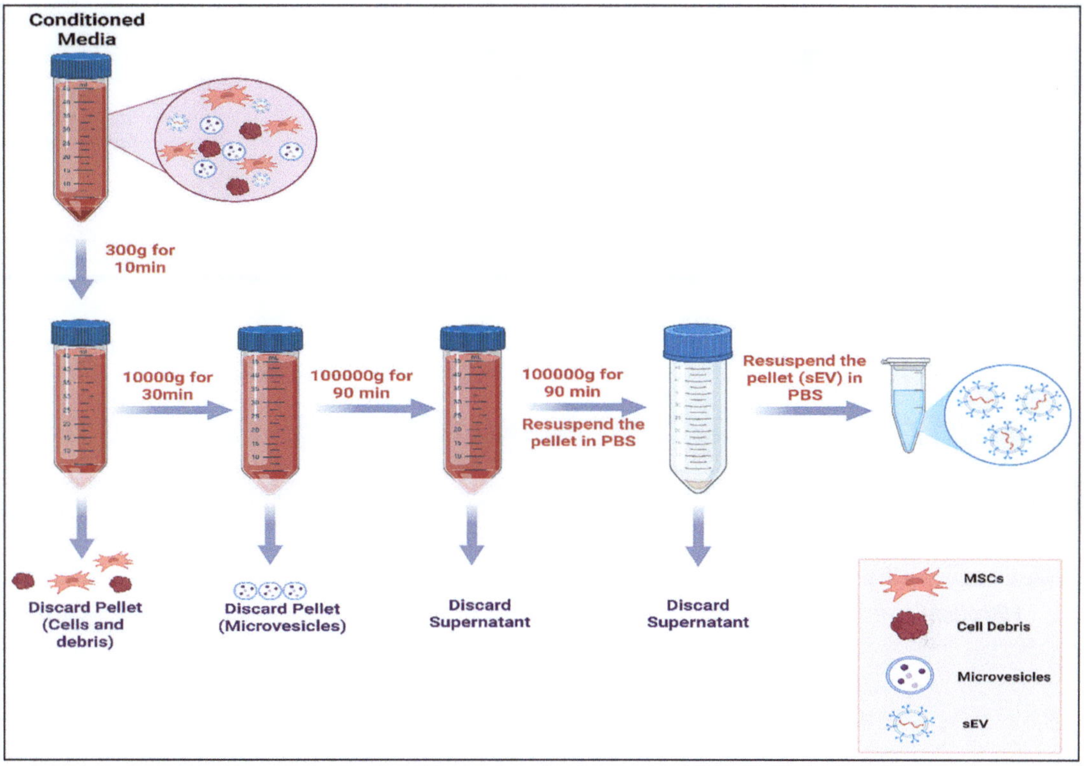

Fig. 2 Isolation of MSC EV by ultracentrifugation

replaced by filtration using a 0.22 μm pore membrane to remove microvesicles. This filter allows only smaller EVs, typically under 200 nm, to pass through, in line with the MISEV Guidelines [8, 24, 25].

It is typically difficult to isolate EVs from MSCs conditioned media, due to a lower yield. Gupta et al., introduced sucrose density gradient ultracentrifugation (SUC) as a method for isolating homogenous populations of EVs from MSC conditioned media [20]. This method follows the same initial steps as traditional UC, but instead of proceeding with high-speed centrifugation, the conditioned media is slowly layered over a 4 mL 30% sucrose solution (prepared in 1× PBS). The mixture is then centrifuged at 100,000 g and 4 °C for 90 min and washed to remove other impurities at 100,000 g for 90 min at 4 °C. The SUC method has been shown to yield a higher particle count than traditional UC when isolating EVs from MSCs. Ultracentrifugation is a simple, cost-effective, and widely used technique that typically produces EVs with less size variability and more uniform shape compared to other methods, such as ultrafiltration and precipitation [14]. However, it is time-consuming, requires specialized equipment, and may result in low yields or contamination with non-exosomal vesicles [26].

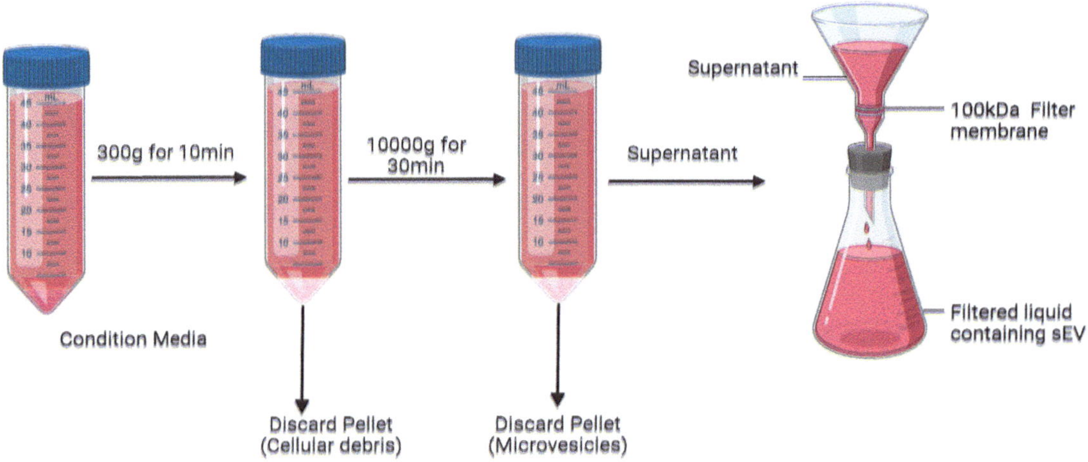

Fig. 3 Isolation of MSC EV by filtration

2.2 Filtration

The ultrafiltration method uses porous membranes/centrifugal filters with different membrane types and pore sizes to trap particles of specific size [14]. One or more filters, most often Amicon filters with a Molecular weight cut-off of 10 kDa or 100 kDa, were used to separate particles larger than 30 nm. In ultrafiltration method, the condition media were centrifugation at 300 × g for 10 min at 4 ° C to remove dead cells. The supernatants were centrifuged at 3000 × g for 30 min at 4 °C to eliminate cellular debris and large microvesicles, and then the media was concentrated and buffer exchanged to PBS using a 100 kDa ultrafiltration column (Fig. 3). The ultrafiltration method isolated vesicles with a wide range of shape and size distribution and lower homogeneity than the samples isolated by the precipitation and ultracentrifugation methods. EV isolated by this method had lower functionality than the EVs purified by other methods. In the view of potential industrialization, this method shows low potential, because of its low scalability and requirement of the use of expensive Amicon filters [14].

Tangential flow filtration uses a membrane filter to separate EVs based on size, while the conditioned media is flowed tangentially over the membrane [27]. Larger particles, like EVs, are retained, while smaller molecules pass through. Conditioned medium was then subjected to ultrafiltration in a TFF system, based on the flow rate and pressure in which the EVs are isolating [13]. Filters were first washed with three times the volume of sterile PBS (pH 7.4) and the biological samples were then processed. The input flow rate and the sheer force need to be maintained to minimize the rupture of EV [28]. Then EVs were concentrated and diafiltrated six times in sucrose buffer (5% sucrose, 50 mM Tris, and 2 mM MgCl). The final EV-solution in sucrose buffer was concentrated to reach a final volume [27, 29]. Concentration was followed by buffer recirculation and diafiltration using an equal volume of PBS as the original conditioned media. Recirculation/diafiltration was repeated for a total of three times [28, 30]. Tangential flow filtration (TFF) offers the advantages of being continuous, scalable, and gentle on EV integrity, making it suitable for isolating EVs from large volumes of conditioned media, but it requires specialized equipment and may not fully eliminate small vesicles or protein contaminants [28].

2.3 Size-Exclusion Chromatography

Size-Exclusion Chromatography (SEC) is a separation technique that isolates EVs based on their size [31]. It uses a column packed with porous beads that allow particles to be separated as they pass through. Larger particles, such as *EVs,* are excluded from entering the pores of the beads and elute first, while smaller particles, such as proteins, lipids, or smaller vesicles, enter the pores and elute later. The separation is based on the *molecular size* and *shape* of the particles, with EVs typically being larger than the other components in conditioned media [32].

The stationary phase or the chromatography column can be packed with a number of gel polymers including crosslinked dextrans (Sephadex), agarose (Sepharose), polyacrylamide (Biogel P), or allyldextran (Sephacryl). The condition media need to be loaded on the top of the column, and elution buffer or PBS is used for the elution. The elute was collected by gravity. The eluate was collected as different fractions (Fig. 4) [16, 33]. The EV were concentrated by using Ultracentrifugation [31, 32]. The column was washed with NaOH for further use.

Size-Exclusion Chromatography (SEC) offers high purity with minimal contamination, preserves EVs integrity without harsh chemicals, and is gentler than ultracentrifugation, but it requires specialized equipment and has lower throughput compared to methods like ultracentrifugation or precipitation [34].

2.4 Precipitation-Based Method

The *precipitation method* for EV isolation is based on the use of chemical agents or polymeric reagents that cause the aggregation and subsequent precipitation of EVs from conditioned media or biological fluids [14]. These reagents work by interacting with the *lipid bilayer* of EVs, facilitating their aggregation into larger particles that can be separated by *centrifugation* [14].

Cell culture supernatants were centrifuged at 300 × g for 10 min at 4 °C to remove dead cells. The supernatants were centrifuged at 3000 × g for 30 min at 4 °C to eliminate cellular debris and large microvesicles, then transferred to a sterile vessel, and an appropriate volume of EVs precipitation solution was added, followed by refrigeration overnight (16 h). After refrigeration, the EVs precipitation/supernatant mixture was centrifuged at 3000 × g for 30 min (Fig. 5). Different types of precipitation agent can used

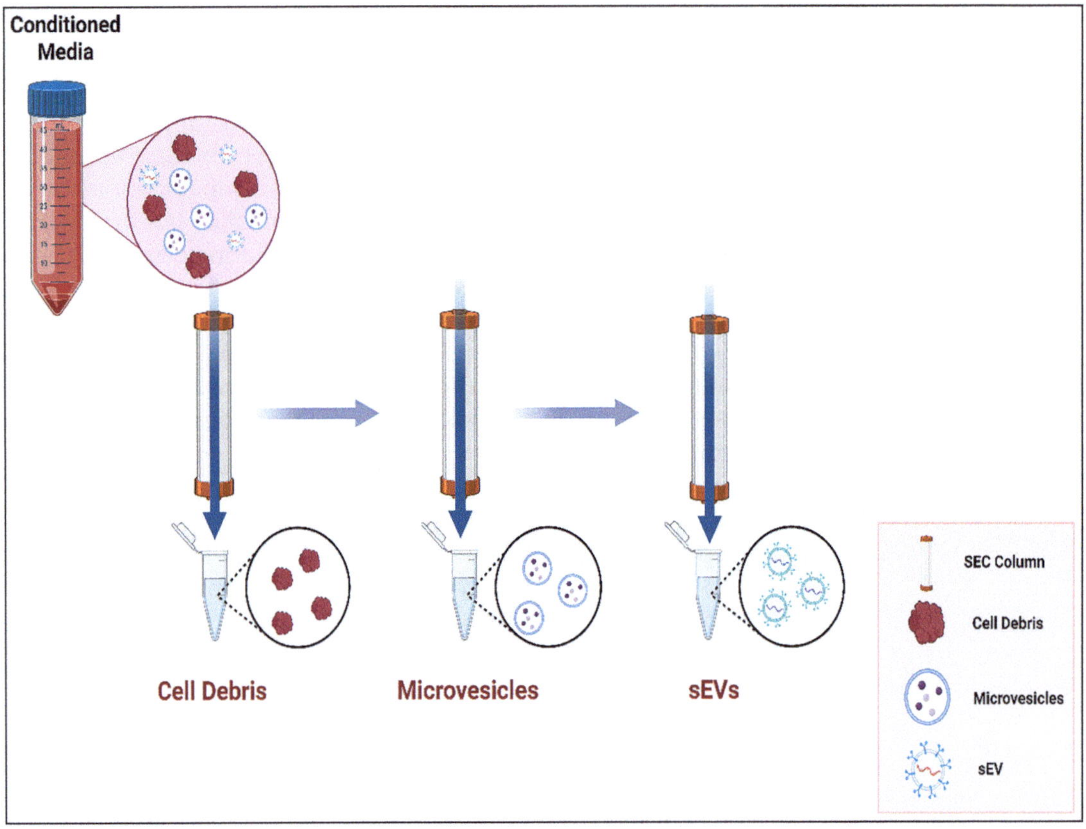

Fig. 4 Isolation of MSC EV by size-exclusion chromatography

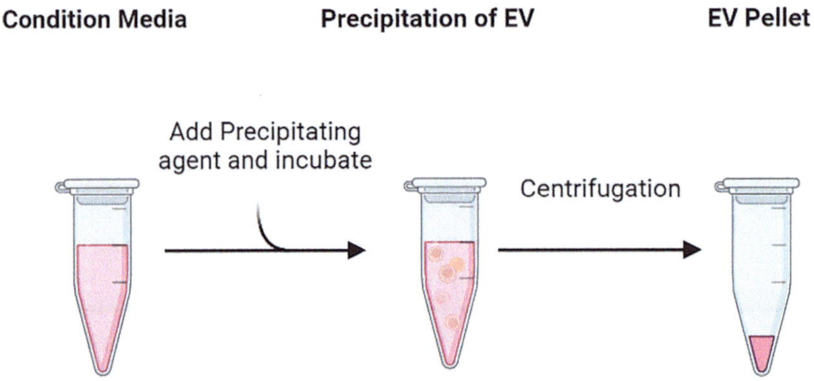

Fig. 5 Isolation of MSC EV by precipitation method

for this method including ExoQuick Exosome Precipitation Solution [35], Polyethylene glycol (PEG) [36], Exocib kit [14], and Polyethylenimine [37].

Precipitation method is typically simpler and faster than other methods like ultracentrifugation, but may also precipitate other

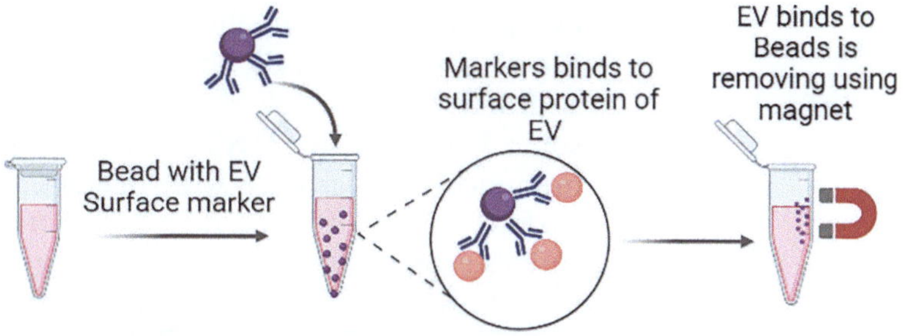

Fig. 6 Isolation of MSC EV by immune affinity capture

non-EV proteins and smaller vesicles, potentially reducing the purity of the EVs preparation [14].

2.5 Immunoaffinity Capture

Immunoaffinity capture for EVs isolation is a highly specific method that utilizes antibodies or ligands that target *surface markers* on EVs. These surface markers, such as *CD9*, *CD63*, *CD81*, or *epithelial cell adhesion molecule (EpCAM)*, are commonly found on EV membranes. By conjugating magnetic beads or other solid supports with these antibodies, EVs can be selectively isolated from complex biological fluids or conditioned media through *affinity binding* [15].

The sEV-containing media were incubated with capture beads for 60 min at room temperature with gentle shaking. Biotinylated antibodies can be used [17]. Using magnets, beads were separated from the supernatant and washed thrice. Further, the beads were gently re-suspended in elution buffer and incubated without mixing for 3 min at RT. Beads were magnetically removed and the supernatant was diluted with PBS and then dialyzed (Fig. 6) [38, 39].

Immunoaffinity capture is highly specific for EVs with known surface markers, allowing for high purity and the isolation of specific subpopulations, but it is limited to EVs that express the targeted markers, can be expensive, requires specialized reagents, and may not capture all EVs types [34, 40].

2.6 Kit Method

Several commercial kits have been developed to simplify EVs isolation, often based on precipitation or immunoaffinity capture techniques. These kits are designed to be user-friendly and reproducible, providing faster isolation protocols compared to traditional methods. Below are some commonly used kits. EVs isolation can be efficiently achieved using various commercial kits, each with distinct advantages and limitations. ExoQuick™ (System Biosciences) and Total Exosome Isolation Reagent (Invitrogen) [12, 41] use a precipitation method, offering simplicity, cost-effectiveness, and fast processing, but they may result in lower purity due to co-precipitation of non-exosomal proteins and lipoproteins. ExoSpin™ (Cell Guidance Systems) utilizes size-exclusion chromatography, providing high-purity EVs without the use of harsh chemicals, though it requires specialized equipment and may be less efficient for large volumes. MagCapture™ Exosome Isolation Kit (Medical & Biological Laboratories) employs immunoaffinity capture with magnetic beads targeting exosomal surface markers, offering high specificity and purity, but it is limited to EVs expressing the targeted markers and can be more expensive [12]. Other kits like miRCURY™ Exosome Isolation Kit (Qiagen) combine precipitation and filtration methods, making them suitable for RNA analysis, but they may sacrifice purity compared to SEC-based methods. Each method or kit is chosen based on the required purity, yield, and intended application, with trade-offs between speed, specificity, and scalability.

2.7 Emerging Methods of EV Isolation

Isolating EVs, including exosomes, from conditioned media poses several challenges, primarily due to contamination with non-EV particles such as proteins, lipoproteins, and microvesicles, which can reduce the purity of the isolated vesicles. The lack of typical surface markers and the existence of other EVs types such as microvesicles commonly lead to co-isolation and impurity of harvested EVs. Methods like precipitation and ultracentrifugation, while commonly used, often result in low yield, loss of EV integrity, and co-isolation of cell debris or aggregates [13]. Additionally, harsh isolation techniques can damage the vesicles, and scalability remains an issue for large-volume isolation [14]. The process can also be time-consuming and labor-intensive, with significant batch-to-batch variability due to differences in media composition or cell culture conditions.

Overcoming these issues requires optimizing protocols, combining different isolation techniques, and developing more efficient, automated, and scalable methods to ensure high-purity and high-quality EVs preparations for research and therapeutic applications [26].

Emerging methods of EV isolation aim to improve purity, yield, scalability, and functionality by utilizing cutting-edge technologies like microfluidics, nanomaterials, acoustic separation, and hybrid methods combining SEC and magnetic capture. While these methods show great potential, they are still under development or require further optimization for widespread adoption.

As the field advances, these technologies are likely to complement or even replace traditional isolation techniques, enabling more efficient and reproducible EVs-based research and therapeutic applications. *Microfluidic-based isolation* uses lab-on-a-chip technology to separate EVs through size-based filtration, immunocapture, or dielectrophoresis, offering high precision and automation with minimal sample volume [42]. *Acoustic-based EVs isolation*, or acoustophoresis, applies ultrasonic waves to separate EVs by size and density in a gentle, non-invasive manner, making it scalable for larger volumes [43, 44]. *Nanomaterials*, such as graphene oxide and gold nanoparticles, selectively capture EVs through electrostatic or chemical interactions, providing high specificity and minimal contamination [45]. *Size-exclusion chromatography (SEC)* combined with *magnetic bead immunocapture* enhances purity by using SEC to filter out larger contaminants before selecting specific EVs populations with surface marker-targeted beads [34]. *Nanoparticle tracking analysis (NTA)* integrated with isolation techniques monitors particle size distribution in real-time to optimize EVs collection [46]. *Magnetic nanoparticles with electrostatic properties* combine magnetic separation and selective binding to EV surface markers, improving scalability and specificity [45]. *Field-flow fractionation (FFF)* isolates EVs based on their size, charge, and shape using a force field, while *SERS-based capture* combines magnetic beads and surface-enhanced Raman spectroscopy for high-sensitivity, rapid detection [47, 48].

Together, these methods offer more efficient, reproducible, and scalable solutions for isolating EVs with minimal contamination, paving the way for advanced diagnostic and therapeutic applications.

3 Characterization of Extracellular Vesicles

Characterizing EVs is crucial for assessing their quantity, confirming their presence, and understanding the contribution of non-EV components within an EV sample [49]. However, this task is made challenging by factors such as the small size and molecular diversity of EVs, the lack of universally accepted identification techniques, and the limitations of many current measurement methods that are not specific to EVs [49]. As a result, no single approach can comprehensively address all the requirements for EV characterization, which calls for the use of complementary techniques.

According to the MISEV Guidelines 2023, various criteria for EV characterization include the quantification of particle number concentration, particle size, total protein, total lipids, total RNA, EV morphology, protein composition, and non-protein markers of EVs [8, 9] (Fig. 7, Table 2).

4 Application of MSC EV in Clinical Research

Numerous studies have investigated the potential of MSC-derived EVs for clinical applications. However, their clinical utility is hindered by several challenges, including low yields from conditioned media, variability in EV populations and cargo influenced by isolation techniques, and inconsistencies in characterization parameters [50]. EVs exhibit heterogeneity in size, content, surface markers, and source, further complicating their isolation and standardization.

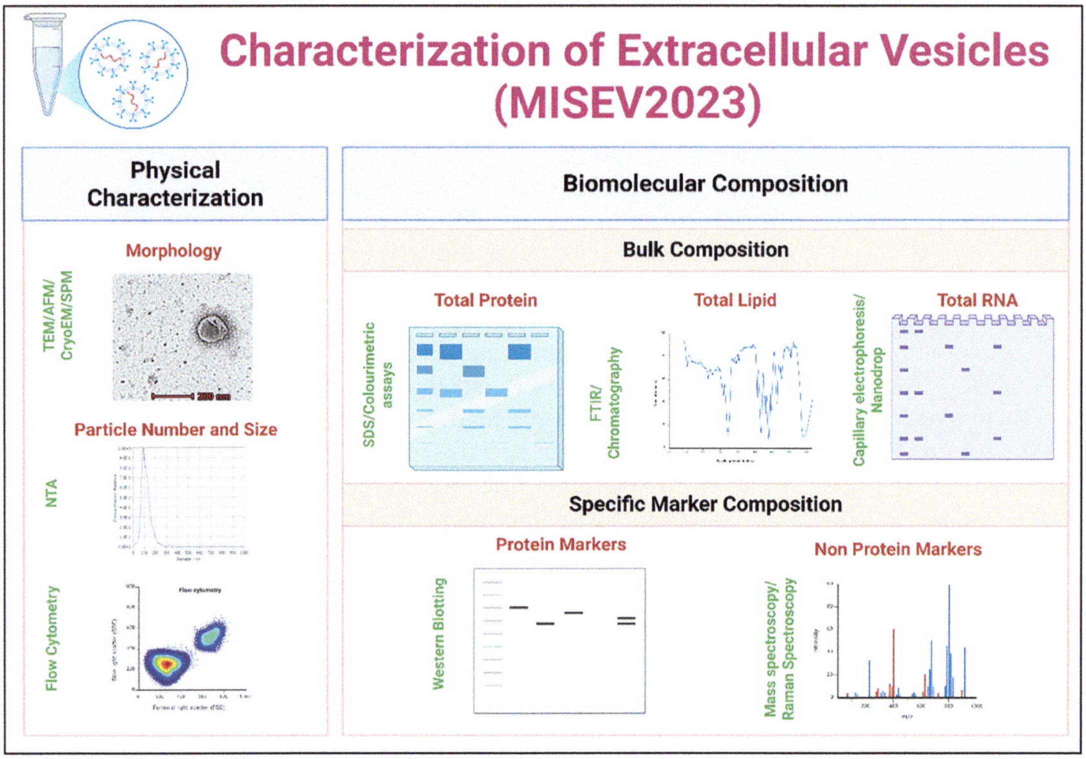

Fig. 7 Characterization of extracellular vesicle as per MISEV2023 guidelines

Table 2
Method of EV characterization

Sl No	Method of characterization	Techniques	Advantage	Limitation
1	Quantification of particle number concentration	Nanoparticle tracking analysis	Precise size distribution and concentration data, real-time particle tracking, nondestructive analysis, high sensitivity for low concentrations, and customizable data collection.	Detecting very small particles, assuming spherical shapes, requires careful sample preparation, struggles with complex samples, lacks compositional information, has difficulty in distinguishing similarly sized particles, and is affected by light scattering interference.

(continued)

Table 2 (continued)

Sl No	Method of characterization	Techniques	Advantage	Limitation
2	Quantification of particle size	Flow cytometry, NTA, RPS, multiangle light scattering, and dynamic light scattering (DLS)	High-throughput, multiplexed analysis of EVs provides detailed information on size, surface markers, and heterogeneity, enabling real-time, highly sensitive measurements of particle size distribution, concentration, molecular weight, and heterogeneity with minimal sample preparation and without the need for labeling or staining.	Complex sample preparation, low sensitivity for small EVs or low concentrations, and challenges in detecting particles smaller than 30 nm, as well as issues with sample heterogeneity, aggregation, and non-exosomal particles, reduced accuracy, particularly for highly polydisperse samples and large aggregates.
3	Quantification of total protein	Colorimetric assays, fluorometric assays, global protein stain on SDS-PAGE	Simple, sensitivity, and the ability to detect a broad range of protein.	Potential interference from sample components, lack of specificity for EVs, and challenges in quantifying heterogeneous populations or small sample amounts limit the accuracy and reliability .
4	Quantification of total lipids	Colorimetric assays, fluorescence of membrane-intercalating dyes, total reflection Fourier-transform infrared spectroscopy (FTIR), or chromatography	Simple and cost-effective, high sensitivity, provides detailed molecular composition data, high resolution for separating EVs.	Lack specificity for EVs, nonspecific binding, requires high concentrations and struggles with complex samples, time-consuming, and less sensitive for low-abundance particles.
5	Quantification of total RNA	Calorimetry and electrophoresis	Simple and offers detailed thermodynamic information about EVs, such as stability, lipid composition, and binding interactions, without the need for labeling.	Requires high-quality RNA extraction, is affected by low RNA yield or contamination, struggles with complex samples, and has limited sensitivity

(continued)

Table 2
(continued)

Sl No	Method of characterization	Techniques	Advantage	Limitation
				for small EVs or low concentrations, while being susceptible to aggregation and interference from non-EV particles.
6	EV morphology	Scanning electron microscopy (SEM), transmission electron microscopy (TEM), cryo-EM, and scanning-probe microscopy (SPM), including atomic force microscopy (AFM)	High-resolution surface morphology, ultrahigh resolution of both surface and internal structures, 3D structural information in the native, hydrated state without fixation, and nondestructive, detailed topographical and mechanical insights with high spatial resolution in solution.	Require extensive and labor-intensive sample preparation, including fixation and thin-sectioning, which can introduce artifacts, are expensive with complex, time-consuming preparation, and have limitations such as slow imaging, limited throughput, and sensitivity to sample preparation and surface properties.
7	Protein composition	Western blotting	High specificity for identifying and quantifying EV-associated proteins.	Time consuming, requires relatively large sample amounts, and can be affected by incomplete transfer, nonspecific binding, or low sensitivity for small EVs.
8	Nonprotein markers	Mass spectrometry or Raman spectroscopy	Proteomic and lipidomic analysis of EVs with high sensitivity, nondestructive, label-free chemical insights with high spatial resolution.	Requires complex preparation, struggles with low-abundance samples, and has limited sensitivity for low-abundance components, while being affected by sample heterogeneity.

Currently available techniques for EV isolation and purification—such as ultracentrifugation, size-exclusion chromatography, and marker-based immunoaffinity methods—yield EVs with varying degrees of purity, specificity, and quantity [49]. For instance, laboratory-scale cultures typically produce approximately 1 μg of EV protein per milliliter of conditioned medium [51]. This yield is insufficient for clinical applications, considering that preclinical studies in rodents often require a dose of 100 μg of EV protein, suggesting an even greater demand for human therapeutic doses [52].

Efforts to enhance MSC-EV production have focused on diverse strategies. For example, Wang et al., 2020, demonstrated that treatments with small molecules such as fenoterol, norepinephrine, N-methyldopamine, mephenesin, and forskolin could increase EV yield by two–six-fold [53]. However, the effects of these treatments on the bioactive molecules encapsulated in the EVs require further investigation to establish their safety and efficacy for human applications. Hypoxic priming has also been shown to enhance EV release by two–five-fold, but no consensus exists regarding the consistent packaging of cargo in hypoxia-primed EVs [53].

Various studies have also evaluated isolation techniques, such as ultracentrifugation, tangential flow filtration, and extrusion, to determine their impact on EV yield and functionality. For instance, Lau et al., 2022, compared ultracentrifugation and extrusion methods for deriving MSC-EVs and vesicles. Their work identified key parameters for GMP-compliant manufacturing, including QA/QC measures such as mycoplasma testing, endotoxin levels, proinflammatory cytokine profiling, and residual benzonase activity [54]. These findings are pivotal in defining standardized clinical processing protocols for MSC-EVs.

A recent study by Nguyen et al., 2024, sought to harmonize MSC-EV production protocols by comparing MSC derivatives from various sources and laboratories [55]. The highlight of this study was the utilization of different protocols and techniques to obtain a standard EV preparation for clinical applications. This study identified CD73, CD90, CD105, and CD44 as robust markers for MSC-derived EVs and highlighted the need for streamlined, reproducible isolation and characterization protocols. These advancements underscore the significant progress being made in the field while highlighting the critical need for further research to standardize protocols, optimize yields, and ensure the safety and efficacy of MSC-derived EVs for clinical applications.

Despite the variability in MSC-derived EV (MSC-EV) production and characterization, numerous clinical trials have demonstrated their potential as safe and effective therapeutic agents. A notable Phase I study by Shi et al. (2021) evaluated the safety profile of allogeneic adipose-derived MSC-EVs in 24 healthy volunteers. Participants received nasal administration of

$2–16 \times 10^8$ EV particles, which were characterized by tetraspanin surface markers and the cytoplasmic marker TSG101 [56]. All subjects tolerated the treatment well, with no adverse reactions reported, establishing the safety of MSC-EVs in healthy individuals.

In another pivotal clinical study addressing severe COVID-19, bone-marrow-derived MSC-EVs (BM-MSC-EVs) were administered intravenously to 24 participants at doses derived from 1 to 10×10^6 MSCs/kg daily over 2 weeks. Remarkably, the study reported a survival rate of 83%, with 71% of patients achieving complete recovery [57]. The therapeutic potential of BM-MSC-EVs has also been investigated in a patient with graft-versus-host disease (GvHD). Treatment resulted in a significant reduction in pro-inflammatory cytokines and notable improvement in clinical symptoms [58].

MSC-EVs have further been evaluated in a variety of clinical contexts, demonstrating promising outcomes across studies. Additional details on these investigations are summarized in Table 3 from clinical trial.gov.

Table 3
Clinical trials using MSC EV

S. No.	NCT ID	Condition/disease	Type of MSC EV	Clinical trial phase	Outcome/current status
1	NCT06245746	Acute myeloid leukemia	UCMSC EV	1	Not yet recruiting.
2	NCT04602104	Acute respiratory distress syndrome	Not mentioned	1&2	Completed, result not published.
3	NCT05354141		BM MSC EV	3	Recruiting.
4	NCT04493242		BM MSC EV	2	The 15 mL dose of ExoFlo was found to be safe in patients with severe or critical COVID-19-associated respiratory failure.
5	NCT06598202	Amyotrophic lateral sclerosis	Umbilical cord-blood–derived MSC EV	1&2	Recruiting.
6	NCT05402748	Anal fistula	Human placental MSC EV	1&2	Not yet recruiting.
7	NCT06568653	Complex perianal fistula	Human placental MSC EV	2	Not yet recruiting.

(continued)

Table 3 (continued)

S. No.	NCT ID	Condition/disease	Type of MSC EV	Clinical trial phase	Outcome/current status
8	NCT04276987	Coronavirus pneumonia	ADMSC EV	1	Completed, result not published.
9	NCT05787288	COVID-19	Not mentioned	1	Recruiting.
10	NCT05191381		Not mentioned	Observational	Recruiting.
11	NCT05871463	Decompensated liver cirrhosis	UCMSC EV	2	Recruiting.
12	NCT05261360	Degenerative meniscal injury	Synovial fluid–derived MSV EV	2	Recruiting.
13	NCT05738629	Dry eye diseases post refractive surgery and associated with blepharospasm	PSC-MSC-EV	1&2	Not yet recruiting.
14	NCT04173650	Dystrophic epidermolysis bullosa	Not mentioned	1&2	Recruiting.
15	NCT06697080	Hair growth in patients with androgenetic alopecia	UCMSC EV	Not mentioned	Completed, result is not published.
16	NCT06677931	Melasma	UCMSC EV	Not mentioned	Recruiting.
17	NCT06599346	Mucosal injuries	Not mentioned	Not mentioned	Recruiting.
18	NCT04356300	Multiple organ dysfunction syndrome after surgical repair of acute type A Aortic dissection	Not mentioned	Not mentioned	Not yet recruiting.
19	NCT06466850	Osteoarthritis	Not mentioned	Not mentioned	Recruiting.
20	NCT06391307	Pilonidal sinus disease	Not mentioned	Not mentioned	Recruiting.
21	NCT05808400	Post-COVID-19 syndrome	UCMSC EV	1	Recruiting.
22	NCT03562715	Pre-eclampsia diagnosis	UCMSC EV	Observational	Completed, results not Published.

(continued)

Table 3 (continued)

S. No.	NCT ID	Condition/disease	Type of MSC EV	Clinical trial phase	Outcome/current status
23	NCT05523011	Psoriasis	Not mentioned	1	Application of 100 μg MSC exosomes/g ointment for 20 days showed better results without adverse effect in all the patients.
24	NCT06536712	Rectal cancer	Human placental MSC EV	1	Not yet Recruiting.
25	NCT05813379	Skin rejuvenation	Not mentioned	1&2	Recruiting.

AD Adipose, *UC* Umbilical Cord, *BM* Bone Marrow, *WJ* Wharton Jelly

These findings underscore the importance of MSC-EVs as a therapeutic tool and pave the way for further research to optimize their clinical translation.

5 Conclusion

MSCs-derived EVs are emerging as leading candidates for cell-free therapeutics, with significant potential in regenerative medicine. These small vesicles have the potential to modulate the physiological state of cells and thereby change the cellular signaling patterns. This chapter has explored the extensive body of research employing diverse protocols for EV isolation and characterization, which underscores their importance as a regenerative modality. These studies also highlight the need for standardized and optimized protocols to ensure uniform isolation and characterization, tailored to both laboratory workflows and clinical applications. Furthermore, these EVs can be incorporated into various formulations by leveraging tissue engineering techniques. For instance, they can be encapsulated into hydrogels or ointments, sprayed as a solution, or freeze-dried for use as a powder, offering versatile approaches to therapeutic delivery.

Acknowledgments

The images are created with Biorender.com. The details of clinical data were derived from clinicaltrials.gov.

Conflict of Interest The authors have no conflicts of interest to declare that are relevant to the content of this chapter.

Financial Support The authors declare no relevance or involvement of any funding source with the subject matter or materials discussed in the manuscript.

References

1. Sharma Y, Mendiratta M, Mohanty S (2024) Therapeutic prospects of stem cell-derived EVs: upcoming avenues, promises, and challenges. In: Ghosh AR (ed) Extracellular vesicles in human health and diseases, vol 1. Springer, Singapore, pp 45–72
2. Sharma Y, Ray S, Mohanty S (2022) MSCs derived extracellular vesicles as a therapeutic paragon for neurodegenerative disorders: a viewpoint. Biocell 46. https://doi.org/10.32604/biocell.2022.018612
3. Sharma Y, Gupta S, Mohanty S (2021) Mesenchymal stem cell-derived exosome as a nano weapon to target the COVID-19 pandemic. Biocell 45. https://doi.org/10.32604/BIOCELL.2021.014621
4. Panda B, Sharma Y, Gupta S, Mohanty S (2021) Mesenchymal stem cell-derived exosomes as an emerging paradigm for regenerative therapy and nano-medicine: a comprehensive review. Life 11:11
5. Sharma Y, Pranshu Rao E, Mohanty S (2022) An insight into the molecular mechanisms of mesenchymal stem cells and their translational approaches to combat COVID-19. In: Stem cells and COVID-19
6. Sharma Y, Mahapatra S, Mohanty S (2025) Tissue-specific mesenchymal stem cells as an emerging modality in cell-based therapy for regenerative medicine. In: Stem cell biology. Cold Spring Harbor Laboratory Press, pp 203–243
7. Sharma Y, Ghatak S, Sen CK, Mohanty S (2024) Emerging technologies in regenerative medicine: the future of wound care and therapy. J Mol Med 102:1425
8. Théry C, Witwer KW, Aikawa E, Alcaraz MJ, Anderson JD, Andriantsitohaina R, Antoniou A, Arab T, Archer F, Atkin-Smith GK et al (2018) Minimal Information for Studies of Extracellular Vesicles 2018 (MISEV2018): a position statement of the International Society for Extracellular Vesicles and Update of the MISEV2014 guidelines. J Extracell Vesicles 7:1535750. https://doi.org/10.1080/20013078.2018.1535750
9. Welsh JA, Goberdhan DCI, O'Driscoll L, Buzas EI, Blenkiron C, Bussolati B, Cai H, Di Vizio D, Driedonks TAP, Erdbrügger U et al (2024) Minimal information for studies of extracellular vesicles (MISEV2023): from basic to advanced approaches. J Extracell Vesicles 13:e12404. https://doi.org/10.1002/jev2.12404
10. Wang BZ, Luo LJ, Vunjak-Novakovic G (2022) RNA and protein delivery by cell-secreted and bioengineered extracellular vesicles. Adv Healthc Mater 11:e2101557
11. Li Y, Chen YH, Liu BY, Nie Q, Li LJ, Duan X, Wu LZ, Chen G (2023) Deciphering the heterogeneity landscape of mesenchymal stem/stromal cell-derived extracellular vesicles for precise selection in translational medicine. Adv Healthc Mater 12. https://doi.org/10.1002/adhm.202202453
12. Patel GK, Khan MA, Zubair H, Srivastava SK, Khushman M, Singh S, Singh AP (2019) Comparative analysis of exosome isolation methods using culture supernatant for optimum yield, purity and downstream applications. Sci Rep 9. https://doi.org/10.1038/s41598-019-41800-2
13. Haraszti RA, Miller R, Stoppato M, Sere YY, Coles A, Didiot MC, Wollacott R, Sapp E, Dubuke ML, Li X et al (2018) Exosomes produced from 3D cultures of MSCs by tangential flow filtration show higher yield and improved

activity. Mol Ther 26. https://doi.org/10.1016/j.ymthe.2018.09.015
14. Ansari FJ, Tafti HA, Amanzadeh A, Rabbani S, Shokrgozar MA, Heidari R, Behroozi J, Eyni H, Uversky VN, Ghanbari H (2024) Comparison of the efficiency of ultrafiltration, precipitation, and ultracentrifugation methods for exosome isolation. Biochem Biophys Rep 38. https://doi.org/10.1016/j.bbrep.2024.101668
15. Greening DW, Xu R, Ji H, Tauro BJ, Simpson RJ (2015) A protocol for exosome isolation and characterization: evaluation of ultracentrifugation, density-gradient separation, and immunoaffinity capture methods. In: Proceedings of the methods in molecular biology, vol 1295. Springer
16. Gámez-Valero A, Monguió-Tortajada M, Carreras-Planella L, Franquesa M, Beyer K, Borràs FE (2016) Size-exclusion chromatography-based isolation minimally alters extracellular vesicles' characteristics compared to precipitating agents. Sci Rep 6. https://doi.org/10.1038/srep33641
17. Sharma P, Ludwig S, Muller L, Hong CS, Kirkwood JM, Ferrone S, Whiteside TL (2018) Immunoaffinity-based isolation of melanoma cell-derived exosomes from plasma of patients with melanoma. J Extracell Vesicles 7. https://doi.org/10.1080/20013078.2018.1435138
18. Börger V, Staubach S, Dittrich R, Stambouli O, Giebel B (2020) Scaled isolation of mesenchymal stem/stromal cell-derived extracellular vesicles. Curr Protoc Stem Cell Biol 55. https://doi.org/10.1002/cpsc.128
19. Antounians L, Tzanetakis A, Pellerito O, Catania VD, Sulistyo A, Montalva L, McVey MJ, Zani A (2019) The regenerative potential of amniotic fluid stem cell extracellular vesicles: lessons learned by comparing different isolation techniques. Sci Rep 9. https://doi.org/10.1038/s41598-018-38320-w
20. Gupta S, Rawat S, Arora V, Kottarath SK, Dinda AK, Vaishnav PK, Nayak B, Mohanty S (2018) An improvised one-step sucrose cushion ultracentrifugation method for exosome isolation from culture supernatants of mesenchymal stem cells. Stem Cell Res Ther 9. https://doi.org/10.1186/s13287-018-0923-0
21. Purushothaman A (2019) Exosomes from cell culture-conditioned medium: isolation by ultracentrifugation and characterization. Methods Mol Biol 1952:233–244
22. Chhoy P, Brown CW, Amante JJ, Mercurio AM (2021) Protocol for the separation of extracellular vesicles by ultracentrifugation from in vitro cell culture models. STAR Protoc 2. https://doi.org/10.1016/j.xpro.2021.100303
23. Chen Y, Qian H, Mak M, Tao Z (2024) Protocol for isolating and identifying small extracellular vesicles derived from human umbilical cord mesenchymal stem cells. STAR Protoc 5: 103197. https://doi.org/10.1016/j.xpro.2024.103197
24. Ramírez-Bajo MJ, Banon-Maneus E, Rovira J, Campistol JM, Diekmann F (2020) Isolation of extracellular vesicles derived from mesenchymal stromal cells by ultracentrifugation. Bio Protoc 10. https://doi.org/10.21769/BioProtoc.3860
25. Morente-López M, Fafián-Labora JA, Carrera M, de Toro FJ, Gil C, Mateos J, Arufe MC (2021) Mesenchymal stem cell-derived extracellular vesicle isolation and their protein cargo characterization. Methods Mol Biol 2259:3–12
26. Théry C, Amigorena S, Raposo G, Clayton A (2006) Isolation and characterization of exosomes from cell culture supernatants and biological fluids. Curr Protoc Cell Biol 30. https://doi.org/10.1002/0471143030.cb0322s30
27. Busatto S, Vilanilam G, Ticer T, Lin WL, Dickson DW, Shapiro S, Bergese P, Wolfram J (2018) Tangential flow filtration for highly efficient concentration of extracellular vesicles from large volumes of fluid. Cells 7. https://doi.org/10.3390/cells7120273
28. Visan KS, Lobb RJ, Ham S, Lima LG, Palma C, Edna CPZ, Wu LY, Gowda H, Datta KK, Hartel G et al (2022) Comparative analysis of tangential flow filtration and ultracentrifugation, both combined with subsequent size exclusion chromatography, for the isolation of small extracellular vesicles. J Extracell Vesicles 11: e12266
29. Chew JRJ, Chuah SJ, Teo KYW, Zhang S, Lai RC, Fu JH, Lim LP, Lim SK, Toh WS (2019) Mesenchymal stem cell exosomes enhance periodontal ligament cell functions and promote periodontal regeneration. Acta Biomater 89. https://doi.org/10.1016/j.actbio.2019.03.021
30. Son JP, Kim EH, Shin EK, Kim DH, Sung JH, Oh MJ, Cha JM, Chopp M, Bang OY (2023) Mesenchymal stem cell-extracellular vesicle therapy for stroke: scalable production and imaging biomarker studies. Stem Cells Transl Med 12. https://doi.org/10.1093/stcltm/szad034
31. Lobb R, Möller A (2017) Size exclusion chromatography: a simple and reliable method for exosome purification. Methods Mol Biol 1660.

https://doi.org/10.1007/978-1-4939-7253-1_9

32. Benedikter BJ, Bouwman FG, Vajen T, Heinzmann ACA, Grauls G, Mariman EC, Wouters EFM, Savelkoul PH, Lopez-Iglesias C, Koenen RR et al (2017) Ultrafiltration combined with size exclusion chromatography efficiently isolates extracellular vesicles from cell culture media for compositional and functional studies. Sci Rep 7. https://doi.org/10.1038/s41598-017-15717-7
33. Contreras H, Alarcón-Zapata P, Nova-Lamperti E, Ormazabal V, Varas-Godoy M, Salomon C, Zuniga FA (2023) Comparative study of size exclusion chromatography for isolation of small extracellular vesicle from cell-conditioned media, plasma, urine, and saliva. Front Nanotechnol 5. https://doi.org/10.3389/fnano.2023.1146772
34. Sidhom K, Obi PO, Saleem A (2020) A review of exosomal isolation methods: is size exclusion chromatography the best option? Int J Mol Sci 21:6466
35. Cheng M, Yang J, Zhao X, Zhang E, Zeng Q, Yu Y, Yang L, Wu B, Yi G, Mao X et al (2019) Circulating myocardial MicroRNAs from infarcted hearts are carried in exosomes and mobilise bone marrow progenitor cells. Nat Commun 10. https://doi.org/10.1038/s41467-019-08895-7
36. Zhang X, Borg EGF, Liaci AM, Vos HR, Stoorvogel W (2020) A novel three step protocol to isolate extracellular vesicles from plasma or cell culture medium with both high yield and purity. J Extracell Vesicles 9. https://doi.org/10.1080/20013078.2020.1791450
37. Djeungoue Petga MA, Taylor C, Macpherson A, Dhadi SR, Rollin T, Roy JW, Ghosh A, Lewis SM, Ouellette RJ (2024) A simple scalable extracellular vesicle isolation method using polyethylenimine polymers for use in cellular delivery. Extracellular Vesicle 3. https://doi.org/10.1016/j.vesic.2023.100033
38. Sharma S, LeClaire M, Wohlschlegel J, Gimzewski J (2020) Impact of isolation methods on the biophysical heterogeneity of single extracellular vesicles. Sci Rep 10. https://doi.org/10.1038/s41598-020-70245-1
39. Hosseinkhani B, van den Akker NMS, Molin DGM, Michiels L (2020) (Sub)populations of extracellular vesicles released by TNF-α –triggered human endothelial cells promote vascular inflammation and monocyte migration. J Extracell Vesicles 9. https://doi.org/10.1080/20013078.2020.1801153
40. Reiner AT, Witwer KW, Van Balkom BWM, De Beer J, Brodie C, Corteling RL, Gabrielsson S, Gimona M, Ibrahim AG, De Kleijn D et al (2017) Concise review: developing best-practice models for the therapeutic use of extracellular vesicles. Stem Cells Transl Med 6:1730
41. Bucan V, Vaslaitis D, Peck CT, Strauß S, Vogt PM, Radtke C (2019) Effect of exosomes from rat adipose-derived mesenchymal stem cells on neurite outgrowth and sciatic nerve regeneration after crush injury. Mol Neurobiol 56. https://doi.org/10.1007/s12035-018-1172-z
42. Gholizadeh S, Shehata Draz M, Zarghooni M, Sanati-Nezhad A, Ghavami S, Shafiee H, Akbari M (2017) Microfluidic approaches for isolation, detection, and characterization of extracellular vesicles: current status and future directions. Biosens Bioelectron 91:588
43. Lee K, Shao H, Weissleder R, Lee H (2015) Acoustic purification of extracellular microvesicles. ACS Nano 9. https://doi.org/10.1021/nn506538f
44. Sridharan B, Lim HG (2023) Exosomes and ultrasound: the future of theranostic applications. Mater Today Bio 19:100556
45. Fang X, Wang Y, Wang S, Liu B (2022) Nanomaterials assisted exosomes isolation and analysis towards liquid biopsy. Mater Today Bio 16. https://doi.org/10.1016/j.mtbio.2022.100371
46. Comfort N, Cai K, Bloomquist TR, Strait MD, Ferrante AW, Baccarelli AA (2021) Nanoparticle tracking analysis for the quantification and size determination of extracellular vesicles. J Vis Exp 2021. https://doi.org/10.3791/62447
47. Guerrini L, Garcia-Rico E, O'loghlen A, Giannini V, Alvarez-Puebla RA (2021) Surface-Enhanced Raman Scattering (Sers) spectroscopy for sensing and characterization of exosomes in cancer diagnosis. Cancers (Basel) 13:2179
48. Giordani S, Marassi V, Placci A, Zattoni A, Roda B, Reschiglian P (2023) Field-flow fractionation in molecular biology and biotechnology. Molecules 28:6201
49. Perocheau D, Touramanidou L, Gurung S, Gissen P, Baruteau J (2021) Clinical applications for exosomes: are we there yet? Br J Pharmacol 178:2375
50. Kim J, Kim EH, Lee H, Sung JH, Bang OY (2023) Clinical-scale mesenchymal stem cell-derived extracellular vesicle therapy for wound healing. Int J Mol Sci 24. https://doi.org/10.3390/ijms24054273
51. Kim JY, Rhim WK, Yoo YI, Kim DS, Ko KW, Heo Y, Park CG, Han DK (2021) Defined

MSC exosome with high yield and purity to improve regenerative activity. J Tissue Eng 12. https://doi.org/10.1177/20417314211008626

52. Jankovic MG, Stojkovic M, Bojic S, Jovicic N, Kovacevic MM, Ivosevic Z, Juskovic A, Kovacevic V, Ljujic B (2023) Scaling up human mesenchymal stem cell manufacturing using bioreactors for clinical uses. Curr Res Transl Med 71:103393

53. Wang J, Bonacquisti EE, Brown AD, Nguyen J (2020) Boosting the biogenesis and secretion of mesenchymal stem cell-derived exosomes. Cells 9. https://doi.org/10.3390/cells9030660

54. Lau H, Han DW, Park J, Lehner E, Kals C, Arzt C, Bayer E, Auer D, Schally T, Grasmann E et al (2022) GMP-compliant manufacturing of biologically active cell-derived vesicles produced by extrusion technology. J Extracell Biol 1. https://doi.org/10.1002/jex2.70

55. Nguyen VVT, Welsh JA, Tertel T, Choo A, van de Wakker SI, Defourny KAY, Giebel B, Vader P, Padmanabhan J, Lim SK et al (2024) Inter-laboratory multiplex bead-based surface protein profiling of MSC-derived EV preparations identifies MSC-EV surface marker signatures. J Extracell Vesicles 13:e12463. https://doi.org/10.1002/jev2.12463

56. Shi M, Yang Q, Monsel A, Yan J, Dai C, Zhao J, Shi G, Zhou M, Zhu X, Li S et al (2021) Preclinical efficacy and clinical safety of clinical-grade nebulized allogenic adipose mesenchymal stromal cells-derived extracellular vesicles. J Extracell Vesicles 10:e12134. https://doi.org/10.1002/jev2.12134

57. Sengupta V, Sengupta S, Lazo A, Woods P, Nolan A, Bremer N (2020) Exosomes derived from bone marrow mesenchymal stem cells as treatment for severe COVID-19. Stem Cells Dev 29. https://doi.org/10.1089/scd.2020.0080

58. Kordelas L, Rebmann V, Ludwig AK, Radtke S, Ruesing J, Doeppner TR, Epple M, Horn PA, Beelen DW, Giebel B (2014) MSC-derived exosomes: a novel tool to treat therapy-refractory graft-versus-host disease. Leukemia 28:970–973

Chapter 9

Large-Scale Isolation of Extracellular Vesicles Using Tangential Flow Filtration Method

Mandeep Kaur, Sowmya Shree Gopal, Zenia Barretto, and Amit K. Srivastava

Abstract

There is an urgent need to develop techniques for isolation to appraise extracellular vesicles (EVs) in diagnostics and therapeutics. One major challenge in the EV field is to improve and standardize methods of EV isolation and characterization to obtain pure and intact EVs on a large scale. Commonly used EV methods, such as ultracentrifugation, size exclusion chromatography, ultrafiltration, immunoaffinity, precipitation, and microfluidics, each have their advantages and limitations. Here, we present a tangential flow filtration (TFF)-based, multistep purification protocol to isolate EVs at high purity from large volumes of cell culture media. Clinical importance necessitates collecting the maximum amount of EVs, and to achieve the diagnostic goal, EVs should be isolated using techniques that provide maximum yield to confirm that the analyzed results are not misleading.

Key words Extracellular vesicles (EVs), Tangential flow filtration (TFF), Conditioned media (CM)-derived EVs, Nanoparticle tracking analysis (NTA)

1 Introduction

Extracellular vesicles (EVs) are a heterogeneous population of small, spherical, bilayer lipid-binding vesicles released into the extracellular environment, which function in intercellular communication [1]. They bear many biological functions and are implicated in several pathologies. EVs also have tremendous potential as biomarkers, therapeutic agents, and vehicles for therapeutic molecules because of their involvement in complex cell-to-cell communication mechanisms [2]. Therefore, the isolation and purification of EVs from biofluids and conditioned media are required.

Despite the considerable potential of EVs, the process of EV isolation poses numerous challenges and remains an area of active research interest. There are several techniques to isolate EVs. The most reliable and widely used method is ultracentrifugation (UC).

Amit K. Srivastava (ed.), *Extracellular Vesicles*, Neuromethods, vol. 228, https://doi.org/10.1007/978-1-0716-4905-3_9,

In differential centrifugation, EVs are isolated based on their size and density through centrifugal force to separate impurities from samples. However, the generation of EV, non-EV aggregates, and EV damage add to its disadvantages [3]. Size-exclusion chromatography (SEC) is based on a single isolation column. Although EV aggregates are absent due to better separation of EVs with high purity and integrity, the contamination by particles of identical size adds to its limitations [4]. Ultrafiltration or microfiltration techniques use porous filters with defined pore sizes and exclusion limits. These methods help eliminate smaller molecules (non-EVs) by flow-through, with minimal aggregation and few ruptured EVs. A major advantage of these methods is production of highly enriched EV fractions at low speeds, while also allowing the collection of smaller EV fractions at higher speeds [5]. However, filter plugging leads to sample loss, and contamination by molecules of the same size is expected [6]. Other affinity-based and precipitation-based techniques also help ensure the purity and integrity of the EV samples [7, 8]. The disadvantages include nonspecific binding, high selectivity, and difficulties with analysis. Protein and organic solvent precipitation methods and commercially available kits such as Total Exosome Isolation Reagent (ThermoFisher Scientific), exoEasy (Qiagen), and miRCURY Exosome Isolation Kit (Exiqon) [9, 10] are also commonly used. However, these methods may not be suitable for large sample volumes because of the high concentration of impurities. Moreover, the kits' components cannot differentiate between different types of EVs, making the analysis challenging. Microfluidics, asymmetrical flow field fractionation, and nanoscale deterministic lateral displacement array are other techniques considered to be reproducible, but their limitation is low yield [11–14]. Therefore, a combination of different techniques is employed to improve EV yield and preservation of integrity.

The objective of EV isolation is to establish a reproducible methodology for generating pure, intact EVs on a large scale. Here, we present tangential flow filtration (TFF)-based protocol, which is a rapid and efficient technique that concentrates and filters EVs simultaneously [15]. TFF has emerged as a powerful tool for the purification and concentration of biomolecules owing to its capacity to efficiently separate components based on size while minimizing sample damage [16]. It is an advanced form of ultrafiltration, utilizing "tangential flow," where the media flows parallel to the membrane, in contrast to the perpendicular flow [17]. This process allows for the effective collection of EVs as filtrate. It uses a semipermeable membrane and a pressure gradient to effectively remove unwanted contaminants while retaining desired EVs [18]. Being a closed system, it gives a high yield of EVs (10–100× more) with more reproducible purity to circumvent problems associated with another techniques [8].

2 Materials

2.1 Conditioned Serum-Free Media

1. 50 mL of cell culture media.
2. 1× Dulbecco's Phosphate Buffered Saline (DPBS) (Gibco, Cat:14190–144): Stored at 4 °C.
3. Syringes.
4. 5 μm Syringe filters.
5. 0.45 μm Syringe filters.
6. Tangential flow filtration (TFF) system.
7. NanoSight NS300 (Malvern Panalytical).

2.2 Tangential Flow Filtration

1. Millipore Labscale Tangential Flow Filtration System.
2. Biomax 300 kDa Pellicon XL filter (Millipore, Billerica, MA).
3. Double-distilled water (ddH_2O).
4. 0.025% Bleach.
5. 1× DPBS: Stored at 4 °C.

3 Methods

3.1 Culture Media (CM) Filtration and Removal of Cell Debris

To make the media free from cells and larger debris, it was divided into equal volumes and centrifuged at 1000 × g for 15 min. The debris and large particles settled at the bottom. The supernatant was first filtered through 0.5 μm syringe filters and then through 0.22 μm syringe filters. The sample should be clear to avoid clogging of the system and filter (*see* **Note 1**).

3.2 Tangential Flow Filtration (TFF)

The system must be cleaned BEFORE and AFTER each sample collection with each of the following solutions in order: 1× DPBS, 0.25% bleach, and double-distilled water. The systematic representation of TFF is shown in Fig. 1.

1. Follow the manufacturer's recommendations and install the filter (300 kDa) to the Millipore-Sigma Flux TFF system. Make sure all connections are tightly closed to prevent leaking (*see* **Note 2**).
2. Precondition the system with 100 mL of 1× DPBS after rinsing the buffer chamber, tubing, and outlets. The lower gauge should be approximately 20 psi; the upper gauge should be 0 psi. Make sure to stop the pump and stirrer at ~15 mL.
3. Add 50 mL of vortexed sample and 450 mL of 1× DPBS into the sample reservoir. Open the valve, turn the pump on and stirrer on. Twist the top gauge clockwise so it reads ~10 psi. It just needs to be greater than 0 psi even if it does not reach

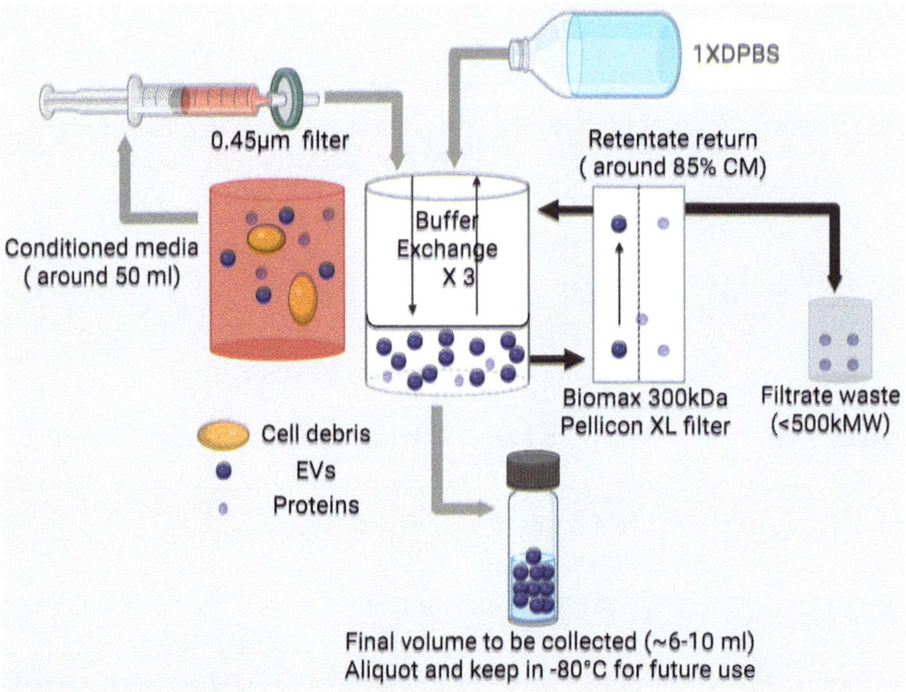

Fig. 1 EV isolation technique—tangential flow filtration and volume reduction

10 psi. But the pressure should be built properly. Once the top pressure gauge reaches 10 psi, the bottom pressure gauge should be increased to 30 psi. Sample must be washed thrice with 1× DPBS.

4. For sample collection, start with the connections. Disconnect connection 3 and place it into a collecting tube. Also, disconnect connection 12 and place it into a waste container, make sure top gauge is completely untwisted (turn counterclockwise). Use a new syringe to collect samples that are stuck in tubes. After collecting the sample (~6–10 mL), reconnect connections. Disconnect connection 2, open valve, and turn on pump and stirrer to collect the remaining sample in the chamber.

5. Once the sample is collected, flush the system. Place 500 mL of distilled water into chamber, open tank valve, disconnect connection 12, and place into waste container. Turn the pump and stirrer on until the feed pressure gauge (bottom pressure) reads 20 psi. Continue pumping into waste container until the volume drops to 350 mL. Turn the pump and stirrer off and reconnect connections. Turn the pump and stirrer back on. Make sure the tank valve is still open and the feed pressure (bottom) should read 20 psi. Twist the top gauge clockwise so

it reads ~10 psi. Continue flushing until the level in the reservoir drops to 50 mL, then drain the system.

6. To clean the system, fill the reservoir with 500 mL of distilled water, open the tank valve, and disconnect connection 12 and place into the waste container. Turn on the pump and stirrer (approximately 20 psi on the bottom pressure gauge). Continue pumping until the reservoir volume reaches 250 mL. Turn off the pump and stirrer, reconnect connection 12. Turn on the pump and stirrer for 15–30 min. If the filtering is slow, this means the filter has been clogged and needs to be cleaned (*see* **Note 3**).
7. Never leave the chamber empty and dry (*see* **Note 4**). For later use, pour 200 mL of double-distilled water into the chamber. Open the valve, turn on the stirrer and pump, let volume drop to 100 mL and leave distilled water in the tank until next use.

3.3 NanoSight NS300

1. The NanoSight NS300 uses nanoparticle tracking analysis (NTA) technology to characterize nanoparticles in liquid. NTA is considered one of the methods to characterize EVs.
2. The concentrated cell culture media-derived EVs were diluted with 1× DPBS with a ratio of 1:10 in a 1 mL syringe and loaded in the instrument. The following parameters were used for measurement: Screen Gain: 7 and Camera Level: 10 in the capture window; Screen Gain:10 and Detection Threshold: 5 in the process window.
3. The infusion rate was set to 1000 for 200 μL and then set to 50 to run the script. Five recordings were taken of 60 s each. After the analysis is completed, the data is exported. The exported data has the excel and pdf file which can be used for the analysis of concentration and size of EVs (*see* Fig. 2).
4. The gasket is cleaned thereafter with around 3 mL of double-distilled water through the machine and with another empty syringe air is passed through gasket to clean the tubing completely (*see* **Note 5**).

4 Notes

1. The sample should be handled carefully while filtering through syringe filters. A large pressure can build up, which can burst the syringe filter. Therefore, it is always good to pass it through a 5 μm filter first and then through a 0.45 μm filter in small volumes.
2. Consult the manufacturer's instructions for proper filter fixing and cleaning procedures. A routine water flux test should be performed to determine filter performance.

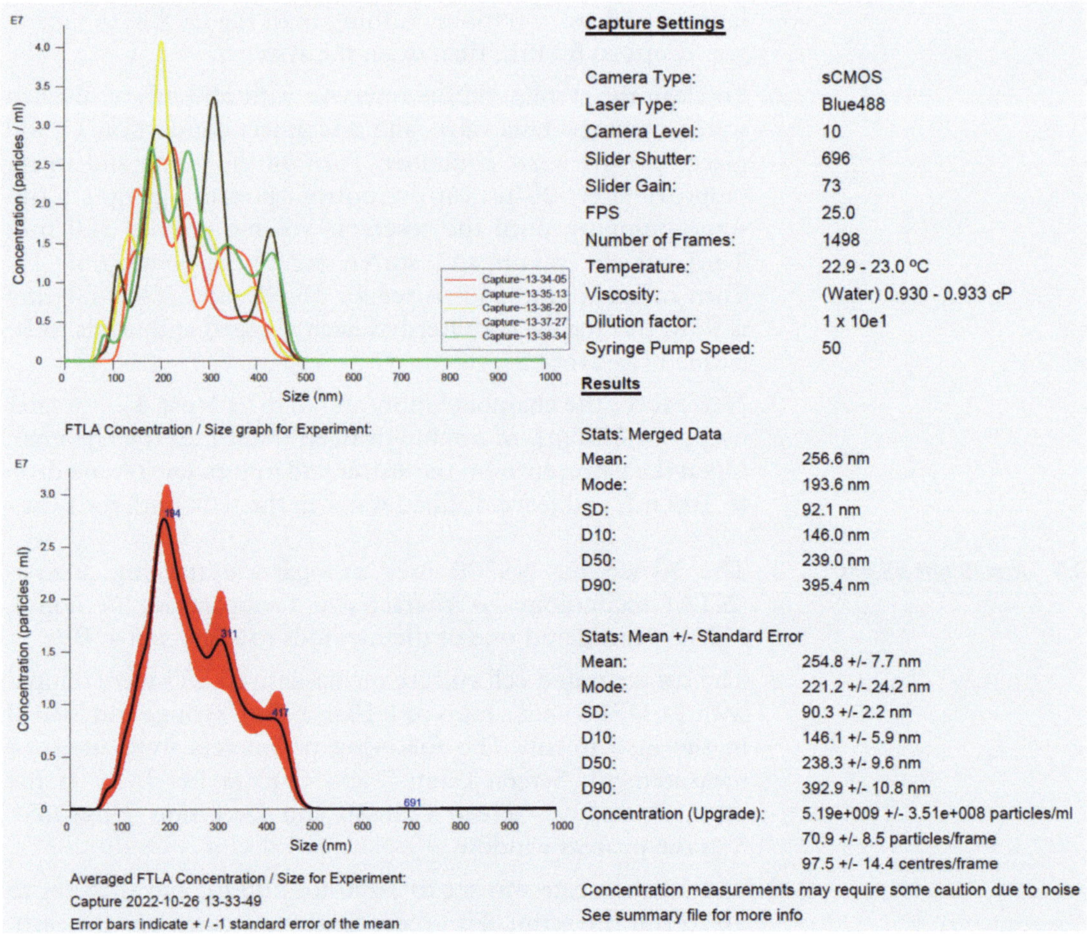

Fig. 2 NTA analysis of conditioned media-derived EVs

3. If the filter becomes clogged, leading to slow filtration, remove filter from the system and inject 20 mL of 0.025% bleach into two opposite diagonal ports. Leave for 30 min then, wash several times with 5 mL of water by injecting into the ports. Also, the system can be cleaned with 500 mL of 0.025% bleach, followed by double-distilled water cleaning.
4. It is good practice to keep the system wet to prevent clogging and maintain its efficiency for long-term use.
5. NTA is a very sensitive machine; therefore, the infusion rate should be in range always. Air should be passed through the machine very carefully and slowly, otherwise it can destroy the whole system.

Acknowledgments

This work was supported by the US Department of Defense award (HT9425-23-1-0138 and W81XWH2110682) to Amit K. Srivastava.

Conflict of Interest None.

References

1. Kaur M, Fusco S, Van den Broek B, Aseervatham J, Rostami A, Iacovitti L, Grassi C, Lukomska B, Srivastava AK (2024) Most recent advances and applications of extracellular vesicles in tackling neurological challenges. Med Res Rev 44:1923
2. van Niel G, Carter DR, Clayton A, Lambert DW, Raposo G, Vader P (2022) Challenges and directions in studying cell–cell communication by extracellular vesicles. Nat Rev Mol Cell Biol 23(5):369–382
3. Lucchetti D, Fattorossi A, Sgambato A (2019) Extracellular vesicles in oncology: progress and pitfalls in the methods of isolation and analysis. Biotechnol J 14(1):1700716
4. Gámez-Valero A, Monguió-Tortajada M, Carreras-Planella L, Franquesa M, Beyer K, Borràs FE (2016) Size-exclusion chromatography-based isolation minimally alters extracellular vesicles' characteristics compared to precipitating agents. Sci Rep 6(1):33641
5. Coughlan C, Bruce KD, Burgy O, Boyd TD, Michel CR, Garcia-Perez JE, Adame V, Anton P, Bettcher BM, Chial HJ (2020) Exosome isolation by ultracentrifugation and precipitation and techniques for downstream analyses. Curr Protoc Cell Biol 88(1):e110
6. Konoshenko MY, Lekchnov EA, Vlassov AV, Laktionov PP (2018) Isolation of extracellular vesicles: general methodologies and latest trends. Biomed Res Int 2018:8545347
7. García-Romero N, Madurga R, Rackov G, Palacín-Aliana I, Núñez-Torres R, Asensi-Puig A, Carrión-Navarro J, Esteban-Rubio S, Peinado H, González-Neira A (2019) Polyethylene glycol improves current methods for circulating extracellular vesicle-derived DNA isolation. J Transl Med 17:1–11
8. Heath N, Grant L, De Oliveira TM, Rowlinson R, Osteikoetxea X, Dekker N, Overman R (2018) Rapid isolation and enrichment of extracellular vesicle preparations using anion exchange chromatography. Sci Rep 8(1): 5730
9. Skottvoll FS, Berg HE, Bjørseth K, Lund K, Roos N, Bekhradnia S, Thiede B, Sandberg C, Vik-Mo EO, Roberg-Larsen H (2018) Ultracentrifugation versus kit exosome isolation: nanoLC–MS and other tools reveal similar performance biomarkers, but also contaminations. Future Sci OA 5(1):FSO359
10. Helwa I, Cai J, Drewry MD, Zimmerman A, Dinkins MB, Khaled ML, Seremwe M, Dismuke WM, Bieberich E, Stamer WD (2017) A comparative study of serum exosome isolation using differential ultracentrifugation and three commercial reagents. PLoS One 12(1): e0170628
11. Zhang H, Lyden D (2019) Asymmetric-flow field-flow fractionation technology for exomere and small extracellular vesicle separation and characterization. Nat Protoc 14(4): 1027–1053
12. Gholizadeh S, Draz MS, Zarghooni M, Sanati-Nezhad A, Ghavami S, Shafiee H, Akbari M (2017) Microfluidic approaches for isolation, detection, and characterization of extracellular vesicles: current status and future directions. Biosens Bioelectron 91:588–605
13. Iliescu FS, Vrtačnik D, Neuzil P, Iliescu C (2019) Microfluidic technology for clinical applications of exosomes. Micromachines 10(6):392
14. Smith JT, Wunsch BH, Dogra N, Ahsen ME, Lee K, Yadav KK, Weil R, Pereira MA, Patel JV, Duch EA (2018) Integrated nanoscale deterministic lateral displacement arrays for separation of extracellular vesicles from clinically-relevant volumes of biological samples. Lab Chip 18(24):3913–3925

15. Yuan R, Zhou Y, Arias GF, Dittmer DP (2023) Extracellular vesicle isolation by a tangential-flow filtration-based large-scale purification method. In: Cell-secreted vesicles: methods and protocols. Springer, pp 45–55
16. Kawai-Harada Y, Nimmagadda V, Harada M (2024) Scalable isolation of surface-engineered extracellular vesicles and separation of free proteins via tangential flow filtration and size exclusion chromatography (TFF-SEC). BMC Methods 1(1):9
17. Visan KS, Lobb RJ, Ham S, Lima LG, Palma C, Edna CPZ, Wu LY, Gowda H, Datta KK, Hartel G (2022) Comparative analysis of tangential flow filtration and ultracentrifugation, both combined with subsequent size exclusion chromatography, for the isolation of small extracellular vesicles. J Extracell Vesicles 11(9):12266
18. Busatto S, Vilanilam G, Ticer T, Lin W-L, Dickson DW, Shapiro S, Bergese P, Wolfram J (2018) Tangential flow filtration for highly efficient concentration of extracellular vesicles from large volumes of fluid. Cells 7(12):273

Chapter 10

Optimized Buffer Conditions for Storage and Recovery of Extracellular Vesicles

André Görgens and Samir El Andaloussi

Abstract

Extracellular vesicles (EVs) play a key role in many physiological and pathophysiological processes and hold both therapeutic and diagnostic promise. However, the broad definition of storage conditions to ensure EV stability remains unclear. In recent years, we and others have explored different concepts and defined protocols aiming to stabilize EVs and ensure reproducible EV recovery. Despite these advances, the assessment of EV stability remains a challenge due to the multimodal composition of EVs and the variety of their biological functions. Here, we summarize and discuss the current state-of-the-art and describe selected experimental procedures that have been utilized to stabilize EVs and to assess their stability and recovery, with a focus on cell-culture-derived EVs.

Key words Extracellular vesicles, Exosomes, Storage, Stability, Formulation, Reference material, Imaging flow cytometry, EV flow cytometry

1 Introduction

Over the past decade, extensive research has explored the biology, roles, and clinical potential of extracellular vesicles (EVs) [1–9]. Despite advances in improving methods, data reporting standardization, and procedures related to EV isolation and characterization in the field, many key aspects concerning EVs remain challenging, including the definition of procedures to stabilize EVs [8, 10–15]. Ensuring EV stability is crucial for consistent data across basic research, therapy development, and diagnostic applications. Instability or loss of EVs during sample storage impacts EV dosing and downstream quantitative and functional assessment in cell-based assays and animal models, as well as the detection of EV subsets in biomarker discovery studies, thereby hindering reliable data interpretation [16, 17].

Initial studies investigating the impact of storage conditions on EV stability and recovery have demonstrated that both the storage conditions and the buffer used can impact the biophysical stability,

Amit K. Srivastava (ed.), *Extracellular Vesicles*, Neuromethods, vol. 228, https://doi.org/10.1007/978-1-0716-4905-3_10,

particle recovery, diameter, and function of EVs [5, 6, 8, 18–28]. The use of phosphate-buffered saline (PBS) for EV storage and freezing EV samples at −80 °C has been commonly reported [24, 29]. However, other studies have also investigated the effects of storage temperature, storage containers, and cryoprotectants like DMSO [30] or trehalose [19], and additional studies have explored lyophilization [31]. Notably, these studies mostly focused on EVs derived from specific biofluid samples and/or used downstream readouts to assess only partial aspects of EVs, such as small RNA quantification or bulk protein content [20, 24, 32–34]. In addition, different EV isolation procedures were used between those studies, which are known to affect EV composition, recovery, and purity [17, 35–38]. Altogether, these studies highlight the need for more systematic studies on EV stability.

In a more recent study, we have compared the effects of various storage buffers supplemented with various additives on short-term and long-term stability and recovery of engineered cell-culture-derived EVs from multiple cellular sources and at different temperatures. For different timepoints up to 2 years, we have compared EV diameter, fluorescence stability as a surrogate for EV intactness, nucleic acid and protein content, EV surface protein composition, and EV function in different readouts. We demonstrated that storage of EVs in PBS leads to reduced EV concentrations in a time-dependent manner, detectable already within 1 day, at all temperatures tested. The highest EV recovery was observed for EVs stored at −80 °C in PBS-HAT buffer (DPBS supplemented with HEPES, human albumin, and trehalose), which also facilitated stable recovery of EVs after several freeze-thaw cycles [17]. Notably, we did not obtain any evidence for EV damage, disruption, or fusion, but identified surface adsorption of EVs to plastic or glass surfaces as the main source for the time-dependent decrease in recovery upon EV storage in buffers devoid of protein and sugar additives. Similar results were obtained when using tubes with functionalized "low-binding" surfaces. These findings were consistent with our observation that storage of more pure and more diluted preparations resulted in generally reduced recovery of EVs and also liposomes [17]. These results have been independently reproduced by another group that was using the same PBS-HAT buffer formulation, with the exception that they replaced human with bovine albumin [39]. These observations imply that EV-containing samples are stable and can be efficiently recovered from containers after being exposed to plastic or glass surfaces as long as the surrounding solution contains a sufficient concentration of sugars and proteins that likely saturate the respective surface of storage containers, thereby minimizing EV adsorption. This hypothesis is further supported by the observation that body fluid EVs appear to be more stable in the body fluid compared to storage as purified EVs over time [14, 15].

Importantly, sensitive single EV analysis methods allowing detection of EVs are crucial to accurately quantify the concentration of EVs in each sample, which in turn is essential information to investigate the loss of EVs in different buffer conditions, at different temperatures, and at different purities and concentrations. In this context, optimized workflows for utilizing single EV imaging flow cytometry (IFCM) have been instrumental for addressing such questions in our lab. Here, we highlight the utility of IFCM in combination with using fluorescently labeled reference EVs and describe selected experimental procedures we have used to stabilize EVs and investigate EV stability and recovery.

2 Materials

2.1 Cell Lines

For production of fluorescently tagged reference EVs, we used HEK293 Freestyle suspension cells (HEK293FS; ThermoFisher Scientific) stably transduced with lentiviral constructs encoding for the CD63-mNeonGreen (CD63-mNG) fusion protein (HEK293FS:CD63mNG) [40]. To produce conditioned medium containing green fluorescent EVs, cells were cultured in FreeStyle 293 Expression Medium (ThermoFisher Scientific) in 125 mL polycarbonate Erlenmeyer flasks (Corning) in a shaking incubator (Infors HT Minitron) according to the manufacturer's instructions. Cells were grown at 37 °C, 5% CO_2 in a humidified atmosphere and regularly tested for the absence of mycoplasma.

2.2 Preparation of EV Storage Buffer (PBS-HAT)

The identification of PBS-HAT as a buffer stabilizing EV preparations has been described previously [17]. PBS-HAT contains Dulbecco's PBS (DPBS, no calcium, no magnesium, ThermoFisher) supplemented with (working concentrations) 25 mM HEPES [4-(2-hydroxyethyl)-1-piperazineethanesulfonic acid, ThermoFisher], 0.2% w/v human serum albumin (HSA, albunorm injection solution, Octapharma AB), and 25 mM trehalose [D-(+)-Trehalose dihydrate, Sigma-Aldrich]. For added experimental flexibility and to avoid diluting samples unnecessarily, we regularly prepare 10× PBS-HAT stock solutions that can be diluted in DPBS to prepare 1x PBS-HAT, i.e., (1×) DBPS supplemented with 250 mM HEPES, 2% HSA, and 250 mM trehalose (commercially available as Vesi-Safe from Vesiculab). The final buffer solution is filtered through 0.22 μm filters (ThermoFisher) and optionally through 0.1 μm PES low protein-binding filters (Sigma-Aldrich) to ensure sterility and to reduce particle counts. Aliquots of 10× PBS-HAT stock solutions appear stable at 4 °C for at least 6 months.

2.3 Preparation of Fluorescent Reference EVs

For specific detection and quantification of EVs by high-sensitivity single EV flow cytometry, it is essential to label them fluorescently to facilitate the identification of fluorescent EVs from non-EV particles. Here, we used stable HEK293FS:CD63mNG cells to prepare green fluorescent CD63-mNG EVs as biological reference material (described in Refs. [40, 41], commercially available as Vesi-Ref CD63-mNG, Vesiculab). Green fluorescent reference EVs can be prepared from conditioned medium of genetically engineered producer cells by the EV isolation method of choice [42]. Here, 500 mL conditioned media from HEK293FS:CD63mNG cells cultured at a density of 0.5–4 × 10^6 cells/mL for 72 h was harvested, precleared from cells and debris by a low-speed centrifugation step (700 × g for 5 min, RT), and supernatant was then subjected to centrifugation at 2000 × *g* for 20 min at RT to remove larger particles and debris. Next, precleared cell-culture supernatant was filtered through 0.22 μm bottle-top vacuum filters (Corning, cellulose acetate, low protein binding) to remove any larger particles. EVs were concentrated by tangential flow filtration (TFF) with a KR2i TFF system (SpectrumLabs) equipped with modified polyethersulfone (mPES) hollow fiber filters with 300 kDa membrane pore size (MidiKros, 370 cm^2 surface area, SpectrumLabs) at a flow rate of 100 mL/min (transmembrane pressure at 3.0 psi and shear rate at 3700 s^{-1}) as previously described [43, 44]. The sample was then loaded onto a bind-elute size exclusion chromatography (BE-SEC) column (HiScreen Capto Core 700 column, GE Healthcare Life Sciences), using an ÄKTAstart chromatography system (GE Healthcare Life Sciences) and collected according to the 280 nm UV absorbance chromatogram [43]. Finally, EVs were concentrated to a final volume of 500 μL by using Amicon Ultra-15 10 kDa molecular weight cut-off spin-filters (Millipore). EVs were immediately supplemented with 10% (v/v) of 10× PBS-HAT stock solution and stored in 1.5 mL polypropylene tubes (Axygen Maxymum Recovery, Corning). Following EV characterization by nanoparticle tracking analysis (NTA) (as described in Ref. [17], estimated mode diameter 138 nm), EV samples were diluted with 1x PBS-HAT buffer to a NTA-based concentration of 2 × 10^{10} particles/mL, and 100 μL aliquots were stored in Maxymum Recovery polypropylene 1.5 mL tubes at −80 °C until usage.

2.4 Flow Cytometer and Preparation for EV Measurements

To facilitate detection of fluorescent EVs by high-sensitivity single EV flow cytometry, we use an imaging flow cytometer (Cytek Amnis Cellstream equipped with 405, 488, 561, and 642 nm lasers) and base our analysis on previously optimized parameters and settings using a Cytek Amnis ImageStream X MkII instrument [42]. IFCM measurements of EVs with a Cellstream instrument are regularly performed in our lab from 96-well U-bottom or V-bottom multiwell plates (ThermoFisher Scientific) by using the instrument's plate reader with flow rate set to slow, FSC turned off,

SSC laser set to 40%, and all other lasers set to 100%. All instrument buffers, including rinse and wash buffers as well as sheath fluid [PBS, Dulbecco's PBS pH 7.4 (Gibco)] are filtered through 0.22 μm and optionally 0.1 μm filters to minimize particle counts and background in the system. All buffers are replaced regularly. Each day, the instrument is prepared for EV measurements by performing two cycles of startup/clean procedures by default. Afterwards, instrument cleanliness and background events are assessed by running filtered DPBS as a sample. Instrument QC is run daily as recommended by the manufacturer, and fluorescence calibration of the instrument is performed every few months, or after lasers have been realigned by service engineers, as described before [42, 45]. In brief, FITC MESF beads (Quantum FITC-5 MESF, Bangs Laboratories) and APC MESF beads (Quantum APC MESF, Bangs Laboratories) with known absolute fluorescence values for each bead population are acquired with the same settings used for EV measurements, with the exception that the SSC laser is turned off to avoid camera saturation, and linear regressions are performed to convert arbitrary fluorescence values into FITC/APC absolute fluorescence values in MESF units, respectively. Flow cytometric dot plots using MESF unit axes are created with FlowJo v10.5.3 (FlowJo, LLC).

3 Methods

3.1 Stabilizing EV Samples with PBS-HAT Buffer

EVs can be prepared by a variety of EV isolation procedures. Isolating EVs at reproducible quality, with high recovery, and at sufficient quantities are the key aims when designing EV preparation experiments [35, 41, 43, 46]. It is also essential that EVs can be quantified postisolation to estimate comparable doses for downstream cell-based assays, animal treatments, or for analytical purposes [16, 17, 45, 47]. One of the most common methods to perform such a quantitative assessment of the sample is nanoparticle tracking analysis (NTA), even though the obtained data are not specific for EVs [25, 48]. However, there is a plethora of other EV characterization methods available, each with its own features and limitations, including single EV flow cytometry [9, 41, 42, 47, 49, 50]. Once EVs have been prepared, aliquoted, and characterized, it is further critical that EVs taken from respective aliquots at later time points have the same quality and quantity as freshly prepared and assessed samples. We have previously shown that the recovery of EVs stored in PBS at different temperatures is decreasing in a time-based manner, and that this decrease critically depends on the concentration of EVs in a sample and their purity, with more pure and more diluted samples showing drastically decreased recovery rates [17]. This time-dependent reduction of EV recovery can be prevented by storing the final samples in PBS-HAT buffer, preferentially at −80 °C [17]. According to our data, even diluted samples at concentrations around 2–3 × 10^6 EVs/mL are stabilized in PBS-HAT, while for the same samples stored in PBS for 7 days the EV recovery is drastically reduced with more than 99% loss of sample due to surface adsorption, also in several "low binding" tubes tested [17]. Based on this data, the tube type and brand has a limited impact on EV recovery, however, we recommend being mindful about not mixing different tube types within one experiment. Practically, PBS-HAT buffer is regularly prepared as 10× concentrated stocks, which provide the flexibility to either use 1× PBS-HAT (e.g., to dissolve EV pellets after ultracentrifugation) or 10× PBS-HAT (e.g., to add 10% volume to EV samples prepared in PBS by ultrafiltration) to stabilize final EV samples (Fig. 1).

3.2 Evaluation of EV Stability and Recovery Using Imaging Flow Cytometry

The stability of an EV sample depends on a variety of different factors, including the surrounding buffer or medium, the storage temperature, the purity of EVs, as well as the concentration of EVs and their surface exposure [17, 38]. It is challenging to test for changes concerning all potential EV features such as biophysical stability, molecular cargo, and function of EVs since this requires several analytical methods. However, given that surface adsorption was identified as the main contributor for reduced EV recovery over time for cell-culture-derived EV samples [17], a good starting point for evaluating the stability of an EV sample is to accurately measure the EV concentration in respectively differently stored samples. While the EV concentration can be estimated by different methods, we regularly use optimized high-sensitivity single EV imaging flow cytometry for quantification of fluorescently labeled EVs [12, 42, 45, 49, 51]. The key advantages comprise sensitive and reproducible quantification of EVs regardless of sample purity as well as very low requirements in terms of EV concentrations required for IFCM analysis, i.e., in the range of 1 × 10^6 to 1 × 10^7 EVs/mL. IFCM requires fluorescent labeling of EVs, which can be achieved by utilizing fluorescently labeled antibodies [45] or fluorescently tagged EVs from engineered cell lines [42]. Here, we use mNeonGreen(mNG)-tagged EVs derived from a suspension cell line genetically engineered to express a CD63-mNG construct, which leads to efficient incorporation of mNG green fluorescent proteins into EVs, which can be utilized as biological reference material and tool to track EVs in different assays and processes [40, 41, 52]. To evaluate the impact of different buffer formulations, here PBS and PBS-HAT, on EV stability and recovery, CD63-mNG EVs were serially diluted 20,000-fold from an NTA-based concentration of 2 × 10^{10} to a final concentration of 1 × 10^6 particles/mL, either in PBS or PBS-HAT. Samples were then incubated either for 30 min or 24 h at 4 °C and analyzed by single EV IFCM (Fig. 2a). Fluorescent EVs can reliably and quantitatively be identified over buffer background for each condition

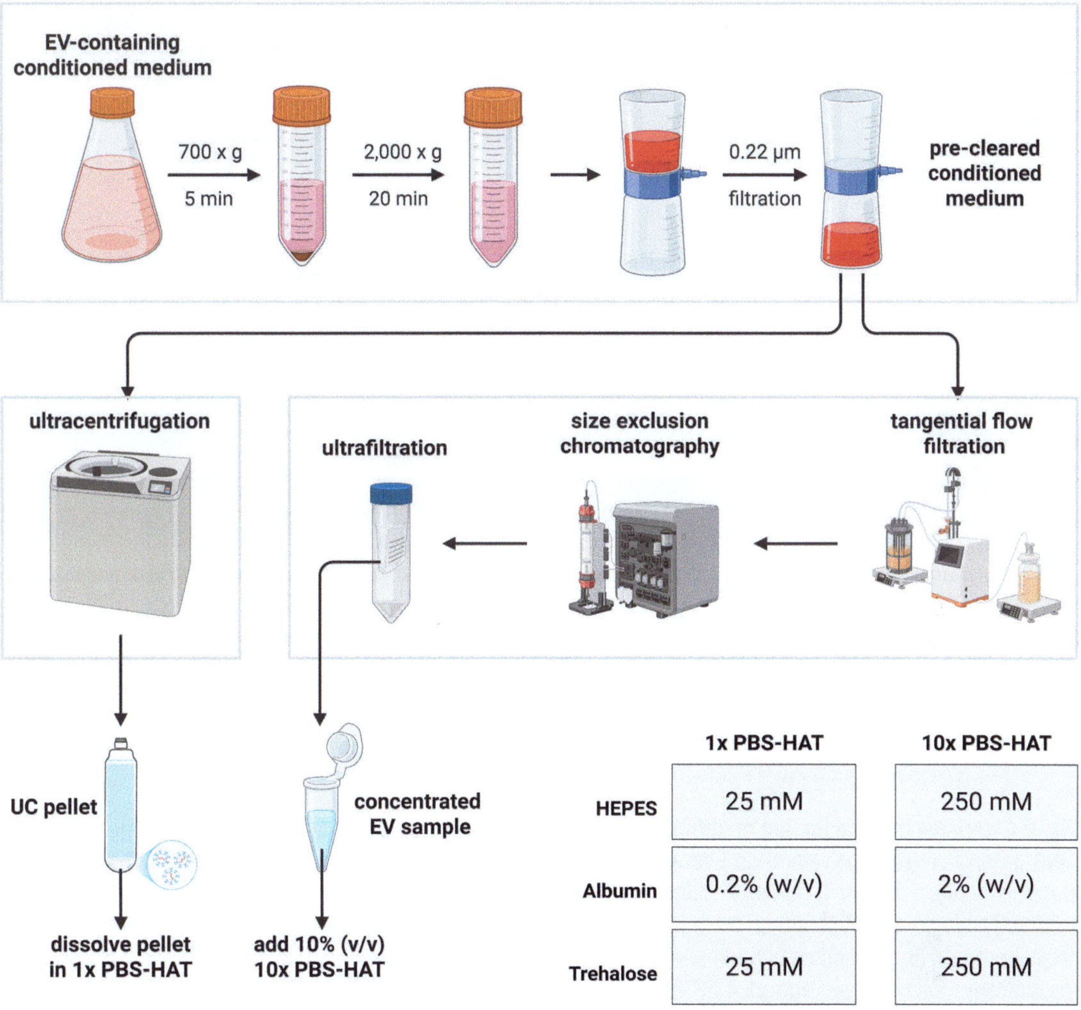

	1x PBS-HAT	10x PBS-HAT
HEPES	25 mM	250 mM
Albumin	0.2% (w/v)	2% (w/v)
Trehalose	25 mM	250 mM

Fig. 1 EVs can be prepared from conditioned media by following different isolation procedures. Here, two example workflows regularly used in our lab are depicted, i.e., differential centrifugation followed by ultracentrifugation (UC) or by a combination of tangential flow filtration, size exclusion chromatography, and ultrafiltration (TFF-SEC) [18, 42, 44, 53, 54]. As a common first step, conditioned media from cell culture is precleared by sequential slow speed centrifugation and subsequent 0.22 μm filtration as indicated. According to our experience, EVs stay functional and stable in conditioned media for 2–3 weeks at 4 °C. Depending on the EV isolation procedure applied, final EV samples are prepared by either dissolving the pellet in a suitable volume of 1× PBS-HAT (UC) or by adding 10% (v/v) of 10× concentrated PBS-HAT to the final sample (TFF-SEC) after the ultrafiltration step. Respective concentrations of additives added to DPBS for 1× and 10× PBS-HAT solutions are provided in the embedded table

and timepoint as SSC (low) fluorescence-positive events (Fig. 2b). Since samples here have been acquired for a defined amount of time, and since the volumetric flow rate of the instrument is known (~3.4 μL/min), IFCM provides accurate concentration measurements for EV subsets, here expressed in mNG+ events/μL (Fig. 2c). In this experiment, we observe a clearly reduced EV

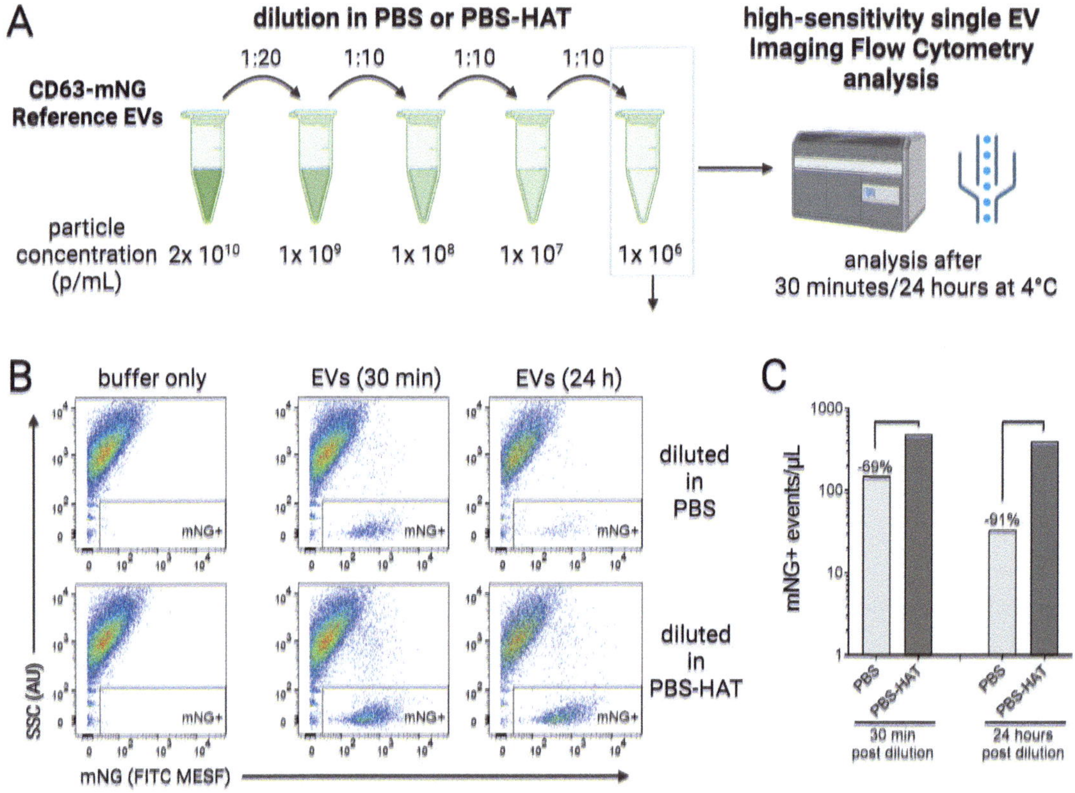

Fig. 2 Comparison of the impact of different dilution buffers on EV recovery and stability over time. (**a**) CD63-mNG green fluorescence reference EVs were diluted serially either in PBS or PBS-HAT in 10–20-fold dilution steps to achieve a final NTA-based concentration of 1 × 10^6 particles/mL, which was analyzed by high-sensitivity single EV IFCM either 30 min or 24 h postdilution. (**b**) Representative dotplots depicting IFCM measurements (5 min acquisition/sample) of buffer controls and CD63-mNG EVs diluted in PBS or PBS-HAT, respectively, following storage at 4 °C for 30 min or 24 h. (**c**) Quantification of data shown in B based on counts recorded per volume for each sample

concentration in the sample diluted in PBS compared to the sample diluted in PBS-HAT already 30 min after dilution, with a reduction of the measured concentration of 69%. After storing the EVs in both buffers for 24 h, an even more drastically reduced recovery was observed with a more than 10-fold reduced EV concentration obtained for the sample diluted and stored in PBS compared to the sample in PBS-HAT (Fig. 2b, c). The results of this dataset are consistent with previous observations [17].

4 Conclusions

The importance of stabilizing EV preparations—and nanoparticle preparations in general—is a priority of the field, and the combination of more quantitative and sensitive methods such as high-sensitivity single EV IFCM with defined biological reference materials such as mNG EVs has facilitated a more detailed experimental assessment of EV stability and recovery within recent years [14, 15]. Our previous comprehensive study [17] and the data shown here highlight that purified EVs are rapidly adsorbed upon surface exposure, which is an important consideration for research and ultimately therapy development. Notably, all components of PBS-HAT are available at clinical grade and/or already used clinically. However, it should be noted that EVs for therapeutic applications would likely be prepared at much higher concentrations, implying that the relative loss of EVs due to surface adsorption will likely be negligible in certain cases. Still, it is crucial to evaluate the stability of potential off-the-shelf therapeutic EV preparations, and more dilute EV formulations might be applicable for certain treatment strategies such as repetitive EV dosing. Nevertheless, we are confident that PBS-HAT is highly suitable and useful to stabilize EV preparations and ensure effective EV recovery, particularly for analytical methods requiring considerable dilution of EV samples before analysis, and in general, in an exploratory EV research environment. PBS-HAT is compatible with most downstream analytical and functional assays, including cellular uptake, in vivo biodistribution [17], and transcriptional profiling [53]. Notably, the albumin in PBS-HAT would interfere with protein measurements and proteomics, which is why we recommend storing EVs prepared for proteomics analysis in separate tubes and performing EV lysis in these tubes directly to avoid sample loss due to poor EV recovery. Alternatively, EVs could be stored in PBS-HAT to ensure stability and be cleaned up using size exclusion chromatography before using samples for proteomics.

Maintaining stable storage of EVs, with preserved function and recovery, is consistently difficult across various cellular systems and sample types, especially when the exact EV functions or their mode of action are not fully understood. Given the multimodal composition and potential functional diversity of EVs in different cellular contexts, it is difficult to broadly evaluate potential impacts of storage conditions on the function of EVs, however, we propose that surrogate approaches such as the assessment of the transcriptional response of different candidate target cells to EV treatments might be a suitable solution in this context [53]. While we have not explored stability and recovery for biological fluid-derived EVs so far, there is evidence that EVs stored in biological fluids or derived from biological fluids appear stable for longer periods of time, especially if not stored highly purified from residual soluble proteins [15, 26–28], holding promise for the future assessment of biobanked samples. In summary, there is still a need for more systematic studies applying high-sensitivity quantitative analytical single EV methods to evaluate EV stability for different sample types and at different storage conditions, covering all areas such as basic EV biology, therapy development, and biomarker discovery.

Acknowledgments

A.G. was an International Society for Advancement of Cytometry (ISAC) Marylou Ingram Scholar 2019–2024 and was funded by Karolinska Institutet (Network Medicine Global Alliance Collaborative Grant), Karolinska Institutet Research Foundation (2024-02858), the European Commission (Innobuyer/EV-NEXT, 101071212), and Eureka (Eurostars, INCANT, 7335). S.E.A. was funded by The European Research Council-H2020 (EXPERT), a Consolidator Grant (DELIVER, 101001374), The Swedish Foundation of Strategic Research (FormulaEx, SM19-0007), The Swedish Cancer Society, The Swedish Research Council (4-258/2021), and Evox Therapeutics. Figures were created using BioRender.com; publication licenses can be supplied upon request.

References

1. Wiklander OPB, Brennan MÁ, Lötvall J, Breakefield XO, EL Andaloussi S (2019) Advances in therapeutic applications of extracellular vesicles. Sci Transl Med 11(492): eaav8521. https://doi.org/10.1126/scitranslmed.aav8521
2. Yanez-Mo M, Siljander PR, Andreu Z, Zavec AB, Borras FE, Buzas EI, Buzas K, Casal E, Cappello F, Carvalho J, Colas E, Cordeiro-da Silva A, Fais S, Falcon-Perez JM, Ghobrial IM, Giebel B, Gimona M, Graner M, Gursel I, Gursel M, Heegaard NH, Hendrix A, Kierulf P, Kokubun K, Kosanovic M, Kralj-Iglic V, Kramer-Albers EM, Laitinen S, Lasser C, Lener T, Ligeti E, Line A, Lipps G, Llorente A, Lotvall J, Mancek-Keber M, Marcilla A, Mittelbrunn M, Nazarenko I, Nolte-'t Hoen EN, Nyman TA, O'Driscoll L, Olivan M, Oliveira C, Pallinger E, Del Portillo HA, Reventos J, Rigau M, Rohde E, Sammar M, Sanchez-Madrid F, Santarem N, Schallmoser K, Ostenfeld MS, Stoorvogel W, Stukelj R, Van der Grein SG, Vasconcelos MH, Wauben MH, De Wever O (2015) Biological properties of extracellular vesicles and their physiological functions. J Extracell Vesicles 4: 27066. https://doi.org/10.3402/jev.v4.27066
3. Sahoo S, Adamiak M, Mathiyalagan P, Kenneweg F, Kafert-Kasting S, Thum T (2021) Therapeutic and diagnostic translation of extracellular vesicles in cardiovascular diseases. Circulation 143(14):1426–1449. https://doi.org/10.1161/CIRCULATIONAHA.120.049254
4. Weng J, Xiang X, Ding L, Wong AL-A, Zeng Q, Sethi G, Wang L, Lee SC, Goh BC (2021) Extracellular vesicles, the cornerstone of next-generation cancer diagnosis? Semin Cancer Biol 74:105–120. https://doi.org/10.1016/j.semcancer.2021.05.011
5. Lener T, Gimona M, Aigner L, Borger V, Buzas E, Camussi G, Chaput N, Chatterjee D, Court FA, Del Portillo HA, O'Driscoll L, Fais S, Falcon-Perez JM, Felderhoff-Mueser U, Fraile L, Gho YS, Gorgens A, Gupta RC, Hendrix A, Hermann DM, Hill AF, Hochberg F, Horn PA, de Kleijn D, Kordelas L, Kramer BW, Kramer-Albers EM, Laner-Plamberger S, Laitinen S, Leonardi T, Lorenowicz MJ, Lim SK, Lotvall J, Maguire CA, Marcilla A, Nazarenko I, Ochiya T, Patel T, Pedersen S, Pocsfalvi G, Pluchino S, Quesenberry P, Reischl IG, Rivera FJ, Sanzenbacher R, Schallmoser K, Slaper-Cortenbach I, Strunk D, Tonn T, Vader P, van Balkom BW, Wauben M, Andaloussi SE, Thery C, Rohde E, Giebel B (2015) Applying extracellular vesicles based therapeutics in clinical trials - an ISEV position paper. J Extracell Vesicles 4:30087. https://doi.org/10.3402/jev.v4.30087
6. Mateescu B, Kowal EJK, van Balkom BWM, Bartel S, Bhattacharyya SN, Buzás EI, Buck

AH, de Candia P, Chow FWN, Das S, Driedonks TAP, Fernández-Messina L, Haderk F, Hill AF, Jones JC, Van Keuren-Jensen KR, Lai CP, Lässer C, di Liegro I, Lunavat TR, Lorenowicz MJ, Maas SLN, Mäger I, Mittelbrunn M, Momma S, Mukherjee K, Nawaz M, Pegtel DM, Pfaffl MW, Schiffelers RM, Tahara H, Théry C, Tosar JP, Wauben MHM, Witwer KW, Nolte-'t Hoen ENM (2017) Obstacles and opportunities in the functional analysis of extracellular vesicle RNA—an ISEV position paper. J Extracell Vesicles 6 (1):1286095. https://doi.org/10.1080/20013078.2017.1286095

7. Buzas EI (2023) The roles of extracellular vesicles in the immune system. Nat Rev Immunol 23(4):236–250. https://doi.org/10.1038/s41577-022-00763-8
8. Coumans FAW, Brisson AR, Buzas EI, Dignat-George F, Drees EEE, El-Andaloussi S, Emanueli C, Gasecka A, Hendrix A, Hill AF, Lacroix R, Lee Y, van Leeuwen TG, Mackman N, Mäger I, Nolan JP, van der Pol E, Pegtel DM, Sahoo S, Siljander PRM, Sturk G, de Wever O, Nieuwland R (2017) Methodological guidelines to study extracellular vesicles. Circ Res 120(10):1632–1648. https://doi.org/10.1161/circresaha.117.309417
9. Welsh JA, Goberdhan DCI, O'Driscoll L, Buzas EI, Blenkiron C, Bussolati B, Cai H, Di Vizio D, Driedonks TAP, Erdbrügger U, Falcon-Perez JM, Fu QL, Hill AF, Lenassi M, Lim SK, Mahoney MG, Mohanty S, Möller A, Nieuwland R, Ochiya T, Sahoo S, Torrecilhas AC, Zheng L, Zijlstra A, Abuelreich S, Bagabas R, Bergese P, Bridges EM, Brucale M, Burger D, Carney RP, Cocucci E, Crescitelli R, Hanser E, Harris AL, Haughey NJ, Hendrix A, Ivanov AR, Jovanovic-Talisman T, Kruh-Garcia NA, Ku'ulei-Lyn Faustino V, Kyburz D, Lässer C, Lennon KM, Lötvall J, Maddox AL, Martens-Uzunova ES, Mizenko RR, Newman LA, Ridolfi A, Rohde E, Rojalin T, Rowland A, Saftics A, Sandau US, Saugstad JA, Shekari F, Swift S, Ter-Ovanesyan D, Tosar JP, Useckaite Z, Valle F, Varga Z, van der Pol E, van Herwijnen MJC, Wauben MHM, Wehman AM, Williams S, Zendrini A, Zimmerman AJ, Théry C, Witwer KW (2024) Minimal information for studies of extracellular vesicles (MISEV2023): from basic to advanced approaches. J Extracell Vesicles 13(2):e12404. https://doi.org/10.1002/jev2.12404
10. Thery C, Witwer KW, Aikawa E, Alcaraz MJ, Anderson JD et al (2018) Minimal information for studies of extracellular vesicles 2018 (MISEV2018): a position statement of the International Society for Extracellular Vesicles and update of the MISEV2014 guidelines. J Extracell Vesicles 7(1):1535750. https://doi.org/10.1080/20013078.2018.1535750
11. Welsh JA, Tang VA, van der Pol E, Gorgens A (2021) MIFlowCyt-EV: the next chapter in the reporting and reliability of single extracellular vesicle flow cytometry experiments. Cytometry A 99(4):365–368. https://doi.org/10.1002/cyto.a.24268
12. Welsh JA, Van Der Pol E, Arkesteijn GJA, Bremer M, Brisson A, Coumans F, Dignat-George F, Duggan E, Ghiran I, Giebel B, Gorgens A, Hendrix A, Lacroix R, Lannigan J, Libregts S, Lozano-Andres E, Morales-Kastresana A, Robert S, De Rond L, Tertel T, Tigges J, De Wever O, Yan X, Nieuwland R, Wauben MHM, Nolan JP, Jones JC (2020) MIFlowCyt-EV: a framework for standardized reporting of extracellular vesicle flow cytometry experiments. J Extracell Vesicles 9(1):1713526. https://doi.org/10.1080/20013078.2020.1713526
13. Ramirez MI, Amorim MG, Gadelha C, Milic I, Welsh JA, Freitas VM, Nawaz M, Akbar N, Couch Y, Makin L, Cooke F, Vettore AL, Batista PX, Freezor R, Pezuk JA, Rosa-Fernandes L, Carreira ACO, Devitt A, Jacobs L, Silva IT, Coakley G, Nunes DN, Carter D, Palmisano G, Dias-Neto E (2018) Technical challenges of working with extracellular vesicles. Nanoscale 10(3):881–906. https://doi.org/10.1039/c7nr08360b
14. Geng T, Ding L, Liu M, Zou X, Gu Z, Lin H, Sun L (2025) Preservation of extracellular vesicles for drug delivery: a comparative evaluation of storage buffers. J Drug Deliv Sci Technol 107:106850. https://doi.org/10.1016/j.jddst.2025.106850
15. Ahmadian S, Jafari N, Tamadon A, Ghaffarzadeh A, Rahbarghazi R, Mahdipour M (2024) Different storage and freezing protocols for extracellular vesicles: a systematic review. Stem Cell Res Ther 15(1):453. https://doi.org/10.1186/s13287-024-04005-7
16. Gupta D, Zickler AM, El Andaloussi S (2021) Dosing extracellular vesicles. Adv Drug Deliv Rev 178:113961. https://doi.org/10.1016/j.addr.2021.113961
17. Gorgens A, Corso G, Hagey DW, Jawad Wiklander R, Gustafsson MO, Felldin U, Lee Y, Bostancioglu RB, Sork H, Liang X, Zheng W, Mohammad DK, van de Wakker SI, Vader P, Zickler AM, Mamand DR, Ma L, Holme MN, Stevens MM, Wiklander OPB, El Andaloussi S (2022) Identification of storage conditions stabilizing extracellular vesicles

preparations. J Extracell Vesicles 11(6): e12238. https://doi.org/10.1002/jev2.12238

18. Yuana Y, Böing AN, Grootemaat AE, van der Pol E, Hau CM, Cizmar P, Buhr E, Sturk A, Nieuwland R (2015) Handling and storage of human body fluids for analysis of extracellular vesicles
19. Bosch S, de Beaurepaire L, Allard M, Mosser M, Heichette C, Chrétien D, Jegou D, Bach J-M (2016) Trehalose prevents aggregation of exosomes and cryodamage. Sci Rep 6:36162. https://doi.org/10.1038/srep36162
20. Jeyaram A, Jay SM (2017) Preservation and storage stability of extracellular vesicles for therapeutic applications. AAPS J 20(1):1. https://doi.org/10.1208/s12248-017-0160-y
21. Lőrincz ÁM, Timár CI, Marosvári KA, Veres DS, Otrokocsi L, Kittel Á, Ligeti E (2014) Effect of storage on physical and functional properties of extracellular vesicles derived from neutrophilic granulocytes. J Extracell Vesicles 3:25465–25465. https://doi.org/10.3402/jev.v3.25465
22. Maroto R, Zhao Y, Jamaluddin M, Popov VL, Wang H, Kalubowilage M, Zhang Y, Luisi J, Sun H, Culbertson CT, Bossmann SH, Motamedi M, Brasier AR (2017) Effects of storage temperature on airway exosome integrity for diagnostic and functional analyses. J Extracell Vesicles 6(1):1359478. https://doi.org/10.1080/20013078.2017.1359478
23. Zhou H, Yuen PS, Pisitkun T, Gonzales PA, Yasuda H, Dear JW, Gross P, Knepper MA, Star RA (2006) Collection, storage, preservation, and normalization of human urinary exosomes for biomarker discovery. Kidney Int 69(8): 1471–1476. https://doi.org/10.1038/sj.ki.5000273
24. Kusuma GD, Barabadi M, Tan JL, Morton DAV, Frith JE, Lim R (2018) To protect and to preserve: novel preservation strategies for extracellular vesicles. Front Pharmacol 9(1199). https://doi.org/10.3389/fphar.2018.01199
25. Sokolova V, Ludwig AK, Hornung S, Rotan O, Horn PA, Epple M, Giebel B (2011) Characterisation of exosomes derived from human cells by nanoparticle tracking analysis and scanning electron microscopy. Colloids Surf B Biointerfaces 87(1):146–150. https://doi.org/10.1016/j.colsurfb.2011.05.013
26. Yang C, Han J, Liu H, He Y, Zhang Z, Liu X, Waqas F, Zhang L, Duan H, He J, Dong L (2024) Storage of plasma-derived exosomes: evaluation of anticoagulant use and preserving temperatures. Platelets 35(1):2337255. https://doi.org/10.1080/09537104.2024.2337255
27. Shen S, Shen Z, Wang C, Wu X, Wang L, Ye L, Zhang S, Cheng X (2023) Effects of lysate/tissue storage at −80°C on subsequently extracted EVs of epithelial ovarian cancer tissue origins. iScience 26(4):106521. https://doi.org/10.1016/j.isci.2023.106521
28. Wright A, Snyder OL, Christenson LK, He H, Weiss ML (2022) Effect of pre-processing storage condition of cell culture-conditioned medium on extracellular vesicles derived from human umbilical cord-derived mesenchymal stromal cells. Int J Mol Sci 23(14). https://doi.org/10.3390/ijms23147716
29. Witwer KW, Buzas EI, Bemis LT, Bora A, Lasser C, Lotvall J, Nolte-'t Hoen EN, Piper MG, Sivaraman S, Skog J, Thery C, Wauben MH, Hochberg F (2013) Standardization of sample collection, isolation and analysis methods in extracellular vesicle research. J Extracell Vesicles 2. https://doi.org/10.3402/jev.v2i0.20360
30. Tegegn TZ, De Paoli SH, Orecna M, Elhelu OK, Woodle SA, Tarandovskiy ID, Ovanesov MV, Simak J (2016) Characterization of procoagulant extracellular vesicles and platelet membrane disintegration in DMSO-cryopreserved platelets. J Extracell Vesicles 5: 30422. https://doi.org/10.3402/jev.v5.30422
31. Trenkenschuh E, Richter M, Heinrich E, Koch M, Fuhrmann G, Friess W (2022) Enhancing the stabilization potential of lyophilization for extracellular vesicles. Adv Healthc Mater 11(5):2100538. https://doi.org/10.1002/adhm.202100538
32. Yuan F, Li Y-M, Wang Z (2021) Preserving extracellular vesicles for biomedical applications: consideration of storage stability before and after isolation. Drug Deliv 28(1): 1501–1509. https://doi.org/10.1080/10717544.2021.1951896
33. Qin B, Zhang Q, Hu X-m, Mi T-y, Yu H-y, Liu S-s, Zhang B, Tang M, Huang J-f, Xiong K (2020) How does temperature play a role in the storage of extracellular vesicles? J Cell Physiol 235(11):7663–7680. https://doi.org/10.1002/jcp.29700
34. Welch JL, Madison MN, Margolick JB, Galvin S, Gupta P, Martinez-Maza O, Dash C, Okeoma CM (2017) Effect of prolonged freezing of semen on exosome recovery and biologic activity. Sci Rep 7:45034. https://doi.org/10.1038/srep45034
35. Nordin JZ, Lee Y, Vader P, Mager I, Johansson HJ, Heusermann W, Wiklander OP,

Hallbrink M, Seow Y, Bultema JJ, Gilthorpe J, Davies T, Fairchild PJ, Gabrielsson S, Meisner-Kober NC, Lehtio J, Smith CI, Wood MJ, El Andaloussi S (2015) Ultrafiltration with size-exclusion liquid chromatography for high yield isolation of extracellular vesicles preserving intact biophysical and functional properties. Nanomedicine 11(4):879–883. https://doi.org/10.1016/j.nano.2015.01.003

36. Willms E, Cabañas C, Mäger I, Wood MJA, Vader P (2018) Extracellular vesicle heterogeneity: subpopulations, isolation techniques, and diverse functions in cancer progression. Front Immunol 9(738). https://doi.org/10.3389/fimmu.2018.00738

37. Allelein S, Medina-Perez P, Lopes ALH, Rau S, Hause G, Kölsch A, Kuhlmeier D (2021) Potential and challenges of specifically isolating extracellular vesicles from heterogeneous populations. Sci Rep 11(1):11585. https://doi.org/10.1038/s41598-021-91129-y

38. van de Wakker SI, van Oudheusden J, Mol EA, Roefs MT, Zheng W, Gorgens A, El Andaloussi S, Sluijter JPG, Vader P (2022) Influence of short term storage conditions, concentration methodsand excipients on extracellular vesicle recovery and function. Eur J Pharm Biopharm 170:59–69. https://doi.org/10.1016/j.ejpb.2021.11.012

39. Kawai-Harada Y, El Itawi H, Komuro H, Harada M (2023) Evaluation of EV storage buffer for efficient preservation of engineered extracellular vesicles. Int J Mol Sci 24(16):12841

40. Cavallaro S, Pevere F, Stridfeldt F, Gorgens A, Paba C, Sahu SS, Mamand DR, Gupta D, El Andaloussi S, Linnros J, Dev A (2021) Multiparametric profiling of single nanoscale extracellular vesicles by combined atomic force and fluorescence microscopy: correlation and heterogeneity in their molecular and biophysical features. Small 17(14):e2008155. https://doi.org/10.1002/smll.202008155

41. Wiklander OPB, Bostancioglu RB, Welsh JA, Zickler AM, Murke F, Corso G, Felldin U, Hagey DW, Evertsson B, Liang XM, Gustafsson MO, Mohammad DK, Wiek C, Hanenberg H, Bremer M, Gupta D, Bjornstedt M, Giebel B, Nordin JZ, Jones JC, El Andaloussi S, Gorgens A (2018) Systematic methodological evaluation of a multiplex bead-based flow cytometry assay for detection of extracellular vesicle surface signatures. Front Immunol 9(1326):1326. https://doi.org/10.3389/fimmu.2018.01326

42. Gorgens A, Bremer M, Ferrer-Tur R, Murke F, Tertel T, Horn PA, Thalmann S, Welsh JA, Probst C, Guerin C, Boulanger CM, Jones JC, Hanenberg H, Erdbrugger U, Lannigan J, Ricklefs FL, El-Andaloussi S, Giebel B (2019) Optimisation of imaging flow cytometry for the analysis of single extracellular vesicles by using fluorescence-tagged vesicles as biological reference material. J Extracell Vesicles 8(1):1587567. https://doi.org/10.1080/20013078.2019.1587567

43. Corso G, Mager I, Lee Y, Gorgens A, Bultema J, Giebel B, Wood MJA, Nordin JZ, Andaloussi SE (2017) Reproducible and scalable purification of extracellular vesicles using combined bind-elute and size exclusion chromatography. Sci Rep 7(1):11561. https://doi.org/10.1038/s41598-017-10646-x

44. Nordin JZ, Bostancioglu RB, Corso G, El Andaloussi S (2019) Tangential flow filtration with or without subsequent bind-elute size exclusion chromatography for purification of extracellular vesicles. Methods Mol Biol 1953: 287–299. https://doi.org/10.1007/978-1-4939-9145-7_18

45. Tertel T, Bremer M, Maire C, Lamszus K, Peine S, Jawad R, Andaloussi SEL, Giebel B, Ricklefs FL, Gorgens A (2020) High-resolution imaging flow cytometry reveals impact of incubation temperature on labeling of extracellular vesicles with antibodies. Cytometry A 97(6):602–609. https://doi.org/10.1002/cyto.a.24034

46. Dhondt B, Pinheiro C, Geeurickx E, Tulkens J, Vergauwen G, Van Der Pol E, Nieuwland R, Decock A, Miinalainen I, Rappu P, Schroth G, Kuersten S, Vandesompele J, Mestdagh P, Lumen N, De Wever O, Hendrix A (2023) Benchmarking blood collection tubes and processing intervals for extracellular vesicle performance metrics. J Extracell Vesicles 12(5): e12315. https://doi.org/10.1002/jev2.12315

47. Sych T, Görgens A, Steiner L, Gucluler G, Huge Y, Alamdari F, Johansson M, Aljabery F, Sherif A, Gabrielsson S, El Andaloussi S, Sezgin E (2025) Imaging single particle profiler to study nanoscale bioparticles using conventional confocal microscopy. Nano Lett 25. https://doi.org/10.1021/acs.nanolett.4c05117

48. Gardiner C, Ferreira YJ, Dragovic RA, Redman CW, Sargent IL (2013) Extracellular vesicle sizing and enumeration by nanoparticle tracking analysis. J Extracell Vesicles 2. https://doi.org/10.3402/jev.v2i0.19671

49. Welsh JA, Arkesteijn GJA, Bremer M, Cimorelli M, Dignat-George F, Giebel B, Gorgens A, Hendrix A, Kuiper M, Lacroix R, Lannigan J, van Leeuwen TG, Lozano-Andres-E, Rao S, Robert S, de Rond L, Tang VA, Tertel T, Yan X, Wauben MHM, Nolan JP,

Jones JC, Nieuwland R, van der Pol E (2023) A compendium of single extracellular vesicle flow cytometry. J Extracell Vesicles 12(2):e12299. https://doi.org/10.1002/jev2.12299

50. Shao H, Im H, Castro CM, Breakefield X, Weissleder R, Lee H (2018) New technologies for analysis of extracellular vesicles. Chem Rev 118(4):1917–1950. https://doi.org/10.1021/acs.chemrev.7b00534

51. Ricklefs FL, Maire CL, Reimer R, Duhrsen L, Kolbe K, Holz M, Schneider E, Rissiek A, Babayan A, Hille C, Pantel K, Krasemann S, Glatzel M, Heiland DH, Flitsch J, Martens T, Schmidt NO, Peine S, Breakefield XO, Lawler S, Chiocca EA, Fehse B, Giebel B, Gorgens A, Westphal M, Lamszus K (2019) Imaging flow cytometry facilitates multiparametric characterization of extracellular vesicles in malignant brain tumours. J Extracell Vesicles 8(1):1588555. https://doi.org/10.1080/20013078.2019.1588555

52. Wiklander OPB, Mamand DR, Mohammad DK, Zheng W, Jawad Wiklander R, Sych T, Zickler AM, Liang X, Sharma H, Lavado A, Bost J, Roudi S, Corso G, Lennaard AJ, Abedi-Valugerdi M, Mager I, Alici E, Sezgin E, Nordin JZ, Gupta D, Gorgens A, El Andaloussi S (2024) Antibody-displaying extracellular vesicles for targeted cancer therapy. Nat Biomed Eng 8. https://doi.org/10.1038/s41551-024-01214-6

53. Hagey DW, Ojansivu M, Bostancioglu BR, Saher O, Bost JP, Gustafsson MO, Gramignoli R, Svahn M, Gupta D, Stevens MM, Gorgens A, El Andaloussi S (2023) The cellular response to extracellular vesicles is dependent on their cell source and dose. Sci Adv 9(35):eadh1168. https://doi.org/10.1126/sciadv.adh1168

Chapter 11

Nanoparticle Tracking Analysis and Its Application to Study the Role of Small Extracellular Vesicles in Health and Disease

Ella Fouhy, Daniel O'Reilly, and Patricia B. Maguire

Abstract

Extracellular vesicles (EVs) are mediators of extracellular communication, present in a diverse range of body fluids, with significant implications for a variety of physiological and pathological processes, including thrombosis, hemostasis, pregnancy, and neonatal health. Nanoparticle tracking analysis has allowed accurate and reliable detection of particles from 50 nm, to detect differences that provide insights into normal and diseased states. Specifically, EVs derived from platelets and endothelial cells play critical roles in vascular biology, sepsis, and inflammatory disorders. Additionally, placental EVs have been shown to influence pregnancy outcomes, with their altered profile contributing to the pathophysiology of complications such as preeclampsia (PE). As the technology for EV detection continues to improve, new avenues for studying their role in diverse clinical conditions are emerging. These advancements hold promise for the identification of novel biomarkers and therapeutic targets, offering the potential for improved diagnostics and personalized treatments in a variety of clinical settings.

Key words Extracellular vesicles (EVs), Nanoparticle tracking analysis (NTA), Exosomes, Small extracellular vesicles (sEVs), Vascular biology, Sepsis, Pregnancy

1 Introduction

The characterization of extracellular vesicles (EVs) using liquid biopsies has provided valuable insights into both physiologic and pathologic processes. These mediators of extracellular communication play important roles in immune response [1], tissue repair [2], hemostasis [3], pregnancy [4], and cancer [5]. Often reflecting the pathophysiological state of their parent cell, these tiny EVs can traverse barriers such as the blood-brain barrier that are otherwise inaccessible to their parent cell. Nanoparticle tracking analysis (NTA) has played a pivotal role in the characterization of these particles and has greatly influenced our understanding of small EVs* (sEVs). The ability to accurately characterize the size and concentration of EVs in a sample, as small as 30 nm, can uncover

Amit K. Srivastava (ed.), *Extracellular Vesicles*, Neuromethods, vol. 228, https://doi.org/10.1007/978-1-0716-4905-3_11,

subtle yet significant changes relevant to specific disease states and normal physiological responses. This chapter provides an overview of NTA and its application to the study of small EVs in health and disease, with a particular focus on the contribution to furthering our understanding of thrombosis and hemostasis, neonatal physiology, sepsis, and pregnancy.

*Earlier literature surrounding extracellular vesicles uses several overlapping terms and existing literature that cites “exosomes” more accurately refers to broader populations of small EVs and not specific EVs released via specific biogenesis pathways [6]. In this chapter, we will interchangeably refer to exosomes and small EVs to reflect the extensive literature to date; however, it is important to note that the International Society of Extracellular Vesicles (https://www.isev.org) discourages the use of biogenesis-based terms, as outlined in the minimal information for studies of extracellular vesicles (MISEV2023) [6].

2 EV Biogenesis and Isolation from Biological Fluids

EVs are membrane-bound vesicles released from cells, further classified based on their specific size, molecular composition, and cellular origin, with exosomes 30–150 nm, microvesicles 100–1000 nm, and apoptotic bodies 1000–5000 nm in size. While blood plasma or serum serves as the primary source of EVs, they are also found in various biological fluids such as urine, synovial fluid, lymph, and saliva.

Exosomes are lipid bilayer particles released from cells ranging from 30 to 150 nm in size with a buoyant density of 1.12–1.19 g/mL. Unlike microvesicles that are formed from direct outward budding of the plasma membrane, exosomes originate from late endosomal multivesicular bodies (MVBs) that fuse with the plasma membrane to release their intraluminal vesicles (ILVs) as exosomes (Fig. 1). Exosomes therefore resemble intraluminal membranes and are highly enriched in certain lipids such as ceramide and cholesterol [7, 8]. This highly regulated cellular process, once viewed as an elimination system for unwanted proteins during reticulocyte maturation, sorts specific bioactive molecules into the vesicle, which is capable of influencing distant recipient cells [9, 10]. Due to the overlap in size, isolation methods for exosomes often include vesicles of non-endocytic origin, formed from plasma membrane budding. Therefore, the term “small EVs (sEVs)” is more appropriate when referring to EVs < 200 nm [6].

EVs contain a rich repertoire of proteins, lipids, metabolites, and nucleic acids protected from degradation from proteases and nucleases by their lipid bilayer and stable when stored at −80 °C [11–14]. This cargo can be delivered to recipient cells, influencing various cellular processes [15, 16]. For instance, Valadi et al.

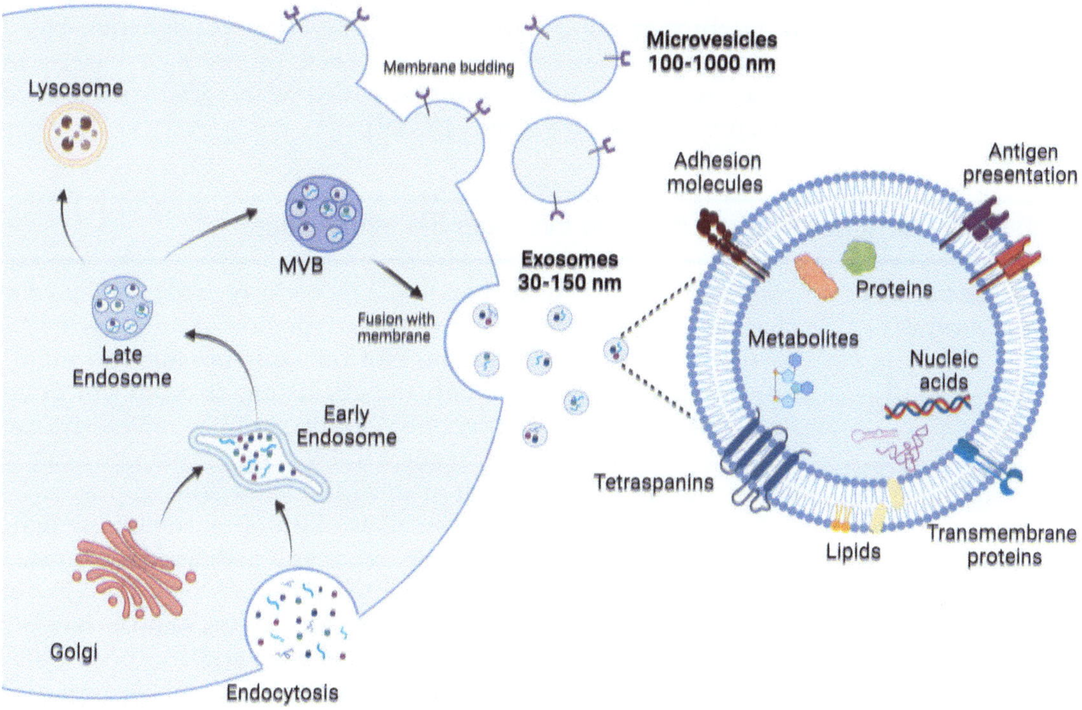

Fig. 1 Extracellular vesicle (EV) biogenesis. EVs are released from a multitude of cell types into extracellular space. The two main types of EVs are exosomes and microvesicles. Exosomes (30–150 nm) originate from late endosomal multivesicular bodies (MVBs) that fuse with the plasma membrane to release their intraluminal vesicles. Exosomes contain a rich content of proteins, lipids, metabolites, and nucleic acids. Microvesicles are formed from direct outward budding of the plasma membrane

demonstrated that mRNA and miRNA molecules within exosomes, derived from mouse and human mast cells, can be translated into functional proteins in recipient cells [17]. Similarly, Escobar et al. showed that EVs derived from human megakaryocytes can drive de novo platelet biogenesis in a murine model, increasing platelet count [16].

Although the cargo of EVs reflects the molecular profile of their parent cell, specific components can be preferentially packaged. In liver cancer, for example, exosomes secreted by gastric adenocarcinoma cells in vitro can transport the epidermal growth factor receptor (EGFR), helping to regulate the tumor microenvironment and promote future metastasis [5]. This dynamic cargo composition highlights the potential of EVs to provide insight into biological processes and their alterations during disease pathogenesis.

EVs represent a highly heterogenous population of vesicles and their isolation is crucial for accurate characterization and removal of contaminant particles. Various techniques are used for isolating EVs from biological fluids, with the most cited methods including size-exclusion chromatography (SEC), sucrose or density gradient ultracentrifugation, and filtration or precipitation-based approaches. These methods have been previously reviewed in detail, with ultracentrifugation regarded as the gold standard for isolation of sEVs in plasma [18, 19].

3 Methods for Quantification of Small EVs from Biological Fluids

3.1 Nanoparticle Tracking Analysis

Nanoparticle tracking analysis (NTA) is a widely used technique for characterizing EVs, providing high-resolution measurements of particle size and concentration. Not only can absolute quantification of EVs lend insights into the circulating environment in an individual but normalization of EV counts for functional assays can help uncover functional differences in EV cargo. It has significantly advanced our understanding of small EVs in health and disease.

NTA detects particles as small as ~50 nm by combining light scattering and Brownian motion [20]. For analysis, the sample is loaded into a syringe and injected into a flow cell where a laser beam (642, 532, 488, or 405 nm) is passed through the sample chamber and particles in the path of the beam scatter and are then illuminated and easily visualized by a microscope. The CCD, EMCCD (electron multiplied charged coupled device), or high sensitivity complementary metal-oxide-semiconductor camera captures a video file of the particles moving under Brownian motion, where the software can then track the center of each particle on a frame-by-frame basis (typically 30 frames per second). Application of the Stokes–Einstein equation allows for determination of size for each tracked particle [20]. Capture of multiple video replicates can improve the variability and accuracy of the data generated, especially in samples with low particle counts [21].

NTA enables both size distribution and concentration analysis, but it is accompanied by some limitations. A significant challenge is the contamination of similar sized particles. NTA does not distinguish between EVs, protein aggregates, or lipoproteins in an unstained sample, so the method of isolation and removal of these particles is a key consideration. This limitation can be overcome if antibody-mediated fluorophores or fluorescent stains are used. NTA instruments also have fluorescent capabilities if appropriate optical filters are used and enable the quantification of specific subpopulations. This chapter, however, will focus on the use of NTA in unlabeled samples for particle size and concentration as most commonly utilized. A detailed methodology for fl-NTA has been previously described [22].

3.2 Flow Cytometry

Extensively utilized in cell biology, flow cytometry has been adapted for EV analysis allowing for quantification of vesicles, particularly those derived from blood cells and the vascular endothelium. Traditionally, this method has been applied to study large vesicles (>200 nm), but recent technical advancements in instrumentation have aimed at improving resolution to enable the detection of smaller EVs.

EVs are detected by flow cytometry based on their light scattering and fluorescence; however, due to their small size, sEVs scatter significantly less light than larger EVs or reference beads, making them difficult to distinguish from background noise and protein aggregates. To address this challenge, improvements in detection optics have enabled the identification of particles as small as ~80 nm using specialized nano-flow cytometers. These instruments aim to characterize EVs in a similar size range as NTA while offering the added advantage of detecting multiple fluorophores simultaneously. Despite these advancements, NTA remains the most used technique to date for characterizing these smaller biological particles.

Flow cytometry comes with distinct limitations compared to NTA, including challenges related to gating strategies, the use of reference beads, and swarming [23]. These factors must be carefully considered when selecting a method for single-particle analysis. To support standardization and ensure accurate reporting of single EV flow cytometry data, the minimum information about a flow cytometry experiment (MIFlowCyt) EV framework has been established [24].

4 Platelet- and Endothelial-Derived EVs in Vascular Biology and Sepsis

Blood-derived EVs are a heterogeneous group originating from a multitude of circulating blood cells, the vascular endothelium, and other distant tissues. The majority of blood-derived EVs originate from erythrocytes, megakaryocytes, and platelets, with an estimated ~25% of circulating EVs being platelet-derived [25], playing a crucial role in maintaining hemostasis and vascular integrity. Dysregulation of this delicate balance in the circulating environment can lead to the hypercoagulable state associated with various thrombotic and inflammatory disorders.

Platelets continuously survey the vasculature, circulating in a quiescent state while receiving inhibitory signals like nitric oxide and prostacyclin from the endothelium. At the site of vascular injury or inflammation, subendothelial matrix components (such as collagen and von Willebrand factor) bind to circulating platelets, triggering activation and cytoskeletal reorganization and subsequent secretion of internal granules. Platelets release a multitude of soluble and cleaved factors along with small and large EVs following activation, of which both content and antigen expression are agonist dependent [26–28]. Platelet exosomes originate from multivesicular compartments and alpha granules within the platelets [29], and can be visualized in platelet progenitor cells, megakaryocytes [30]. Notably, in a mouse model of gray platelet syndrome (*Nbeal*$^{-/-}$), characterized by the absence of platelet alpha granules cargo, NTA showed a 50% reduction of circulating exosomal vesicles when compared to wild-type mice, demonstrating a functional decrease in concentration of platelet-derived EVs (PEVs) [30].

Supporting evidence from NTA and TEM has demonstrated that the majority of PEVs secreted are 100–250 nm in size, with 90% of total EVs secreted measuring <500 nm [3]. These EVs share their platelet-specific parent antigen CD41 (glycoprotein IIb) on their surface, and early work by Heijen et al. demonstrated that exosomes released from activated platelets are also enriched in CD63 and exhibited minimal procoagulant activity, indicated by low Annexin-V binding and absence of factor X and prothrombin binding. The opposite is true for large PEVs, which are strongly associated with coagulation due to their negatively charged phospholipid phosphatidylserine and tissue factor rich surface that provide a site for coagulation factor binding and thrombin generation [29, 31]. Together, small and large PEVs are significantly increased in conditions associated with inflammation and platelet activation, such as cardiovascular disease (CVD) and cancer. Total sEVs in plasma, as measured by NTA, correlate with the overall risk for CVD as well as the individual risk factors like BMI, blood pressure, and plasma triacylglycerol levels with an association between higher circulating EVs and increased thrombin generation [32].

As a key contributor to thrombosis, PEVs have also been implicated in sepsis, a dysregulated host response to infection that can lead to multiorgan failure and death. Part of the dysregulated host response is believed to be endothelial dysfunction and hypercoagulability in the context of inflammation, which can progress to a consumptive coagulopathy, which is initially non-overt and referred to as sepsis-induced coagulopathy (SIC) but can progress to an overt process, referred to as disseminated intravascular coagulation (DIC). PEVs have been demonstrated to contribute to this phenomenon by infiltrating the bone marrow during inflammation in vivo and in vitro causing functional reprogramming in cells furthering their role in the regulation of hematopoiesis [15]. Brain-derived EVs, defined by the presence of neuron-specific enolase (NSE) and glial fibrillary protein (GFAP), have also been suggested to drive the development of a consumptive coagulopathy in a Cecal ligation and puncture (CLP) rat model. Isolated BDEVs (100–500 nm particles confirmed by NTA and TEM) caused apoptosis in the liver, lung, and kidney tissue as identified by TUNEL staining, suggesting they are key mediators of organ dysfunction and death in this setting [33].

Evidence is more limited in clinical cohorts, where the literature is further confused by differing clinical definitions of sepsis. Some research uses Sepsis 2 criteria, based on the systemic inflammatory response syndrome (SIRS), while others apply the more modern Sepsis 3 criteria, which requires evidence of organ dysfunction as defined by Sequential Organ Failure Assessment. In a Japanese study classifying patients into the now defunct septic shock, sepsis, and non-sepsis categories, the vast majority of EVs were <600 nm in size and there was a statistically significant difference in PECAM$^+$ and VE-cadherin$^+$ EVs in those with septic shock as opposed to sepsis [34]. This underpins the key role endothelial dysfunction has in separating those with organ dysfunction versus those who do not. Further evidence has examined the impact of endothelial EVs expressing vascular cell adhesion molecule 1 (VCAM-1) and the development of acute respiratory distress syndrome (ARDS), a type of respiratory failure caused by inflammation, a recognized complication of sepsis. They demonstrated, using both a rodent model and from examining human plasma samples from patients who were diagnosed as septic according to sepsis 3 criteria, that VCAM-1+ endothelial cell EVs were highly expressed in patients with ARDS [35]. Using an in vitro model, they demonstrated that EVs released by endothelial cells treated with LPS stimulate monocytes in the lung to differentiate into an M1(CD80$^+$/CD86$^+$) or pro-inflammatory phenotype via initial uptake of the EVs, then activation of ITGA4, and subsequent stimulation of the NF-κB pathway [35]. These M1 monocytes then propagate inflammation through release of pro-inflammatory cytokines worsening lung injury. The importance of endothelial cell-derived EVs in the development of ARDS has been further underlined by work using plasma from septic patients (albeit without sepsis clearly defined). These EVs, characterized by NTA, were 30–120 nm in size, tested positive for EV markers CD9, CD63, and CD81, and showed lower transendothelial resistance, a measure of vascular permeability, in those with ARDS compared to those without. Total EVs displayed a higher caspase-1 activity and were significantly associated with the development of ARDS [36].

Evidence to date supports the key role of EVs in mediating the organ dysfunction associated with sepsis, including ARDS and sepsis-induced coagulopathy, and work remains to be done to examine how these processes may be interfered with therapeutically.

5 EVs in Pregnancy and Pregnancy-Related Complications

Human pregnancy is a tightly regulated physiological process that requires an intricate balance between the developing embryo and the maternal uterus. EVs play an important role in the complex bidirectional communication essential in pregnancy from implantation of the embryo in early pregnancy, to signals aiding in parturition [37, 38]. In addition to the previously mentioned biological fluids that contain EVs, they are also found in the umbilical cord, placenta, amniotic fluid, and amniotic membranes in pregnancy. Here, they are pivotal players in angiogenesis and immunomodulation and aid in the transfer of bioactive molecules vital for a successful pregnancy [39].

Coordination between the maternal endometrium and the embryo in early pregnancy is a highly complex process which involves a myriad of growth factors, hormones, chemokines, cytokines, and immune cells [40–43]. Exosomes fulfill critical functions in this paracrine communication and exosomal cargo can be transferred bidirectionally from the fetus to the mother throughout pregnancy [44]. Maternal exosomes isolated from the maternal cord serum display up to a 16-fold increase than those isolated from the umbilical cord serum when measured by NTA, with umbilical EVs expressing specific and relevant miRNAs specific to the pluripotency of embryonic stem cells [45]. NTA of plasma-derived EVs has demonstrated that pregnancy is associated with a significant increase in total circulating exosomes when compared to nonpregnant controls, of which the concentration increases with each trimester of pregnancy, peaking at term [4, 14, 46].

Specific to pregnancy are placental-derived EVs, which can be detected in maternal plasma from as early as 6 weeks' gestation [14]. These placental EVs are released from syncytiotrophoblast (STB) cells that line the surface of the placenta, contain cytokines and growth factors such as RANTES and VEGF, and are positive for placental alkaline phosphatase (PLAP) and often HLA-G and synctin-1 [47, 48]. STBs secrete miRNAs into maternal plasma with evidence supporting that specific placental miRNAs such as MIR517A and MIR21 are transferred into maternal circulation in an exosome-mediated fashion [49]. PLAP positive exosomes mirror the same gestational increase in maternal plasma as total EVs, and at term positively correlate with placental weight [4]. Placental small EVs have been associated with vasculogenesis and angiogenesis [4], processes crucial for a developing placenta, and are therefore being exploited to further our understanding in pathological processes. Interestingly, a study that injected spontaneously hypertensive mice with placental EVs from normotensive term placentae found a long-lasting protective effect, reducing hypertension in vivo and providing evidence for a role in cardioprotection [50].

Much of the research on EVs in reproductive cells has been concentrated on preeclampsia (PE), a disorder associated with improper remodeling of maternal spiral arteries and high perfusion pressure in the intervillous space [51, 52]. Circulating EVs in plasma are significantly increased in PE compared to healthy pregnancies at the same gestation [53], and may contribute to the endothelial dysfunction associated with the disorder by increasing reactive oxygen species, VWF and VCAM-1 [54–56]. Functional miRNAs packaged in STBEVs can be transferred to endothelial cells in culture and cause a downregulation of the PE associated gene fms-related tyrosine kinase 1 (FLT1) [57], furthering their role in PE pathogenesis. Isolated exosomes analyzed by NTA confirm a significant increase in total small EVs in PE, with a higher contribution of placental exosomes in those that developed early onset PE (diagnosis <33 + 6 weeks of gestation) in comparison to those with late onset PE (diagnosis >34 weeks of gestation) [58]. Placental exosomes in PE can be enriched with neprilysin, with elevated levels contributing to hypertension—the hallmark of PE [59]. This enrichment suggests that exosomes may play a direct role in the pathogenesis of the disorder, rather than acting merely as indirect communicators or mediators.

Subtle changes in EVs can unravel disease-specific alterations, and even in presymptomatic women who later develop PE, there is a marked increase in total and placental exosomes compared to those that do not develop PE in their pregnancies [46]. Ultimately, the dynamic changes in circulating and placental-derived EVs not only enhance our understanding of normal pregnancy physiology but also offer promising avenues for identifying biomarkers that could lead to earlier diagnosis and better management of pregnancy-related disorders such as PE.

6 EVs in Neonatal Complications

The role of EVs in neonatal complications has been explored extensively in recent studies, including those conducted in our lab as part of the extracellular vesicles in early preterm neonates and thrombin generation (EVENT) study. Here, there were significant variations in platelet-derived and endothelial-derived EV content between days 1 and 3 of preterm birth which are not recapitulated in term neonates [60].

Although research into EVs as diagnostic biomarkers for neonatal complications is still in its early stages, prediction of chronic lung disease (CLD) and bronchopulmonary dysplasia (BPD) has been a particular area of focus. Given the high burden of CLD/BPD and the limited progress in both prevention and cure of this condition in preterm newborns, there is an urgent need to identify a specific biomarker which may help both an understanding of this condition and pick out a "high risk" group beyond the current parameters of lower gestational age and birthweight. EVs have been isolated from amniotic fluid, tracheal aspirates, serum, and plasma in neonates at risk of CLD/BPD, with NTA showing that tracheal aspirates from infants diagnosed with severe BPD, requiring ventilation at 36 weeks corrected gestational age, exhibit a smaller modal EV size (65 nm compared to 105 nm in controls) [61].

Early lung development is associated with changes in EV profiles, with tracheal aspirates showing specific EV associated markers (CD9, CD63, and CD81) and differing expression of specific markers (CD24, CD14) across the late canalicular stage of lung development (defined as 22–26 weeks and 5 days corrected gestational age) and saccular stage of development (defined as 27–34 weeks and 6 days corrected gestational age). Additionally, high expression of CD24, an epithelial marker, and CD14, a macrophage specific marker, on aspirate EVs was associated with eventual development of CLD/BPD. EV size, as defined by NTA, demonstrated larger mean and median diameter in the late canalicular phase as opposed to the saccular phase [61]. In a hyperoxia model of BPD/CLD, circulating EVs were increased compared to neonatal rats maintained at room air, and NTA revealed that most EVs were 50–150 nm in size [62]. Based on these, EVs carrying surfactant protein C (SPC) were determined to be released by type II pneumocyte/alveolar type II epithelial cells. They also were demonstrated to contain Gasdermin-D, an inflammasome associated protein key for pyroptosis. Adoptive transfer of these circulating EVs from hyperoxic rat pups led to increased lung inflammation in normoxic neonatal rats with higher concentrations of circulating neutrophils, lymphocytes, and macrophages and lung-specific increases in the expression of IL-18, TNF-α, CXCL1, CTGF, and TGF-α [62]. Pathological samples from neonatal rats that received EVs from hyperoxic rats showed simplified alveolar structures and reduced vascular density, hallmarks of BPD/CLD in humans.

Overall, the evidence points to EVs being both generated by the processes which contribute to BPD/CLD, such as high oxygen and inflammation, as well as sustaining the process through direct lung damage themselves. Further work on human neonates to establish how and when this occurs could provide not only greater insights into the pathology of BPD/CLD in a modern cohort but also novel therapeutic and diagnostic avenues for this substantial cause of neonatal morbidity and mortality.

7 Summary

EVs have emerged as critical mediators of intercellular communication, influencing a range of physiological processes such as thrombosis, hemostasis, pregnancy, and neonatal health. NTA has allowed for quantification and size determination of sEVs, uncovering subtle yet distinct differences relevant to physiologic and pathologic response. Platelet- and endothelial-derived EVs are highly implicated in vascular biology and sepsis, contributing to systemic inflammation and coagulability associated with disorders like ARDS. PEVs can influence distant cells, offer insights into clinical outcomes, and shed light on potential molecular mechanisms contributing to differing prognoses. EVs are vital for a successful pregnancy, mediating communication and transfer of bioactive molecules bidirectionally from the growing fetus and mother. This intricate balance can be disrupted in complications of pregnancy such as PE, where EVs from the placenta drive systemic inflammation and carry antiangiogenic molecules that are hallmarks of the disorder. The increasing capabilities of instrumentation for detection of EVs make the field exciting to explore their roles in various clinical contexts and uncover novel insights into health and disease.

Conflict of Interest

This study was funded by Taighde Éireann—Research Ireland, AI for Good Challenge Fund (19/FIP/AI/7490R2).

References

1. Robbins PD, Morelli AE (2014) Regulation of immune responses by extracellular vesicles. Nat Rev Immunol 14(3):195–208
2. Lai RC, Arslan F, Lee MM, Sze NSK, Choo A, Chen TS et al (2010) Exosome secreted by MSC reduces myocardial ischemia/reperfusion injury. Stem Cell Res 4(3):214–222
3. Aatonen MT, Öhman T, Nyman TA, Laitinen S, Grönholm M, Siljander PR-M (2014) Isolation and characterization of platelet-derived extracellular vesicles. J Extracell Vesicles 3(1):24692
4. Salomon C, Torres MJ, Kobayashi M, Scholz-Romero K, Sobrevia L, Dobierzewska A et al (2014) A gestational profile of placental exosomes in maternal plasma and their effects on endothelial cell migration. PLoS One 9(6): e98667
5. Zhang H, Deng T, Liu R, Bai M, Zhou L, Wang X et al (2017) Exosome-delivered EGFR regulates liver microenvironment to promote gastric cancer liver metastasis. Nat Commun 8(1):15016
6. Welsh JA, Goberdhan DC, O'Driscoll L, Buzas EI, Blenkiron C, Bussolati B et al (2024) Minimal information for studies of extracellular vesicles (MISEV2023): from basic to advanced approaches. J Extracell Vesicles 13(2):e12404
7. Subra C, Grand D, Laulagnier K, Stella A, Lambeau G, Paillasse M et al (2010) Exosomes account for vesicle-mediated transcellular transport of activatable phospholipases and prostaglandins [S]. J Lipid Res 51(8): 2105–2120
8. Skotland T, Sandvig K, Llorente A (2017) Lipids in exosomes: current knowledge and the way forward. Prog Lipid Res 66:30–41

9. Harding C, Heuser J, Stahl P (1983) Receptor-mediated endocytosis of transferrin and recycling of the transferrin receptor in rat reticulocytes. J Cell Biol 97(2):329–339
10. Raposo G, Nijman HW, Stoorvogel W, Liejendekker R, Harding CV, Melief C et al (1996) B lymphocytes secrete antigen-presenting vesicles. J Exp Med 183(3): 1161–1172
11. Pathan M, Fonseka P, Chitti SV, Kang T, Sanwlani R, Van Deun J et al (2019) Vesiclepedia 2019: a compendium of RNA, proteins, lipids and metabolites in extracellular vesicles. Nucleic Acids Res 47(D1):D516–D5D9
12. Keerthikumar S, Chisanga D, Ariyaratne D, Al Saffar H, Anand S, Zhao K et al (2016) ExoCarta: a web-based compendium of exosomal cargo. J Mol Biol 428(4):688–692
13. Cheng L, Sharples RA, Scicluna BJ, Hill AF (2014) Exosomes provide a protective and enriched source of miRNA for biomarker profiling compared to intracellular and cell-free blood. J Extracell Vesicles 3(1):23743
14. Sarker S, Scholz-Romero K, Perez A, Illanes SE, Mitchell MD, Rice GE et al (2014) Placenta-derived exosomes continuously increase in maternal circulation over the first trimester of pregnancy. J Transl Med 12:1–19
15. French SL, Butov KR, Allaeys I, Canas J, Morad G, Davenport P et al (2020) Platelet-derived extracellular vesicles infiltrate and modify the bone marrow during inflammation. Blood Adv 4(13):3011–3023
16. Escobar C, Kao C-Y, Das S, Papoutsakis ET (2020) Human megakaryocytic microparticles induce de novo platelet biogenesis in a wild-type murine model. Blood Adv 4(5):804–814
17. Valadi H, Ekström K, Bossios A, Sjöstrand M, Lee JJ, Lötvall JO (2007) Exosome-mediated transfer of mRNAs and microRNAs is a novel mechanism of genetic exchange between cells. Nat Cell Biol 9(6):654–659
18. Konoshenko MY, Lekchnov EA, Vlassov AV, Laktionov PP (2018) Isolation of extracellular vesicles: general methodologies and latest trends. Biomed Res Int 2018(1):8545347
19. Zhang Q, Jeppesen DK, Higginbotham JN, Franklin JL, Coffey RJ (2023) Comprehensive isolation of extracellular vesicles and nanoparticles. Nat Protoc 18(5):1462–1487
20. Kestens V, Bozatzidis V, De Temmerman P-J, Ramaye Y, Roebben G (2017) Validation of a particle tracking analysis method for the size determination of nano-and microparticles. J Nanopart Res 19:1–16
21. Parsons ME, McParland D, Szklanna PB, Guang MHZ, O'Connell K, O'Connor HD et al (2017) A protocol for improved precision and increased confidence in nanoparticle tracking analysis concentration measurements between 50 and 120 nm in biological fluids. Front Cardiovasc Med 4:68
22. Carnell-Morris P, Tannetta D, Siupa A, Hole P, Dragovic R (2017) Analysis of extracellular vesicles using fluorescence nanoparticle tracking analysis. In: Kuo WP, Jia S (eds) Extracellular vesicles: methods and protocols. Springer New York, New York, pp 153–173
23. Buntsma NC, Shahsavari M, Gąsecka A, Nieuwland R, van Leeuwen TG, van der Pol E (2023) Preventing swarm detection in extracellular vesicle flow cytometry: a clinically applicable procedure. Res Pract Thromb Haemost 7(4):100171
24. Welsh JA, Van Der Pol E, Arkesteijn GJ, Bremer M, Brisson A, Coumans F et al (2020) MIFlowCyt-EV: a framework for standardized reporting of extracellular vesicle flow cytometry experiments. J Extracell Vesicles 9(1):1713526
25. Arraud N, Linares R, Tan S, Gounou C, Pasquet JM, Mornet S et al (2014) Extracellular vesicles from blood plasma: determination of their morphology, size, phenotype and concentration. J Thromb Haemost 12(5):614–627
26. Coppinger JA, Cagney G, Toomey S, Kislinger T, Belton O, McRedmond JP et al (2004) Characterization of the proteins released from activated platelets leads to localization of novel platelet proteins in human atherosclerotic lesions. Blood 103(6):2096–2104
27. Milioli M, Ibáñez-Vea M, Sidoli S, Palmisano G, Careri M, Larsen MR (2015) Quantitative proteomics analysis of platelet-derived microparticles reveals distinct protein signatures when stimulated by different physiological agonists. J Proteome 121:56–66
28. Palviainen M, Puutio J, Østergaard RH, Eble JA, Maaninka K, Butt U et al (2024) Beyond basic characterization and omics: immunomodulatory roles of platelet-derived extracellular vesicles unveiled by functional testing. J Extracell Vesicles 13(10):e12513
29. Heijnen HF, Schiel AE, Fijnheer R, Geuze HJ, Sixma JJ (1999) Activated platelets release two types of membrane vesicles: microvesicles by surface shedding and exosomes derived from exocytosis of multivesicular bodies and alpha-granules. Blood 94(11):3791–3799
30. Parsons ME, Szklanna PB, Guerrero JA, Wynne K, Dervin F, O'Connell K et al (2018) Platelet releasate proteome profiling reveals a core set of proteins with low variance between healthy adults. Proteomics 18(15):1800219

31. Sinauridze EI, Kireev DA, Popenko NY, Pichugin AV, Panteleev MA, Krymskaya OV et al (2007) Platelet microparticle membranes have 50-to 100-fold higher specific procoagulant activity than activated platelets. Thromb Haemost 97(03):425–434
32. Zhou R, Bozbas E, Allen-Redpath K, Yaqoob P (2022) Circulating extracellular vesicles are strongly associated with cardiovascular risk markers. Front Cardiovasc Med 9:907457
33. Lin H, Chen H, Qi B, Jiang Y, Lian N, Zhuang X et al (2021) Brain-derived extracellular vesicles mediated coagulopathy, inflammation and apoptosis after sepsis. Thromb Res 207:85–95
34. Takei Y, Yamada M, Saito K, Kameyama Y, Aihara T, Iwasaki Y et al (2024) Endothelium-derived extracellular vesicles expressing intercellular adhesion molecules reflect endothelial permeability and sepsis severity. Anesth Analg 139(2):385–396
35. Wang L, Tang Y, Tang J, Liu X, Zi S, Li S et al (2024) Endothelial cell-derived extracellular vesicles expressing surface VCAM1 promote sepsis-related acute lung injury by targeting and reprogramming monocytes. J Extracell Vesicles 13(3):e12423
36. Li P, Wu Y, Goodwin AJ, Wolf B, Halushka PV, Wang H et al (2023) Circulating extracellular vesicles are associated with the clinical outcomes of sepsis. Front Immunol 14:1150564
37. Tan Q, Shi S, Liang J, Zhang X, Cao D, Wang Z (2020) MicroRNAs in small extracellular vesicles indicate successful embryo implantation during early pregnancy. Cells 9(3):645
38. Sheller-Miller S, Trivedi J, Yellon SM, Menon R (2019) Exosomes cause preterm birth in mice: evidence for paracrine signaling in pregnancy. Sci Rep 9(1):608
39. Sabapatha A, Gercel-Taylor C, Taylor DD (2006) Specific isolation of placenta-derived exosomes from the circulation of pregnant women and their immunoregulatory consequences 1. Am J Reprod Immunol 56(5–6): 345–355
40. Hannan NJ, Jones RL, White CA, Salamonsen LA (2006) The chemokines, CX3CL1, CCL14, and CCL4, promote human trophoblast migration at the feto-maternal interface. Biol Reprod 74(5):896–904
41. Ledee-Bataille N (2004) Secreted cytokines in the uterine lumina are predictive of subsequent implantation. Presence of IL18 in the uterine flushing. J Gynecol Obstet Biol Reprod (Paris) 33(1 Pt 2):S29–S32
42. Aluvihare VR, Kallikourdis M, Betz AG (2004) Regulatory T cells mediate maternal tolerance to the fetus. Nat Immunol 5(3):266–271
43. Griffith OW, Chavan AR, Protopapas S, Maziarz J, Romero R, Wagner GP (2017) Embryo implantation evolved from an ancestral inflammatory attachment reaction. Proc Natl Acad Sci 114(32):E6566–E6E75
44. Sheller-Miller S, Lei J, Saade G, Salomon C, Burd I, Menon R (2016) Feto-maternal trafficking of exosomes in murine pregnancy models. Front Pharmacol 7:432
45. Jia L, Zhou X, Huang X, Xu X, Jia Y, Wu Y et al (2018) Maternal and umbilical cord serum-derived exosomes enhance endothelial cell proliferation and migration. FASEB J 32(8): 4534–4543
46. Salomon C, Guanzon D, Scholz-Romero K, Longo S, Correa P, Illanes SE et al (2017) Placental exosomes as early biomarker of preeclampsia: potential role of exosomal microRNAs across gestation. J Clin Endocrinol Metabol 102(9):3182–3194
47. Helige C, Ahammer H, Hammer A, Huppertz B, Frank H-G, Dohr G (2008) Trophoblastic invasion in vitro and in vivo: similarities and differences. Hum Reprod 23(10): 2282–2291
48. Ouyang Y, Bayer A, Chu T, Tyurin VA, Kagan VE, Morelli AE et al (2016) Isolation of human trophoblastic extracellular vesicles and characterization of their cargo and antiviral activity. Placenta 47:86–95
49. Luo S-S, Ishibashi O, Ishikawa G, Ishikawa T, Katayama A, Mishima T et al (2009) Human villous trophoblasts express and secrete placenta-specific microRNAs into maternal circulation via exosomes. Biol Reprod 81(4): 717–729
50. Feng Y, Lau S, Chen Q, Oyston C, Groom K, Barrett CJ et al (2024) Normotensive placental extracellular vesicles provide long-term protection against hypertension and cardiovascular disease. Am J Obstet Gynecol 231(3):350. e1–e24
51. Brosens I, Robertson W, Dixon H (1967) The physiological response of the vessels of the placental bed to normal pregnancy. J Pathol Bacteriol 93(2):569–579
52. Steegers EA, Von Dadelszen P, Duvekot JJ, Pijnenborg R (2010) Pre-eclampsia. Lancet 376(9741):631–644
53. Chen Y, Huang Y, Jiang R, Teng Y (2012) Syncytiotrophoblast-derived microparticle shedding in early-onset and late-onset severe pre-eclampsia. Int J Gynecol Obstet 119(3): 234–238
54. Knight M, Redman CW, Linton EA, Sargent IL (1998) Shedding of syncytiotrophoblast microvilli into the maternal circulation in

pre-eclamptic pregnancies. BJOG Int J Obstet Gynaecol 105(6):632–640

55. Goswami D, Tannetta D, Magee L, Fuchisawa A, Redman C, Sargent I et al (2006) Excess syncytiotrophoblast microparticle shedding is a feature of early-onset pre--eclampsia, but not normotensive intrauterine growth restriction. Placenta 27(1):56–61
56. Ramos A, Youssef L, Molina P, Torramadé-Moix S, Martinez-Sanchez J, Moreno-Castaño AB et al (2024) Circulating extracellular vesicles and neutrophil extracellular traps contribute to endothelial dysfunction in pre-eclampsia. Front Immunol 15:1488127
57. Cronqvist T, Tannetta D, Mörgelin M, Belting M, Sargent I, Familari M et al (2017) Syncytiotrophoblast derived extracellular vesicles transfer functional placental miRNAs to primary human endothelial cells. Sci Rep 7(1):4558
58. Pillay P, Maharaj N, Moodley J, Mackraj I (2016) Placental exosomes and pre-eclampsia: maternal circulating levels in normal pregnancies and, early and late onset pre-eclamptic pregnancies. Placenta 46:18–25
59. Gill M, Motta-Mejia C, Kandzija N, Cooke W, Zhang W, Cerdeira AS et al (2019) Placental syncytiotrophoblast-derived extracellular vesicles carry active NEP (neprilysin) and are increased in preeclampsia. Hypertension 73(5):1112–1119
60. Murphy CA, Daniel OR, Luisa W, Stephen M, Hayley M, Ana LC, Elaine N et al (2025) Unique patterns of circulating extracellular vesicles in preterm infants during adaptation to extra-uterine life. J Extracell Vesicles 14(4): e70064. https://doi.org/10.1002/jev2.70064
61. Lal CV, Olave N, Travers C, Rezonzew G, Dolma K, Simpson A et al (2018) Exosomal microRNA predicts and protects against severe bronchopulmonary dysplasia in extremely premature infants. JCI Insight 3(5):e93994
62. Ali A, Zambrano R, Duncan MR, Chen S, Luo S, Yuan H et al (2021) Hyperoxia-activated circulating extracellular vesicles induce lung and brain injury in neonatal rats. Sci Rep 11(1):8791

Chapter 12

Innovative Methods to Label, Track, and Quantitate Extracellular Vesicle Uptake

Amritha Achuthkumar, Peter J. Wermuth, and Mỹ G. Mahoney

Abstract

Extracellular vesicles (EVs) have garnered tremendous attention for their role in facilitating intercellular communication by transporting biomolecules and chemical compounds to recipient cells. This has sparked interest in their potential as targeted delivery systems for therapeutic agents. Researchers are exploring ways to enhance these capabilities by engineering the composition of EVs to ensure they are taken up by specific subpopulations of recipient cells. For EVs to serve effectively as delivery vehicles for targeted therapies, it is crucial that they are internalized by the intended cells, allowing them to release their cargo. However, despite extensive research, the factors that regulate the entry of EVs and synthetic lipid nanoparticles (LNPs) into target cells, as well as the mechanisms that enable them to evade lysosomal degradation and release their contents, remain largely unclear. One challenge contributing to this knowledge gap is the difficulty in tracking the binding, internalization, and intracellular movement of these particles. To address this issue, significant efforts are being made to develop innovative techniques for monitoring and quantifying the internalization and intracellular dynamics of EVs and other nanoparticles. In this context, we will discuss the key factors to consider when designing such methods and review five innovative strategies for labeling, tracking, and quantifying the internalization of EVs and synthetic LNPs.

Key words Extracellular vesicles, Uptake, Methods, Internalization

Abbreviations

AKT	AK strain-transforming protein
AP1	activator Protein 1
AP2	activator Protein 2
AP3	activator Protein 3
AP180	activator Protein 180
ARP2/3	actin-related Protein 2/3
BF	bright field
CD9	Cluster of Differentiation Protein 9
CD63	Cluster of Differentiation Protein 63

This study was supported by grants from the National Institutes of Health (NIH) NIAMS-R01-AR074314, and NIDCR-R21-DE033558-01A1 to Mỹ Mahoney.

Amit K. Srivastava (ed.), *Extracellular Vesicles*, Neuromethods, vol. 228, https://doi.org/10.1007/978-1-0716-4905-3_12,

CDC42	cell division cycle 42
CFSE	carboxyfluorescein succinimidyl ester
CCD	charge-coupled device
CLIC	clathrin-independent carriers
CPZ	chlorpromazine
Cyt D	cytochalasin D
DIBO-NHS	dibenzocyclooctyne *N*-hydroxysuccinimide ester
DiD, 1	1′-dioctadecyl-3, 3, 3′, 3′-tetramethylindodicarbocyanine, 4-chlorobenzenesulfonate salt
Dil, 1	1′-dioctadecyl-3, 3, 3′, 3′-tetramethylindocarbocyanine perchlorate
DiR, 1	1′-dioctadecyl-3, 3, 3′, 3′-tetramethylindotricarbocyanine iodide
DMEM	Dulbecco's Modified Eagle Medium
DMSO	dimethyl sulfoxide
DSP1	Dual Split Protein 1
DSP2	Dual Split Protein 2
Dyn	dynasore
EEA1	early endosome antigen-1
EDC	*N*-(3-(dimethylamino)propyl-*N′*-ethylcarbodiimide hydrochloride
EGFP	enhanced green fluorescent protein
EGFR	epidermal growth factor receptor
EIPA	envelope integrity Protein A
EV	extracellular vesicle
FBS	fetal bovine serum
GEEC	GPI-AP-enriched compartments
GPCR	G protein-coupled receptor
GPI-AP	glycosylphosphatidylinositol-anchored proteins
GTPase	guanosine triphosphatase
HEK	human embryonic kidney
HER-2	human epidermal growth factor receptor 2
ICP-MS	inductively coupled mass spectrometry
IEDDA	inverse electron demand Diels-Alder
lEV	large extracellular vesicle
LNP	lipid nanoparticle
MALS	medium angle light scatter
mAb	monoclonal antibody
MAPK	mitogen-activated protein kinase
MFI	mean fluorescence intensity
MHz	megahertz
MUNC-18	Sec1/Mammalian homolog of Caenorhabditis elegans uncoordinated gene 18
MV	microvesicle
MWCO	molecular weight cutoff
NDZ	nocodazole
PALM	photoactivated localization-based microscopy
PBS	phosphate buffered saline
PKH26	Paul Karl Hogan 26 dye
PKH67	Paul Karl Hogan 67 dye
PLB	passive lysis buffer
PMT	photomultiplier tube
RAB4	RAS-associated binding Protein 4

RAB5	RAS-associated binding Protein 5
RAB7	RAS-associated binding Protein 7
RAB11	RAS-associated binding Protein 11
RAB14	RAS-associated binding Protein 14
RAB22	RAS-associated binding Protein 22
RAB34	RAS-associated binding Protein 34
RAC1	RAS-related C3 botulinum toxin substrate 1
RAC2	RAS-related C3 botulinum toxin substrate 2
RAS	rat sarcoma
RMS	root mean square
SDS-PAGE	sodium dodecyl sulfate-polyacrylamide gel electrophoresis
sEV	small extracellular vesicle
SHIP	specific hybridization internalization probe
shRNA	short hairpin RNA
SIM	structured illumination microscopy
siRNA	short interfering RNA
SNARE	soluble N-ethylmaleimide-sensitive factor attachment protein receptor
SRC	proto-oncogene tyrosine-protein kinase
SSC	side scatter
STED	stimulated depletion microscopy
STORM	stochastic optical reconstruction microscopy
TBS	Tris-buffered saline
TCO	*Trans*-cyclooctane
TDI	time delay integration
Tet	tetrazine
VEGF	vascular endothelial growth factor
VSV-G	vesicular stomatitis virus glycoprotein G
WGA	wheat germ agglutinin
Wort	wortmannin

1 Introduction

Extracellular vesicles (EVs), lipid-bound particles that have been released into the extracellular space, have generated intense interest for their crucial roles in intercellular communication [1, 2], regulation of immune responses [3–6], modification of the tumor microenvironment [7, 8], disease progression [9–12], and their potential to serve as biomarkers for disease prognosis and response to therapeutic interventions [13–18]. The ability of EVs to modify the phenotype of recipient cells has produced copious studies to exploit their enormous potential to function as natural therapeutic nanovesicles [19–21]. EVs can be classified based on distinct biogenic pathways, composition, and functions into three major categories: microvesicles, exosomes, and apoptotic bodies. Due to the heterogeneity of these EV populations, further subclassifications of these three main groups based on size have been proposed [22, 23].

Initial studies on the biological effects of small EVs (sEVs) secreted by cancer cells identified their luminal cargo as the mediators of these effects [13, 24–27]. Modification of recipient cell physiology by intraluminal cargo molecules requires EVs to be internalized by the recipient cell, followed by the release of the vesicle contents into the cell. Internalization of EVs is accomplished by employing one or more endocytosis pathways. Each of these pathways utilizes distinct molecular mechanisms that can affect cellular function. The major endocytic pathways include clathrin-mediated endocytosis, caveolin-mediated endocytosis, lipid raft-mediated endocytosis, glycosylphosphatidylinositol (GPI)-anchored proteins/glycosylphosphatidylinositol (GPI)-anchored protein-enriched compartment (CLIC/GEEC) endocytosis, macropinocytosis, phagocytosis, and direct membrane fusion as depicted in Fig. 1 [28, 29].

1.1 EV Uptake Pathways

Clathrin-mediated endocytosis begins with clustering of the molecular scaffold protein clathrin and accessory adapter proteins like Activator Protein-1 (AP1), AP2, AP3, and AP180 on the cytoplasmic side of the plasma membrane [30, 31]. In addition to promoting clathrin coat assembly, these adapter proteins may play a crucial role in recruiting EVs to the nascent clathrin-coated pit by binding to proteins displayed on the EV membrane or their biomolecular corona [32]. Dynamin-2 is recruited to the clathrin structure, which mediates invagination of the membrane that pinches off to form the clathrin-coated vesicle [33]. Following the uncoating of clathrin, the vesicles with their EV cargo bind to the early endosome, also termed the sorting endosome, which is positive for the RAS-associated binding protein 5 (RAB5) guanine triphosphatase (GTPase) [34]. In the sorting endosome, G-protein-coupled receptor (GPCR) cargos are sorted for recycling to the plasma membrane by either a fast-RAB4-mediated pathway [35] or a slow RAB11-mediated pathway [36]. The $RAB5^+$ recycling endosome matures into a $RAB7^+$ late endosome, which fuses with or matures into the lysosome, where the remaining cargo is degraded [37].

Caveolin-mediated endocytosis begins with the formation of 50–60 nm invaginations of the plasma membrane called caveolae [38–41]. Caveolae formation originates in the endoplasmic reticulum with the formation of hetero-oligomers of the caveolin-1 and caveolin-2 proteins. These complexes are sorted to the Golgi complex, where they bind cholesterol, followed by export to the plasma membrane. Cavin protein family members, which are abundant and essential components of caveolae, form multimeric complexes in the cytoplasm composed of ~50 cavin monomers [42]. Caveolin-

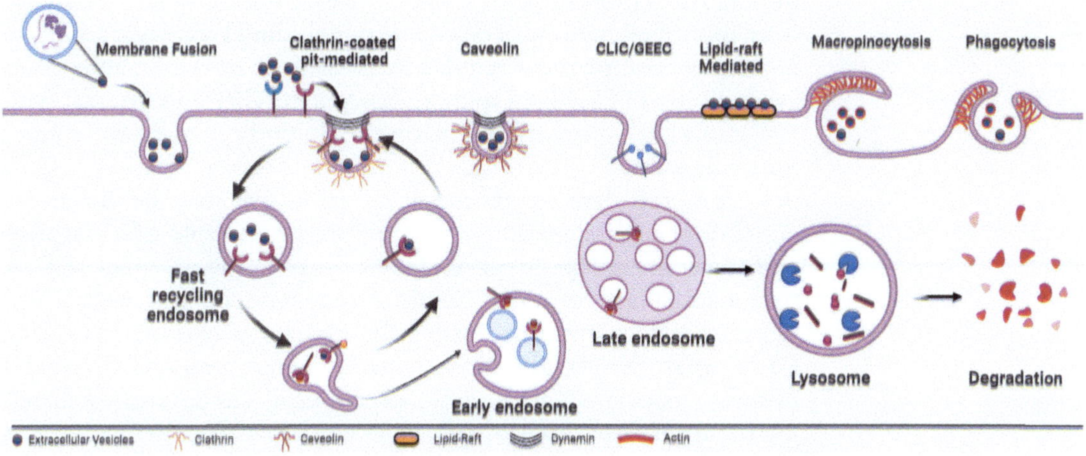

Fig. 1 Overview of endocytosis pathways. Cells can employ multiple mechanisms to internalize particles. Clathrin-mediated endocytosis is a sophisticated process involving the formation of clathrin-coated pits that capture extracellular particles and facilitate their internalization and transport within the cell. Lipid raft-mediated and caveolin-dependent endocytosis occurs when particles associate with caveolin-1 protein, a key component of caveolae. Both clathrin-mediated and caveolin-mediated endocytosis require the protein dynamin. There are additional various pathways for clathrin- and caveolin-independent endocytosis, some of which depend on dynamin, while others, like the clathrin-independent carrier (CLIC) and glycosylphosphatidylinositol-anchored protein-enriched endosomal compartments (GEEC) pathways, do not. Macropinocytosis is characterized by the engulfment and uptake of larger particles, whereas phagocytosis pertains to the uptake of smaller ones. Both processes necessitate actin-mediated extension of the plasma membrane to surround and internalize particles. During transcytosis, extracellular vesicles (EVs) taken up through clathrin-mediated endocytosis, caveolin-mediated endocytosis, or macropinocytosis are transported within the cell from one luminal surface to another before being released. This process occurs in endothelial cells of the central nervous system and enables EVs to traverse the blood–brain barrier. After being internalized through any of these pathways, the cargo is transported intracellularly and ultimately fuses with either early endosomes or lysosomes. (Figure designed using BioRender)

cholesterol complexes, multimeric cavin complexes, and sphingomyelin associate at the plasma membrane to form caveolae. Other accessory proteins are recruited that stabilize the nascent caveolae and that can bind to ligands in the EV membrane or biomolecular corona. Dynamin-2 recruitment [33] and tyrosine phosphorylation of caveolin proteins by Src family kinases [43] facilitate the inward budding of mature caveolae. Once internalized, caveolae and their cargo are shuttled to the early endosome via the actin cytoskeleton [41, 44].

Lipid raft-mediated endocytosis is a broad category of clathrin- and caveolin-independent, cholesterol-sensitive endocytic pathways [45–47]. As the name implies, lipid raft domains are plasma membrane microdomains enriched in various lipids, particularly sphingolipids and cholesterol. These microdomains become associated with integral cell membrane proteins called glycosylphosphatidylinositol (GPI)-anchored proteins (GPI-APs). Since the protein moiety lacks a transmembrane domain, its association with the membrane requires the GPI anchor. More than 150 GPI-Aps have been identified, with varied functions including receptors, adhesion molecules, complement regulators, hydrolytic enzymes, and protease inhibitors [48–50]. Lipid raft-associated GPI-APs interact with EV-associated proteins, mediating EV binding and subsequent internalization.

Lipid raft-dependent endocytic pathways may be dynamin-dependent or -independent. An important example of a clathrin- and caveolin-independent endocytosis pathway is mediated by tubulovesicular "uncoated" clathrin-independent carriers (CLICs) generated from invaginations of the plasma membrane [51]. The CLICs mature into GPI-AP-enriched compartments (GEECs) and become associated with actin-dependent, cholesterol-mediated microclusters of GPI-APs [52, 53]. The CLIC/GEEC pathway, a major contributor to membrane dynamics that enables rapid membrane turnover for plasma membrane homeostasis and repair [54], can be further characterized by the presence of various accessory proteins such as GTPase regulator associated with focal adhesion kinase-1 (GRAF1) [55], RAS homolog A RHOA [56], flotillin 1 (FLOT1) [57–59], or adenosine diphosphate ribosylation factor 6 (ARF6) [60].

Macropinocytosis is a clathrin- and caveolin-independent endocytosis pathway. Extension of the plasma membrane, which is mediated by cholesterol, actin, and microtubules, generates membrane ruffles that fold over and fuse to the plasma membrane [61, 62]. This results in the formation of macropinosomes, which are large (>250 nm) intracellular vacuoles. Macropinocytosis occurs independently of specific membrane receptor binding and activation but instead, it is mediated by phosphatidylinositol-3 kinase (PI3-K) [63, 64] and by various small non-RHO GTPases such as cell division cycle 42 (CDC42), Ras-related C3 botulinum toxin substrate 1 (RAC1), and RAC2 that promote actin-related protein 2/3 (ARP2/3)-dependent branched actin polymerization [65, 66]. Macropinosomes fuse to the early endosome as mediated by RAB5, RAB14, and RAB22 [67], or to the late endosome or lysosome as mediated by RAB7 and RAB34 [68–70].

Phagocytosis is an endocytosis pathway employed by cells for the engulfment of large particles, primarily pathogens and cellular debris. Phagocytic cells such as monocytes, macrophages, neutrophils, and dendritic cells, as well as some non-phagocytic cells such as fibroblasts, endothelial cells, and epithelial cells, have been reported to internalize EVs via phagocytosis [71, 72]. Phagosome formation begins with membrane receptor activation that triggers an actin-dependent extension of membrane-derived pseudopods that surround the particle and fuse together, forming the phagosome [61, 71, 73]. As with macropinosomes, phagosomes have been reported to fuse with both early endosomes and lysosomes [67–70].

In addition to endocytosis-mediated EV uptake, EVs can also be internalized following the fusion of EV membranes to the plasma membrane, followed by the release of the EV luminal cargo directly into the cytoplasm. Following the hemifusion of the outer leaflets of the EV and plasma membrane, a fusion pore can open that allows the two hydrophobic leaflets to form a single structure [73–75]. The participation of several protein families, including soluble N-ethylmaleimide-sensitive factor attachment protein receptors (SNAREs), RAB family proteins, and Secl/ Mammalian homolog of *Caenorhabditis elegans* uncoordinated gene 18 (MUNC-18)-related proteins, is required to mediate this process [76–78].

1.2 Uptake-Independent EV Signaling

The importance of EV intraluminal cargo in mediating phenotypic changes in recipient cells has been called into question based on a quantitative stoichiometric analysis of the distribution of miRNAs in EVs [79], and studies have shown that most EV cargo is degraded after EV internalization due to fusion of late endosomes with lysosome [80]. Increasingly, the role of proteins displayed on the outer leaflet of the EV membrane in EV-mediated modulation of recipient cell phenotypes, independent of EV internalization, is being investigated.

Analyses of the protein components of EVs isolated from multiple cell types and bodily fluids have identified thousands of proteins [81, 82]. A 2014 analysis of the 3027 EV proteins listed in the Vesiclepedia database reported that ~21% of EV-associated proteins were plasma membrane proteins, with diverse classifications that included adhesion molecules, tetraspanins, membrane proteases, heparan sulfate proteoglycans, cytokines, chemokines, cell surface receptors, and their ligands [83].

It has been reported that many cytokines are released from cells both as free proteins and associated with EVs [84, 85]. For example, a study quantifying the amount of vascular endothelial growth factor (VEGF) released by cancer cells found that nearly 35% was associated with a population of EVs <200 nm. Another study that measured the proportion of 33 secreted and EV-associated cytokines from eight different cultured cells, tissue explants, or bodily fluids reported that all of the cytokines measured were associated to some degree with EVs and that the magnitude of this association depended on the cell type or context [84]. By extending their analysis to determine whether the EV-associated cytokines could be localized to the EV membrane or lumen, they reported that 18 of the 33 cytokines were predominantly localized in the lumen, while the others were more often membrane-associated.

One of the advantages of secreting soluble factors encapsulated in the lumen of vesicles is that they would be protected during transport from proteases and other enzymes present in bodily fluids. Such an advantage was not thought to apply to integral EV membrane proteins or to proteins displayed on the EV surface. The half-life of soluble secreted cytokines in plasma is typically less than 1 h [85]; however, the stability of the ovarian, colorectal, and renal cancer cell EV-associated $VEGF_{189}$ isoform was profoundly increased, with no significant change in the levels measured after 24 h [84]. In addition, the association of proteins with EV membranes can affect cytokine potency, possibly due to conformational changes. For example, the levels of $VEGF_{90k}$, which is composed of cross-linked $VEGF_{165}$, stimulates twofold greater levels of endothelial tube formation than free $VEGF_{165}$ [86], while the epidermal growth factor receptor (EGFR) ligand amphiregulin associated with the membrane of EVs produced by colorectal cancer cells stimulates fivefold more invasiveness than soluble amphiregulin [87].

Multiple recent studies have reported evidence that EV-mediated phenotype modification in recipient cells can be induced independently of EV internalization and cargo release. Muller et al. reported that exosomes secreted by tumor cells upregulated the expression of multiple immune response-related genes in activated T cells [5]. Since neither resting nor activated T-cells had internalized exosomes that were labeled with the red fluorescent dye PKH26, the signals responsible for activating gene expression were proposed to be achieved by the activation of T-cell surface receptors by macromolecules on the surface of the tumor-derived exosomes [5]. The previously discussed ovarian, colorectal, and renal cancer cell EV-associated $VEGF_{189}$ isoform was able to exert its effects on endothelial cell tube formation independently of EV uptake [84]. Lung, colorectal, and skin cancer cell microvesicles stimulated autocrine VEGF signaling as well as the activation of mitogen-associated protein kinase (MAPK) and AK strain-transforming protein (AKT) in endothelial cells by transferring oncogenic epidermal growth factor receptor (EGFR) from the EV membranes to the recipient cell plasma membranes [88].

1.3 EV Uptake Assays

Exploiting the capability of LNPs to serve as targeted delivery systems for chemical and biomolecular therapeutic agents for cancer and other diseases is a major research focus [89]. LNPs can be engineered to target recipient cell populations, allowing them to transport encapsulated insoluble chemical agents or various biomolecular therapeutics through the aqueous circulatory microenvironment with minimal toxicity, protected from chemical or enzymatic degradation, compared with systemic administration of non-encapsulated "naked" therapeutics [90–94]. Conversely, the effectiveness of synthetic LNPs as targeted therapeutic delivery modalities is limited by several factors, such as difficulty loading high concentrations of therapeutic agents [95]; nanoparticle instability during storage due to lyophilization, pH, and temperature conditions [96]; induction of a persistent immune response mediated by the complement system and the innate immune response [97, 98]; and rapid nanoparticle accumulation in the liver [99].

1.4 Association, Internalization, and Uptake

Given the interest in exploiting EVs as targeted delivery vehicles for therapeutic agents, understanding the intricacies of particle uptake and internalization will be necessary to modify them to maximize their uptake by recipient cells. The ability to measure EV uptake is vital to be able to understand the roles that properties such as particle size, shape, composition, and surface modifications play in the efficient internalization of EVs and synthetic LNPs [100, 101]. Identifying the endocytic pathways responsible for EV uptake and the mechanistic differences between particle binding to the membrane and their internalization into the endocytic compartments is critical for a number of bioconjugated therapies that are engineered with pH-sensitive labile linkers that will release biomolecules in the acidic environment found in endosomes [102].

Before providing an overview of several novel strategies used to quantify cellular uptake of EVs and other particles, it is important to define some of the common terms to more clearly understand the advantages and disadvantages of these approaches for tracking and measuring EV uptake. One of the most common mistakes made in this area is the failure to differentiate between particle uptake or internalization and particle association [103]. Although often used interchangeably, they are not, in fact, the same. Very often, the uptake process begins with the binding or tethering of the particle to a molecule on the cell membrane. This can be an actual binding of a ligand to a transmembrane receptor or a more transient association that brings the particle into close association with the membrane until the uptake process is initiated. Once the associated particle has been taken into the cell, the particle has been internalized. Many studies lump together "association" and "internalization" as the more general term "uptake."

1.5 Synchronization

Due to the small size of EVs and synthetic LNPs, distinguishing between associated and internalized particles can be challenging, especially immediately after internalization has occurred but before the particles are shuttled away from the cell membrane. This is common when attempting to quantify internalization by fluorescence-based mechanisms using microscopy or flow cytometry [104]. The values reported are calculated from the mean fluorescence intensity (MFI), which is a measure of association. MFI values are indicative of the average amount of fluorescent particles bound to a sample. However, MFI values cannot distinguish between a sample with a small cluster of cells that bind numerous particles and another sample that has low but uniform binding by most cells in the sample. Therefore, an effective strategy to accurately quantify EV or LNP uptake must be able to discriminate between association and internalization.

Some proposed methods for measuring uptake attempt to decouple particle binding from particle internalization by incubating cells at 4 °C for ~15 min prior to introducing EVs and LNPs to inhibit endocytosis [105]. This occurs as a result of plasma membrane lipids undergoing a phase transition to a crystal gel state, rigidifying the membrane and interfering with vesicle budding but still allowing particle binding to membrane proteins [106, 107]. This can be visualized in fluorescently labeled particles as a ring of particles at the cell surface [108, 109]. After binding has occurred, the cells would be washed to remove unbound particles and then returned to 37 °C to reverse the lipid phase transition, allowing internalization and particle endocytosis to resume.

Although this method is effective and greatly simplifies kinetic analyses of particle internalization, it has significant limitations. The kinetics of the recovery of the various endocytosis pathways differ considerably. For example, clathrin-mediated endocytosis recovers rapidly within minutes, whereas the CLIC/GEEC pathway may take longer than 4 h to recover [110]. Therefore, any kinetic measurements of particle internalization may differ substantially because the uptake has been shifted toward fast-recovering pathways whose kinetics may diverge substantially from the predominant pathways normally used by cells that are maintained at 37 °C.

1.6 Endocytosis Inhibitors

As discussed previously, multiple pathways can be employed by cells to internalize EVs, synthetic LNPs, and other particles [28, 29]. The pathways responsible for internalizing specific classes of particles may be determined by variations in their size, shape, and composition and may also differ based on the target recipient cell. The endocytosis pathway employed may also influence the intracellular tracking of the particles following their internalization [111, 112]. It is also the case that several of these pathways can be operating simultaneously in the same cell, complicating an already complex process.

Endocytosis inhibitors are commonly used tools to study intercellular uptake and internalization pathways by interfering with specific pathways [112–116]. Some of the more commonly employed endocytosis inhibitors are listed in Table 1. While the use of chemical inhibitors of endocytosis is attractive since they can be administered acutely to reveal direct inhibition of a particular pathway, their use must be carefully analyzed.

As is evident in Fig. 1, although several distinct endocytosis pathways have been identified, they often share components. As an example, cytochalasin D or latrunculin block actin polymerization, thereby blocking both actin-dependent phagocytosis and macropinocytosis [117]. Additionally, many of these chemical inhibitors can have serious toxic off-target effects, and their effectiveness in

Table 1
Commonly used endocytosis inhibitors and their mechanisms of action

Pathway inhibited	Inhibitors	Mechanism	References
Clathrin-mediated endocytosis	Chlorpromazine	Inhibits clathrin-coated pit assembly at the plasma membrane	[256]
	Bolinaquinone (BLQ)	Anti-inflammatory; inhibits serum phospholipase A2 by chelating Ca^{2+}	[118, 257]
	Hydroxy chloroquine	Inhibits endosomal acidification	[258]
	Pitstop2	Inhibits clathrin-adaptor interactions	[259]
	Transmembrane domain derived (TDD)	Blocks clathrin-AP2/SNX9 interaction	[260]
Caveolin-mediated endocytosis	Filipin	Disrupts noncoated vesicle structure	[261]
	Methyl-β-cyclodextrin (MβCD)	Depletes plasma membrane cholesterol	[262]
	Nystatin	Sequesters cholesterol	[263]
	Genistein	Tyrosine kinase inhibitor; disrupts caveolar structure	[264]
Dynamin-mediated pathways	Dynasore	Blocks vesicle scission of vesicles from the plasma membrane	[265]
	Myristyl trimethyl ammonium bromide (MiTMAB)	Blocks dynamin recruitment to membranes	[266, 267]
MAcropinocytosis	Rottlerin	Inhibits PKC activity, disrupting cytoskeletal dynamics	[268]
	NSC23766	Blocks Rac1, disrupting actin polymerization	[269]
	Amiloride	Blocks Na+/H+ exchange, reducing cellular pH and altering actin polymerization	[200]
	Wortmannin	Inhibits PI3K, crucial for membrane ruffling	[270]
Actin polymerization	Cytochalasin	Inhibits actin polymerization	[271]
	CK-636	Inhibits Arp2/3 complex, *actin polymerization* and *stress fiber formation*	[272]
Microtubule polymerization	Nocodazole	Binds β-tubulin; destabilizes microtubules	[273]
	Vinblastine	Disrupts the microtubule network	[274]
	Colchicine	Binds to tubulin dimers	[274]
Endosomal acidification	Chloroquine Bafilomycin A	Disrupts endosomal maturation/ lysosomal degradation	[258, 275]
Others	Heparin	Competes with proteoglycan for EV binding	[234]

blocking specific endocytic pathways can vary among different cell types [113]. Dynasore, for example, was identified as an inhibitor of the GTPase of dynamin and an inhibitor of clathrin-mediated endocytosis [118]. However, it has subsequently been found to disrupt lipid rafts [119], destabilize F-actin, and interfere with membrane ruffling [120], which would inhibit macropinocytosis and lipid raft-mediated endocytosis pathways. Dynasore also depletes labile cholesterol from the plasma membrane, accumulates cholesterol in late endosomal compartments [121, 122], inhibits Golgi vesiculation [123, 124], and vacuolar H^{+}-ATPase [125].

Because of these potential drawbacks, genetic approaches to inhibit endocytosis pathways by knocking down the levels of proteins specific to each pathway employing short hairpin RNA (shRNA) or short interfering RNA (siRNA) [126–129] have been employed. A disadvantage of these approaches are that to fully deplete the cell of the targeted protein may require 3–7 days, during which time the cell may compensate for loss of the targeted pathway by upregulating other endocytic pathways.

1.7 Designing EV Uptake Assays to Quantitate EV Internalization

To accurately differentiate between particle association and internalization, the approach employed must be carefully considered by understanding the capabilities and the limitations of the various available detection methods. To enable the tracking and quantitation of EVs and other nanosized particles, they must be labeled in some manner. The choice of label will obviously be influenced by the detection method to be employed. The most common labeling method involves the use of fluorescent dyes, such as hydrophobic, lipophilic, or amphipathic dyes that associate with EV membrane lipid bilayers, including 1, 1′-dioctadecyl-3, 3, 3′, 3′--tetramethylindodicarbocyanine, 4-chlorobenzenesulfonate salt (DiD), 1′-dioctadecyl-3, 3, 3′, 3′-tetramethylindocarbocyanine perchlorate (Dil), 1, 1′-dioctadecyl-3, 3, 3′, 3′--tetramethylindotricarbocyanine iodide (DiR), Paul Karl Hogan dye 26 (PKH26), PKH67, carboxyfluorescein succinimidyl ester (CFSE), or rhodamine B [130–133].

Direct bioconjugation of the fluorescent tag to components of the plasma membrane [134–138] or engineering a fusion protein consisting of a fluorescent tag fused to a membrane protein [139–141] is a common but more complicated fluorescence-based approach. The presence of the fluorescent tag permits the tracking of EVs and other nanoparticles by microscopy and flow cytometry. Another approach has been to label EVs by engineering a fusion between an EV membrane protein and functional luciferase, allowing quantitation of EV internalization by a luminescence assay [137, 140, 142, 143]. A novel approach involving labeling EVs with the lanthanide europium to allow detection by inductively coupled mass spectrometry has recently been proposed [133].

Once labeled, the particles must be detected and quantified. Ideally, a method should be amenable to high-throughput analyses that can reproducibly and actively quantitate particles that have been internalized by recipient cells. Unfortunately, most of the methods for the detection and quantification of internalized particles are limited by the resolution capability of the detectors. Three types of detection methods will be considered: confocal microscopy-based detection, flow cytometry-based detection, and mass spectrometry-based detection. These methods have been used to quantify nanoparticle "uptake"; however, as previously discussed, the nanometer sizes of these particles make it extremely challenging to differentiate and localize distinct particles.

Microscopy approaches are limited in their resolution by the diffraction properties of light [144]. Epifluorescence microscopy overcomes limitations imposed by brightness at high (40–100×) magnification, allowing localization of distinct particles at the full diffraction-limited resolution (~200 nm). The best resolution can be achieved at shorter wavelengths; however, dyes that fluoresce within this range have a lower quantum yield and often are poorly photostable. Worse, proteins and nucleic acids absorb in the UV portion of the spectrum, limiting the resolution achievable in the XY axis, and the resolution in the Z dimension is worse, approaching only 500–700 nm [145, 146].

Four recent approaches to super-resolution microscopy, namely structured illumination microscopy (SIM), photoactivated localization microscopy (PALM), stochastic optical reconstruction microscopy (STORM), and stimulated electron depleted (STED), have been developed that use mathematical approaches to extend the resolution capability beyond the diffraction limit [147–150]. SIM improves XY resolution to 100 nm and Z resolution to 250 nm; PALM/STORM techniques improve XY resolution to 20 nm and Z resolution to 50 nm, while STED improves XY resolution to <50 nm under optimal conditions. None of these super-resolution strategies have been applied to EV internalization studies. However, although they offer the potential to discriminate between particles, they do not address the other major limitation of microscopy-based analyses, i.e., the low-throughput nature of these techniques.

Flow cytometry has been a routine method for the detection of µm-sized cells and particles; however, adapting it for the analysis of nanosized particles based on their light scattering signal has been a major challenge. The lower detection limit for standard flow cytometers is between 300 and 500 nm, although by optimizing triggering strategies, a minimal detection limit of 150 nm is achievable [151–153]. Another challenge for the use of flow cytometry in quantifying EV internalization is the possibility of "swarm" detection, where detection of multiple EVs is counted as a single event that occurs since the fluidics are optimized for the detection of much larger (2–30 µm) particles [154–165].

The advent of high-resolution flow cytometers has attempted to address these challenges by replacing photodiodes on light scatter channels with photomultiplier tubes (PMTs), adding reduced wide-angle forward scatter/medium angle light scatter (MALS) collection, increasing the power of the lasers, and decreasing sample and sheath flow rates [157–160], which has lowered the resolution limit rendering their use in EV detection increasingly prevalent [150, 157, 159–161].

The development of imaging flow cytometry allowed the detection of synthetic nanospheres as small as 20 nm [162]. This can be attributed to several improvements, such as the use of charge-coupled device (CCD) cameras with a larger dynamic range and a lower signal-to-noise ratio than PMTs, and the use of time delay integration (TDI) of pixel intensity for each particle to detect dim signals [163–166]. Also, unlike traditional flow cytometers, images of each particle are saved, which permits more powerful data analysis and event classification to reduce swarm detection.

More recently, nanoflow cytometers specifically designed for the detection and analysis of nanoparticles have been developed [167, 168]. Nanoflow cytometry is a high-throughput method capable of preparative sorting for various biological or inorganic nanoparticles that also preserves their functional activity. These systems incorporate the use of a high-sensitivity scatter parameter that allows a more complete representation of the particle population, optimized fluidic conditions that permit visualization of single particles as small as 40 nm, and methods to evaluate dimly scattered light signals.

Mass spectrometry has been used extensively in analyses of the proteins found in EVs isolated from numerous cells and bodily fluids [169–172]. Interest in using technologies like inductively coupled plasma mass spectrometry (ICP-MS), for the multiparametric analysis of complex samples has grown significantly. A method using metal-tagged antibodies for the simultaneous detection of multiple proteins in acidified biological samples was reported [173–178]. ICP-MS is widely used in industrial applications that demand precise quantification of elemental abundance. Samples are atomized and ionized in plasma at temperatures approaching 7000 K. The mass spectrometer resolves and quantifies the various elemental isotopes [179]. This approach has been adapted to spatial proteomics using antibodies tagged with metal isotopes [180, 181]. Lanthanides are often used to tag antibodies and are useful for ICP-MS studies because of their very low background levels in biological samples [182]. ICP-MS holds enormous potential for the multiparametric analysis of complex heterogeneous populations of EVs.

As discussed previously, given the promise of EVs to serve as targeted delivery vehicles for chemical and biomolecular therapeutics, it is crucial to quantify their efficiency in delivering these agents to recipient cells. This information could help in the design of new strategies for the optimization of EVs as therapeutic delivery vehicles. We will now present in depth five methodologies for tracking and quantifying the uptake of EVs or other nanoparticles. These strategies employ a variety of labeling and detection methods that could be adapted for use in many experimental systems. The studies reported here are not intended to be comprehensive of all possible strategies for measuring EV internalization but are presented instead as representing innovative approaches to address the complexities involved in assays of this nature.

2 Paper Summaries

2.1 Development of a Quantitative Method to Measure EV Uptake

Most techniques for tracking EV uptake and internalization involve labeling EVs with fluorescent or luminescent markers. The most straightforward and widely used techniques utilize hydrophobic, lipophilic, or amphipathic fluorescent dyes such as Dil, DiD, DiR, PKH67, PKH26, or rhodamine B that associate with EV membrane lipid bilayers [183]. However, these dyes have short half-lives, can interfere with normal EV behavior, are susceptible to micelle formation, and may generate nonspecific signals in recipient cells due to direct dye transfer between the EV and recipient cell membranes [184, 185]. Additionally, because they stain any lipid bilayer membrane indiscriminately, they cannot differentiate between specific EV subpopulations [183].

As an alternative, palmitoylated fluorescent dyes have been developed to anchor the dye within the plasma membrane, preventing dye exchange with recipient cell membranes [186, 187]. More frequently, researchers have focused on engineering producer cells to express membrane-bound fusion proteins that incorporate fluorescent tags [188, 189]. This approach allows for tracking these proteins throughout the early stages of EV biogenesis to their release and subsequent uptake and internalization by recipient cells. These fusion proteins also have a longer half-life compared to nonspecific fluorescent labeling dyes, rendering them suitable for in vivo tracking studies. Furthermore, EVs have been engineered that incorporate luciferase enzymes that can produce luminescent signals, primarily for in vivo uptake imaging [190–193].

These fluorescence-based labeling methods are semiquantitative, relying on estimates of particle numbers derived from the intensity of the fluorescence signal. They also face challenges in distinguishing between EV particles that are bound to the membrane and those that have been internalized. Furthermore, the ability of these methods to track individual EVs is limited because the size range of EVs falls below the resolution limit of the optical techniques used for their detection and analysis, necessitating detailed, time-intensive, low-throughput confocal imaging.

The aim of this study was to design a novel, quantitative, and highly sensitive assay for EV internalization. The authors developed an assay inspired by techniques used to investigate virus-induced cell fusion or viral entry into recipient cells [194, 195]. This assay utilizes a pair of chimeric reporter proteins that label EVs with *Renilla reniformis* luciferase and enhanced green fluorescent protein (EGFP) simultaneously. Dual Split Protein 1 (DSP1) encodes amino acids (aa) 1–229 of *Renilla* luciferase and β-sheets 1–7 of EGFP, while DSP2 encodes aa 230–311 of *Renilla* luciferase and β-sheets 8–11 of EGFP [195].

These fusion proteins, when present separately, are nonluminescent and nonfluorescent. However, when they are in the same compartment, they self-reassociate, restoring both the luciferase and green fluorescence activities. Producer cells were engineered to express a fusion protein comprising an EV-associated tetraspanin protein, such as Cluster of Differentiation Protein 63 (CD63) with DSP1 that would be incorporated into EVs. Recipient cells were engineered to express DSP2. This high-throughput assay enables quick and quantitative evaluation of EV uptake and internalization, aids in identifying key functional molecular pathways involved in this process, and supports the screening of potential treatments aimed at modifying EV uptake and internalization in pathological conditions.

2.1.1 Methods

Antibodies

The primary monoclonal antibodies (mAbs) utilized in this study included: anti-CD63 (Tea 3/10) and anti-CD9 (V11/20) (Immunostep), conditioned media from mouse hybridoma [196, 197]; polyclonal anti-early endosome antigen-1 (EEA-1) (Santa Cruz) at a dilution of 1:50 for immunofluorescence, and polyclonal anti-GFP (Living Colors, Clontech) at a dilution of 1:1000 for immunoblotting, or 1:200 for immunofluorescence.

The secondary antibodies used included Donkey-α-Rabbit Alexa488, Donkey-α-Goat Alexa647, Phalloidin-Alexa647 (Invitrogen), and Goat-α-Mouse Alexa647 (Life Technologies), at a dilution of 1:200 for immunofluorescence; Goat-α-Rabbit HRP and Goat-α-Mouse HRP (Thermo Scientific) at 1:5000 for immunoblotting.

Cell Culture

The breast cancer cell line SUM159 was cultured in Dulbecco's modified Eagle medium (DMEM) F-12 culture medium, supplemented with 5% fetal bovine serum (FBS), 1% penicillin/streptomycin, 80 mg/mL nonessential amino acids (HyClone GE Healthcare), 10 mM HEPES, 5 μg/mL insulin, and 1 μg/mL hydrocortisone. The human embryonic kidney 293 (HEK293) cell line was cultured in DMEM supplemented with 10% FBS, 1% penicillin/streptomycin, and 10 mM HEPES. To prepare EV-depleted media, DMEM F-12 was supplemented with FBS that had been depleted of bovine EVs by ultracentrifugation at 120,000 × *g* overnight.

Generation of CD9 and CD63-DSP Constructs

The coding sequences for the CD9 and CD63 tetraspanin proteins were amplified via PCR and subsequently subcloned into the C-terminal end of the Dual Split Protein 1 (DSP1–7) construct, which was generously provided by Dr. Zane Matsuda [198]. The complete coding sequences for the fusion protein, as well as for DSP1–7 (DSP1) and DSP8–11 (DSP2), were amplified again by PCR, sequenced, and subcloned into either the pcDNA3 or PCR2.1 vectors using the TOPO system from Invitrogen. The PCR2.1 vector was digested with EcoRI to facilitate the subcloning of the DSP constructs into a lentiviral PLVX Puro vector, enabling the creation of stably transduced cell lines.

Cell Transfection and Transduction

SUM159 cells were transfected with DSP1 and DSP2 constructs cloned into a pcDNA3 vector. The cells were trypsinized (1/2 P100 culture plate at 80% confluence), pelleted at 400 × g, and then resuspended in 200 μL of incomplete DMEM F-12 culture medium supplemented with 1.5 M NaCl.

For each transfection point, 20 μg of plasmid DNA was used. Electroporation was performed at 200 V and 975 μF in a 4 mm width cuvette using a Bio-Rad Gene Pulser XCell. After electroporation, the cells were placed on ice for 10–15 min and then seeded into a p6 culture plate.

To produce lentivirus, HEK293 packaging cells were transfected with 5 μg of PLP1, 3.24 μg of PLP2, 2.16 μg of VSVG vector encoding the vesicular stomatitis virus glycoprotein G (VSV-G), 2.76 μg of the plasmid encoding one of the DSP1 constructs, and the DSG2 construct in the PLVX vector, using lipotransfectine (Solmeglas) as per the manufacturer's instructions. The culture supernatants were collected 48 h post-transfection and incubated with SUM159 cells in the presence of 8 μg/ mL polybrene (Santa Cruz). Infected cells were selected using 1 μg/mL puromycin. To assess the infection efficiency of single DSP constructs, transient co-transfection of the complementary construct was performed, followed by flow cytometry analysis.

SUM159 cell lines that stably expressed DSP1-CD9/DSP2 or DSP1-CD63/DSP2 were selected based on their green fluorescence signal at 488 nm through cell sorting. Cells transduced with DSP1-CD9 or DSP1-CD63 were labeled with VJ1/20 anti-CD9 or Tea 3/10 anti-CD63 primary mAbs, respectively, along with goat-α-mouse Alexa 647 secondary antibody. Cells exhibiting higher expression levels of the corresponding tetraspanin were sorted based on their fluorescence at 647 nm using a FACSAria Fusion cell sorter (BD Biosciences).

EV Enrichment

EVs were isolated from the conditioned media of SUM159 cells through ultracentrifugation. Secretory SUM159 cells were cultured to confluence in p150 culture plates containing 20 mL of DMEM F-12, supplemented with 5% EV-depleted FBS, for

5–7 days. The supernatants were collected and centrifuged for 30 min at 4000 × *g* to remove cellular debris. Subsequently, the supernatants underwent a second centrifugation for 2 h at 100,000 × *g* to obtain a concentrated pellet of EVs. To distinguish between microvesicles and exosomes, the supernatants were first ultracentrifuged at 10,000 × *g* for 1 h to collect a pellet rich in microvesicles. The remaining supernatants were then ultracentrifuged at 100,000 × *g* for 1 h to yield an exosome-enriched pellet. All ultracentrifugation steps were performed using an L8-70M ultracentrifuge (Beckman) equipped with an SW-Bucket AH-627 rotor. The pellets obtained from the p150 culture plates were resuspended in 500 μL of 0.22 μM filtered phosphate buffered saline (PBS).

BCA Protein Analysis

Protein concentration was determined following the BCA Protein Assay Kit protocol (Pierce). The absorbance of duplicate samples at 540 nm was measured in a GENios microplate reader (Tecan).

Nanoparticle Tracking Analysis

A 1/100 dilution of the sample in filtered PBS was measured with a Nanosight NS500 (Malvern) with the following parameters: camera level: 10, detection threshold: 10, and three captures of 1 min. Data were analyzed with NTA 3.1 Build 3.1.54 software.

Western Blot

Cells or EVs were lysed using Tris-buffered saline (TBS) that contained 1% Triton X-100 and protease inhibitors (Roche) at 4 °C for 30 min. The lysates were then heated in nonreducing Laemmli buffer at 96 °C for 5 min, and 40 μL of the lysates were loaded onto 10% polyacrylamide SDS-PAGE gels. Following electrotransfer with a Transfer-Blot Turbo system (Bio-Rad), the membranes were blocked with 5% skim milk in TBS and 0.1% Tween-20 for 20 min. Immunoblots were developed using SuperSignal West Femto HRP substrate (Thermo Scientific), and images were captured with a LAS 4000 minisystem (General Electric).

Immunofluorescence Microscopy

Coverslips were coated with 5 μg/mL fibronectin (Sigma-Aldrich) for 30 min at 37 °C, after which 30,000 cells per well were seeded in p24 well plates and cultured overnight at 37 °C with 5% CO_2. When indicated, EVs were added to the cells in a final volume of 400 μL of complete culture medium and incubated for 5 h at 37 °C. The samples were washed with PBS and fixed with 4% paraformaldehyde for 10 min at room temperature. Coverslips were washed several times with TBS, and, where indicated, permeabilized by incubating with TBS and 1% Triton X-100 for 5 min, followed by washing with TBS and blocking with TNB blocking solution (100 mM Tris–HCl, 150 mM NaCl, 0.5% blocking reagent; PerkinElmer FP1020) for 30 min at room temperature. Samples were stained with primary antibodies overnight in a wet chamber at 4 °C, followed by appropriate secondary antibodies for 30 min at 37 °C, and then mounted with Fluoromount-G Aqueous Mounting Medium (Sigma) containing 0.1 mM DAPI. The samples were analyzed using an Axiovert 200 inverted microscope (Zeiss) coupled to a CCD monochromatic camera (C9100–2 Hamamatsu) with a 63X/1.4 oil Plan-Apochromat Ph3 objective. Filters used included GFP 503–538 nm (Semrock) and Cy5 665 LP (Chroma). Confocal microscopy images were acquired with a TCS-SP5 confocal laser scanning unit attached to a DMIRBE inverted epifluorescence microscope (Leica Microsystems, Heidelberg, Germany) using an HCX PL APO CS 100.0 × 1.40 OIL objective. Excitation was achieved using a 405 nm laser for UV, a He/Ne laser at 488 nm for EGFP, and a diode 633 nm laser for Alexa 647. Emission wavelengths were set for UV (422–477 nm), green (494–563 nm), and far-red (726–800 nm). The effective pixel size was 0.3 μm per pixel. EGFP and Alexa 647 signals were acquired simultaneously, but sequentially with the UV signal.

Renilla Luciferase Assay from Lysates

Two 90% confluent p100 plates, each of SUM159 cells expressing DSP1, tetraspanin-DSP1, or DSP2 proteins, were lysed using 500 μL of Passive Lysis Buffer (PLB) from the Dual-Glo Luciferase System (Promega), along with protease inhibitors (Roche) at 4 °C. EVs were isolated as previously described from 40 mL of conditioned media from cells expressing either DSP1 or tetraspanin-DSP1 and lysed in 500 μL of PLB. A total of 125 μL of EVs or cell lysates from DSP1-expressing cells was combined with 75 μL of cell lysates from DSP2-expressing cells and co-incubated for the specified durations in a clear-bottom p96 well plate (Corning Incorporated Costar). The luminescent signal was produced by adding 50 μL of the *Renilla* luciferase substrate Stop&Glo from the Dual-Glo Luciferase Assay System and measured using a GENios microplate reader (Tecan) with an exposure time of 5000 ms and a gain setting of 100.

EV Uptake Luciferase Assay

The EV uptake luciferase assay is illustrated in Fig. 2. A total of 32,000 recipient cells per well were cultured in a clear-bottom p96 well plate for 24 h using EV-depleted culture medium. The cells were then co-incubated for 2 h with 100 μL of a 60 μM Enduren luciferase substrate (Promega) mixed in EV-depleted culture medium. Once the co-incubation period was completed, the supernatant was discarded. When indicated, some wells of recipient cells were fixed with 4% paraformaldehyde after loading with Enduren but prior to the addition of EVs.

Endocytosis inhibitors (all obtained from Sigma-Aldrich) were diluted from stock solutions to the specified concentrations in EV-depleted culture medium: 7.5 μM chlorpromazine (CPZ), 10 μM nocodazole (NDZ), 1 μg/mL cytochalasin D (Cyt D), 200 μM genistein, 1 μM wortmannin (Wort), and 80 μM Dynasore (Dyn) stocks prepared in dimethyl sulfoxide (DMSO), 75 μM EIPA

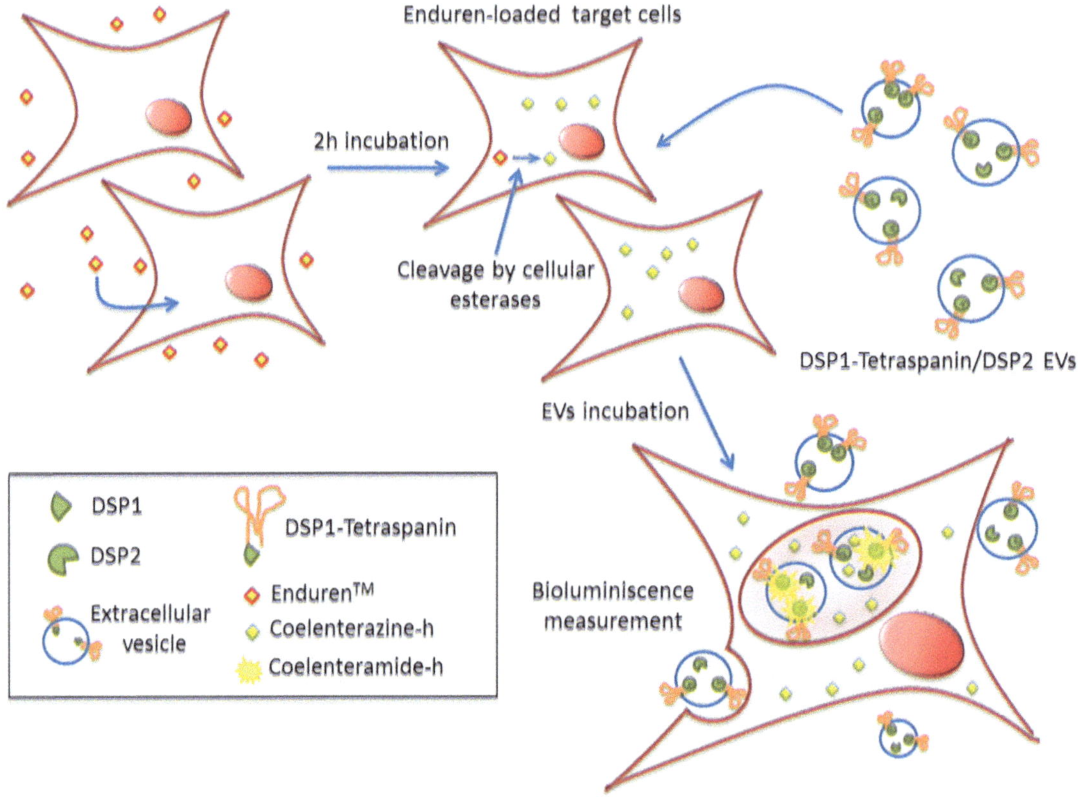

Fig. 2 Scheme of the EV uptake assay. First, receptor cells are preloaded with Enduren, a cytopermeable protected *Renilla* luciferase substrate. Once internalized, Enduren is processed by intracellular esterases, rendering coelenterazine-h *Renilla* luciferase substrate. EVs are isolated from cells that stably express both DSP2 and either DSP1-CD9 or DSP1-CD63, and thus carrying fully competent *Renilla* enzyme. Only those EVs that are taken up and internalized by target cells will gain access to the deprotected *Renilla* luciferase substrate and will transform it into coelenteramide-h, which emits a luminescence signal. (Reproduced with permission. Under a Creative Commons license (https:/creativecommons.org/licenses/by/4.0/) from {*Sci Rep.* 2019, 9, 1, 10522})

stock prepared in methanol, and 10 μg/mL heparin stock prepared in water.

To assess the effects of various inhibitors, recipient cells preloaded with Enduren were treated for 30 min with 50 μL of the diluted inhibitors following the Enduren supernatant removal. Following this, 4–8 μg of EVs were introduced to a total volume of 100 μL of EV-depleted medium. In wells receiving both inhibitors and EVs, they were added simultaneously with the inhibitors.

When using the Dual-Glo Luciferase Assay System, samples were lysed with PLB prior to adding the *Renilla* luciferase substrate Stop&Glo. All measurements were conducted in duplicate and measured at various times using a GENios microplate reader (Tecan), with an exposure time of 5000 ms and a gain setting of 150.

To explore the mechanisms involved in the acquisition and processing of the Enduren substrate by EVs, SUM159 cells were incubated for either 2 or 6 h with 100 μL of 60 μM Enduren diluted in EV-depleted culture medium. Supernatants from six wells of a p96 well plate were collected and pooled. The cells were washed with PBS, and 100 μL of PBS containing protease inhibitors was added to each well. The cells were mechanically lifted and lysed using sonication with a LABSONIC M ultrasonic processor (Sartorius) set to 80% amplitude and 0.6 cycles at 4 °C. The lysates were centrifuged at 21,000 × *g* to eliminate cellular debris. A total of 100 μL of either the supernatant or cell lysates preloaded with Enduren for 2 or 6 h was co-incubated in duplicate with 3 μg of DSP1-CD9/DSP2 EVs, and the luminescence signal was measured at various time points.

Annexin V-PE/7AAD Viability Protocol

Cells were cultured in p24 well culture plates and treated with inhibitors at the previously specified concentrations. After trypsinization, the cells were collected and resuspended in annexin binding buffer (50 mM HEPES, 700 mM NaCl, 12.5 mM $CaCl_2$, pH 7.4) at a concentration of 1×10^6 cells/mL. A total of 100 μL of this cell solution was stained for 15 min at room temperature with 2.5 μL of Annexin V-PE and 5 μL of 7AAD (BD Pharmingen 559,763). The cells were then analyzed by flow cytometry using a FACSCalibur flow cytometer (Becton Dickinson). For PE detection, the emission wavelength was set at 488 nm with detection at 588/42 nm, while for 7AAD detection, the emission wavelength was also 488 nm but with detection at 610/13 nm. The results were analyzed using the FlowJo V-10 program.

Statistical Analyses

Statistical analyses were performed using GraphPad Prism version 6.0 (GraphPad Software Inc., CA).

2.1.2 Results Summary

Toribio et al. developed a highly sensitive and quantitative method to measure the internalization of EVs, enabling the identification of the mechanisms that regulate this process. They designed an assay employing two tags, DSP1 and DSP2, which encode nonfunctional split segments of EGFP and *Renilla* luciferase. When these tagged fusion proteins are expressed separately, they do not emit fluorescence or luminescence; however, when present in the same cellular compartment, they can associate to form a complete EGFP/luciferase complex, restoring green fluorescence and luciferase activity. To ensure that the DSP1 tag is targeted to EVs, a fusion protein was engineered that attaches the DSP1 tag to the C-terminal tail of EV-associated tetraspanin proteins, specifically CD9 or CD63. Four distinct constructs (DSP1, DSP1-CD9, DSP1-CD63, and DSP2) were designed and subcloned into both the pc3DNA-TOPO vector for transfection and transient expression, and the pLVX-Puro lentivirus vector for viral transduction and stable expression in producer cells.

DSP localization was assessed using fluorescence microscopy following transient transfection into SUM159 breast cancer cell line, employing an anti-GFP antibody. The DSP1 and DSP2 constructs exhibited a diffuse distribution throughout the cytoplasm, while the DSP1-CD9 and DSP1-CD63 fusion protein constructs were specifically localized to plasma membrane and vesicular membranes, consistent with tetraspanin localization. Notably, DSP1-CD63 displayed a higher localization to the intracellular vesicle compartments. When the DSP1, DSP1-CD9, or DSP1-CD63 constructs were co-transfected with the DSP2 construct, green fluorescence was recovered, indicating an association between the two components.

After confirming the anticipated transient expression patterns, SUM159 cells were transduced with one of the four single lentiviral constructs (DSP1, DSP1-CD9, DSP1-CD63, or DSP2) or with one of the three double lentiviral construct combinations (DSP1/DSP2, DSP1-CD9/DSP2, or DSP1-CD63/DSP2). The expression of the single constructs was evaluated as previously described using fluorescence microscopy. To measure the recovery of *Renilla* luciferase activity in the single constructs, cell lysates from cells expressing the DSP1 constructs were mixed 1:1 with lysates from the transduced cells expressing the DSP2 constructs, followed by the addition of luciferase substrate. The doubly transduced cells exhibited the highest luciferase activity. EVs were isolated from the conditioned culture medium of single DSP1-expressing cells and DSP2-expressing cells through ultracentrifugation. The EVs were lysed, and the resulting lysates were mixed and analyzed in the same manner as previously described. When mixed with DSP2-carrying EV lysates, the lysates from EVs containing DSP1-CD9 or DSP1-CD63 produced stronger signals than the lysates containing DSP1 alone, likely due to the enriched sorting of the tetraspanin fusion proteins into the EVs.

To assess EV uptake and internalization, DSP1-CD9-EVs were incubated with lung carcinoma A549 epithelial recipient cells expressing DSP2, which had been preloaded with one of two *Renilla* substrates to measure the luminescent signal. The substrates used were either the traditional Stop&Glo reagent, which requires cell lysis, or Enduren, a cytopermeable substrate that permits detection without cell lysis, enabling a kinetic assessment of EV internalization. No luminescent signal was detected after either 6 or 24 h incubation in the presence of either substrate, indicating that DSP1/DSP2 reassociation had not taken place.

EVs isolated from the conditioned media of cells stably cotransduced with both DSP1-CD9 and DSP2, containing a fully functional intraluminal *Renilla* luciferase, served as a positive control. Unexpectedly, these EVs generated a strong signal with the classical Stop&Glo substrate following EV lysis, but no signal was detected using the Enduren substrate in intact EVs. However, if the EVs were placed in contact with and internalized by Enduren-loaded recipient cells, the luciferase signal was readily detected.

Enduren is permeable through the plasma membrane but cannot be processed directly by *Renilla* luciferase. Enduren first needs to be deprotected by cellular esterases to generate the active *Renilla* substrate coelenterazine-h. Isolated EVs containing fully competent *Renilla* luciferase did not emit a signal when incubated with Enduren, suggesting that either Enduren cannot penetrate the EV membrane or the esterase necessary to process Enduren is absent or inaccessible in the EV lumen. When EVs from DSP1-CD9/DSP2-expressing SUM159 cells were incubated with either substrate, no signal was detected. However, incubating intact EVs with lysates, obtained by detergent-free mechanical scraping and sonication, from A549 recipient cells that had been preloaded with Enduren for 2 or 6 h resulted in a robust signal, with the most robust signal coming from lysates preloaded for 6 h. Conditioned media from A549 recipient cells preloaded with Enduren for 2 or 6 h did not induce a luminescent signal from the intact EVs. These results suggest that cytoplasmic cellular esterases can convert Enduren to coelenterazine-h, which can easily traverse the EV membrane, but that the coelenterazine-h either cannot leave the cell or is unstable in solution outside the cell. When EVs expressing the DSP1-CD9/DSP2 *Renilla* enzyme were exposed to live A549 recipient cells preloaded with Enduren, an increasing signal was observed, indicating gradual exposure of the substrate to processed Enduren due to time-dependent EV uptake and internalization by the recipient cells.

Based on the finding that cellular esterases are necessary for the deprotection of Enduren, the experimental strategy was adjusted as depicted in Fig. 2. The revised assay protocol was focused on the inherent luciferase activity of EVs isolated from DSP1-CD9/DSP2-expressing SUM159 cells, rather than relying on DSP reassociation. In this updated method, recipient cells would first be preloaded with the cytopermeable *Renilla* luciferase substrate Enduren. After removing any excess reagent, EVs isolated from the conditioned medium of DSP1-CD9/DSP2 SUM159 producer cells would be introduced. The signal detected would originate solely from the EVs that were internalized, as only those would have access to the deprotected Enduren substrate within the recipient cells.

A comprehensive kinetic analysis was conducted to measure luminescence at various time points: 0, 0.5, 2, 4, 6, and 24 h. No luminescent signal was observed when intact EVs were incubated directly with Enduren, which is the undeprotected substrate, nor when EVs were incubated with fixed cells, as they could not internalize the EVs. The deprotected substrate became accessible to the

EV-encapsulated *Renilla* substrate only after the EVs were internalized by recipient cells, resulting in a measurable luminescent signal after 30 min. Subsequently, the impact of temperature was evaluated by incubating cells at either 4 or 37°C for 0, 2, 4, or 6 h. No signal was produced by EVs incubated with cells at 4°C because EV internalization was inhibited, while a strong signal was observed when EVs were incubated with recipient cells at 37°C. Exosomes (EXs) and microvesicles (MVs) isolated from producer cells transduced with DSP1-CD9/DSP2 or with DSP1-CD63/DSP2 were incubated with recipient cells. The MVs from DSP1-CD9/DSP2-expressing cells showed the highest signal, followed by the DSP1-CD63/DSP2 MVs, then the DSP1-CD63/DSP2 EXs, with the lowest signal coming from the DSP1-CD9/DSP2 EXs. None of these MVs or EXs generated a detectable signal when incubated directly with the Enduren substrate.

To demonstrate the quantitative assay, the effect of different doses of EVs on the luminescent signal was evaluated. DSP1-CD9/DSP2 EVs were diluted in ratios of 1:2, 1:4, 1:10, and 1:100, followed by incubation with recipient cells. The luminescent signal was measured at 0, 2, 4, and 6 h (Fig. 3a). The luciferase signal showed a decrease that closely matched the dilution factor, indicating a proportional relation that aligns with linear regression. At the 1:100 EV dilution, the signal was no longer statistically significant compared to the EV background, with a detection limit established at $9.35 \pm 2.46 \times 10^8$ EVs/well. Subsequently, recipient cells were treated with various drugs that inhibit or disrupt uptake/internalization pathways (Fig. 3b). The drugs included CPZ, a clathrin inhibitor; NDZ and Cyt D, which block tubulin or actin dynamics, respectively; genistein, a general tyrosine kinase inhibitor that indirectly affects dynamin2; Wort, a PI3K inhibitor that hinders macropinocytosis; EIPA, a Na^+/H^+ inhibitor that also impacts macropinocytosis; heparin, which was used as a lectin competitor; and Dyn, a dynamin2 inhibitor. The most significant reduction in EV internalization in the SUM159 cells occurred with the inhibition of dynamin2 function by genistein, while a lesser reduction was observed with the clathrin inhibitor CPZ. To confirm that these reductions were not due to toxicity of the inhibitors, the viability of the recipient cells treated with these drugs was assessed in parallel (Fig. 3c). Only CPZ displayed minor toxicity at the doses and durations employed in the assay.

In addition to measuring luminescence from *Renilla* luciferase activity, the association of DSP1 and DSP2 also labels the EVs with green fluorescence, permitting detection via confocal fluorescence microscopy. EVs containing DSP1-CD9/DSP2, isolated from SUM159 producer cells, were incubated for 5 h with A549 recipient cells. Analysis through confocal fluorescence microscopy showed that most cells exhibited a concentrated GFP signal that

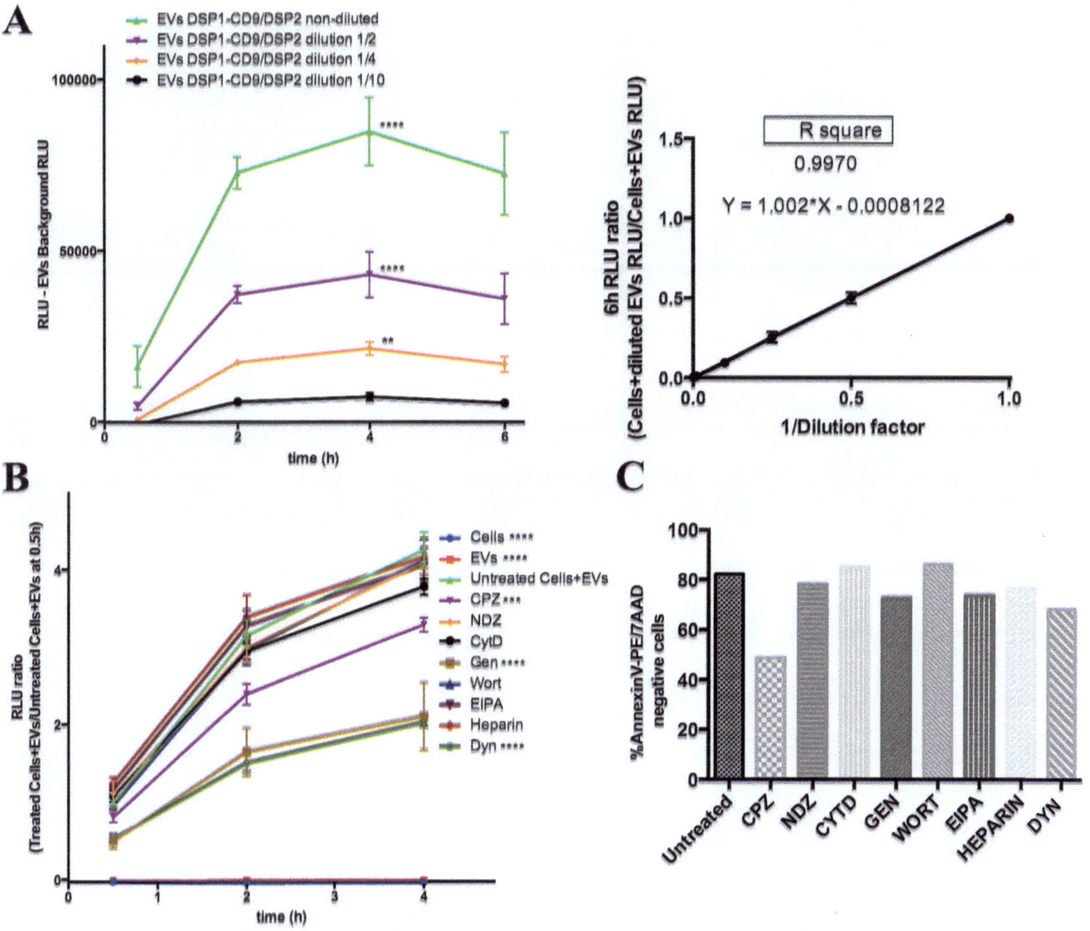

Fig. 3 DSP1 and DSP2 in vitro re-association assessment. **a** DSP1 and DSP2 reassociation kinetics in cell lysates. SUM159 stably expressing DSP1, DSP1-CD9 or DSP1-CD63 were lysed, and each lysate was mixed with a cell lysate from DSP2-expressing cells. Renilla luciferase activity was measured with Stop&Glo Renilla reagent at the indicated times. $n = 2$. **b** Comparison of the signal obtained with Stop&Glo reagent at 24 h from combinations of cell lysates of cells expressing DSP1 constructs with cell lysates from cells expressing DSP2 (DSP1 + DSP2, DSP1-CD9 + DSP2 and DSP1-CD63 + DSP2) and with cell lysates from cells that coexpressed both constructs (DSP1/DSP2, DSP1-CD9/DSP2, and DSP1-CD63/DSP2). Negative controls of independent lysates from cells expressing DSP1, DSP2, DSP1-CD9, or DSP1-CD63 are depicted. **c** Reassociation kinetics with EV lysates. EVs were isolated from conditioned media of SUM159 stably expressing DSP1, DSP1-CD9, or DSP1-CD63 by ultracentrifugation at 100,000 × g, lysed, and each lysate was mixed with a cell lysate from DSP2-expressing cells. Renilla luciferase activity was measured with Stop& Glo Renilla reagent at the indicated times. (Reproduced with permission. Under a Creative Commons license (https:/creativecommons.org/licenses/by/4.0/) from {*Sci Rep.* 2019, 9, 1, 10522})

was primarily perinuclear, although some cells displayed a more scattered GFP distribution (Fig. 4). Staining for CD63 expression indicated partial colocalization with the GFP signal associated with DSP1-CD9/DSP2-EVs in some A549 target cells. However, no colocalization of the DSP1-CD9/DSP2-EV-associated GFP signal

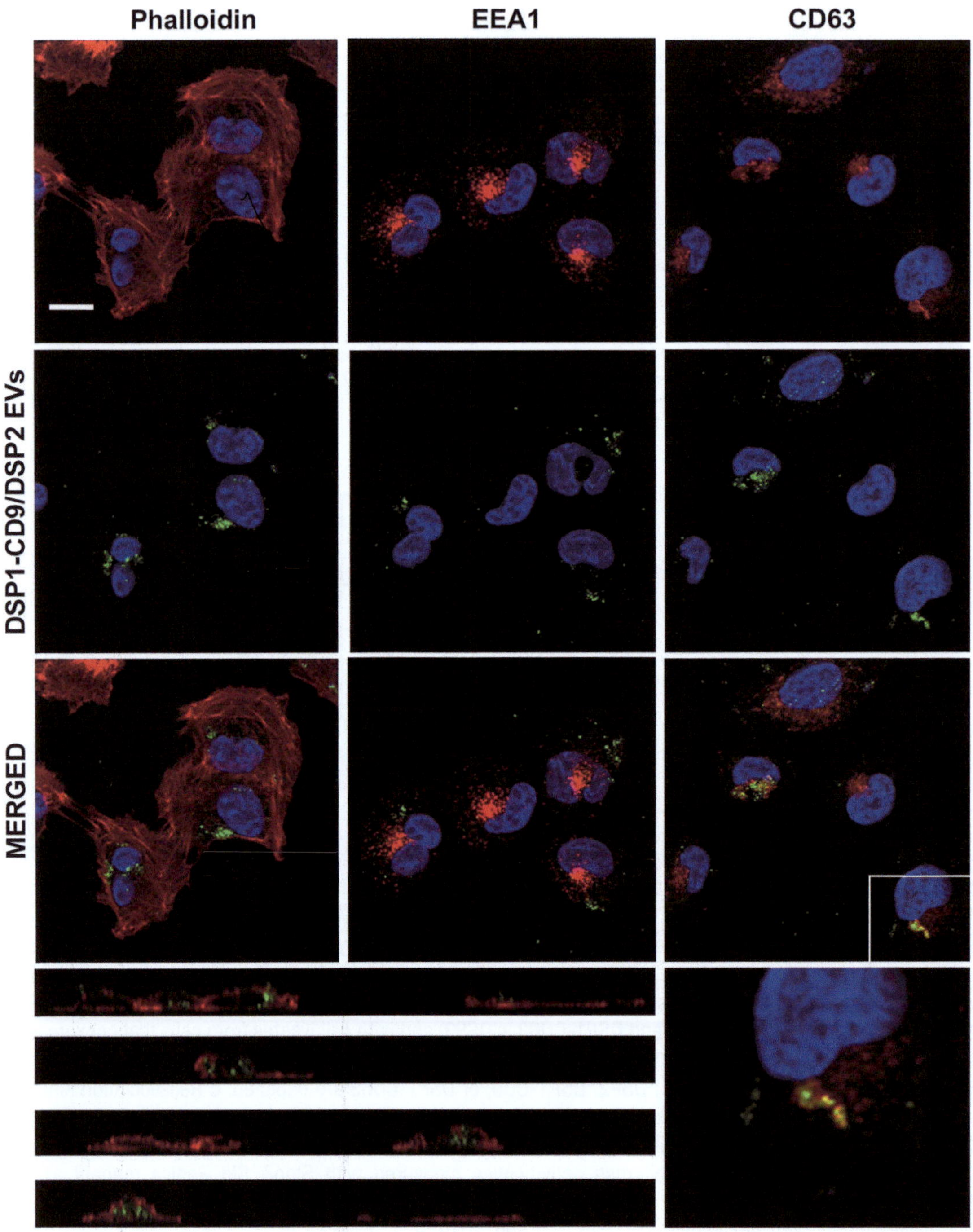

Fig. 4 DSP1-CD9/DSP2 EVs uptake detected by fluorescence confocal microscopy. SUM159 target cells were incubated with DSP1-CD9/DSP2 EVs isolated by ultracentrifugation. Cells were fixed and permeabilized after 5 h of incubation with EVs. Samples were stained with Phalloidin-647, anti-EEA1 antibody, or anti-CD63 (Tea3/10) mAb. All samples were co-stained with DAPI and analyzed by confocal fluorescence microscope. A maximal projection of central optical sections is shown. Below, a series of vertical sections of Phalloidin stained samples are depicted, clearly showing the intracellular localization of uptaken EVs. On the right, a zoomed image in which colocalization of EVs with CD63 labeling is demonstrated on a single confocal plane. Bar = 10 μm. (Reproduced with permission. Under a Creative Commons license (https:/creativecommons.org/licenses/by/4.0/) from {*Sci Rep.* 2019, 9, 1, 10522})

was detected when staining for EEA-1 or with the actin cytoskeleton after staining with phalloidin-647.

In conclusion, Toribio et al. developed an extremely sensitive and quantitative method for measuring EV internalization, which allows for evaluating the efficiency of this process and facilitates investigation into the underlying controlling mechanisms. They designed an assay utilizing two tags, DSP1 and DSP2, which encode nonfunctional split portions of EGFP and *Renilla* luciferase. When the encoded proteins associate, both green fluorescence and luciferase activity are restored. The quantification of EV internalization is accomplished by measuring the luminescent signal produced by the luciferase processing of its substrate.

Their initial approach involved incubating EVs containing the DSP1-tetraspanin fusion protein with recipient cells expressing the DSP2 protein, but they found that no reassociation occurred between the split proteins. In contrast, EVs that carried both the DSP1 fusion protein and the DSP2 protein produced a strong luminescent signal. The authors did not explain why there was no DSP1/DSP2 association after internalization, but it is likely from the DSP2 protein's inability to access the DSP1 protein, as the EV-bound DSP1 could not escape the endosomal compartment. This means that the internalized EVs remained intact and would eventually be shuttled to the lysosome for degradation.

They encountered an additional barrier with their strategy: the EV-encapsulated DSP1/DSP2 did not generate a luminescence signal when directly incubated with the luciferase substrate Enduren. Further experiments revealed that Enduren needs cellular esterases to convert it into coelenterazine-h, which can then function as a substrate for luciferase. By modifying their original strategy, they were able to quantify EV internalization in live cells by measuring luminescence generated by luciferase and to visually track the intracellular distribution of EVs using confocal fluorescence microscopy to detect EGFP. This capability enables the analysis of the kinetics and dose dependence of EV internalization and allows for monitoring the effects of endocytosis inhibitors.

2.2 Quantitative Analysis of Extracellular Vesicle Uptake and Fusion with Recipient Cells

For EV luminal cargo to impact the phenotype of recipient cells, it must successfully exit both the EV lumen and the endosomal compartment into which it had been endocytosed, as depicted in Fig. 5a. Gaining a deeper understanding of the specific mechanisms that govern EV endocytic uptake and endosomal escape in recipient cells is essential for assessing their biological relevance and improving their therapeutic applications [199].

While measuring EV uptake can yield important insights into the factors influencing the targeting of EVs to particular recipient cell populations and the pathways involved in their internalization, it does not provide information about the capacity of bioactive drugs and biomolecules to escape the EV transport vehicle and

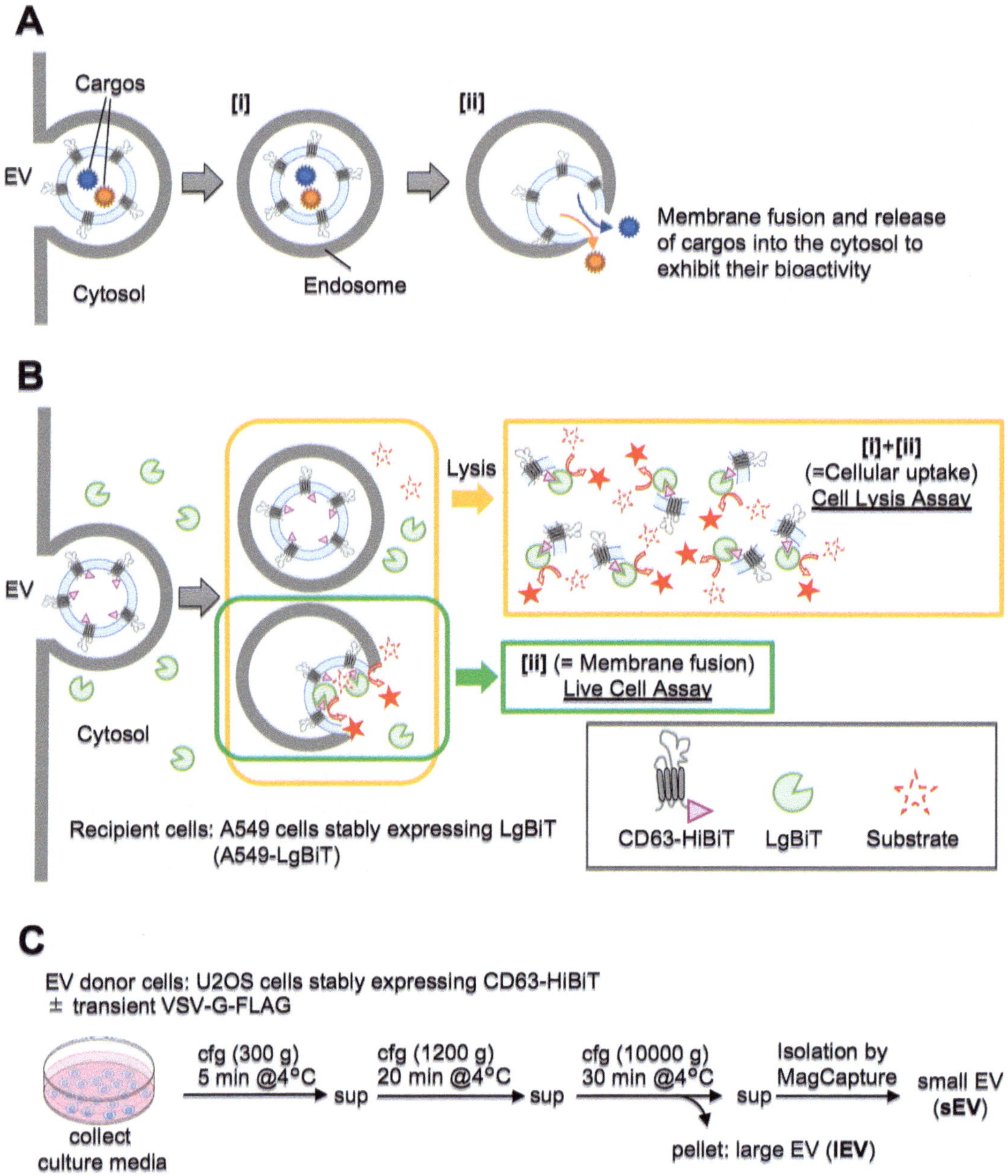

Fig. 5 The concept of the study and isolation of EVs. Prior to obtaining cargo bioactivities **a** EVs are taken up by cells and encapsulated in their vesicular compartment (i.e., endosome) (i), and subsequently fuse with endosomal membrane to release cargo into the cytosol (ii). **b** Quantification of total uptake and membrane fusion of EVs using the NanoBiT system. In this study, "cellular uptake" of EVs is defined to include both EVs encapsulated in endosomes (i) and those fused with endosomal membranes (ii). **c** Isolation of sEVs and lEVs. EVs, extracellular vesicles; sEVs, small extracellular vesicles; lEVs, large extracellular vesicles. (Reprinted with permission from {*Bioconjugate Chem.* 2022, 33, 10, 1852–1859}. Copyright {2022})

affect cell physiology. Consequently, there is an urgent need to develop methods that accurately quantify the portion of internalized EVs that either fuse with or otherwise breach the endosomal membrane barrier.

As previously highlighted in the study by Toribio et al., the use of lipophilic dyes, luciferases, and fluorescent proteins to assess the total number of EVs associated with recipient cells is widely accepted [140, 192, 200, 201]. However, there are still limited methods available for accurately assessing the fusion of EVs with either the plasma or endosomal membranes. Two recent studies have proposed methods for detecting membrane fusion. The first approach involved tagging GFP to the EV lumen in producing cells, followed by EV uptake by recipient cells engineered to overexpress an anti-GFP fluobody [28]. This strategy allowed for cytoplasmic interaction between GFP and the anti-GFP fluobody after membrane fusion, resulting in a dual fluorescent signal. A second study introduced a quantification assay for EV–recipient cell membrane fusion using a split nanoluciferase assay system [202]. While these methods enable the quantification of membrane fusion, they do not allow for the determination of total EV internalization. In this study, Hirose et al. present an alternative method to quantify both aspects of the EV uptake process [143]. They utilize the NanoBiT system (Fig. 5b), which is a split nanoluciferase (NLuc) system consisting of an 18 kDa large subunit (LgBiT) and a small 11-mer peptide fragment (HiBiT) that binds to LgBiT with high affinity [203]. EV-producing cells are engineered to express a fusion transmembrane protein tagged with the HiBiT fragment, ensuring that HiBiT is oriented towards the EV lumen. The isolated EVs are then co-incubated with an engineered recipient cell line that expresses the LgBiT fragment. The HiBiT-tagged protein can only interact with LgBiT after EV–endosomal membrane fusion occurs, which exposes the HiBiT tag to the cytoplasm. When a luciferase substrate like furimazine is present, a luminescent signal is produced by the LgBiT/HiBiT complex, allowing for the tracking and quantitation of the EV-associated HiBiT-tagged protein of interest. By combining a cell lysis luciferase assay with a live cell assay, it becomes possible to quantify both the total amount of internalized EVs and the fraction of internalized EVs that have fused to the endosomal membranes, respectively.

2.2.1 Methods

Antibodies

The primary antibodies utilized in this study included: mouse anti-CD63 mAb (clone 3–13) from FUJIFILM Wako at a dilution of 1: 1000; mouse anti-CD9 mAb (10626D) from ThermoFisher at a dilution of 1:500; rabbit anti-syntenin mAb (EPR8102) from Abcam at a dilution of 1:1000; rabbit anti-annexin A1 mAb (EPR19342) from Abcam at a dilution of 1:2000; and mouse anti-FLAG mAb (clone D2) from Sigma-Aldrich at a dilution of 1:400. The secondary antibodies conjugated to horseradish peroxidase used in this study were: sheep-α-mouse IgG (NA931) from Cytiva at a dilution of 1:5000; and goat-α-rabbit (7074) from Cell Signaling Technology at a dilution of 1:1000.

Oligonucleotides

The DNA oligonucleotides used in this study were:

HiBiT_Fw: 5′-AAT TCG GTG GAT CAG GAG GAG GTA GTG TTA GTG GAT GGA GAC TAT TCA AGA AGA TCT CGT GAG GGC C-3′.

HiBiT_Rv: 5′-CTC ACG AGA TCT TCT TGA ATA GTC TCC ATC CAC TAA CAC TAC CTC CTC CTG ATC CAC CAC CG -3′.

VSV-G-Fw: 5′-TGA CGA CAA GTA ACT CAA ATC CTG CAC AAC AGA TTC-3′.

VSV-G_TV: 5′-CCT TAT AGT CCT TTC CAA GTC GGT TCA TC-3′.

FLAG_Fw: 5′-AGA TGA ACC GAC TTG GAA AGG ACT ATA AGG ACG ATG ACG ACA AGT AAC TCA AAT CCT GCA CAA C-3′.

FLAG_Rv: 5′-GTT GTG CAG GAT TTG AGTTAC TTG TCG TCA TCG TCC TTA ATA TAG TCC TTT CCA AGT CGG TTC ATC T-3′.

HiBiT_Fw: 5′-GGT GGA TCA GGA GGA GGT AG-3′.

HiBiT_Rv: 5′-GGT GCC AAG CTT AAG TTT AAA CGC-3′.

Halo Tag_Fw: 5′- AGC GTT TAA ACT TAA GCT TGC CAC CAT GAG AGG ATC GCA TCA CCA TCA C-3′.

Halo Tag_Rv: 5′-TAA CAC TAC CTC CTC CTG ATC CAC CAC CGG AAA TCT CGA GCG TCG-3′.

Plasmid Construction

To generate the plasmid encoding the CD63-HiBiT fusion protein, an expression plasmid encoding the tetraspanin protein CD63 (CD63-L7Ae, provided by Dr. Martin Fussenegger) was incubated with the restriction enzymes *EcoRI* and *ApaI* [204]. Oligonucleotides HiBiT_Fw and HiBiT_Rv were phosphorylated at the 5′ end using T4 polynucleotide kinase (New England Biolabs). The phosphorylated oligonucleotides were annealed to generate the (GGGS)2-HiBiT coding sequence, which was then ligated into the previously digested CD63-L.7Ae plasmid.

To incorporate a HaloTag into the CD63-HiBiT fusion protein plasmid, the HaloTag sequence was amplified from the pHis-MCP-HaloTag plasmid, provided by Dr. Timothy Stasevich [205], using the HaloTag_Fw and HaloTag_Rv primer oligonucleotide primers. The CD63-HiBiT plasmid was also amplified by PCR using the HiBiT_Fw and HiBiT_Rv oligonucleotide primers. The amplified

HaloTag fragment was assembled with NEBuilder HiFi DNA assembly master mix following the manufacturer's instructions.

To generate a plasmid encoding a VSV-G protein tagged with the FLAG protein tag (DYKDDDDK), the VSV-G sequence was amplified by PCR from the VSV-G-HiBiT plasmid, supplied by Dr. Masaharu Somiya [203], using the VSV-G-Fw and VSV-G_Rv primer oligonucleotides. After that, the amplified VSV-G sequence was used to create A DNA duplex using the NEBuilder HiFi DNA assembly master mix (New England Biolabs) by annealing the FLAG_Fw and FLAG_Rv oligonucleotides with the amplified VSV-G fragments, resulting in the VSV-G-FLAG plasmid.

Establishment of U2OS Cells Stably Expressing CD63-HiBiT

U2OS cells (1.2×10^6 cells) were plated in a 100 mm dish and cultured for 1 day. After this incubation, 11 μg of CD63-HiBiT plasmid, complexed with 33 μL of Lipofectamine LTX reagent (Invitrogen) and 11 μL of PLUS reagent (Invitrogen) in DMEM with high glucose (HG) and 10% FBS were added. After 3 h, the cells were washed with PBS and cultured for another day. Following this, 1 mg/mL of G418 (FUJIFILM Wako) was added to generate a polyclonal pool of CD63-HiBiT stably-expressing U2OS cells. These cells were trypsinized and reseeded into a 96-well plate in DMEM HG with 10% FBS and G418 by limiting dilution. Ultimately, a cell line capable of producing EVs stably expressing CD63-HiBiT (U2OS-CD63-HiBiT cells) was established.

Establishment of A549 Cells Stably Expressing LgBiT

As previously reported, A549-LgBiT cells, a human Caucasian lung carcinoma cell line stably expressing LgBiT, were used as recipient cells [206]. This cell line was established by subcloning the LgBiT coding region from pBiT1.1-C[TK/LgBiT] (Promega) vector into CSII-CMV-MCS-IRES2-Bsd (RIKEN BRC), generating the CSII-CMV-LgBiT-IRES2-Bsd plasmid. Cotransfecting CSII-CMV-MCS-IRES2-Bsd with the Lentiviral High Titer Packaging Mix (Takara Bio) produced a self-inactivating lentiviral vector. A549 cells were inoculated with this vector, followed by selection with 20 μg/mL of blasticidin for cells stably expressing LgBiT. Single-cell clones were obtained by limiting dilution and screened for LgBiT expression by Western blot. A single clone was expanded, resulting in the A549-LgBiT cell line.

Cell Culture and Transfection

All cells were cultured in 100 mm dishes at 37°C with 5% CO_2, unless stated otherwise. U2OS cells and U2OS-CD63-HiBiT cells were cultured in DMEM HG supplemented with 10% FBS. In addition, A549-LgBiT cells were cultured in DMEM low glucose (LG) containing 10% FBS. U2OS or U2OS-CD63-HiBiT cells (1.2×10^6 cells) were plated and cultured for 1 day. Following this, 11 μg of a plasmid complexed with 33 μL of the Lipofectamine LTX reagent and 11 μL PLUS reagent in DMEM HG containing

10% FBS were added. After 3 h, the cells were washed 4 times with 4 mL of Ca^{2+}- and Mg^{2+}-free PBS, i.e., PBS(−), to eliminate excess plasmid-LTX complexes.

Isolation of Large EVs (lEVs) and Small EVs (sEVs)

To produce EVs, transfected and control (non-transfected) U2OS-CD63-HiBiT cells were cultured for 2 days in 100 mm dishes containing DMEM HG with 10% exosome-depleted FBS (System Biosciences). After the incubation period, the cell culture medium was removed and centrifuged at 300 × *g* for 5 min at 4 °C. The supernatant was sequentially centrifuged, first at 1200 × *g* for 20 min at 4 °C, then at 10,000 × *g* for 30 min at 4 °C. The resulting pellet was identified as the lEV fraction, which was resuspended in 100 μL PBS(−), while the sEV fraction remained in the supernatant. Amicon Ultra-15,100 kDa devices (Merck Millipore) were employed to concentrate the sEV fraction to a 1 mL volume, followed by sEV isolation using the MagCapture Kit (FUJIFILM Wako) and elution twice in 50 μL of elution buffer, for a final volume of approximately 100 μL. Protein concentrations were measured by a bicinchoninic acid (BCA)-enhanced protocol using the Pierce BCA protein assay kit (ThermoFisher Scientific).

Detection of HiBiT Peptide

Western blotting was performed by transferring isolated EV proteins to a PVDF membrane, followed by incubation in Tris-buffered saline with 0.1% Triton X-100 (TBST) for 4–5 h at 25°C to solubilize the HiBiT tag. The HiBiT tag was detected using the Nano-Glo HiBiT Blotting System (Promega) by incubating the membrane in Nano-Glo blotting buffer containing LgBiT at 4°C overnight and imaged using LAS3000 mini (FUJIFILM).

Quantification of Cellular Uptake of EVs into Recipient Cells (Cell Lysis Assay)

A549-LgBiT cells (9 × 10^4 cells/well) plated in a 24-well plate in DMEM LG supplemented with 10% FBS were cultured for 1 day. Following two washes with 500 μL of PBS(−), the cells were treated with isolated EVs in DMEM LG containing 10% exosome-depleted FBS for specified time points (2, 4, and 24 h). Reconstituted luciferase activity from internalized CD63-HiBiT, as a result of the total cellular uptake of EVs, was measured employing the Nano-Glo HiBiT Lytic Detection System (Promega). After treatment, to eliminate any non-internalized EVs, the cells were washed twice with 500 μL PBS(−), then trypsinized with 100 μL of 0.01% trypsin in PBS(−) for 10 min at 37 °C. The cells were resuspended in 100 μL of DMEM LG with 10% FBS and collected on ice in a 1.5 mL tube, followed by centrifugation at 800 × *g* for 5 min at 4 °C.The supernatant was discarded, and the EV-treated cells were resuspended in 50 μL of a mixture of PBS(−) and Opti-MEM for analysis (ThermoFisher Scientific).

To establish a calibration curve (x-axis, EV amount, y-axis, luminescence intensity), control non-treated cells were resuspended in Opti-MEM with specific amounts of EVs in PBS(−) at a final volume of 50 μL. Subsequently, the resuspended cells were combined in a 96-well optical bottom white plate (ThermoFisher Scientific Nunc) with 50 μL HiBiT lytic detergent buffer containing the luciferase substrate for 10 min. The luminescence intensity (counts per second) was recorded at 25°C using a Nivo plate reader (PerkinElmer) with an integration time of 1 s per well.

Investigation of Cellular Uptake Mechanism of EVs into Recipient Cells

To investigate the uptake mechanisms for EVs, a cell lysis assay was conducted by plating A549-LgBiT cells (9×10^4 cells/well) in a 24-well plate with DMEM LG supplemented with 10% FBS and culturing for 1 day. Prior to the addition of EVs (300 ng protein per well), the cells were preincubated at either 4 °C, to block energy-dependent endocytosis, or at 37 °C for 30 min. Non-internalized EVs were removed by washing the cells twice with 500 μL of either cold or warm PBS(−), respectively, followed by incubation with the isolated EVs in DMEM LG containing 10% exosome-depleted FBS for 4 h at 4 °C or 37 °C. After the treatment, the cells were harvested, and luminescence measurements were conducted as previously described.

To inhibit the endocytic pathways, the cells were preincubated with 15 μM Pitstop 2, 20 μM CPZ, 20 μM EIPA, 2 μM Wort, or 20 μM Nystatin, prepared using less than 0.5% v/v dimethyl sulfoxide (DMSO), for 30 min in DMEM LG without FBS. DMSO (0.5% v/v) served as the vehicle control. The cells were then incubated with isolated EVs (300 ng of total protein) in DMEM LG containing 10% exosome-depleted FBS along with each inhibitor for 4 h. The cells were then collected, and luminescence was measured as previously outlined.

Determination of a Correction Factor Between Cell Lysis and Live Cell Assay

A correction factor between the cell lysis and live cell assays was established by measuring the reconstituted luciferase activity in A549 cells that were transiently expressing HaloTag-HiBiT employing the Nano-Glo HiBiT Lytic Detection System (Promega) and the Nano-Glo Live Cell Assay System (Promega). A549-LgBiT cells (4.5×10^6 cells/well) were seeded into a 24-well plate in DMEM LG containing 10% FBS and cultured for 1 day. The culture medium was removed, and cells were washed twice with 500 μL of PBS(−), followed by transfection with HaloTag-HiBiT plasmid at varying concentrations (0, 0.01, 0.1, 1, 5, 10, 50, or 100 ng/well) using 0.6 μL Lipofectamine LTX reagent mixed with 0.4 μL PLUS reagent in DMEM LG with 10% FBS. After 3 h, the cells were washed 4 times with PBS(−) and cultured in DMEM LG with 10% FBS for 1 day. The cells were then washed twice with 500 μL of PBS(−), trypsinized using 100 μL of 0.01% trypsin in PBS(−) for 10 min, and resuspended in 100 μL of DMEM LG with 10% FBS on ice in a 1.5 mL tube. To lyse the cells, they were pelleted by centrifuging at 800 × g for 5 min at 4°C and resuspended in 50 μL Opti-MEM and 50 μL of HiBiT lytic detergent buffer containing the substrate for 10 min. For the live cell assay, the cells were combined with 50 μL of non-lytic assay reagent containing the substrate in a 96-well optical bottom white plate (Thermo Scientific Nunc). Finally, the luminescence intensity (counts per second) was measured at 25 °C using a Nivo plate reader (PerkinElmer) with an integration time of 1 s per well. A correction factor (F, lytic/live cell) was calculated using the slope of the calibration curves (x-axis, transfected plasmid amount; y-axis, luminescence intensity).

$$F = \frac{\text{slope of lytic assay}}{\text{slope of live cell assay}} = 0.72$$

Quantification of Membrane Fusion of EVs with Recipient Cells (Live Cell Assay)

The quantification of EVs that fused with the recipient cells was measured by assessing the restored luciferase activity employing the Nano-Glo Live Cell Assay System (Promega). A549-LgBiT cells (9×10^4 cells/well) were plated into a 24-well plate in DMEM LG containing 10% FBS and cultured for 1 day followed by incubation with EVs as previously described. Non-internalized EVs were removed by washing the cells twice with 500 μL of PBS(−) and trypsinizing with 100 μL of 0.01% trypsin in PBS(−) for 10 min. The cells were resuspended in 100 μL of DMEM LG with 10% FBS and collected on ice in a 1.5 mL tube. The cells were pelleted by centrifugation at 800 × *g* for 5 min at 4 °C, then resuspended in 50 μL of Opti-MEM mixed with 50 μL of a live cell assay reagent containing the substrate and plated in a 96-well optical bottom plate (Thermo Scientific Nunc) and the luminescence intensity (counts per second) was measured at 25 °C using a Nivo plate reader with an integration time of 1 s per well. The luminescence intensity in live cell buffer (L) was converted to the luminescence intensity (L') using the correction factor (0.72 as mentioned above) between the two conditions.

$$L' = 0.72 \times L$$

The quantity of EVs fused with the recipient cells was determined using the lytic assay calibration curve.

Statistical Analysis

Unless specified otherwise, all data were presented as the mean + standard deviation (SD) of three independent biological experiments ($n = 3$). Statistical analyses were performed using JMP Pro 15 software (JMP 15.1.0). The calculated p-values are shown in the figure legends, and significance was defined at $p < 0.05$.

2.2.2 Results Summary

A significant body of research has been dedicated to enhancing the natural properties of EVs that enable them to function as targeted delivery systems for biomolecular and chemical therapies. A crucial step in translating this potential into clinical applications is the creation of high-throughput methods to accurately and efficiently measure the effective dose delivered to recipient cells. For these therapeutic agents to access the cellular machinery, they must either escape from or be released from the confines of the EV membrane. Identifying, tracking, and quantifying both the total number of EVs taken up by recipient cells and the specific subset that has released their therapeutic contents will be essential for understanding the mechanisms that control EV cargo release. By comprehending this process, EVs could be engineered to optimize cargo release, thereby improving their therapeutic specificity and effectiveness.

To distinguish between the overall population of internalized EVs and the subset that has released their cargo into the cell, a sensor is required that can be detected only after the EV membrane barrier has been breached. One proposed method to differentiate internalized EVs from those merely attached to the recipient cell surface involves using a fusion protein tagged with a pH-sensitive marker. This fusion protein fluoresces at the neutral pH outside the cell but does not fluoresce once inside the cell [138, 139].

In this study, a two-component complementary nano-luciferase system was designed, allowing EVs produced by U2OS osteosarcoma producer cells to carry one component (HiBiT) of the nano-luciferase, which is fused to the CD63 tetraspanin protein, a known marker for EVs (Fig. 5b). The A549 adenocarcinoma recipient cells were modified to express the second component of the nano-luciferase (LgBiT) in their cytoplasm. Both components remain inactive until they can self-associate to form a functional luciferase enzyme. The LgBiT-expressing A549 recipient cells were preloaded with the furimazine luciferase substrate and then incubated with the HiBiT-carrying EVs isolated from the U2OS producer cells. The normal self-association between HiBiT and LgBiT is blocked by the intact EV membrane. It is only when the EV membrane fuses with the endosomal membrane that the cytoplasmic LgBiT can self-associate with HiBiT, resulting in an active nano-luciferase. This active enzyme can then process the furimazine substrate, producing a measurable luminescent signal. By combining a live cell assay measuring luminescence generated solely by the subpopulation of internalized EVs that have released their contents into the cytoplasm, along with a cell lysis assay that measures the total internalized EV population, researchers can evaluate the percentage of internalized EVs capable of releasing their therapeutic cargo to alter the recipient cell phenotype.

Two subpopulations of EVs were isolated from the U2OS producer cells. The conditioned media from U2OS cell cultures underwent sequential centrifugation at $300 \times g$ and $1200 \times$ g. The resulting pellet was resuspended and centrifuged at $10{,}000 \times g$ for 30 min. This pellet was then resuspended and identified as the lEV fraction, which is enriched in microparticles. The supernatant was treated with MagCapture beads, which enable Ca^{2+}-dependent purification of exosomes through their interaction with the phosphatidylserine present on the exosome surface [207]. This exosome-enriched fraction was labeled as the sEV fraction.

Characterization of both the lEV and sEV fractions through Western blot analysis revealed comparable expression levels of the tetraspanin proteins CD9 and CD63. Additionally, the levels of Syntenin 1, another marker for exosomes, and Annexin A1, a marker for microvesicles, were assessed using Western blotting [208, 209].

The sEV fraction showed high levels of Syntenin 1 and low levels of Annexin A1, while the lEV fraction displayed low levels of Syntenin 1 and high levels of Annexin A1. The presence of the CD63-HiBiT fusion protein was confirmed in both the sEV and lEV subpopulations using the HiBiT detection system.

To create a standard curve of luminescence, different concentrations of HiBiT-containing sEVs and lEVs were incubated with A549 cell lysates expressing LgBiT and containing the furimazine substrate. The total uptake of sEVs (Fig. 6a) and lEVs (Fig. 6b) was then analyzed by measuring uptake at 2, 4, and 24 h, at either 4 or 37 °C. The quantification of cellular uptake indicated that approximately 4.5% of the incubated sEVs and nearly 25% of the lEVs were internalized after 24 h in A549 cells, suggesting a higher uptake efficiency for lEVs compared with sEVs. Notably, cellular uptake was significantly reduced in both EV subpopulations at 4 °C compared to the levels observed at 37 °C, which supports the idea that cellular EV uptake is facilitated by energy-dependent endocytosis pathways. To investigate the pathways involved in EV uptake by the A549 recipient cells, individual wells were preincubated with specific chemical endocytosis inhibitors before co-incubation with either lEVs or sEVs. The uptake of both sEVs and lEVs was significantly inhibited by the clathrin-mediated endocytosis inhibitors CPZ and Pitstop2, and by the macropinocytosis inhibitors Wort and EIPA. However, nystatin, an inhibitor of caveolin-dependent endocytosis, did not impact the cellular uptake of either subpopulation.

The total particle uptake by the cell lysis assay for each subpopulation was determined using previously established luminescence standard curves. This analysis showed that approximately 6% of the total sEVs and about 18% of the total lEVs added were taken up by A549 recipient cells, translating to roughly 90 sEV particles and 270 lEV particles internalized per cell.

Following this, the number of EVs that fused with recipient cells was assessed using a live cell assay. Interestingly, the live cell assay indicated very low levels of EV fusogenic activity. No fusion events for sEVs and only around 0.02% fusion events for lEVs were detected under the experimental conditions employed. One

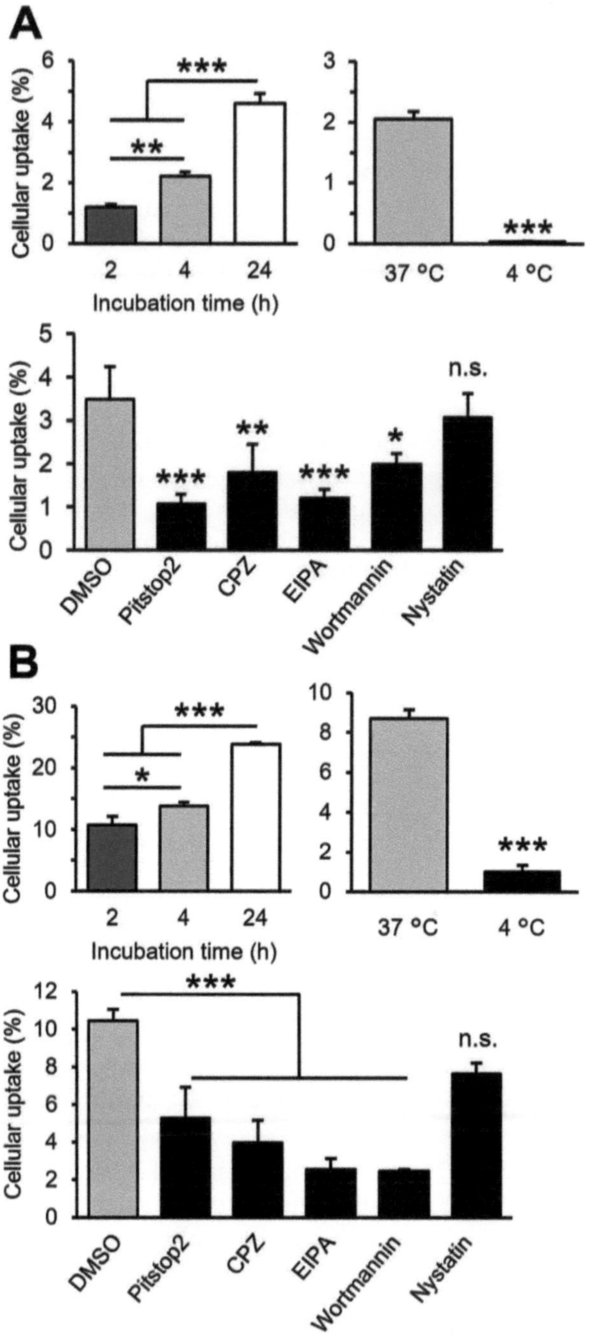

Fig. 6 Evaluation of EV uptake into A549-LgBiT cells. **a** sEV uptake data. **b** lEV uptake data. The results are presented as the mean ± standard deviation (SD) ($n = 3$). $^{**}P < 0.01$; $^{***}P < 0.001$ (one-way analysis of variance [ANOVA] followed by Tukey–Kramers honestly significant difference test for time-course uptake data, ANOVA followed by an unpaired *t*-test for temperature-dependent uptake data, and Dunnett's post hoc test vs dimethyl sulfoxide (DMSO; vehicle control) for endocytosis inhibiting data). EVs, extracellular vesicles; sEVs, small extracellular vesicles; lEVs, large extracellular vesicles. (Reprinted with permission from {*Bioconjugate Chem.* 2022, 33, 10, 1852–1859}. Copyright {2022})

possible explanation for this could be that the design of the assay did not allow for the detection of EV fusion.

Decorating EV membranes with the fusogenic envelope glycoprotein of the vesicular stomatitis virus (VSV-G) significantly improves EV membrane fusion [202]. To assess this, the U2OS-CD63-HiBiT producer cells were transiently transfected with an expression vector that encodes the VSV-G glycoprotein along with a FLAG tag for detection by Western blotting. The sEVs and lEVs were isolated from the conditioned culture medium, and the presence of the VSV-G protein was confirmed. The sEVs and lEVs were then co-incubated with A549-LgBiT recipient cells, and EV fusion was quantified using a live cell assay as previously described. The presence of the VSV-G fusogenic protein improved the efficiency of membrane fusion of sEVs and lEVs to the recipient cells, resulting in the fusion of ~3% and ~7% of total sEVs and lEVs, respectively, or around 40 sEV and 100 lEV particles per cell. These numbers indicate that the assay system is capable of detecting EV fusion and further suggest the potential advantages of lEVs as intracellular transport vehicles. The positive effect of VSV-G on EV cargo delivery implies that incorporating fusogenic proteins or peptides will be essential for the effective use of EVs and LNPs as therapeutic delivery vehicles.

In conclusion, the NanoBiT system provided a successful method to detect and quantify both the total number of internalized EVs and those EVs that had undergone membrane fusion. The uptake numbers were low for both sEVs and lEVs, with negligible levels of EV fusion in A549 recipient cells. The addition of the VSV-G fusogenic glycoprotein increased the frequency of fusion events, suggesting that strategies involving fusogenic proteins or peptides will enhance the efficiency of EV cargo release.

2.3 Quantifying Cellular Internalization with a Fluorescent Click Sensor

A significant drawback of many uptake studies is that the labeling strategies used often fail to clearly differentiate between particle association with recipient cells and the actual internalization of those particles [103]. The ability to distinguish between these two events is crucial for developing EVs and synthetic LNPs to realize their potential as effective delivery vehicles for therapeutic drugs and biomolecules. Most uptake studies utilize fluorescent tags to identify particles, but they vary in how they measure or quantify the fluorescence present. Typically, these measurements include the fluorescence signals from both internalized particles and those merely attached to the cell surface. Many studies that quantify internalization rely on nonspecific techniques to remove bound particles, such as acid washing [210, 211] or trypan blue quenching [212, 213], or use confocal microscopy to assess uptake based on fluorescence intensity [214].

As an alternative, researchers have proposed using pH-sensitive dyes to exploit the acidification of the endosomal compartment, which can assist in differentiating between particles bound to the plasma membrane and those that have been endocytosed [138, 139, 215]. Although these dyes significantly reduce their fluorescence at the appropriate pH, they still retain some measurable fluorescence, resulting in a high background and low signal-to-noise ratio.

Selby et al. describe the design of a fluorescent sensor that is comparable in size to conventional fluorophores. This sensor was designed to enable the tagging of materials of interest through an inverse electron demand Diels–Alder (IEDDA) click-type reaction between tetrazine and *trans*-cyclooctene groups (Fig. 7b and c). The bioconjugation of this sensor allows for the visualization of particles using fluorescence microscopy and the quantification of the amount of internalized material through flow cytometry [108, 137].

2.3.1 Methods

Oligonucleotide Sensor

A previously published [108] oligonucleotide internalization sensor (FIP) and its complementary quencher (QComp) were purchased (Integrated DNA Technologies) with the following custom sequences: FIP-5′-Cy5-TCA GTT CAG GAC CCT CGG CT-azide-3′; QComp-5′-AGC CGA GGG TCC TGA ACT GA-BHQ2-3′. The oligonucleotides were reconstituted in Milli-Q water to prepare 150 or 600 μM stock solutions.

Nanobody Expression

A histidine-tagged antihuman epidermal growth factor receptor 2 (anti-HER2) nanobody clone 2D3 [216] was expressed in *E. coli* BL21(DE3) and purified through immobilized metal affinity chromatography followed by size exclusion chromatography. Nanobody protein concentration was measured by UV–vis absorbance at 280 nm using an extinction coefficient of 35,075 M/cm, calculated using ProtParam [217].

DNA Sensor Bioconjugation

The labeling reactions used 50 μL of transferrin at a concentration of 5.0 mg/mL in phosphate buffered saline (PBS). An excess (20 equiv) of 1 mg/mL dibenzocyclooctyne *N*-hydroxysuccinimide ester (DIBO-NHS), dissolved in dimethyl sulfoxide (DMSO), was added and allowed to react for 2 h at 4°C. To remove any unreacted DIBO-NHS, a Zeba Spin Desalting column with a 7 kDa molecular weight cutoff (MWCO) was employed. The column was equilibrated by washing it 3 times with 300 μL PBS and spinning at 1500 × *g* for 1 min. The sample was then added and spun at 1500 × *g* for 1 min. Following this, an excess of 1.5 equiv. of FIP-azide oligonucleotide sensor was introduced and allowed to react overnight at 4 °C. Excess FIP-azide was removed using an Amicon Ultra 0.5 mL 30 kDa MWCO centrifugal filter [108, 137].

The filter was rinsed with 500 μL PBS by spinning at 13,000 × *g* for 30 s. After adding the sample, PBS was added to reach a total volume of 0.5 mL, spun at 13,000 × *g* for 3 min, and

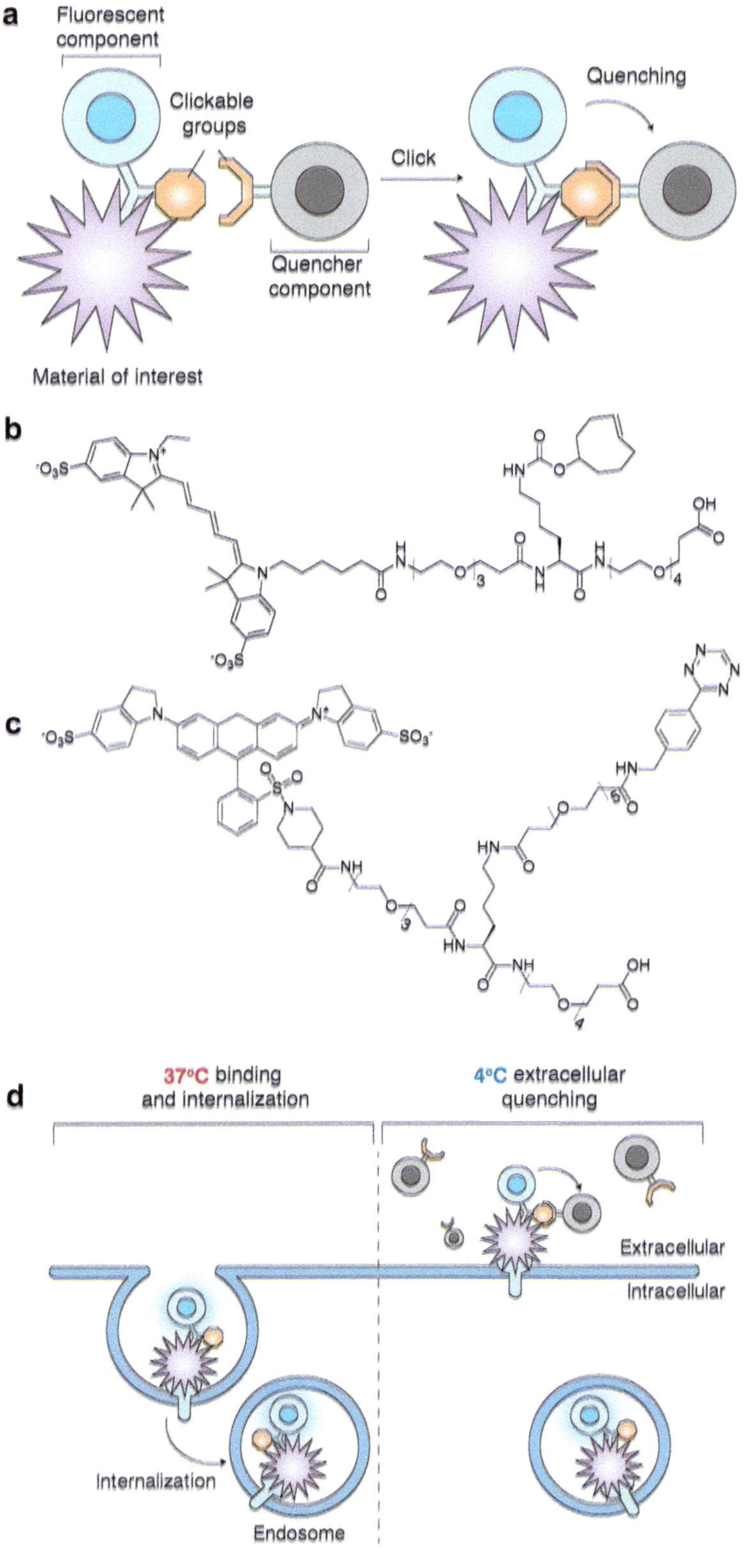

Fig. 7 Scheme of click internalization sensor. **a** The fluorescent component of the sensor is attached to a material of interest and contains a click group compatible with the quencher unit. Chemical structures of the **b** fluorescent

this process was repeated until the filtrate was clear. The unit was then inverted and spun at 1000 × *g* for 2 min to collect the bioconjugated product. The concentrations of the protein and label were calculated using an extinction coefficient and molecular weight (MW) for transferrin of 87,000 cm/M and 81 kDa, respectively [218], and an extinction coefficient of 250,000 cm/M for Cy5 [219]. Absorbance measurements were taken using a NanoDrop Spectrophotometer ND-1000, and the degree of labeling was determined as previously described [220].

pHrodo Red Bioconjugation

pHrodo Red Ester (1.5 equiv., 3.3 μg, 5 nmol) at a concentration of 1 mg/mL was added to 5 mg/mL human holo transferrin (0.25 mg, 3.1 nmol) and incubated at 4 °C overnight before purifying using a Zeba Spin Desalting column, 7 kDa MWCO. The concentration and degree of labeling were obtained using the NanoDrop Spectrophotometer ND-1000 by measuring the absorbance at 560 nm and using an extinction coefficient of 65,000 M/cm.

Click Sensor Bioconjugation

Click chemistry bioconjugation reactions were used to label 50 μL of 5.0 mg/mL transferrin, 38 μL of 0.9 mg/mL 2D3 nanobody (0.035 mg), or 14 μL of 14 mg/mL trastuzumab (0.2 mg) in phosphate buffered saline (PBS). An excess of the click sensor (sCy5-TCO), ranging from 2–4 equiv. of the protein amount, was added to the solution. This was followed by the addition of *N*-3-(3-diethylaminopropyl)-*N′*-carbodiimide hydrochloride (EDC) and NHS, both dissolved in PBS, to achieve final concentrations of 0.1 M and 5 mM, respectively. For trastuzumab labeling, the sensor was preincubated with 10, 50, or 100 mM EDC and 5 mM NHS for 15 min before this mixture was added to the antibody along with Alexa Fluor 488-NHS dissolved in DMSO (2.5 equiv). Additionally, sulfo-Cy5 was conjugated to trastuzumab with sCy5-NHS (2.5 equiv) for comparison with samples treated with EDC/NHS to evaluate cross-linking. All reactions were allowed to incubate overnight at 4 °C. Unreacted dye was removed using a Zeba Spin Desalting column, 7 kDa MWCO. The concentration and degree of labeling were obtained on the NanoDrop Spectrophotometer ND-1000 using an extinction coefficient of 250,000 M/ cm for Alexa Fluor 488 at 495 nm. An extinction coefficient of 250,000 M/cm was used for sCy5-TCO. For trastuzumab, an

Fig. 7 (continued) component (sCy5-TCO) and **c** quencher component sQSY-Tet. **d** Following binding and internalization of the protein or material into cells at 37 °C, uptake is halted by lowering the temperature to 4 °C. Fluorescence of any material on the surface is selectively removed by adding the quencher component of the sensor, leaving intracellular signal unaffected. (Reprinted with permission from {*ACS Sens.* 2018, 3, 6, 1182–1189}. Copyright {2018}) extinction coefficient of 225,000 M/cm at 280 nm and an approximate MW of 150 kDa were used.

Cell Culture

All cells were maintained at 37 °C and 5% CO_2. The nonadherent human lymphoblast cell line C1R (ATCC CRL-1993) was cultured in DMEM with GlutaMax supplemented with 10% FBS, 100 U/mL penicillin, and 100 μg/mL streptomycin. The adherent human ovarian epithelial cell line SKOV-3 (ATCC HTB-77) was cultured in McCoy's 5A modified medium with 10% FBS.

Cell Surface Quenching and Internalization of Transferrin

C1R cells were split into tubes with 4×10^5 cells per sample and cooled on ice for 10 min, followed by the addition of 15 μg/mL of the sensor-labeled transferrin and incubation on ice for an additional 15 min to allow transferrin binding to the cell surface. To remove unbound transferrin, cells were washed twice in cold PBS before incubating the cells at either 4 °C or 37°C for 15 min. In experiments using wheat germ agglutinin-Alexa Fluor 488 (WGA-AF488), the lectin was added at 1 μg/mL for the final 5 min of binding time. The cells were then split into a 96-well V-bottom plate, centrifuged at 300 × *g* for 5 min to pellet the cells, which were then resuspended in 200 μL of PBS with or without quencher (4 equiv. for the click sensor, 5 equiv. for the DNA sensor) relative to the total amount of the sensor added, calculated using the degree of labeling. The samples were mixed and left on ice for approximately 15 min to prevent further transferrin internalization before analysis by flow cytometry.

For microscopy, following incubation with 50 μg/mL of transferrin at 4 °C or 37 °C for 25 min with 1 μg/L WGA-AF488 added for the final 5 min of binding as described above, cooled C1R cells (4×10^5 cells per sample) were washed twice in cold PBS by spinning at 300 × *g* for 5 min. Samples were split into two and resuspended in PBS with or without 4 equiv. of QSY-Tet, based on the degree of labeling of transferrin with sCy5-TCO, before imaging by fluorescence microscopy.

Trastuzumab and 2D3 Binding and Internalization

SKOV-3 cells were plated at 1×10^5 cells in 400 μL of McCoy's 5a modified medium with 10% FBS per well in two 24-well plates 1 day prior to the experiment. One plate of cells was cooled to 4 °C on ice for 15 min, while the other was maintained at 37 °C. Aliquots of either 4 nM trastuzimab antibody or 8 nM 2D3 nanobody labeled with sCy-TCO were added per well for 0.5, 1, 2, or 4 h. The cells were washed twice in cold PBS and detached with 200 μL TrypLE for 10 min. Aliquots of 100 μL of 1% bovine serum albumin in PBS were added to each sample, and the entire contents were transferred to a 96-well V-bottom plate. The cells were spun at 400 × *g* for 5 min and resuspended in 200 μL PBS with or without 0.5 μM sQSY-Tet before analysis by flow cytometry.

Quenching Efficiency and Internalization

To calculate internalization and quenching efficiency, the average mean fluorescence intensity (MFI) of the background (PBS for solution phase, control cells for flow cytometry experiments) was subtracted from each sample. Quenching efficiency was then calculated as a percentage from the following equation using the MFI:

$$n_Q = \left[1 - \left(\frac{Q_0}{N_0}\right)\right] \times 100 \quad (1)$$

where $n_Q =$ quenching efficiency, $N_0 =$ MFI of the unquenched sample at 4 °C, and $Q_0 =$ MFI of the quenched sample at 37 °C.

The percentage of material internalized, compensating for incomplete surface quenching, was calculated from the following equation:

$$I_f = \left[1 - \frac{N_1 - Q_1}{N_1 - \frac{N_1 Q_1}{N_0}}\right] \quad (2)$$

where $I_f =$ fraction internalized, $N_1 =$ MFI of the unquenched sample at 37 °C, $Q_1 =$ MFI of the quenched sample at 37°C, and $N_0 =$ MFI of the unquenched sample at 4 °C.

For the trastuzumab and 2D3 experiment, the average quenching efficiency determined at 30 min and 4 h was used to compensate for incomplete surface quenching.

2.3.2 Results Summary

Two key attributes are essential for creating an effective internalization sensor: the sensor's specificity for its molecular target and the speed at which it can detect the event. For applications such as microscopy and flow cytometry involving live cells, the quenching of surface fluorescence needs to occur within a timeframe of seconds to minutes. In the authors' previously published DNA-based Specific Hybridization Internalization Probe (SHIP) assay, this process happens very quickly, with a second-order rate constant of 10^4–10^5 M/ s for 15–20 base pair oligonucleotides [108]. The aim here was to develop a sensor with similar reaction kinetics but with a lower MW. The sensor design is based on the bio-orthogonal IEDDA reaction between tetrazine (Tet) and *trans*-cyclooctene (TCO) groups, which exhibit high specificity and a rapid reaction rate.

The internalization sensor is made up of two components (Fig. 7a). The first component includes a fluorescent dye (sCy5), a click group (TCO), and an anchor point for attaching the sensor to a molecule or particle of interest, with short polyethylene glycol (PEG) linkers connecting these functional groups to minimize steric hindrance (Fig. 7b). The MW of the sCy5-TCO component is approximately 1390 g/mol, slightly larger than the widely used Alexa Fluor 647 dye.

The second component consists of a quencher dye that has an absorbance matched to the emission of sCy5 (BHQ2 or sQSY21) along with a compatible click group (Fig. 7c). They assessed the quenching abilities of both Tet and mTet, which has a methyl group at the sixth position of the 1,2,4,5-tetrazine ring. This modification enhances stability under biological conditions but is over 30-fold less reactive than Tet [221].

In the experiments designed to quantify internalization, the fluorescent component was initially conjugated to the material of interest (transferrin, the anti-HER2 nanobody 2D3, or the anti-HER2 monoclonal antibody trastuzumab) and incubated with recipient cells for the specified duration. The cells were subsequently cooled to 4°C to halt further endocytosis [105, 108] and to provide the quencher adequate time to interact with the remaining sensors attached to surface-bound particles (Fig. 7d). It is essential that the quencher is impermeable to the plasma membrane and is confined to the extracellular space.

The initial test of the system was to evaluate the quenching efficiency in solution. Both the sQSY-Tet and BHQ2-Tet quenchers resulted in an immediate (<5 min) decrease in fluorescence intensity to the sCy5-TCO fluorescent tag in solution. The reaction between the BHQ-mTet quencher and sCy5-TCO took approximately 1 h to complete and therefore it was excluded from further testing. The signal-to-noise ratio of all quencher components was better than that of the pHrodo red dye, which retained more than 50% of its intensity when transitioning from a pH 7.5 environment to a pH 4.5 environment.

The ability of the quenchers to suppress the signal when the fluorescent component was linked to transferrin bound to the cell surface in vitro was assessed. The click sensor was attached to transferrin using EDC/NHS chemistry [222, 223]. sCy5-TCO conjugated to transferrin and incubated with C1R recipient cells was quenched by BHQ2-Tet and sQSY-Tet. The transferrin-sCy5-TCO formed a ring around the cell that colocalized with the cell membrane stain wheat germ agglutinin (WGA) labeled with Alexa Fluor 488 (WGA-AF488) and appeared as punctate structures within the cell. It was observed that BHQ2-Tet and sQSY-Tet quenched only the cell membrane signal; however, BHQ2-Tet also quenched the internal punctate fluorescence, indicating that it is membrane-permeable, rendering it unsuitable for this assay.

Transferrin bound to the cell surface at 4°C was visualized by fluorescence microscopy as a ring around the cell membrane that colocalized with WGA-AF488 (Fig. 8a) that vanished when quenched with sQSY-Tet (Fig. 8b). At 37°C both internalized and extracellular transferrin were visible (Fig. 8c). The addition of sQSY-Tet quenched the extracellular fluorescence, but the endocytosed fraction remained unaffected (Fig. 8d).

The internalization and quenching of transferrin-sCy5-TCO were quantified by flow cytometry. Addition of sQSY-Tet quencher

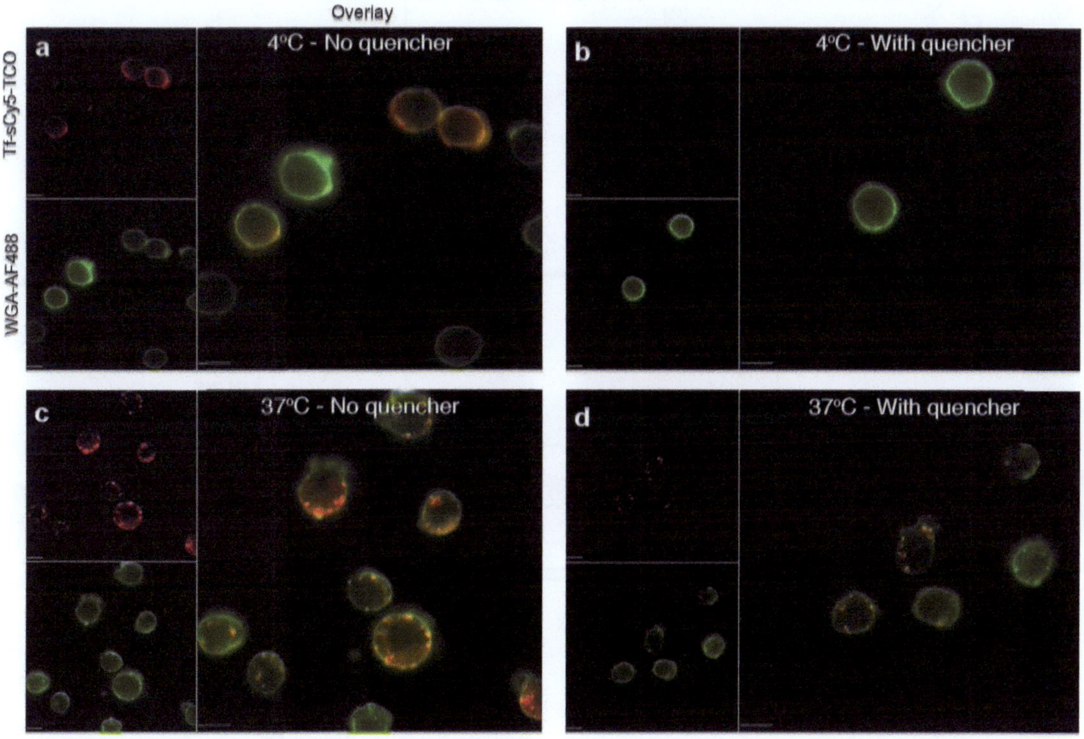

Fig. 8 Extracellular transferrin labeled with click sensor is quenched by sQSY-Tet. C1R cells incubated with 50 μg/mL transferrin labeled with sCy5-TCO click sensor at 4 °C for 10 min, washed twice, and then incubated for a further 25 min at **a** 4 °C or **c** 37 °C. Addition of 1.8 μM sQSYTet removes external fluorescence at **b** 4 °C and **d** 37 °C, but does not quench intracellular signal. Scale bar = 10 μm. (Reprinted with permission from {*ACS Sens.* 2018, 3, 6, 1182–1189}. Copyright {2018})

at 4°C resulted in a shift in the fluorescence histogram, corresponding to a quenching efficiency of 76% (Fig. 9a). At 37° C, the sQSY-Tet quenched the extracellular fluorescence but did not impact the internalized fluorescence, leading to a minimal shift in fluorescence intensity (Fig. 9b). A comparison of the transferrin uptake calculated from Eq. (1) (71 ± 2%) was comparable to the internalization determined using the FIP internalization sensor (67 ± 4%) with no statistically significant difference as determined by an unpaired *t*-test, two-tailed *P*-value = 0.15 (Fig. 9c).

The internalization sensors were linked to the 2D3 anti-HER2 nanobody and the anti-HER2 monoclonal antibody trastuzumab, then incubated with SKOV-3 cells for 0.5, 1, 2, or 4 h. The sQSY-Tet quencher demonstrated quenching efficiencies at 4 °C of 79% for the 2D3 nanobody and 71% for trastuzumab. The percentage of internalization was dependent on incubation time, with only 5% of both molecules internalized after 0.5 h. After 4 h, the final uptake percentage of the 2D3 nanobody reached 30%, while 22% of the trastuzumab had been internalized.

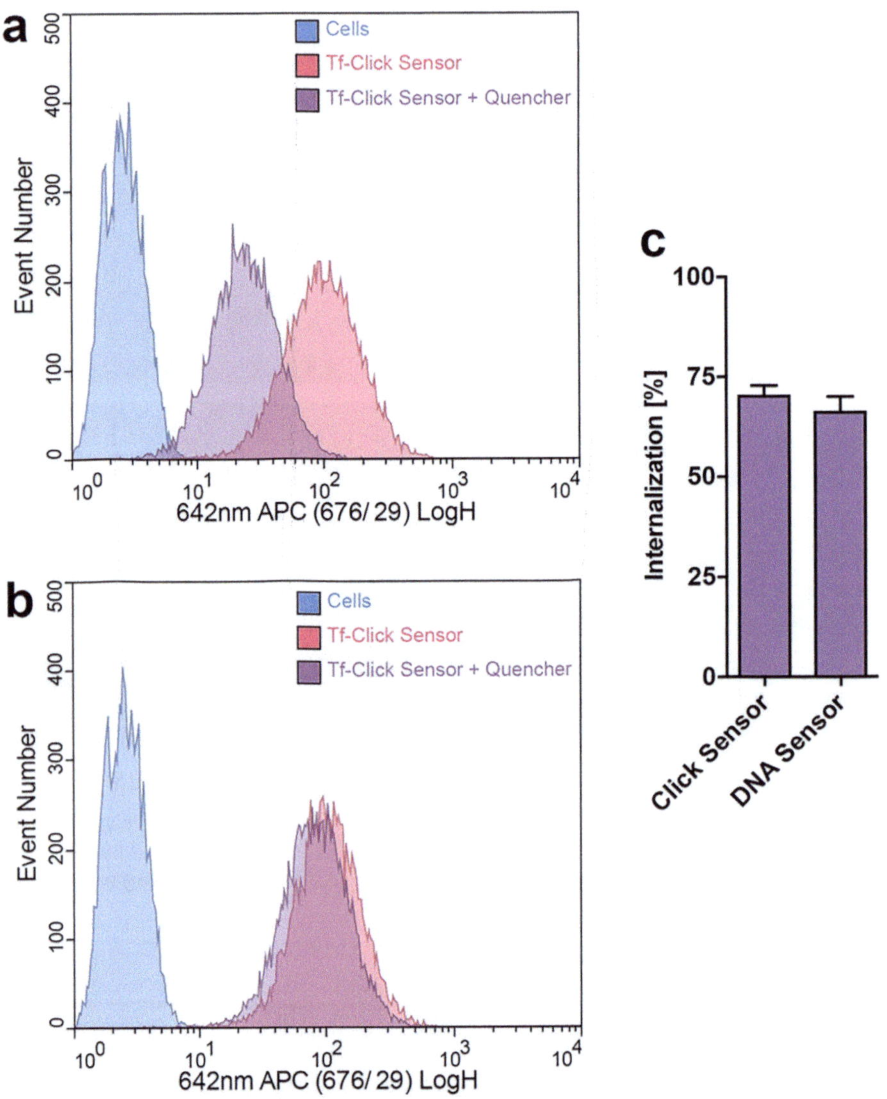

Fig. 9 Flow cytometry analysis of transferrin quenching and internalization with sQSY-Tet in C1R cells. Representative histogram of cells incubated with Tf-sCy5-TCO (15 g $^{-1}$mL) at 4°C for 15 min, washed twice, and then incubated for a further 15 min at **a** 4 °C or **b** 37 °C before quenching with sQSY-Tet. **c** Internalization using the click or DNA sensor. Quenching or internalization efficiency calculated from duplicate binding samples quenched in triplicate, from $n = 4$ (click sensor) and $n = 2$ (DNA sensor) separate experiments. (Reprinted with permission from {*ACS Sens.* 2018, 3, 6, 1182–1189}. Copyright {2018})

In summary, Selby et al. successfully developed a two-component internalization sensor capable of click chemistry-mediated conjugation to various cellular biomolecules [137, 185] The external membrane-bound fluorescence can be quenched by the second component of the sensor, enabling more accurate quantification of the percentage of internalization of the labeled proteins through flow cytometry analysis. Although this approach was not specifically designed to label EVs for internalization studies, it could be easily adapted to label any surface protein with appropriate click chemistry reactive groups.

2.4 Kinetics and Specificity of HEK293T Extracellular Vesicle Uptake Using Imaging Flow Cytometry

An alternative approach for quantifying EV uptake and internalization is imaging flow cytometry (IFC). In the study discussed here, the researchers rigorously evaluated various parameters related to EV uptake by examining the effects of EV dosage, internalization kinetics, and the differences in internalization specificity among different recipient cell types [141]. A recent position paper from ISEV emphasized the urgent need for more meticulous analyses focusing on the impact of experimental parameters such as time and dose on EV biogenesis, uptake, stability, and cargo transfer [224]. These concerns stem from conflicting findings in the literature, which may be attributed to a lack of standardization in the measurement platforms employed. For instance, varying incubation times of EVs with recipient cells, from 15 min to 4 h, influenced uptake measurements [225–229]. Additionally, the uptake of serum EVs can change based on the dose administered [230]. Another study indicated that the biodistribution of internalized human embryonic kidney 293 T (HEK293T EV) cell-derived EVs was affected by increasing the dose [231]. Consequently, establishing a reliable standard to determine time and dose curves for identifying the minimum effective dose and incubation times should lead to more robust and reproducible results in future research.

The use of flow cytometry to assess the internalization of EVs and synthetic LNPs has typically been paired with low-throughput microscopy techniques, such as confocal microscopy, to confirm EV uptake [229, 232]. However, these methods face limitations imposed by the subjective nature and time-consuming aspects of microscopy-based approaches, and due to the challenges in adapting conventional flow cytometers to accurately detect single nanometer-sized particles and distinguish them from EV aggregates or background fluorescence [163, 165, 233].

Jurgielewicz and colleagues have developed an elegant and sophisticated approach to quantify EV internalization in recipient cells by counting the internalized EVs employing IFC [141].

2.4.1 Methods

Cell Culture

HEK293T cells (ATCC), human liver epithelial cells (C3A), human neural stem cells (hNSC), SH-SY5Y neural cells, neurons, and human umbilical vein vascular endothelial cells were cultured and maintained in complete culture medium (DMEM supplemented with 10% fetal bovine serum (FBS), 100 U/mL penicillin, and 100 μg/mL streptomycin) at 37 °C in a humidified incubator. For all procedures involving EV isolation and for all EV coculture experiments, cells were cultured in FBS-free culture medium consisting of DMEM supplemented with 100 U/mL penicillin and 100 μg/mL streptomycin.

EV Labeling, Isolation, and Characterization

HEK293T cells on 10-cm dishes in complete DMEM culture medium were cultured to 70% confluency. Cells were transfected using Lipofectamine 2000 according to the manufacturer's protocol with 10 μg of CD63-eGFP (Addgene #62962) or CD63-pEGFP C2 (Addgene #62964) plasmid DNA. Control transfections were conducted by replacing the plasmid aliquot with water. After 24 h, complete culture medium was removed and replaced with FBS-depleted culture medium. Media was collected daily and replaced with fresh media for 3 days. Cellular debris in HEK293T-conditioned media was removed by filtering through a 22 μm filter, and EVs isolated by ultrafiltration employing a 100 kDa regenerated cellulose Amicon centrifugal filter unit that had been washed twice with PBS. EVs that were isolated from each HEK293T 100 cm culture dish were considered separate biological replicates. EVs were concentrated to a 1 mL volume. Particle concentration and size distributions were measured on Nanosight NS300 (Malvern, UK) according to the manufacturer's protocols.

EV Uptake Assays

Recipient cell lines at 60% confluency were seeded into a 6-well plate in complete culture medium and cultured for 24 h under standard conditions at 37 °C. Media was removed and replaced with FBS-depleted culture medium. CD63-EGFP-tagged EVs were administered at varying doses and time points at 37 °C, which has been the standard temperature used for uptake experiments since it is permissive for active cellular uptake pathways [130, 227, 234–236]. After co-culture, cells were suspended in 5% trypsin and concentrated to approximately 10^6 cells/50 μL for analysis by IFC.

EV Uptake Inhibition Assays

Cold Assay

Recipient cell lines at 60% confluency were seeded into a 6-well plate in complete culture medium and cultured for 24 h under standard conditions at 37 °C. Media was removed and replaced with FBS-depleted culture medium. CD63-EGFP-tagged EVs were administered at varying doses and time points at 37 °C, which has been the standard temperature used for uptake experiments since it is permissive for active cellular uptake pathways [130, 227, 234–236].

Fixation Assay

To inhibit all active endocytic uptake, the media were removed from recipient cells and replaced with 4% paraformaldehyde for 30 min on ice prior to coculture with EVs. Fixed cells were washed with PBS and FBS-depleted culture media and CD63-EGFP EVs were added at various doses and time points.

ImageStreamX Acquisition Software

Measurement of EV uptake was performed using the ImageStreamX Mark II Imaging Flow Cytometer (Luminex Corporation, Seattle, WA) using INSPIRE software by acquiring a minimum of 500–10,000 cell events per sample. Three biological replicates for each dose or time point were aliquoted into three separate technical wells each and individually acquired ($n = 9$). Brightfield images were acquired on channel one, and side scatter (785 nm) was collected on channel six. GFP excitation was performed employing a 488 nm argon laser at 200 mW, and the fluorescent signal was collected on channel two (480–560 nm). A magnification of 60× was employed for every sample with a low acquisition rate to achieve high sensitivity.

Image Data Exploration and Analysis (IDEA) Software Analysis

Data and image analyses were performed using the IDEA software (Luminex) with the following gating strategy:

1. To eliminate cells not in the field of focus, the focus gate was determined using the gradient root mean square (RMS) value, which detects changes in pixel values in the image, thereby measuring the sharpness quality of the brightfield image.
2. Gating of the focused cells to eliminate doublets and debris using brightfield vs aspect ratio bright field was performed. The gated data were used to generate histograms and statistical references that measure fluorescent intensity (the sum of all pixels in an image), maximum pixel intensity (intensity of the brightest pixels in an image), and spot count values via internal algorithms for each sample using the applicable IDEAS Wizards. Spot counts, mean intensity, and maximum pixel ratios were calculated using the formula (Output value with EVs/Output values without EVs).

Statistics

Quantitative data were analyzed using GraphPad Prism 8.1.2 (San Diego, CA) and performed in triplicate. Data are presented as mean ± standard error of the mean (SEM). Statistical significance was determined using an unpaired *t*-test or one-way analysis of variance (ANOVA) with Tukey's or Dunnett's multiple comparison post hoc compared to controls when appropriate. Significance was defined as $p < 0.05$.

2.4.2 Results Summary

Fluorescently labeled EVs were produced by transfecting HEK293T cells with a plasmid that encodes a CD63-eGFP fusion protein. At 24 h post-transfection, the complete culture medium was replaced with FBS-depleted culture medium. The culture medium was collected daily over a period of 3 days and pooled together. Each transfected plate was treated as a separate biological replicate. EVs were isolated using ultrafiltration techniques. The concentrations of EVs and their particle size distributions were analyzed using nanoparticle tracking analysis on the Nanosight

NS300 instrument. No significant differences were found in either the median particle diameter or the size distribution of EVs isolated from control-transfected vs CD63-eGFP-transfected HEK293T cells.

To assess whether CD63-eGFP fluorescence was associated with the isolated EVs, IFC analysis was performed. Nonfluorescent 1.34 μm beads in a buffer solution did not exhibit fluorescence when excited by the 488 nm excitation wavelength, but they were visible in both brightfield (BF) and side scatter (SSC). Untagged HEK293T EVs showed negative results in the BF, GFP, and SSC channels, which was consistent with the absence of fluorescence and a size below the BF threshold (<300 nm). The lack of detectable particles in the BF channel indicates minimal aggregation or "swarming" of the EVs. In contrast, CD63-eGFP EVs were negative in BF but positive in the GFP channel, indicating that CD63-eGFP fluorescence was indeed associated with these EVs.

The presence of common EV tetraspanin protein markers (CD9, CD63, and CD81), along with CD29, which has been previously reported on HEK293T-derived EVs, was assessed using a commercially available flow cytometry-based method. Both untagged and CD63-eGFP EVs showed similar levels of these protein markers. This led to the conclusion that the labeling strategies did not alter the physical characteristics of HEK293T EVs.

Two inhibitory internalization assays were conducted: one involved coculturing EVs with recipient cells at 4 °C, and the other involved fixing recipient cells with paraformaldehyde prior to coculturing with EVs. These methods resulted in a decreased presence of eGFP-labeled EVs compared to unfixed recipient cells cocultured at 37 °C. The cold and fixed inhibitory assays led to reductions in maximum pixel intensity, fluorescence intensity, and spot count, indicating decreased cellular uptake. This suggests that increases in these output parameters would reflect EV internalization in subsequent uptake assays.

A standard dose curve was constructed using the IFC platform by coculturing HEK293T recipient cells with varying doses of EVs, ranging from 0 to a maximum of 20,000 EVs per cell at 37 °C. A total of eight doses (0, 100, 1000, 2000, 4000, 6000, 10,000, and 20,000 EVs/cell) were tested. A dose-dependent increase in fluorescence was observed, as shown in Fig. 10. The lowest detectable EV dose was 6000 EVs per cocultured HEK293T cell (Fig. 10a). At this dose, measurements of spot count, fluorescence intensity, and maximum pixel intensity were significantly increased compared to recipient cells without EVs (Fig. 10b–d). The higher doses of 10,000 or 20,000 EVs/cell also showed significant increases in all three measurements compared to cells cultured without EVs. However, no significant differences were observed between the three

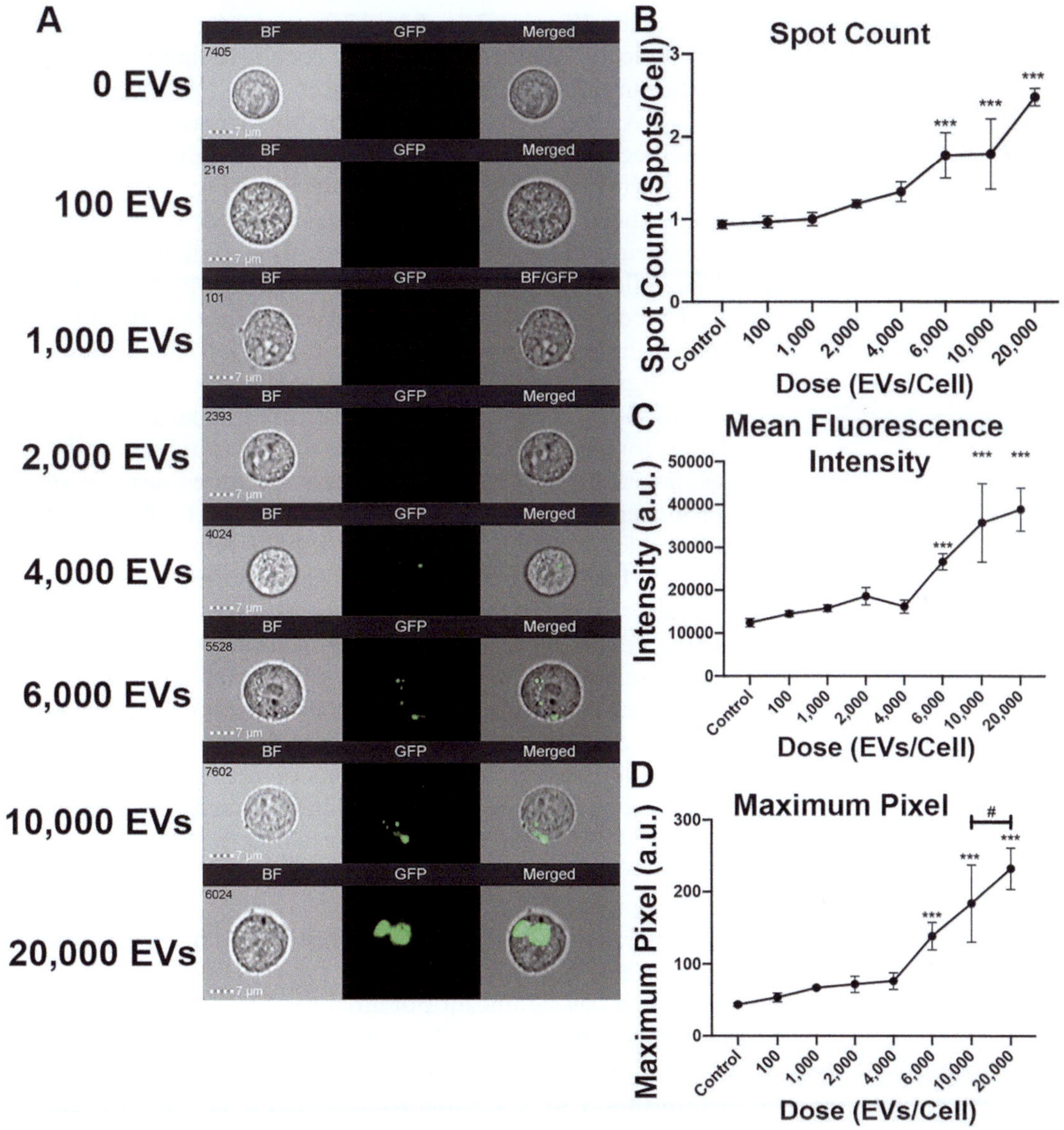

Fig. 10 HEK293T EV uptake has a dose effect with a minimum threshold of 6000 EVs. HEK293T cells were co-cultured with HEK293T EVS at increasing doses from 0 to 20,000/cell. **a** Representative IFC images of recipient cells with respective EV doses. GFP localization signifies HEK293T EV uptake. (**b**–**d**) Quantification of dose assays compared with controls and each group via spot count, mean fluorescence intensity, and maximum pixel ratios. Bars represent mean ± SEM; $N = 3$; one-way ANOVA followed with Tukey's post hoc test. *$p < 0.05$; ***$p < 0.01$. (Reproduced with permission. Under a Creative Commons license (https:/creativecommons.org/licenses/by/4.0/) from {*Nanoscal Res Lett.* 2020, 15, 1, 170})

detectable doses (i.e., 6000 vs 10,000; 10,000 vs 20,000; or 6000 vs 20,000) in any parameters, except for maximum pixel intensity for 6000 vs 20,000 EVs per recipient cell comparison.

Next, the kinetics of EV uptake by HEK293T cells were evaluated by coculturing them with 6000 EVs per cell across six time points (5 min, 30 min, 2 h, 4 h, 12 h, and 24 h) at 37 °C. The findings indicated that the duration of coculture significantly influenced the amount of visible fluorescence in the recipient cells (Fig. 11a). At the 30 min mark, a notable increase in spot count was observed, although other parameters remained unchanged compared with recipient cells without EVs (Fig. 11b–d). From 2 to 24 h, there were significant increases in spot counts, fluorescence intensity, and maximum pixel intensity compared to control recipient cells. While no differences in uptake parameters were noted between 4 and 12 h, both fluorescence intensity and maximum pixel intensity showed a decline from 12 to 24 h. Based on these findings, a dose of 6000 EVs/recipient cell and a coculture duration of 4 h were standardized for future uptake assays.

To explore the hypothesis that EV uptake is influenced by cell lineage, favoring recipient cells that share the same lineage with the EV producer cells, 6000 HEK293T CD63-eGFP EVs were co-cultured for 4 h at 37 °C with HEK293T cells, C3A liver cells (epithelial), HUVECs, or SH-SY5Y glioblastoma cells (neural). The results showed that eGFP fluorescence was highest in HEK293T cells compared to the other three cell types, supporting the idea of selective uptake of EVs by cells of similar lineage.

The effect of cellular differentiation state on EV uptake was evaluated by co-culturing 6000 EVs per recipient cell for 4 h at 37 ° C, using either mature human neurons or neural stem cells. The findings indicated that both cell types internalized EVs, but neural stem cells exhibited higher uptake than mature human neurons. This supports the hypothesis that the state of cellular differentiation affects EV uptake.

In conclusion, the authors successfully developed a high-throughput and quantitative method for assessing HEK293T EV uptake, which has significant potential for analyzing the uptake of EVs from other producer cells, as well as uptake by other recipient cell types. Notably, this method enables rigorous and quantitative evaluation of key experimental factors such as EV dose, coculture duration, and selective uptake. This approach will be of immense value for assessing and refining potential therapeutic protocols based on synthetic lipid nanoparticles and EVs for in vitro preclinical testing.

2.5 Absolute Quantification of Dynamic Cellular Uptake of Small Extracellular Vesicles via Lanthanide Element Labeling and ICP-MS

In the studies previously discussed, the binding, internalization, and intracellular movement of EVs have been monitored using various fluorescent labeling techniques. One common approach involves genetically engineering EV-producing cells to overexpress an intrinsic EV protein that is fused to a fluorescent tag [134–141], while another method uses chemical ligation to attach a fluorescent dye to an EV biomolecule [237, 238]. Although these strategies are widely used for tracking and quantifying EVs, they can be complex, time-consuming, labor-intensive, and may inadvertently affect cellular uptake or alter the biological activity of EVs.

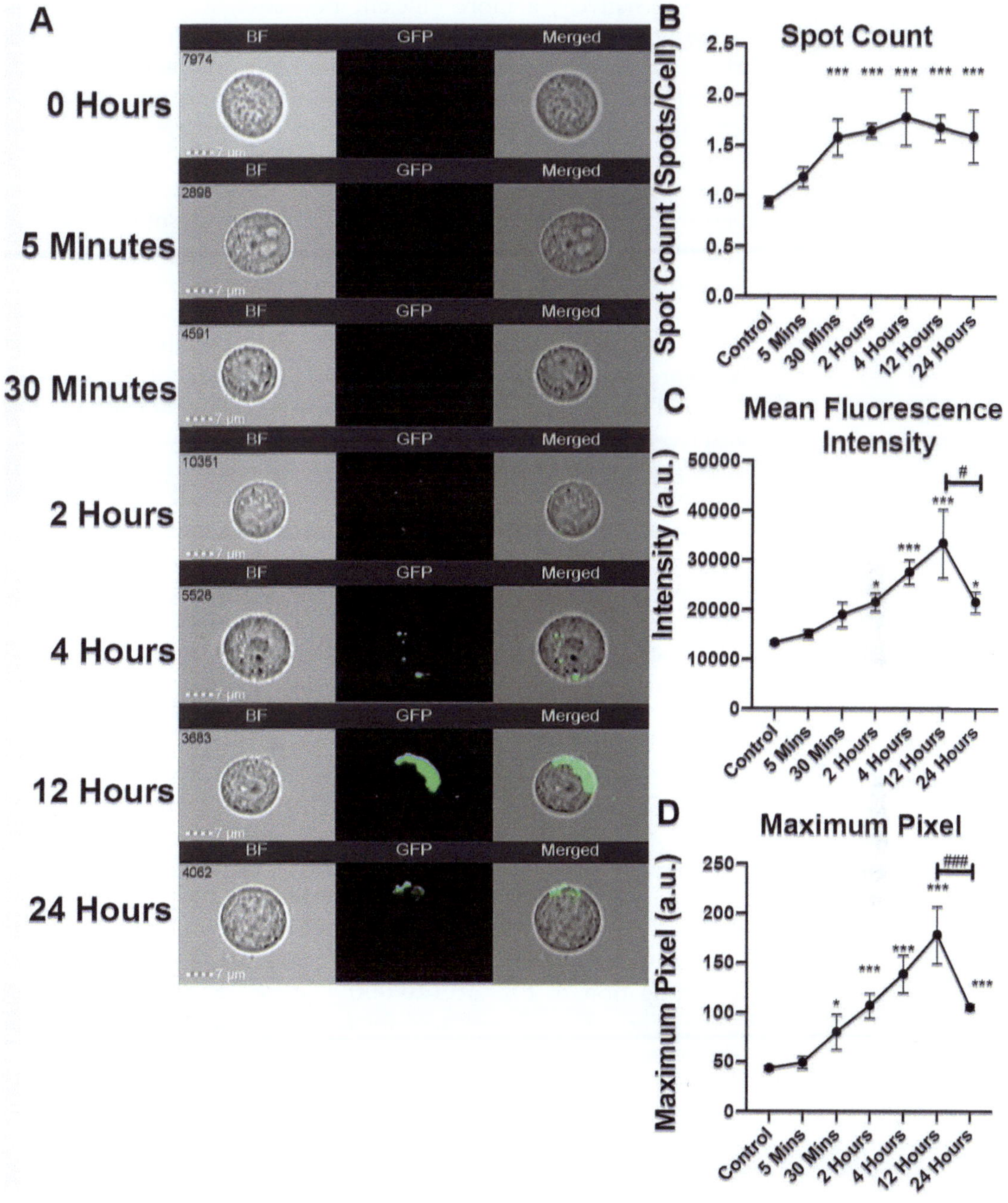

Fig. 11 HEK293T EV uptake is time-dependent. HEK293T cells were co-cultured with 6000 HEK293T EVs/cell for increasing lengths of time. **a** Representative IFC images of recipient cells. GFP localization signifies increased HEK293T EV uptake. **b–d** Quantification of time course assays compared with controls and each group via spot count, mean fluorescence intensity, and maximum pixel ratios. Bars represent mean ± SEM; $N = 3$; one-way ANOVA followed with Tukey's post hoc test. $*p < 0.05$; $***p < 0.01$. (Reproduced with permission. Under a Creative Commons license (https:/creativecommons.org/licenses/by/4.0/) from {*Nanoscal Res Lett.* 2020, 15, 1, 170})

Alternatively, a more efficient non-covalent labeling can be achieved by using hydrophobic, lipophilic, or amphipathic dyes that interact with the lipid bilayers of EV membranes. However, despite the simplicity and ease of this labeling technique, these dyes can aggregate or form micelles, which complicates the accurate quantification of the labeled particles [184, 185]. Additionally, fluorescence-based labeling methods tend to be semiquantitative, relying on fluorescence intensity that can overestimate particle numbers due to the challenges in distinguishing between EVs bound to the membrane and those that have been internalized.

To overcome these limitations, Yang and colleagues have developed a sophisticated and elegant method that enables the precise absolute quantitative assessment of EV internalization by recipient cells. This approach utilizes a one-step, biocompatible, and interference-free labeling strategy for EVs incorporating the lanthanide compound europium trichloride ($EuCl_3$). The internalized particles are quantified using ICP-MS, which measures the EV europium content [133]. The labeling of EVs with $EuCl_3$ occurs through a chelation interaction between Eu^{3+} and the phosphate groups in the EV lipid bilayer. Thanks to the high sensitivity of ICP-MS and the negligible background levels of lanthanides in biological systems, this method can detect and quantify internalized EVs across a linear range of three orders of magnitude, with the capacity to detect as few as 1.12×10^7 particles [239, 240].

2.5.1 Methods

Isolation of HeLa Cell-Derived Small EVs (sEVs)

HeLa cells were cultured and maintained in complete culture medium, specifically DMEM supplemented with 10% FBS, 100 U/mL penicillin, and 100 μg/mL streptomycin, at 37 °C in a humidified incubator. For the isolation of sEVs, HeLa cells were seeded into 15 cm Petri dishes. After achieving 80% confluence, the media was removed, and the cells were cultured in DMEM supplemented with 10% EV-depleted FBS, which was obtained by ultracentrifugation of FBS at 120,000 × *g* for 18 h at 4 °C [241–243]. The cells were cultured for an additional 24 h, after which the conditioned cell culture medium was collected for sEV isolation.

Isolation of sEVs from the collected HeLa cell culture medium was accomplished according to the ultracentrifugation protocol of Théry et al. [244]. Briefly, the medium underwent a series of centrifugations at progressively higher speeds to sequentially eliminate contaminating dead cells, cell debris, and large EVs. A final ultracentrifugation of the supernatant at 110,000 × *g* using an SW 32 Ti rotor, Beckman Coulter, for 70 min at 4 °C produced a crude sEV pellet. The pellet was washed twice with Tris-buffered saline (150 mM NaCl, 10 mM Tris–HCl, pH 6), repelleted by ultracentrifugation, and the purified sEV pellet was resuspended in 1× Tris buffer.

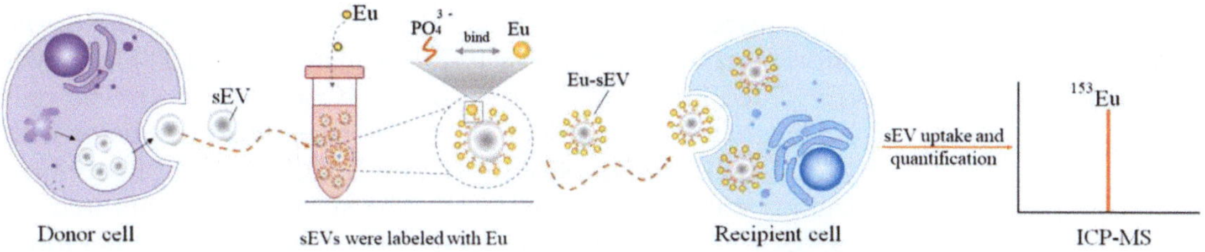

Fig. 12 Illustration of the one-step europium labeling strategy for absolute quantification of cellular sEV uptake. (Reprinted with permission from {*Anal. Chem.* 2023, 95, 32, 11,934–11,942}. Copyright {2023})

Europium Labeling of sEVs

Figure 12 illustrates the one-step Europium (Eu) labeling strategy. In summary, a chelation reaction occurs when the isolated sEVs are incubated with an $EuCl_3$ solution, through the binding of the Europium with the lipid phosphate groups. The efficiency of Eu^{3+} labeling was enhanced by adjusting reaction parameters such as the concentration of $EuCl_3$, the duration of incubation, the number of washes required to eliminate unincorporated Eu^{3+}, and the temperature during incubation. The labeling efficiency was assessed by measuring the Eu^{3+} content by ICP-MS. In the resulting optimized method, a 20 μg aliquot of sEVs was incubated with a 0.1 mM $EuCl_3$ solution for 30 min at 37 °C. Any excess unincorporated Eu^{3+} was removed by washing 4 times with 1× Tris–HCL using a 100 kDa Millipore ultrafiltration unit.

Fluorescent Labeling of sEVs

To enable visual tracking, in addition to Eu^{3+} labeling, sEVs were also labeled with the membrane-permeable fluorescent dye carboxyfluorescein succinimidyl ester (CFSE). In brief, a 20 μg aliquot of Eu^{3+}-labeled sEVs was incubated for 10 min with 10 μM CFSE at 37 °C. The excess unincorporated CFSE was removed by washing 4 times using a 100 kDa Millipore ultrafiltration unit.

Characterization of sEVs

Various methods were employed to assess the potential impacts of Eu^{3+}-labeling on the morphology, stability, and functional characteristics of sEVs. The protein concentrations of both labeled and unlabeled sEVs were measured using the microBCA kit (ThermoFisher, USA). Nanoflow cytometry (N30E, Xiamen, China) was employed to measure the concentration and particle size distribution of labeled and unlabeled sEVs. Additionally, the morphology of Eu^{3+}-labeled sEVs was assessed by transmission electron microscopy (JEOL JEM, Japan). Western blotting was employed to analyze the expression levels of known sEV marker proteins. The zeta potentials of both labeled and unlabeled sEVs were measured with a Zetasizer Nano ZS Zen3600 (Malvern Instruments, UK). The chemical stability of labeled and unlabeled sEVs was evaluated by monitoring Eu^{3+} signals by ICP-MS after incubation in EV-depleted culture medium at 37 °C for either 0 or 9 h.

Effects of Temperature or Incubation Duration on Eu-sEV Uptake by HeLa Cells

For fluorescent imaging, 8 × 10^4 HeLa cells were plated into 6-well confocal imaging dishes. For ICP-MS analysis, 1 × 10^6 HeLa cells were seeded into 6-well culture dishes (Corning, USA). The cells were incubated overnight at 37 °C in a humidified incubator with 5% CO_2. A 20 μg aliquot of Eu-sEVs (CFSE-labeled for fluorescent imaging) was added to each well. The sEVs were co-cultured for 1, 3, 6, or 9 h at 37 °C to assess the effects of incubation duration, or at 4 °C or 37 °C to assess the effect of temperature on sEV uptake, before being analyzed by fluorescent imaging or ICP-MS.

Competitive Inhibition of Eu-sEV Uptake by HeLa Cells

HeLa cells (1 × 10^6) were plated into 6-well culture plates and cultured overnight. A 10 μg aliquot of Eu^{3+}-sEVs, either alone or combined with a 20 μg aliquot of native sEVs, was added to each well and cocultured for 3 h at 37 °C before analysis using ICP-MS.

Inhibition of HeLa Cell Uptake of Eu-sEVs by Endocytosis Inhibitors

HeLa cells (1 × 10^6) were seeded into 6-well culture plates and cultured overnight. The cells were preincubated with or without 10 μM chlorpromazine (CPZ), which inhibits clathrin-coated pit formation, or 50 μM IPA-3, a highly specific inhibitor of p21-activated kinase-1 (Pak-1) that blocks both clathrin- and caveolin-independent endocytosis, for 30 min. Afterward, a 20 μg aliquot of Eu^{3+}-sEVs was added to each well and cocultured for 6 h at 37 °C before analysis by ICP-MS.

Fluorescence Imaging of CFSE-Labeled Eu-EV Uptake

After incubating 8 × 10^4 HeLa cells with Eu^{3+}-sEVs for the specified times and temperatures, the culture medium was removed, and the cells were washed 3 times in 1X phosphate buffered saline (PBS). The cells were then fixed for 30 min in freshly prepared 4% paraformaldehyde (PFA) at room temperature. Nuclear staining was performed by incubating the cells with 1 μg/μL Hoechst 33342 for 10 min at room temperature. Excess stain was removed by washing 3 times with 1X PBS. Uptake was assessed by confocal microscopy with a 20× objective lens (Zeiss 880, Germany).

Wound Healing Assay

HeLa cells (1 × 10^6 cells) were seeded into 6-well plates and cultured until the cell monolayer reached confluence. The HeLa monolayer was scratched using a sterile 10 μL pipette tip and then washed 3 times with 1× PBS to remove any detached cells. The remaining cells were incubated with or without 50 μg of either native-sEVs or Eu^{3+}-sEVs. Wound closure was assessed by capturing images at 0 and 18 h of incubation in fresh culture medium using an inverted microscope (Nikon, Japan) with a 10× objective lens. The extent of wound closure was analyzed with ImageJ software.

Detection and Quantification of Eu-Labeled sEVs by ICP-MS

A standard curve was created based on the Eu^{3+} signal intensities obtained from a series of diluted Eu^{3+} standard solutions (0, 5, 10, 20, 50, 100, 200, and 500 μg/L). The concentrations of Eu solutions served as the horizontal coordinates, while the Eu signal intensities measured by ICP-MS (Agilent 7800, USA) were plotted as the vertical coordinates. To determine the Eu^{3+} content in sEVs or in cell-related samples, the sEVs or cell samples were predigested in nitric acid in a water bath for 1 h.

The Eu^{3+} content was subsequently measured using the Agilent 7800 with the following key parameters: radiofrequency power = 1500 W; carrier gas (Ar) flow rate = 1 L/min; auxiliary gas (Ar) flow rate = 0.9 L/min; coolant gas (Ar) flow rate = 15 L/min; dwell time = 300 ms; scanning mode = He; pump rate = 29 r/min; monitored isotope = ^{153}Eu. The actual Eu^{3+} y content was then calculated using the standard curve.

2.5.2 Results Summary

HeLa cell-derived sEVs were isolated from the cell culture medium using ultracentrifugation at 110,000 × *g* for 70 min at 4 °C. A 20 μg aliquot of sEVs was labeled by incubating with 0.1 mM $EuCl_3$ for 30 min at 4 °C, and any excess $EuCl_3$ was removed by washing 4 times using a 100 kDa ultrafiltration unit. The measurement of Eu^{3+} incorporation into the sEVs using ICP-MS found that the sEV sample contained 11 times more Eu^{3+} than the residual Eu^{3+} amount measured in the fourth wash buffer. Experiments were performed to identify the optimal labeling conditions, which were established as 0.1 mM $EuCl_3$ for 30 min at 37 °C. These labeling conditions resulted in a saturation of approximately 14 ng Eu^{3+} per 20 μg sEVs, which had a size range of 30–150 nm.

The impact of Eu^{3+}-labeling on the physical characteristics of sEVs was assessed using various methods. The stability of Eu^{3+} incorporation was evaluated by incubating Eu^{3+}-labeled sEVs for 9 h in HeLa culture medium supplemented with EV-depleted FBS, followed by the re-isolation of the sEVs from the culture medium and comparing the Eu^{3+} levels to a 0 h sEV sample. No statistically significant loss of sEV Eu^{3+} content was detected. The particle size distribution of Eu^{3+}-sEVs was unchanged compared to native sEVs, as measured by Nanoflow cytometry (NanoFCM). The zeta potential of Eu^{3+}-sEVs showed a slight decrease in charge density compared to native sEVs, but it remained negatively charged, indicating no significant alteration in sEV surface properties. Visual examination by fluorescent confocal microscopy of CFSE-labeled Eu^{3+}-sEVs and native sEVs did not show any differences in particle aggregation. Additionally, no differences in the typical cup-shaped morphology of sEVs as evaluated by TEM were observed between Eu^{3+}-sEVs and native sEVs.

Confocal fluorescence microscopy of HeLa cells incubated with CFSE-native sEVs and of CFSE-Eu^{3+}-sEVs displayed similar strong fluorescent patterns within the cells, suggesting that the Eu^{3+}-

labeling of sEVs did not hinder their internalization. To further investigate, a wound healing experiment was performed, which demonstrated that Eu^{3+}-labeling did not compromise the ability of sEVs to facilitate cell migration. Consequently, the authors concluded that their Eu^{3+}-labeling strategy is a viable method for efficiently labeling sEVs without altering their typical physical characteristics and biological functions.

The number of internalized sEV particles was quantified by NanoFCM, while the amount of Eu^{3+} was measured by ICP-MS. A linear relationship ($R = 0.99$) was established between the number of sEV particles and the Eu^{3+} content, covering a range of three orders of magnitude from 1.12×10^7 to 5.24×10^{19} particles. Given that most experiments utilize between 1.0×10^8 and 1.9×10^9 particles to investigate the cellular uptake of EVs, it was concluded that ICP-MS-based quantification of Eu^{3+}-sEVs can detect subtle changes induced by experimental conditions.

The ability to quantify the cellular uptake and internalization of Eu^{3+}-sEVs was evaluated. The findings from these experiments are shown in Fig. 13. Initially, HeLa cells were incubated with Eu^{3+}-sEVs for 1, 3, 6, and 9 h at 37 °C for analysis by ICP-MS. A second parallel experiment involved incubating HeLa cells with CFSE-Eu^{3+}-sEVs for analysis via NanoFCM. As illustrated in Fig. 13a and b, the internalization of sEVs was evident through scattered punctate green fluorescence within the HeLa cells, and the associated Eu^{3+} signal was detectable by ICP-MS. The internalization process was time-dependent, as both the Eu^{3+} signal and the associated fluorescence signal increased gradually over the incubation period. Notably, the Eu^{3+} signal specifically provided a more accurate and quantitative measure of the actual amount of internalized sEVs compared to merely estimating based on a few scattered particles seen on a confocal image.

The effect of temperature on internalization was assessed next. Parallel plates were incubated with Eu^{3+}-sEVs or CFSE-Eu^{3+}-sEVs for 3 h at either 4 °C or 37 °C. As shown in Fig. 13c, internalization of Eu^{3+}-sEVs at 4 °C was significantly lower, as measured by both NanoFCM and ICP-MS. The ICP-MS analysis revealed a 157% increase in Eu^{3+}-sEV internalization at 37 °C compared to the levels measured at 4 °C. This suggests that the internalization of Eu^{3+}-sEVs occurs through an energy-dependent endocytosis pathway rather than passive membrane fusion. Additionally, the introduction of native sEVs led to a 40% reduction in Eu^{3+}-sEV internalization (Fig. 13d), indicating that the Eu^{3+}-labeling does not change the endocytosis pathway utilized for internalization.

To determine the endocytic pathways utilized by HeLa cells for sEV internalization, the Eu^{3+} signal was measured in HeLa cells that had been preincubated with IPA-3 or chlorpromazine for 30 min, followed by a 9 h incubation with Eu^{3+}-sEVs. After

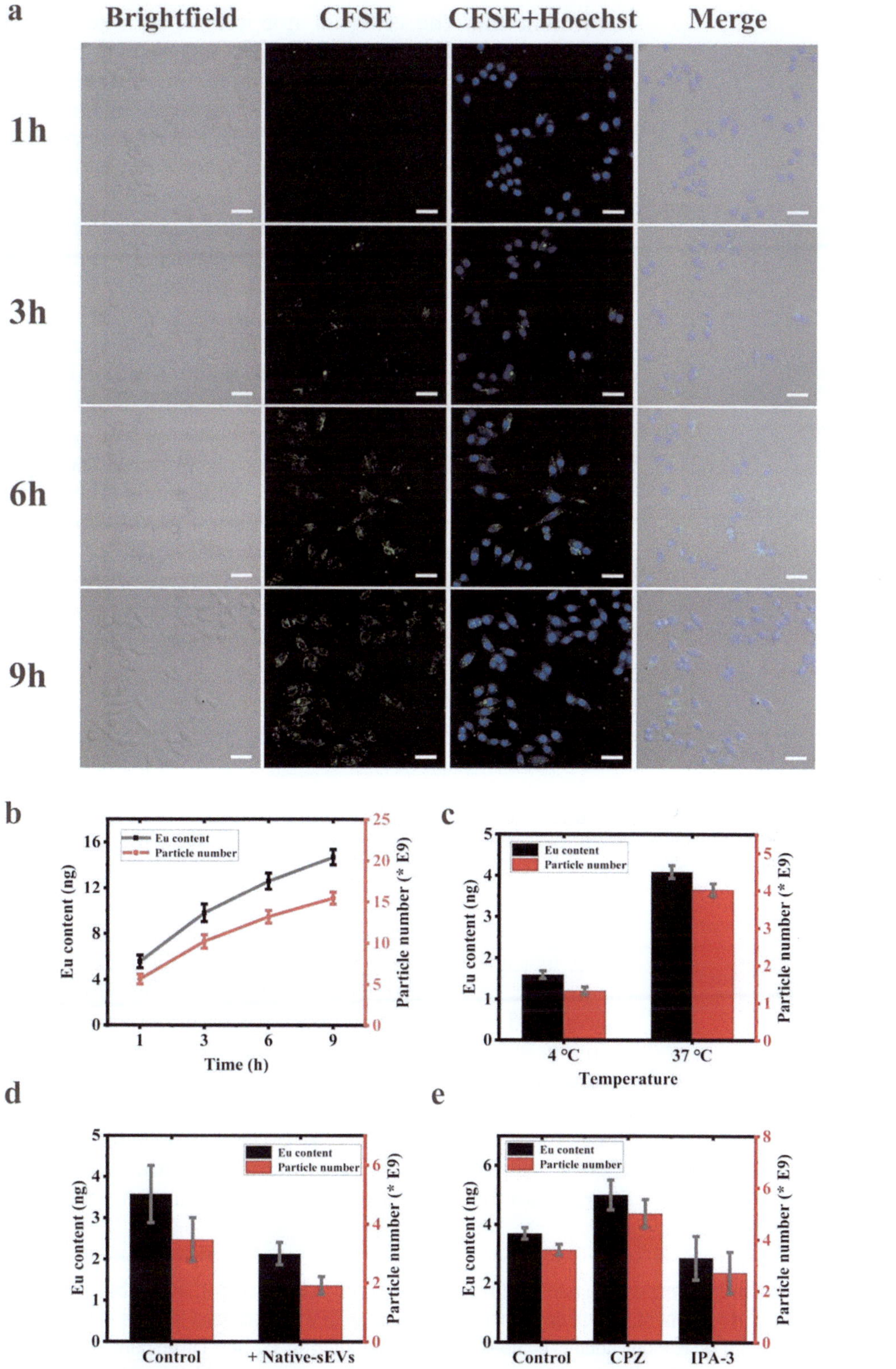

Fig. 13 Quantification of the cellular uptake of sEVs. **a** Confocal fluorescence images of CFSE-Eu-sEV uptake by HeLa cells after incubation for 1, 3, 6, and 9 h. Scale bar: 50 μm. **b** Amount of Eu and corresponding number of sEVs internalized in HeLa cells after incubation with Eu-sEVs for 1, 3, 6, and 9 h. **c** Amount of Eu and

confirming that the cells did not induce cytotoxicity, the Eu^{3+} signal was quantified using ICP-MS, and particle numbers were evaluated with NanoFCM. IPA-3, which inhibits macropinocytosis by blocking PAK-1, resulted in a 23% decrease in Eu^{3+}-sEV internalization (Fig. 13e). This indicated that macropinocytosis is one of the mechanisms HeLa cells use to internalize sEVs, although other endocytosis pathways are also active. In contrast, CPZ, which inhibits the formation of clathrin-coated pits, led to a slight increase in sEV incorporation. This could be explained by a compensatory upregulation of clathrin-independent pathways in response to the blockage of clathrin-mediated endocytosis [229, 245].

The authors found that their Eu^{3+}-labeling strategy did not change the physical properties or the ability to alter the phenotype of recipient cells. By using ICP-MS to measure the Eu^{3+}-signal and NanoFCM to count the number of internalized sEVs simultaneously, they were able to quantify sEV internalization across a range of three orders of magnitude. Additionally, these techniques were sensitive enough to detect minor variations in sEV uptake influenced by experimental parameters. Therefore, this approach can be a valuable tool for research involving not only sEV internalization but also in studies of cellular communication and for verifying the delivery of targeted chemical and biomolecular therapeutics.

3 Conclusions

Accurately quantifying EVs and synthetic LNPs presents daunting challenges. The primary obstacles stem from the tiny size of these particles. EVs and other nanoparticles are significantly smaller than the resolution limit of instruments designed to analyze particles an order of magnitude larger in size. Consequently, despite advancements in these technologies, it remains a significant challenge to ensure that single particles are counted rather than aggregates or swarms. Additionally, it is necessary to understand that the first event in several endocytosis pathways for particle internalization involves the interaction between the EV membrane and the recipient cell plasma membrane. Particle binding to the plasma membrane keeps the particle in close proximity until the internalization is complete. A successful quantification method must be able to discriminate between bound particles and internalized particles.

Fig. 13 (continued) corresponding number of sEVs internalized in HeLa cells after incubation with Eu-sEVs at 37 °C and 4 °C. **d** Amount of Eu and corresponding number of sEVs internalized in HeLa cells after incubation with Eu-sEVs with or without adding unlabeled sEVs. **e** Amount of Eu and corresponding number of sEVs internalized in HeLa cells after incubation with Eu-sEVs with or without CPZ (10 μM) and IPA-3 (50 μM). (Reprinted with permission from {*Anal. Chem.* 2023, 95, 32, 11,934–11,942}. Copyright {2023})

The engineering of EVs and synthetic particles aimed at targeting specific subpopulations of recipient cells has led to the design of a wide variety of strategies to tag these nanoparticles to facilitate their detection and permit tracking of their movement within cells. The choice of tag and labeling method largely depends on the technology used to detect and quantify their internalization. Most tags utilize fluorescent dyes to stain the EV membranes, but there are methods to attach tags directly to specific components of the EV membrane or protein components. This can be achieved through bioconjugation reactions or by designing fusion proteins that incorporate fluorescent tags. Alternative tagging methods, such as luciferase tags or isotopic tags, are becoming increasingly common, as discussed in the method papers described here.

Fluorescence microscopy and flow cytometry have been widely utilized for analyzing how nanoparticles are taken up by recipient cells, even though these methods have limitations in resolution. The development of high-resolution fluorescence microscopes, imaging flow cytometers, and specialized nanoflow cytometers specifically designed to detect nanoparticles represents a significant improvement in overcoming the challenges posed by older technologies, enabling more accurate and high-throughput analysis of nanoparticle internalization. Additionally, the growing application of luminescence assays and ICP-MS offers promising opportunities to advance our understanding of the factors that regulate nanoparticle internalization. The ability to accurately evaluate the impact of modifications to EVs and other nanoparticles will be crucial for realizing their therapeutic potential.

To understand the in vivo biological and clinical effects of EVs, it is crucial to be able to visualize and analyze their distribution and dynamic movements in situ in the tissue or tumor microenvironment. Four of the five methodologies for visualizing, tracking, and quantification of EV internalization previously discussed in detail rely on tagging EV marker molecules with fluorescence-based labels or the use of hydrophobic, lipophilic, or amphipathic fluorescent dyes that associate with EV membrane lipid bilayers [137, 140, 141, 143]. These marker-based approaches are fundamentally limited by the tissue distribution of the labeled EV marker proteins. The labels used may also interfere with the normal physiological functions of the EV markers or result in nonspecific false-positive binding of the fluorescent dyes [246, 247].

An intriguing new label-free approach has been developed that exploits the intrinsic optical structural and metabolic contrast of EVs to permit the analysis of individual nanoparticles in the context of their spatial distribution by their optical signature [248]. This was accomplished by the simultaneous excitation of multiharmonic and autofluorescent signals [249–252]. Using a tailored optical fiber source with a 1080–1140 nm wavelength, pulse shaping, and a repetition rate of 10 MHz allows the acquisition of EV structural properties of the EV lipid-water interface by third-generation harmonic microscopy [253]. The EV metabolic signatures were captured by 2-photon fluorescence of FAD and by 3-photon fluorescence of NAD(P)H [249–251, 254, 255]. The combined effect of these two components overcomes the challenges of low specificity and contrast achievable using standard light microscopy to visualize nanometer-sized EVs and complex tissue distribution, allowing for the label-free detection and analysis of EVs from such diverse sources as fresh human tissues, living cells, living animals, or isolated EV samples. This technique holds great potential for future studies of EV and synthetic LNP binding, internalization, and intracellular tracking.

References

1. Choi D, Lee TH, Spinelli C, Chennakrishnaiah S, D'Asti E, Rak J (2017) Extracellular vesicle communication pathways as regulatory targets of oncogenic transformation. Semin Cell Dev Biol 67:11–22
2. Tkach M, Théry C (2016) Communication by extracellular vesicles: where we are and where we need to go. Cell 164:1126–1232
3. Robbins PD, Morelli AE (2014) Regulation of immune responses by extracellular vesicles. Nat Rev Immunol 14(3):195–208. https://doi.org/10.1038/nri3622
4. Qian K, Fu W, Li T, Zhao J, Lei C, Hu S (2022) The roles of small extracellular vesicles in cancer and immune regulation and translational potential in cancer therapy. J Exp Clin Cancer Res 41(1):286
5. Muller L, Mitsuhashi M, Simms P, Gooding WE, Whiteside TL (2016) Tumor-derived exosomes regulate expression of immune function-related genes in human T cell subsets. Sci Rep 6:20254
6. Chen G, Huang AC, Zhang W, Zhang G, Wu M, Xu W, Yu Z, Yang J, Wang B et al (2018) Exosomal PD-L1 contributes to immunosuppression and is associated with anti-PD-1 response. Nature 560(7718): 82–386
7. Chen W, Jiang J, Xia W, Huang J (2017) Tumor-related exosomes contribute to tumor-promoting microenvironment: an immunological perspective. J Immunol Res:1073947
8. Bhatta B, Cooks T (2020) Reshaping the tumor microenvironment: extracellular vesicles as messengers of cancer cells. Carcinogenesis 41(11):1461–1470
9. Lattmann E, Levesque MP (2022) The role of extracellular vesicles in melanoma progression. Cancers (Basel) 14:3086
10. Peinado H, Alečković M, Lavotshkin S, Matei I, Costa-Silva B, Moreno-Bueno G, Hergueta-Redondo M (2012) Melanoma exosomes educate bone marrow progenitor cells toward a pro-metastatic phenotype through MET. Nat Med 18:883–891
11. Saman S, Kim W, Raya M, Visnick Y, Miro S, Saman S, Jackson B, McKee AC, Alvarez VE, Lee NC, Hall GF (2012) Exosome-associated tau is secreted in tauopathy models and is selectively phosphorylated in cerebrospinal fluid in early Alzheimer disease. J Biol Chem 287(6):3842–3849
12. Candelario KM, Steindler DA (2014) The role of extracellular vesicles in the progression of neurodegenerative disease and cancer. Trends Mol Med 20(7):368–374
13. Skog J, Würdinger T, van Rijn S, Meijer DH, Gainche L, Sena-Esteves M, Curry WTJ, Carter BS, Krichevsky AM, Breakefield XO (2008) Glioblastoma microvesicles transport RNA and proteins that promote tumour growth and provide diagnostic biomarkers. Nat Cell Biol 10:1470–1476
14. Xu K, Liu Q, Wu K, Liu L, Zhao M, Yang H, Wang X, Wang W (2020) Extracellular vesicles as potential biomarkers and therapeutic approaches in autoimmune diseases. J Transl Med 18:432
15. Zocco D, Ferruzzi P, Cappello F, Kuo WP, Fais S (2014) Extracellular vesicles as shuttles of tumor biomarkers and anti-tumor drugs. Front Oncol 4:267
16. Palviainen M, Saraswat M, Varga Z, Kitka D, Neuvonen M, Puhka M, Joenväärä S, Renkonen R, Nieuwland R, Takatalo M, Siljander PRM (2020) Extracellular vesicles from human plasma and serum are carriers of extravesicular cargo-implications for biomarker discovery. PLoS One 15:e0236439

17. Williams HL, Frei AL, Koessler T, Berger MD, Dawson H, Michielin O, Zlobec I (2024) The current landscape of spatial biomarkers for prediction of response to immune checkpoint inhibition. NPJ Precis Oncol 8:178
18. Zhang Q, Jeppesen DK, Higginbotham JN, Graves-Deal R, Trinh VQ, Ramirez MA et al (2021) Supermeres are functional extracellular nanoparticles replete with disease biomarkers and therapeutic targets. Nat Cell Biol 23:1240–1254
19. Chen Z, Wang H, Xia Y, Yan F, Lu Y (2018) Therapeutic potential of mesenchymal cell-derived miRNA-150-5p-expressing exosomes in rheumatoid arthritis mediated by the modulation of MMP14 and VEGF. J Immunol 201:2472–2482
20. Liu S, Wu X, Chandra S, Lyon C, Ning B, Jiang L, Fan J, Hu TY (2022) Extracellular vesicles: emerging tools as therapeutic agent carriers. Acta Pharm Sin B 12:3822–3842
21. Kawikova I, Askenase PW (2015) Diagnostic and therapeutic potentials of exosomes in CNS diseases. Brain Res 1617:63–71
22. Théry C, Witwer KW, Aikawa E, Alcaraz MJ, Anderson JD, Riantsitohaina R et al (2018) Minimal information for studies of extracellular vesicles 2018 (MISEV2018): a position statement of the International Society for Extracellular Vesicles and update of the MISEV2014 guidelines. J Extracell Vesicles 7:1535750
23. Welsh JA, Goberdhan DCI, O'Driscoll L, Buzas EI, Blenkiron C, Bea B (2024) Minimal information for studies of extracellular vesicles (MISEV2023): from basic to advanced approaches. J Extracell Vesicles 13:e12404
24. Zhou W, Fong MY, Min Y, Somlo G, Liu L, Palomares MR et al (2014) Cancer-secreted miR-105 destroys vascular endothelial barriers to promote metastasis. Cancer Cell 25(4):501–515
25. Paggetti J, Haderk F, Seiffert M, Janji B, Distler U, Ammerlaan W et al (2015) Exosomes released by chronic lymphocytic leukemia cells induce the transition of stromal cells into cancer-associated fibroblasts. Blood 126(9):1106–1117
26. Hu S, Ma J, Su C, Chen Y, Shu Y, Qi Z, Zhang B, Shi G, Zhang Y, Zhang Y, Huang A, Kuang Y, Cheng P (2021) Engineered exosome-like nanovesicles suppress tumor growth by reprogramming tumor microenvironment and promoting tumor ferroptosis. Acta Biomater 135:567–581
27. Fang T, Lv H, Lv G, Li T, Wang C, Han Q et al (2018) Tumor-derived exosomal miR-1247-3p induces cancer-associated fibroblast activation to foster lung metastasis of liver cancer. Nat Commun 9(1):191
28. Joshi BS, de Beer MA, Giepmans BNG, Zuhorn IS (2020) Endocytosis of extracellular vesicles and release of their cargo from endosomes. ACS Nano 14:4444–4455
29. Mulcahy LA, Pink RC, Carter DR (2014) Routes and mechanisms of extracellular vesicle uptake. J Extracell Vesicles 4:3
30. Kirchhausen T (2000) Clathrin. Annu Rev Biochem 69:699–727
31. Kirchhausen T, Owen D, Harrison SC (2014) Molecular structure, function, and dynamics of clathrin-mediated membrane traffic. Cold Spring Harb Perspect Biol 6:a016725
32. Maldonado-Báez L, Wendland B (2006) Endocytic adaptors: recruiters, coordinators and regulators. Trends Cell Biol 16(10):505–513
33. Oh P, McIntosh DP, Schnitzer JE (1998) Dynamin at the neck of caveolae mediates their budding to form transport vesicles by GTP-driven fission from the plasma membrane of endothelium. J Cell Biol 141(1):101–114. https://doi.org/10.1083/jcb.141.1.101
34. Woodman PG (2000) Biogenesis of the sorting endosome: the role of Rab5. Traffic 1(9):695–701
35. van der Sluijs P, Hull M, Webster P, Mâle P, Goud B, Mellman I (1992) The small GTP-binding protein rab4 controls an early sorting event on the endocytic pathway. Cell 70(5):729–740
36. Schonteich E, Wilson GM, Burden J, Hopkins CR, Anderson K, Goldenring JR, Prekeris R (2008) The Rip11/Rab11-FIP5 and kinesin II complex regulates endocytic protein recycling. J Cell Sci 121(Pt 22):3824–3833. https://doi.org/10.1242/jcs.032441
37. Rink J, Ghigo E, Kalaidzidis Y, Zerial M (2005) Rab conversion as a mechanism of progression from early to late endosomes. Cell 122(5):735–749
38. Kiss AL, Botos E (2009) Endocytosis via caveolae: alternative pathway with distinct cellular compartments to avoid lysosomal degradation? J Cell Mol Med 13:1228–1237
39. Simón L, Campos A, Leyton L, Quest AFG (2020) Caveolin-1 function at the plasma membrane and in intracellular compartments in cancer. Cancer Metastasis Rev 39:435–453
40. Dalton CM, Schlegel C, Hunter CJ (2023) Caveolin-1: a review of intracellular functions, tissue-specific roles, and epithelial tight junction regulation. Biology (Basel) 12:1402
41. D'Alessio A (2023) Unraveling the cave: a seventy-year journey into the caveolar

network, cellular signaling, and human disease. Cells 12:2680

42. Gambin Y, Ariotti N, McMahon K-A, Bastiani M, Sierecki E, Kovtun O et al (2014) Single-molecule analysis reveals self assembly and nanoscale segregation of two distinct cavin subcomplexes on caveolae. elife 3:e01434
43. Zimnicka AM, Husain YS, Shajahan AN, Sverdlov M, Chaga O, Chen Z et al (2016) Src-dependent phosphorylation of caveolin-1 Tyr-14 promotes swelling and release of caveolae. Mol Biol Cell 27(13):2090–2106
44. Pelkmans L, Püntener D, Helenius A (2002) Local actin polymerization and dynamin recruitment in SV40-induced internalization of caveolae. Science 296:535–539
45. Lajoie P, Nabi IR (2007) Regulation of raft-dependent endocytosis. J Cell Mol Med 11: 644–653
46. Sezgin E, Levental I, Mayor S, Eggeling C (2017) The mystery of membrane organization: composition, regulation and roles of lipid rafts. Nat Rev Mol Cell Biol 18:361–374
47. Li Y, Gao L, Tan X, Li F, Zhao M, Peng S (2016) Lipid rafts-mediated endocytosis and physiology-based cell membrane traffic models of doxorubicin liposomes. Biochim Biophys Acta 1858:1801–1811
48. Chatterjee S, Mayor S (2001) The GPI-anchor and protein sorting. Cell Mol Life Sci 58(14):1969–1987
49. Fivaz M, Vilbois F, Thurnheer S, Pasquali C, Abrami L, Bickel PE et al (2002) Differential sorting and fate of endocytosed GPI-anchored proteins. EMBO J 21(15): 3989–4000
50. Kinoshita T (2020) Biosynthesis and biology of mammalian GPI-anchored proteins. Open Biol 10(3):190290
51. Kalia M, Kumari S, Chadda R, Hill MM, Parton RG, Mayor S (2006) Arf6-independent GPI-anchored protein-enriched early endosomal compartments fuse with sorting endosomes via a Rab5/phosphatidylinositol-3′-kinase-dependent machinery. Mol Biol Cell 17(8):3689–3704
52. Goswami D, Gowrishankar K, Bilgrami S, Ghosh S, Raghupathy R, Chadda R et al (2008) Nanoclusters of GPI-anchored proteins are formed by cortical actin-driven activity. Cell 135(6):1085–1097
53. Chadda R, Howes MT, Plowman SJ, Hancock JF, Parton RG, Mayor S (2007) Cholesterol-sensitive Cdc42 activation regulates actin polymerization for endocytosis via the GEEC pathway. Traffic 8(6):702–717
54. Howes MT, Kirkham M, Riches J, Cortese K, Walser PJ, Simpson F et al (2010) Clathrin-independent carriers form a high capacity endocytic sorting system at the leading edge of migrating cells. J Cell Biol 190(4):675–691
55. Doherty GJ, Lundmark R (2009) GRAF1-dependent endocytosis. Biochem Soc Trans 37:1061–1065
56. Lamaze C, Dujeancourt A, Baba T, Lo CG, Benmerah A, Dautry-Varsat A (2001) Interleukin 2 receptors and detergent-resistant membrane domains define a clathrin-independent endocytic pathway. Mol Cell Biol 7(3):661–671
57. Dermine JF, Duclos S, Garin J, St-Louis F, Rea S, Parton RG, Desjardins M (2001) Flotillin-1-enriched lipid raft domains accumulate on maturing phagosomes. J Biol Chem 276:18507–18512
58. Glebov OO, Bright NA, Nichols BJ (2006) Flotillin-1 defines a clathrin-independent endocytic pathway in mammalian cells. Nat Cell Biol 8(1):46–54
59. Frick M, Bright NA, Riento K, Bray A, Merrified C, Nichols BJ (2007) Coassembly of flotillins induces formation of membrane microdomains, membrane curvature, and vesicle budding. Curr Biol 17(13):1151
60. Damm EM, Pelkmans L, Kartenbeck J, Mezzacasa A, Kurzchalia T, Helenius A (2005) Clathrin- and caveolin-1-independent endocytosis: entry of simian virus 40 into cells devoid of caveolae. J Cell Biol 168:477–488
61. Swanson JA (2008) Shaping cups into phagosomes and macropinosomes. Nat Rev Mol Cell Biol 9:639–649
62. Grimmer S, van Deurs B, Sandvig K (2002) Membrane ruffling and macropinocytosis in A431 cells require cholesterol. J Cell Sci 115(14):2953–2962
63. Hoeller O, Bolourani P, Clark J, Stephens LR, Hawkins PT, Weiner OD, Weeks G, Kay RR (2013) Two distinct functions for PI3-kinases in macropinocytosis. J Cell Sci 126(Pt 18): 4296–4307
64. Stephens L, Ellson C, Hawkins P (2002) Roles of PI3Ks in leukocyte chemotaxis and phagocytosis. Curr Opin Cell Biol 14:203–213
65. Ridley AJ (2006) Rho GTPases and actin dynamics in membrane protrusions and vesicle trafficking. Trends Cell Biol 16:522–529
66. Ahram M, Sameni M, Qiu RG, Linebaugh B, Kirn D, Sloane BF (2000) Rac1-induced endocytosis is associated with intracellular proteolysis during migration through a

three-dimensional matrix. Exp Cell Res 260: 292–303

67. Gutierrez MG (2013) Functional role(s) of phagosomal Rab GTPases. Small GTPases 4: 148–158
68. Harrison RE, Bucci C, Vieira OV, Schroer TA, Grinstein S (2003) Phagosomes fuse with late endosomes and/or lysosomes by extension of membrane protrusions along microtubules: role of Rab7 and RILP. Mol Cell Biol 23: 6494–6506
69. Kasmapour B, Cai L, Gutierrez MG (2013) Spatial distribution of phagolysosomes is independent of the regulation of lysosome position by Rab34. Int J Biochem Cell Biol 45:2057–2065
70. Kasmapour B, Gronow A, Bleck CK, Hong W, Gutierrez MG (2012) Size-dependent mechanism of cargo sorting during lysosome-phagosome fusion is controlled by Rab34. Proc Natl Acad Sci USA 109: 20485–20490
71. Feng D, Zhao WL, Ye YY, Bai XC, Liu RQ, Chang LF, Zhou Q, Sui SF (2010) Cellular internalization of exosomes occurs through phagocytosis. Traffic 11:675–687
72. Kovács ÁF, Láng O, Turiák L, Ács A, Kőhidai L, Fekete N (2018) The impact of circulating preeclampsia-associated extracellular vesicles on the migratory activity and phenotype of THP-1 monocytic cells. Sci Rep 8(1):5426
73. Jahn R, Lang T, Südhof TC (2003) Membrane fusion. Cell 112:519–533
74. Chernomordik LV, Melikyan GB, Chizmadzhev YA (1987) Biomembrane fusion: a new concept derived from model studies using two interacting planar lipid bilayers. Biochim Biophys Acta 906:309–352
75. Chernomordik LV, Kozlov MM (2008) Mechanics of membrane fusion. Nat Struct Mol Biol 15:675–683
76. Jahn R, Scheller RH (2006) SNAREs—engines for membrane fusion. Nat Rev Mol Cell Biol 7:631–643
77. Han J, Pluhackova K, Böckmann RA (2017) The multifaceted role of SNARE proteins in membrane fusion. Front Physiol 8:5
78. Südhof TC, Rothman JE (2009) Membrane fusion: grappling with SNARE and SM proteins. Science 323:474–477
79. Chevillet JR, Kang Q, Ruf IK, Briggs HA, Vojtech LN, Hughes SM et al (2014) Quantitative and stoichiometric analysis of the microRNA content of exosomes. Proc Natl Acad Sci USA 111(41):14888–14893
80. Kanada M, Bachmann MH, Hardy JW, Frimannson DO, Bronsart L, Wang A et al (2015) Differential fates of biomolecules delivered to target cells via extracellular vesicles. Proc Natl Acad Sci USA 112(12): E1433–E1442
81. Kim DK, Lee J, Kim SR, Choi DS, Yoon YJ, Kim JH et al (2015) EVpedia: a community web portal for extracellular vesicles research. Bioinformatics 31(6):933–939
82. Pathan M, Fonseka P, Chitti SV, Kang T, Sanwlani R, Van Deun J et al (2019) Vesiclepedia 2019: a compendium of RNA, proteins, lipids and metabolites in extracellular vesicles. Nucleic Acids Res 47(D1):D516–D519
83. Ung TH, Madsen HJ, Hellwinkel JE, Lencioni AM, Graner MW (2014) Exosome proteomics reveals transcriptional regulator proteins with potential to mediate downstream pathways. Cancer Sci 105(11): 1384–1392. https://doi.org/10.1111/cas.12534
84. Ko SY, Lee W, Kenny HA, Dang LH, Ellis LM, Jonasch E et al (2019) Cancer-derived small extracellular vesicles promote angiogenesis by heparin-bound, bevacizumab-insensitive VEGF, independent of vesicle uptake. Commun Biol 2:-386
85. Tarrant JM (2010) Blood cytokines as biomarkers of in vivo toxicity in preclinical safety assessment: considerations for their use. Toxicol Sci 117:4–16
86. Feng Q, Zhang C, Lum D, Druso JE, Blank B, Wilson KF et al (2017) A class of extracellular vesicles from breast cancer cells activates VEGF receptors and tumour angiogenesis. Nat Commun 8:14450
87. Higginbotham JN, Demory Beckler M, Gephart JD, Franklin JL, Bogatcheva G, Kremers GJ et al (2011) Amphiregulin exosomes increase cancer cell invasion. Curr Biol 21: 779–786
88. Al-Nedawi K, Meehan B, Kerbel RS, Allison AC, Rak J et al (2009) Endothelial expression of autocrine VEGF upon the uptake of tumor-derived microvesicles containing oncogenic EGFR. Proc Natl Acad Sci USA 106:3794–3799
89. Hoffman AS (2008) The origins and evolution of "controlled" drug delivery systems. J Control Release 132:153–163
90. ten Tije AJ, Verweij J, Loos WJ, Sparreboom A (2003) Pharmacological effects of formulation vehicles: implications for cancer chemotherapy. Clin Pharmacokinet 42:665–685

91. Erikson A, Tufto I, Bjønnum AB, Bruland ØS, Davies CL (2008) The impact of enzymatic degradation on the uptake of differently sized therapeutic molecules. Anticancer Res 28(6A):3557–3566
92. Krishna M, Nadler SG (2016) Immunogenicity to biotherapeutics—the role of anti-drug immune complexes. Front Immunol 7:21
93. Lin A, Giuliano CJ, Palladino A, John KM, Abramowicz C, Yuan ML, Sausville EL, Lukow DA, Liu L, Chait AR, Galluzzo ZC, Tucker C, Sheltzer JM (2019) Off-target toxicity is a common mechanism of action of cancer drugs undergoing clinical trials. Sci Transl Med 11:eaaw8412
94. Li M, Mei S, Yang Y, Shen Y, Chen L (2022) Strategies to mitigate the on- and off-target toxicities of recombinant immunotoxins: an antibody engineering perspective. Antib Ther 5:164–176
95. Yoon G, Park JW, Yoon IS (2013) Solid lipid nanoparticles (SLNs) and nanostructured lipid carriers (NLCs): recent advances in drug delivery. J Pharm Investig 43:353–362
96. Young RE, Hofbauer SI, Riley RS (2022) Overcoming the challenge of long-term storage of mRNA-lipid nanoparticle vaccines. Mol Ther 30:1792–1793
97. Pondman K, Le Gac S (2023) Nanoparticle-induced immune response: health risk versus treatment opportunity? Immunobiology 228: 152317
98. Thompson EA, Sayers BC, Glista-Baker EE, Shipkowski KA, Taylor AJ, Bonner JC (2014) Innate immune responses to nanoparticle exposure in the lung. J Environ Immunol Toxicol 1:150–156
99. Di J, Du Z, Wu K, Jin S, Wang X, Tea L (2022) Biodistribution and non-linear gene expression of mRNA LNPs affected by delivery route and particle size. Pharm Res 39: 105–114
100. Albanese A, Tang PS, Chan WCW (2012) The effect of nanoparticle size, shape, and surface chemistry on biological systems. Annu Rev Biomed Eng 14:1–16
101. Wang W, Gaus K, Tilley RD, Chan WCW (2019) The impact of nanoparticle shape on cellular internalization and transport: what do the different analysis methods tell us? Mater Horizons 6:1538–1547
102. Senter PD (2009) Potent antibody drug conjugates for cancer therapy. Curr Opin Chem Biol 13:235–244
103. Johnston APR (2017) Life under the microscope: quantifying live cell interactions to improve nanoscale drug delivery. ACS Sens 2:4–9
104. Overton WR (1988) Modified histogram subtraction analysis for flow cytometry data. Cytometry 9:619–626
105. Goldenthal KL, Pastan I, Willingham MC (1984) Initial steps in receptor-mediated endocytosis. The influence of temperature on the shape and distribution of plasma membrane clathrin-coated pits in cultured mammalian cells. Exp Cell Res 152(2):558–564. https://doi.org/10.1016/0014-4827(84)90658-x
106. Wisniewski BJ, Parkes JG, Huang YO, Fox CF (1974) Physical and physiological evidence for two phase transitions in cytoplasmic membranes of animal cells. Proc Natl Acad Sci USA 71:4381–4385
107. Ghetler Y, Yavin S, Shalgi R, Arab A (2005) The effect of chilling on membrane lipid phase transition in human oocytes and zygotes. Hum Reprod 20(12):3385
108. Liu H, Johnston APR (2013) A programmable sensor to probe the internalization of proteins and nanoparticles in live cells. Agnew Chem Int Ed Engl 52(22):5744–5748
109. Jensen DD, Halls ML, Murphy JE, Canals M, Cattaruzza F, Poole DP et al (2014) Endothelin-converting enzyme 1 and β-arrestins exert spatiotemporal control of substance P-induced inflammatory signals. J Biol Chem 289(29):20283–20294
110. Thottacherry JJ, Kosmalska AJ, Kumar A, Vishen AS, Elosegui-Artola A, Pradhan S et al (2018) Mechanochemical feedback control of dynamin independent endocytosis modulates membrane tension in adherent cells. Nat Commun 9(1):4217
111. Duncan R, Richardson SC (2012) Endocytosis and intracellular trafficking as gateways for nanomedicine delivery: opportunities and challenges. Mol Pharm 9(9):2380–2402. https://doi.org/10.1021/mp300293n
112. Rennick JJ, Johnston APR, Parton RG (2021) Key principles for studying the endocytosis of biological and nanoparticle therapeutics. Nat Nanotechnol 16:266–276
113. Vercauteren D, Vandenbroucke RE, Jones AT, Rejman J, Demeester J, De Smedt SC et al (2010) The use of inhibitors to study endocytic pathways of gene carriers: optimization and pitfalls. Mol Ther Nucleic Acids 18:561–569
114. Szewczyk-Roszczenko OK, Roszczenko P, Shmakova A, Finiuk N, Holota S, Lesyk R et al (2023) The chemical inhibitors of endocytosis: from mechanisms to potential clinical applications. Cells 12:2312
115. Dutta D, Donaldson JG (2012) Search for inhibitors of endocytosis: intended specificity

and unintended consequences. Cell Logist 2: 203–208
116. Ivanov AI (2008) Pharmacological inhibition of endocytic pathways: is it specific enough to be useful? Methods Mol Biol 440:15–33
117. Stockinger W, Zhang SC, Trivedi V, Jarzylo LA, Shieh EC, Lane WS et al (2006) Differential requirements for actin polymerization, calmodulin, and Ca2+ define distinct stages of lysosome/phagosome targeting. Mol Biol Cell 17:1697–1710
118. Margarucci L, Monti MC, Fontanella B, Riccio R, Casapullo A (2011) Chemical proteomics reveals bolinaquinone as a clathrin-mediated endocytosis inhibitor. Mol BioSyst 7:480–485
119. Macia E, Ehrlich M, Massol R, Boucrot E, Brunner C, Kirchhausen T (2006) Dynasore, a cell-permeable inhibitor of dynamin. Dev Cell 10:839–850
120. Hill TA, Gordon CP, McGeachie AB, Venn-Brown B, Odell LR, Chau N et al (2009) Inhibition of dynamin mediated endocytosis by the dynoles—synthesis and functional activity of a family of indoles. J Med Chem 52:3762
121. Park RJ, Shen H, Liu L, Liu X, Ferguson SM, De Camilli P (2013) Dynamin triple knockout cells reveal off target effects of commonly used dynamin inhibitors. J Cell Sci 126: 5305–5312
122. Girard E, Paul JL, Fournier N, Beaune P, Johannes L, Lamaze C et al (2011) The dynamin chemical inhibitor dynasore impairs cholesterol trafficking and sterol-sensitive genes transcription in human HeLa cells and macrophages. PLoS One 6:e29042
123. Robinet P, Fradagrada A, Monier MN, Marchetti M, Cogny A, Moatti N et al (2006) Dynamin is involved in endolysosomal cholesterol delivery to the endoplasmic reticulum: role in cholesterol homeostasis. Traffic 7:811–823
124. Grimmer S, Ying M, Wälchli S, van Deurs B, Sandvig K (2005) Golgi vesiculation induced by cholesterol occurs by a dynamin- and cPLA2-dependent mechanism. Traffic 6: 144–156
125. Weller SG, Capitani M, Cao H, Micaroni M, Luini A, Sallese M et al (2010) Src kinase regulates the integrity and function of the Golgi apparatus via activation of dynamin 2. Proc Natl Acad Sci USA 107:5863–5868
126. Sakai H, Moriura Y, Notomi T, Kawawaki J, Ohnishi K, Kuno M (2010) Phospholipase C-dependent Ca2+-sensing pathways leading to endocytosis and inhibition of the plasma membrane vacuolar H+-ATPase in osteoclasts. Am J Physiol Cell Physiol 299(3): C570–C578
127. Al Soraj M, He L, Peynshaert K, Cousaert J, Vercauteren D, Braeckmans K et al (2012) siRNA and pharmacological inhibition of endocytic pathways to characterize the differential role of macropinocytosis and the actin cytoskeleton on cellular uptake of dextran and cationic cell penetrating peptides octaarginine (R8) and HIV-Tat. J Control Release 161: 132–141
128. Hinrichsen L, Harborth J, Andrees L, Weber K, Ungewickell EJ (2003) Effect of clathrin heavy chain- and alpha-adaptin-specific small inhibitory RNAs on endocytic accessory proteins and receptor trafficking in HeLa cells. J Biol Chem 278:45160–45170
129. Motley A, Bright NA, Seaman MN, Robinson MS (2003) Clathrin-mediated endocytosis in AP-2-depleted cells. J Cell Biol 162:909–918
130. Escrevente C, Keller S, Altevogt P, Costa J (2011) Interaction and uptake of exosomes by ovarian cancer cells. BMC Cancer 11:108
131. Ofir-Birin Y, Abou Karam P, Rudik A, Giladi T, Porat Z, Regev-Rudzki N (2018) Monitoring extracellular vesicle cargo active uptake by imaging flow cytometry. Front Immunol 9:1011
132. Loconte L, Arguedas D, El R, Zhou A, Chipont A, Guyonnet L et al (2023) Detection of the interactions of tumour derived extracellular vesicles with immune cells is dependent on EV-labelling methods. J Extracell Vesicles 12:e12384
133. Yang N, Zhao C, Kong L, Zhang B, Han C, Zhang Y et al (2023) Absolute quantification of dynamic cellular uptake of small extracellular vesicles via lanthanide element labeling and ICP-MS. Anal Chem 95:11934–11942
134. Ma D, Luo Y, Zhang X, Xie Z, Yan Y, Ding C, F. (2024) A highly sensitive and selective fluorescent biosensor for breast cancer derived exosomes using click reaction of azide-CD63 aptamer and alkyne-polymer dots. Anal Methods 16:2751–2759
135. Ciferri MC, Bruno S, Rosenwasser N, Gorgun C, Reverberi D, Gagliani MC et al (2024) Standardized method to functionalize plasma-extracellular vesicles via copper-free click chemistry for targeted drug delivery strategies. ACS Appl Bio Mater 7:827–838
136. Salunkhe S, Dheeraj BM, Chitkara D, Mittal A (2020) Surface functionalization of exosomes for target-specific delivery and in vivo imaging & tracking: strategies and significance. J Control Release 326:599–614

137. Selby LI, Aurelio L, Yuen D, Graham B, Johnston APR (2018) Quantifying cellular internalization with a fluorescent click sensor. ACS Sens 3:1182–1189
138. Sung BH, von Lersner A, Guerrero J, Krystofiak ES, Inman D, Pelletier R et al (2020) A live cell reporter of exosome secretion and uptake reveals pathfinding behavior of migrating cells. Nat Commun 11:2092
139. Sung BH, Weaver AM (2023) Visualization of exosome release and uptake during cell migration using the live imaging reporter pHluorin_M153R-CD63. Methods Mol Biol 2608:83–96
140. Toribio V, Morales S, López-Martín S, Cardeñes B, Cabañas C, Yáñez-Mó M (2019) Development of a quantitative method to measure EV uptake. Sci Rep Sci Rep 9:10522
141. Jurgielewicz BJ, Yao Y, Stice SL (2020) Kinetics and specificity of HEK293T extracellular vesicle uptake using imaging flow cytometry. Nanoscale Res Lett 15:170
142. Bonsergent E, Grisard E, Buchrieser J, Schwartz O, Théry C, Lavieu G (2021) Quantitative characterization of extracellular vesicle uptake and content delivery within mammalian cells. Nat Commun 12:1864
143. Hirose H, Hirai Y, Sasaki M, Sawa H, Futaki S (2022) Quantitative analysis of extracellular vesicle uptake and fusion with recipient cells. Bioconjug Chem 33:1852–1859
144. Murphy DB, Davidson MW (2012) Fundamentals of light microscopy and electronic imaging, 2nd edn. Wiley
145. Hell SW, Dyba M, Jakobs S (2004) Concepts for nanoscale resolution in fluorescence microscopy. Curr Opin Neurobiol 14:599–609
146. Cavanagh HD, Petroll WM, Jester JV (1993) The application of confocal microscopy to the study of living systems. Neurosci Biobehav Rev 17:483–498
147. Gustafsson MG (2000) Surpassing the lateral resolution limit by a factor of two using structured illumination microscopy. J Microsc 198:82–87
148. Betzig E, Patterson GH, Sougrat R, Lindwasser OW, Olenych S, Bonifacino JS et al (2006) Imaging intracellular fluorescent proteins at nanometer resolution. Science 313:1642–1645
149. Rust MJ, Bates M, Zhuang X (2006) Sub-diffraction-limit imaging by stochastic optical reconstruction microscopy (STORM). Nat Methods 3:793–795
150. Hell SW, Wichmann J (1994) Breaking the diffraction resolution limit by stimulated emission: stimulated-emission-depletion fluorescence microscopy. Opt Lett 19:780–7826
151. van der Pol E, Coumans FA, Grootemaat AE, Gardiner C, Sargent IL, Harrison P et al (2014) Particle size distribution of exosomes and microvesicles determined by transmission electron microscopy, flow cytometry, nanoparticle tracking analysis, and resistive pulse sensing. J Thromb Haemost 12:1182–1192
152. Arraud N, Gounou C, Turpin D, Brisson AR (2016) Fluorescence triggering: a general strategy for enumerating and phenotyping extracellular vesicles by flow cytometry. Cytometry A 89(2):184–195. https://doi.org/10.1002/cyto.a.22669
153. Arraud N, Gounou C, Linares R, Brisson AR (2015) A simple flow cytometry method improves the detection of phosphatidylserine-exposing extracellular vesicles. J Thromb Haemost 13(2):237–247. https://doi.org/10.1111/jth.12767
154. Nolan JP, Stoner SA (2013) A trigger channel threshold artifact in nanoparticle analysis. Cytometry A 83(3):301–305. https://doi.org/10.1002/cyto.a.22255
155. van der Pol E, van Gemert MJ, Sturk A, Nieuwland R, van Leeuwen TG (2012) Single vs. swarm detection of microparticles and exosomes by flow cytometry. J Thromb Haemost 10(5):919–930. https://doi.org/10.1111/j.1538-7836.2012.04683.x
156. Libregts S, Arkesteijn GJA, Németh A, Nolte-'t Hoen ENM, Wauben MHM (2018) Flow cytometric analysis of extracellular vesicle subsets in plasma: impact of swarm by particles of non-interest. J Thromb Haemost 16(7):1423–1436. https://doi.org/10.1111/jth.14154
157. van der Vlist EJ, Nolte-'t Hoen EN, Stoorvogel W, Arkesteijn GJ, Wauben MH (2012) Fluorescent labeling of nano-sized vesicles released by cells and subsequent quantitative and qualitative analysis by high-resolution flow cytometry. Nat Protoc 7(7):1311–1326. https://doi.org/10.1038/nprot.2012.065
158. Groot Kormelink T, Arkesteijn GJ, Nauwelaers FA, van den Engh G, Nolte-'t Hoen EN, Wauben MH (2016) Prerequisites for the analysis and sorting of extracellular vesicle subpopulations by high-resolution flow cytometry. Cytometry A 89(2):135–147. https://doi.org/10.1002/cyto.a.22644
159. Stoner SA, Duggan E, Condello D, Guerrero A, Turk JR, Narayanan PK, Nolan

JP (2016) High sensitivity flow cytometry of membrane vesicles. Cytometry A 89(2): 196–206. https://doi.org/10.1002/cyto.a.22787
160. Zhu S, Ma L, Wang S, Chen C, Zhang W, Yang L, Hang W, Nolan JP, Wu L, Yan X (2014) Light-scattering detection below the level of single fluorescent molecules for high-resolution characterization of functional nanoparticles. ACS Nano 8(10): 10998–11006. https://doi.org/10.1021/nn505162u
161. Lacroix R, Robert S, Poncelet P, Dignat-George F (2010) Overcoming limitations of microparticle measurement by flow cytometry. Semin Thromb Hemost 36(8):807–818. https://doi.org/10.1055/s-0030-1267034
162. Headland SE, Jones HR, D'Sa AS, Perretti M, Norling LV (2014) Cutting-edge analysis of extracellular microparticles using ImageStream(X) imaging flow cytometry. Sci Rep 4:5237. https://doi.org/10.1038/srep05237
163. Erdbrügger U, Rudy CK, Etter ME, Dryden KA, Yeager M, Klibanov AL, Lannigan J (2014) Imaging flow cytometry elucidates limitations of microparticle analysis by conventional flow cytometry. Cytometry A 85(9):756–770. https://doi.org/10.1002/cyto.a.22494
164. Lannigan J, Nolan JP, Zucker R (2016) Measurement of extracellular vesicles and other submicron size particles by flow cytometry. Cytometry A 89(2):109–110. https://doi.org/10.1002/cyto.a.22814
165. Lannigan J, Erdbruegger U (2017) Imaging flow cytometry for the characterization of extracellular vesicles. Methods 112:55–67. https://doi.org/10.1016/j.ymeth.2016.09.018
166. Görgens A, Bremer M, Ferrer-Tur R, Murke F, Tertel T, Horn PA, Thalmann S, Welsh JA, Probst C, Guerin C, Boulanger CM, Jones JC, Hanenberg H, Erdbrügger U, Lannigan J, Ricklefs FL, El-Andaloussi S, Giebel B (2019) Optimisation of imaging flow cytometry for the analysis of single extracellular vesicles by using fluorescence-tagged vesicles as biological reference material. J Extracell Vesicles 8(1): 1587567. https://doi.org/10.1080/20013078.2019.1587567
167. Morales-Kastresana A, Musich TA, Welsh JA, Telford W, Demberg T, Wood JCS, Bigos M, Ross CD, Kachynski A, Dean A, Felton EJ, Van Dyke J, Tigges J, Toxavidis V, Parks DR, Overton WR, Kesarwala AH, Freeman GJ, Rosner A, Perfetto SP, Pasquet L, Terabe M, McKinnon K, Kapoor V, Trepel JB, Puri A, Kobayashi H, Yung B, Chen X, Guion P, Choyke P, Knox SJ, Ghiran I, Robert-Guroff-M, Berzofsky JA, Jones JC (2019) High-fidelity detection and sorting of nanoscale vesicles in viral disease and cancer. J Extracell Vesicles 8(1):1597603. https://doi.org/10.1080/20013078.2019.1597603
168. Friedrich R, Block S, Alizadehheidari M, Heider S, Fritzsche J, Esbjörner EK, Westerlund F, Bally M (2017) A nano flow cytometer for single lipid vesicle analysis. Lab Chip 17(5):830–841. https://doi.org/10.1039/c6lc01302c
169. De Giorgis V, Barberis E, Manfredi M (2024) Extracellular vesicles proteins for early cancer diagnosis: from omics to biomarkers. Semin Cancer Biol 104-105:18–31. https://doi.org/10.1016/j.semcancer.2024.07.003
170. Gao Y, Kim H, Kitata RB, Lin TT, Swensen AC, Shi T, Liu T (2024) Multiplexed quantitative proteomics in prostate cancer biomarker development. Adv Cancer Res 161: 31–69. https://doi.org/10.1016/bs.acr.2024.04.003
171. Bandu R, Oh JW, Kim KP (2024) Extracellular vesicle proteins as breast cancer biomarkers: mass spectrometry-based analysis. Proteomics 24(11):e2300062. https://doi.org/10.1002/pmic.202300062
172. Anastasi F, Botto A, Immordino B, Giovannetti E, McDonnell LA (2023) Proteomics analysis of circulating small extracellular vesicles: focus on the contribution of EVs to tumor metabolism. Cytokine Growth Factor Rev 73:3–19. https://doi.org/10.1016/j.cytogfr.2023.08.003
173. Baranov VI, Quinn Z, Bandura DR, Tanner SD (2002) A sensitive and quantitative element-tagged immunoassay with ICPMS detection. Anal Chem 74(7):1629–1636. https://doi.org/10.1021/ac0110350
174. Zhang C, Zhang Z, Yu B, Shi J, Zhang X (2002) Application of the biological conjugate between antibody and colloid Au nanoparticles as analyte to inductively coupled plasma mass spectrometry. Anal Chem 74(1):96–99. https://doi.org/10.1021/ac0103468
175. Zhang S, Zhang C, Xing Z, Zhang X (2004) Simultaneous determination of alpha-fetoprotein and free beta-human chorionic gonadotropin by element-tagged immunoassay with detection by inductively coupled plasma mass spectrometry. Clin Chem 50(7): 1214–1221. https://doi.org/10.1373/clinchem.2003.029850
176. Bettmer J, Jakubowski N, Prange A (2006) Elemental tagging in inorganic mass spectrometric bioanalysis. Anal Bioanal Chem

386(1):7–11. https://doi.org/10.1007/s00216-006-0557-4

177. Careri M, Elviri L, Mangia A, Mucchino C (2007) ICP-MS as a novel detection system for quantitative element-tagged immunoassay of hidden peanut allergens in foods. Anal Bioanal Chem 387(5):1851–1854. https://doi.org/10.1007/s00216-006-1091-0
178. Careri M, Elviri L, Mangia A (2009) Element-tagged immunoassay with inductively coupled plasma mass spectrometry for multianalyte detection. Anal Bioanal Chem 393(1):57–61. https://doi.org/10.1007/s00216-008-2419-8
179. Tanner S, Shen Z, Ng J, Florea L, Guigó R, Briggs SP, Bafna V (2007) Improving gene annotation using peptide mass spectrometry. Genome Res 17(2):231–239. https://doi.org/10.1101/gr.5646507
180. Becker JS, Jakubowski N (2009) The synergy of elemental and biomolecular mass spectrometry: new analytical strategies in life sciences. Chem Soc Rev 38(7):1969–1983. https://doi.org/10.1039/b618635c
181. Vaníčková L, Do T, Vejvodová M, Horák V, Hubálek M, Emri G, Zemánková K, Pavelicová K, Křížková S, Faltusová V, Pompeiano A, Vaculovičová M, Zítka O, Vaculovič T, Adam V (2022) Mapping of MeLiM melanoma combining ICP-MS and MALDI-MSI methods. Int J Biol Macromol 203:583–592. https://doi.org/10.1016/j.ijbiomac.2022.01.139
182. Ornatsky OI, Kinach R, Bandura DR, Lou X, Tanner SD, Baranov VI, Nitz M, Winnik MA (2008) Development of analytical methods for multiplex bio-assay with inductively coupled plasma mass spectrometry. J Anal At Spectrom 23(4):463–469. https://doi.org/10.1039/b710510j
183. Carpintero-Fernández P, Fafián-Labora J, O'Loghlen A (2017) Technical advances to study extracellular vesicles. Front Mol Biosci 4:79. https://doi.org/10.3389/fmolb.2017.00079
184. Simonsen JB (2019) Pitfalls associated with lipophilic fluorophore staining of extracellular vesicles for uptake studies. J Extracell Vesicles 8(1):1582237. https://doi.org/10.1080/20013078.2019.1582237
185. Pužar Dominkuš P, Stenovec M, Sitar S, Lasič E, Zorec R, Plemenitaš A, Žagar E, Kreft M (1860) Lenassi M (2018) PKH26 labeling of extracellular vesicles: characterization and cellular internalization of contaminating PKH26 nanoparticles. Biochim Biophys Acta Biomembr 6:1350–1361. https://doi.org/10.1016/j.bbamem.2018.03.013
186. Wąchalska M, Rychłowski M, Grabowska K, Kowal K, Narajczyk M, Bieńkowska-Szewczyk K, Lipińska AD (2020) Palmitoylated mNeonGreen protein as a tool for visualization and uptake studies of extracellular vesicles. Membranes (Basel) 10(12). https://doi.org/10.3390/membranes10120373
187. Sawada S, Yoshikawa M, Tsutsui K, Miyazaki T, Kano K, Mishiro-Sato E, Tsukiji S (2023) Palmitoylation-dependent small-molecule fluorescent probes for live-cell Golgi imaging. ACS Chem Biol 18(5):1047–1053. https://doi.org/10.1021/acschembio.3c00046
188. Lo Cicero A, Delevoye C, Gilles-Marsens F, Loew D, Dingli F, Guéré C, André N, Vié K, van Niel G, Raposo G (2015) Exosomes released by keratinocytes modulate melanocyte pigmentation. Nat Commun 6:7506. https://doi.org/10.1038/ncomms8506
189. Verweij FJ, Bebelman MP, Jimenez CR, Garcia-Vallejo JJ, Janssen H, Neefjes J, Knol JC, de Goeij-de Haas R, Piersma SR, Baglio SR, Verhage M, Middeldorp JM, Zomer A, van Rheenen J, Coppolino MG, Hurbain I, Raposo G, Smit MJ, Toonen RFG, van Niel G, Pegtel DM (2018) Quantifying exosome secretion from single cells reveals a modulatory role for GPCR signaling. J Cell Biol 217(3):1129–1142. https://doi.org/10.1083/jcb.201703206
190. Baglio SR, Lagerweij T, Pérez-Lanzón M, Ho XD, Léveillé N, Melo SA, Cleton-Jansen AM, Jordanova ES, Roncuzzi L, Greco M, van Eijndhoven MAJ, Grisendi G, Dominici M, Bonafede R, Lougheed SM, de Gruijl TD, Zini N, Cervo S, Steffan A, Canzonieri V, Martson A, Maasalu K, Köks S, Wurdinger T, Baldini N, Pegtel DM (2017) Blocking tumor-educated MSC paracrine activity halts osteosarcoma progression. Clin Cancer Res 23(14):3721–3733. https://doi.org/10.1158/1078-0432.Ccr-16-2726
191. Hikita T, Miyata M, Watanabe R, Oneyama C (2018) Sensitive and rapid quantification of exosomes by fusing luciferase to exosome marker proteins. Sci Rep 8(1):14035. https://doi.org/10.1038/s41598-018-32535-7
192. Lai CP, Tannous BA, Breakefield XO (2014) Noninvasive in vivo monitoring of extracellular vesicles. Methods Mol Biol 1098:249–258. https://doi.org/10.1007/978-1-62703-718-1_19
193. Kondo N, Miyauchi K, Meng F, Iwamoto A, Matsuda Z (2010) Conformational changes of the HIV-1 envelope protein during membrane fusion are inhibited by the replacement of its membrane-spanning domain. J Biol

Chem 285(19):14681–14688. https://doi.org/10.1074/jbc.M109.067090
194. Atanasiu D, Saw WT, Gallagher JR, Hannah BP, Matsuda Z, Whitbeck JC, Cohen GH, Eisenberg RJ (2013) Dual split protein-based fusion assay reveals that mutations to herpes simplex virus (HSV) glycoprotein gB alter the kinetics of cell-cell fusion induced by HSV entry glycoproteins. J Virol 87(21): 11332–11345. https://doi.org/10.1128/jvi.01700-13
195. Ishikawa H, Meng F, Kondo N, Iwamoto A, Matsuda Z (2012) Generation of a dual-functional split-reporter protein for monitoring membrane fusion using self-associating split GFP. Protein Eng Des Sel 25(12): 813–820. https://doi.org/10.1093/protein/gzs051
196. Suárez H, Gámez-Valero A, Reyes R, López-Martín S, Rodríguez MJ, Carrascosa JL, Cabañas C, Borràs FE, Yáñez-Mó M (2017) A bead-assisted flow cytometry method for the semi-quantitative analysis of Extracellular Vesicles. Sci Rep 7(1):11271. https://doi.org/10.1038/s41598-017-11249-2
197. Longo N, Yáñez-Mó M, Mittelbrunn M, de la Rosa G, Muñoz ML, Sánchez-Madrid F, Sánchez-Mateos P (2001) Regulatory role of tetraspanin CD9 in tumor-endothelial cell interaction during transendothelial invasion of melanoma cells. Blood 98(13): 3717–3726. https://doi.org/10.1182/blood.v98.13.3717
198. Aoki N, Aoyama K, Nagata M, Matsuda T (2001) A growing family of dual specificity phosphatases with low molecular masses. J Biochem 130(1):133–140. https://doi.org/10.1093/oxfordjournals.jbchem.a002952
199. van Niel G, D'Angelo G, Raposo G (2018) Shedding light on the cell biology of extracellular vesicles. Nat Rev Mol Cell Biol 19:213–228
200. Nakase I, Futaki S (2015) Combined treatment with a pH-sensitive fusogenic peptide and cationic lipids achieves enhanced cytosolic delivery of exosomes. Sci Rep 5:10112
201. Murphy DE, de Jong OG, Evers MJW, Nurazizah M, Schiffelers RM, Vader P (2021) Natural or synthetic RNA delivery: a stoichiometric comparison of extracellular vesicles and synthetic nanoparticles. Nano Lett 21(4):1888–1895. https://doi.org/10.1021/acs.nanolett.1c00094
202. Somiya M, Kuroda S (2021) Reporter gene assay for membrane fusion of extracellular vesicles. J Extracell Vesicles 10(13):e12171. https://doi.org/10.1002/jev2.12171
203. Somiya M, Kuroda S (2021) Real-time luminescence assay for cytoplasmic cargo delivery of extracellular vesicles. Anal Chem 93(13): 5612–5620. https://doi.org/10.1021/acs.analchem.1c00339
204. Kojima R, Bojar D, Rizzi G, Hamri GC, El-Baba MD, Saxena P, Ausländer S, Tan KR, Fussenegger M (2018) Designer exosomes produced by implanted cells intracerebrally deliver therapeutic cargo for Parkinson's disease treatment. Nat Commun 9(1):1305. https://doi.org/10.1038/s41467-018-03733-8
205. Morisaki T, Lyon K, DeLuca KF, DeLuca JG, English BP, Zhang Z, Lavis LD, Grimm JB, Viswanathan S, Looger LL, Lionnet T, Stasevich TJ (2016) Real-time quantification of single RNA translation dynamics in living cells. Science 352(6292):1425–1429. https://doi.org/10.1126/science.aaf0899
206. Dixon AS, Schwinn MK, Hall MP, Zimmerman K, Otto P, Lubben TH, Butler BL, Binkowski BF, Machleidt T, Kirkland TA, Wood MG, Eggers CT, Encell LP, Wood KV (2016) NanoLuc complementation reporter optimized for accurate measurement of protein interactions in cells. ACS Chem Biol 11(2):400–408. https://doi.org/10.1021/acschembio.5b00753
207. Nakai W, Yoshida T, Diez D, Miyatake Y, Nishibu T, Imawaka N, Naruse K, Sadamura Y, Hanayama R (2016) A novel affinity-based method for the isolation of highly purified extracellular vesicles. Sci Rep 6:33935
208. Kugeratski FG, Hodge K, Lilla S, McAndrews KM, Zhou X, Hwang RF, Zanivan S, Kalluri R (2021) Quantitative proteomics identifies the core proteome of exosomes with syntenin-1 as the highest abundant protein and a putative universal biomarker. Nat Cell Biol 23(6):631–641. https://doi.org/10.1038/s41556-021-00693-y
209. Jeppesen DK, Fenix AM, Franklin JL, Higginbotham JN, Zhang Q, Zimmerman LJ, Liebler DC, Ping J, Liu Q, Evans R, Fissell WH, Patton JG, Rome LH, Burnette DT, Coffey RJ (2019) Reassessment of exosome composition. Cell 177(2):428–445.e418. https://doi.org/10.1016/j.cell.2019.02.029
210. Göstring L, Chew MT, Orlova A, Höidén-Guthenberg I, Wennborg A, Carlsson J, Frejd FY (2010) Quantification of internalization of EGFR-binding Affibody molecules: methodological aspects. Int J Oncol 36(4): 757–763. https://doi.org/10.3892/ijo_00000551

211. Kondapalli KC, Hack A, Schushan M, Landau M, Ben-Tal N, Rao R (2013) Functional evaluation of autism-associated mutations in NHE9. Nat Commun 4:2510. https://doi.org/10.1038/ncomms3510
212. Nuutila J, Lilius EM (2005) Flow cytometric quantitative determination of ingestion by phagocytes needs the distinguishing of overlapping populations of binding and ingesting cells. Cytometry A 65(2):93–102. https://doi.org/10.1002/cyto.a.20139
213. Van Amersfoort ES, Van Strijp JA (1994) Evaluation of a flow cytometric fluorescence quenching assay of phagocytosis of sensitized sheep erythrocytes by polymorphonuclear leukocytes. Cytometry 17(4):294–301. https://doi.org/10.1002/cyto.990170404
214. Gottstein C, Wu G, Wong BJ, Zasadzinski JA (2013) Precise quantification of nanoparticle internalization. ACS Nano 7(6):4933–4945. https://doi.org/10.1021/nn400243d
215. Muñoz Javier A, del Pino P, Bedard MF, Ho D, Skirtach AG, Sukhorukov GB, Plank C, Parak WJ (2008) Photoactivated release of cargo from the cavity of polyelectrolyte capsules to the cytosol of cells. Langmuir 24(21):12517–12520. https://doi.org/10.1021/la802448z
216. Vaidyanathan G, McDougald D, Choi J, Koumarianou E, Weitzel D, Osada T, Lyerly HK, Zalutsky MR (2016) Preclinical evaluation of 18F-labeled anti-HER2 nanobody conjugates for imaging HER2 receptor expression by Immuno-PET. J Nucl Med 57(6):967–973. https://doi.org/10.2967/jnumed.115.171306
217. Gasteiger E, Gattiker A, Hoogland C, Ivanyi I, Appel RD, Bairoch A (2003) ExPASy: the proteomics server for in-depth protein knowledge and analysis. Nucleic Acids Res 31(13):3784–3788. https://doi.org/10.1093/nar/gkg563
218. Turcot I, Stintzi A, Xu J, Raymond KN (2000) Fast biological iron chelators: kinetics of iron removal from human diferric transferrin by multidentate hydroxypyridonates. J Biol Inorg Chem 5(5):634–641. https://doi.org/10.1007/s007750000149
219. McNamara G, Difilippantonio M, Ried T, Bieber FR (2017) Microscopy and image analysis. Curr Protoc Hum Genet 94:4.4.1–4.4.89. https://doi.org/10.1002/cphg.42
220. Dumont C, Czuba E, Chen M, Villadangos JA, Johnston AP, Mintern JD (2017) DNA-based probes for flow cytometry analysis of endocytosis and recycling. Traffic 18(4):242–249. https://doi.org/10.1111/tra.12466
221. Karver MR, Weissleder R, Hilderbrand SA (2011) Synthesis and evaluation of a series of 1,2,4,5-tetrazines for bioorthogonal conjugation. Bioconjug Chem 22(11):2263–2270. https://doi.org/10.1021/bc200295y
222. Staros JV, Wright RW, Swingle DM (1986) Enhancement by N-hydroxysulfosuccinimide of water-soluble carbodiimide-mediated coupling reactions. Anal Biochem 156(1):220–222. https://doi.org/10.1016/0003-2697(86)90176-4
223. Sehgal D, Vijay IK (1994) A method for the high efficiency of water-soluble carbodiimide-mediated amidation. Anal Biochem 218(1):87–91. https://doi.org/10.1006/abio.1994.1144
224. Russell AE, Sneider A, Witwer KW, Bergese P, Bhattacharyya SN, Cocks A, Cocucci E, Erdbrügger U, Falcon-Perez JM, Freeman DW, Gallagher TM, Hu S, Huang Y, Jay SM, Kano SI, Lavieu G, Leszczynska A, Llorente AM, Lu Q, Mahairaki V, Muth DC, Noren Hooten N, Ostrowski M, Prada I, Sahoo S, Schøyen TH, Sheng L, Tesch D, Van Niel G, Vandenbroucke RE, Verweij FJ, Villar AV, Wauben M, Wehman AM, Yin H, Carter DRF, Vader P (2019) Biological membranes in EV biogenesis, stability, uptake, and cargo transfer: an ISEV position paper arising from the ISEV membranes and EVs workshop. J Extracell Vesicles 8(1):1684862. https://doi.org/10.1080/20013078.2019.1684862
225. Sancho-Albero M, Navascués N, Mendoza G, Sebastián V, Arruebo M, Martín-Duque P, Santamaría J (2019) Exosome origin determines cell targeting and the transfer of therapeutic nanoparticles towards target cells. J Nanobiotechnol 17(1):16. https://doi.org/10.1186/s12951-018-0437-z
226. Morelli AE, Larregina AT, Shufesky WJ, Sullivan ML, Stolz DB, Papworth GD, Zahorchak AF, Logar AJ, Wang Z, Watkins SC, Falo LDJ, Thomson AW (2004) Endocytosis, intracellular sorting, and processing of exosomes by dendritic cells. Blood 104:3257–3266
227. Tian T, Zhu YL, Zhou YY, Liang GF, Wang YY, Hu FH, Xiao ZD (2014) Exosome uptake through clathrin-mediated endocytosis and macropinocytosis and mediating miR-21 delivery. J Biol Chem 289:22258–22267

228. Mastoridis S, Bertolino GM, Whitehouse G, Dazzi F, Sanchez-Fueyo A, Martinez-Llordella M (2018) Multiparametric analysis of circulating exosomes and other small extracellular vesicles by advanced imaging flow cytometry. Front Immunol 9:1583. https://doi.org/10.3389/fimmu.2018.01583
229. Verdera HC, Gitz-Francois JJ, Schiffelers RM, Vader P (2017) Cellular uptake of extracellular vesicles is mediated by clathrin-independent endocytosis and macropinocytosis. J Control Release 266:100–108
230. Busatto S, Giacomini A, Montis C, Ronca R, Bergese P (2018) Uptake profiles of human serum exosomes by murine and human tumor cells through combined use of colloidal nanoplasmonics and flow cytofluorimetric analysis. Anal Chem 90(13):7855–7861. https://doi.org/10.1021/acs.analchem.7b04374
231. Wiklander OP, Nordin JZ, O'Loughlin A, Gustafsson Y, Corso G, Mäger I, Vader P, Lee Y, Sork H, Seow Y, Heldring N, Alvarez-Erviti L, Smith CI, Le Blanc K, Macchiarini P, Jungebluth P, Wood MJ, Andaloussi SE (2015) Extracellular vesicle in vivo biodistribution is determined by cell source, route of administration and targeting. J Extracell Vesicles 4:26316
232. Hazan-Halevy I, Rosenblum D, Weinstein S, Bairey O, Raanani P, Peer D (2015) Cell-specific uptake of mantle cell lymphoma-derived exosomes by malignant and non-malignant B-lymphocytes. Cancer Lett 364(1):59–69. https://doi.org/10.1016/j.canlet.2015.04.026
233. Gomes J, Lucien F, Cooper TT, Kim Y, Williams KC, Liao X, Kaufman L, Lagugné-Labarthet F, Kenyon O, Boysen J, Kay NE, McIntyre CW, Leong HS (2018) Analytical considerations in nanoscale flow cytometry of extracellular vesicles to achieve data linearity. Thromb Haemost 118(9):1612–1624. https://doi.org/10.1055/s-0038-1668544
234. Christianson HC, Svensson KJ, van Kuppevelt TH, Li JP, Belting M (2013) Cancer cell exosomes depend on cell-surface heparan sulfate proteoglycans for their internalization and functional activity. Proc Natl Acad Sci USA 110:17380–17385
235. Temchura V, Tenbusch M, Nchinda G, Nabi G, Tippler B, Zelenyuk M, Wildner O, Uberla K, Kuate S (2008) Enhancement of immunostimulatory properties of exosomal vaccines by incorporation of fusion-competent G protein of vesicular stomatitis virus. Vaccine 26:3662–3672
236. Delenclos M, Trendafilova T, Mahesh D, Baine AM, Moussaud S, Yan IK, Patel T, McLean PJ (2017) Investigation of endocytic pathways for the internalization of exosome-associated oligomeric alpha-synuclein. Front Neurosci 11:172. https://doi.org/10.3389/fnins.2017.00172
237. Jiang Y, Wang L, Zhang P, Liu X, Di H, Yang J, Liu SL, Pang DW, Liu D (2020) Chemoenzymatic labeling of extracellular vesicles for visualizing their cellular internalization in real time. Anal Chem 92(2):2103–2111. https://doi.org/10.1021/acs.analchem.9b04608
238. Smyth T, Petrova K, Payton NM, Persaud I, Redzic JS, Graner MW, Smith-Jones P, Anchordoquy TJ (2014) Surface functionalization of exosomes using click chemistry. Bioconjug Chem 25:1777–1784
239. Yan X, Luo Y, Zhang Z, Li Z, Luo Q, Yang L, Zhang B, Chen H, Bai P, Wang Q (2012) Europium-labeled activity-based probe through click chemistry: absolute serine protease quantification using (153)Eu isotope dilution ICP/MS. Angew Chem Int Ed Engl 51(14):3358–3363. https://doi.org/10.1002/anie.201108277
240. Luo Y, Yan X, Huang Y, Wen R, Li Z, Yang L, Yang CJ, Wang Q (2013) ICP-MS-based multiplex and ultrasensitive assay of viruses with lanthanide-coded biospecific tagging and amplification strategies. Anal Chem 85(20):9428–9432. https://doi.org/10.1021/ac402446a
241. Shelke GV, Lässer C, Gho YS, Lötvall J (2014) Importance of exosome depletion protocols to eliminate functional and RNA-containing extracellular vesicles from fetal bovine serum. J Extracell Vesicles 3. https://doi.org/10.3402/jev.v3.24783
242. Li J, Yao Z, Xiong H, Cui H, Wang X, Zheng W, Qian Y, Fan C (2020) Extracellular vesicles from hydroxycamptothecin primed umbilical cord stem cells enhance anti-adhesion potential for treatment of tendon injury. Stem Cell Res Ther 11(1):500. https://doi.org/10.1186/s13287-020-02016-8
243. Vu DM, Nguyen VT, Nguyen TH, Do PTX, Dao HH, Hai DX, Le NT, Nguyen XH, Than UTT (2022) Effects of extracellular vesicles secreted by TGFβ-stimulated umbilical cord mesenchymal stem cells on skin fibroblasts by promoting fibroblast migration and ECM protein production. Biomedicine 10(8). https://doi.org/10.3390/biomedicines10081810

244. Théry C, Amigorena S, Raposo G, Clayton A (2006) Isolation and characterization of exosomes from cell culture supernatants and biological fluids. Curr Protoc Cell Biol 3:22
245. Adler M, Mayo A, Zhou X, Franklin RA, Jacox JB, Medzhitov R, Alon U (2018) Endocytosis as a stabilizing mechanism for tissue homeostasis. Proc Natl Acad Sci USA 115(8):E1926–e1935. https://doi.org/10.1073/pnas.1714377115
246. Fortunato D, Mladenović D, Criscuoli M, Loria F, Veiman KL, Zocco D, Koort K, Zarovni N (2021) Opportunities and pitfalls of fluorescent labeling methodologies for extracellular vesicle profiling on high-resolution single-particle platforms. Int J Mol Sci 22(19). https://doi.org/10.3390/ijms221910510
247. Takov K, Yellon DM, Davidson SM (2017) Confounding factors in vesicle uptake studies using fluorescent lipophilic membrane dyes. J Extracell Vesicles 6(1):1388731. https://doi.org/10.1080/20013078.2017.1388731
248. You S, Barkalifa R, Chaney EJ, Tu H, Park J, Sorrells JE, Sun Y, Liu YZ, Yang L, Chen DZ, Marjanovic M, Sinha S, Boppart SA (2019) Label-free visualization and characterization of extracellular vesicles in breast cancer. Proc Natl Acad Sci USA 116(48):24012–24018. https://doi.org/10.1073/pnas.1909243116
249. Sun Y, You S, Tu H, Spillman DR Jr, Chaney EJ, Marjanovic M, Li J, Barkalifa R, Wang J, Higham AM, Luckey NN, Cradock KA, George Liu Z, Boppart SA (2018) Intraoperative visualization of the tumor microenvironment and quantification of extracellular vesicles by label-free nonlinear imaging. Sci Adv 4(12):eaau5603. https://doi.org/10.1126/sciadv.aau5603
250. You S, Sun Y, Chaney EJ, Zhao Y, Chen J, Boppart SA, Tu H (2018) Slide-free virtual histochemistry (Part I): development via nonlinear optics. Biomed Opt Express 9(11):5240–5252. https://doi.org/10.1364/boe.9.005240
251. You S, Sun Y, Chaney EJ, Zhao Y, Chen J, Boppart SA, Tu H (2018) Slide-free virtual histochemistry (Part II): detection of field cancerization. Biomed Opt Express 9(11):5253–5268. https://doi.org/10.1364/boe.9.005253
252. You S, Tu H, Chaney EJ, Sun Y, Zhao Y, Bower AJ, Liu YZ, Marjanovic M, Sinha S, Pu Y, Boppart SA (2018) Intravital imaging by simultaneous label-free autofluorescence-multiharmonic microscopy. Nat Commun 9(1):2125. https://doi.org/10.1038/s41467-018-04470-8
253. Weigelin B, Bakker GJ, Friedl P (2016) Third harmonic generation microscopy of cells and tissue organization. J Cell Sci 129(2):245–255. https://doi.org/10.1242/jcs.152272
254. Tu H, Liu Y, Turchinovich D, Marjanovic M, Lyngsø J, Lægsgaard J, Chaney EJ, Zhao Y, You S, Wilson WL, Xu B, Dantus M, Boppart SA (2016) Stain-free histopathology by programmable supercontinuum pulses. Nat Photonics 10(8):534–540. https://doi.org/10.1038/nphoton.2016.94
255. Tu H, Liu Y, Marjanovic M, Chaney EJ, You S, Zhao Y, Boppart SA (2017) Concurrence of extracellular vesicle enrichment and metabolic switch visualized label-free in the tumor microenvironment. Sci Adv 3(1):e1600675. https://doi.org/10.1126/sciadv.1600675
256. Wang LH, Rothberg KG, Anderson RG (1993) Mis-assembly of clathrin lattices on endosomes reveals a regulatory switch for coated pit formation. J Cell Biol 123:1107–1117
257. Monti MC, Chini MG, Margarucci L, Tosco A, Riccio R, Bifulco G et al (2009) The molecular mechanism of human group IIA phospholipase A2 inactivation by bolinaquinone. J Mol Recognit 22:530–537
258. Kuhn Y, Rohrbach P, Lanzer M (2007) Quantitative pH measurements in Plasmodium falciparum-infected erythrocytes using pHluorin. Cell Microbiol 7:1004–1013
259. von Kleist L, Stahlschmidt W, Bulut H, Gromova K, Puchkov D, Robertson MJ et al (2011) Role of the clathrin terminal domain in regulating coated pit dynamics revealed by small molecule inhibition. Cell 146:471–484
260. Chen Z, Mino RE, Mettlen M, Michaely P, Bhave M, Reed DK et al (2020) A clathrin terminal domain-derived peptide inhibitor of clathrin-mediated endocytosis. J Cell Biol 219:e201908189
261. Schnitzer JE, Oh P, Pinney E, Allard J (1994) Filipin-sensitive caveolae-mediated transport in endothelium: reduced transcytosis, scavenger endocytosis, and capillary permeability of select macromolecules. J Cell Biol 127(5):1217–1232
262. Rodal SK, Skretting G, Garred O, Vilhardt F, van Deurs B, Sandvig K (1999) Extraction of cholesterol with methyl-beta-cyclodextrin perturbs formation of clathrin-coated endocytic vesicles. Mol Biol Cell 10(4):961–974

263. Chen Y, Wang S, Lu X, Zhang H, Fu Y, Luo Y (2011) Cholesterol sequestration by nystatin enhances the uptake and activity of endostatin in endothelium via regulating distinct endocytic pathways. Blood 117:6392–6403
264. Lamaze C, Baba T, Redelmeier TE, Schmid SL (1993) Recruitment of epidermal growth factor and transferrin receptors into coated pits in vitro: differing biochemical requirements. Mol Biol Cell 4:715–727
265. Kirchhausen T, Macia E, Pelish HE (2008) Use of dynasore, the small molecule inhibitor of dynamin, in the regulation of endocytosis. Methods Enzymol 438:77–93
266. Quan A, McGeachie AB, Keating DJ, van Dam EM, Rusak J, Chau N et al (2007) Myristyl trimethyl ammonium bromide and octadecyl trimethyl ammonium bromide are surface-active small molecule dynamin inhibitors that block endocytosis mediated by dynamin I or dynamin II. Mol Pharmacol 72: 1425–1439
267. Linares-Clemente P, Rozas JL, Mircheski J, García-Junco-Clemente P, Martínez-López JA et al (2015) Different dynamin blockers interfere with distinct phases of synaptic endocytosis during stimulation in motoneurones. J Physiol 593:2867–2888
268. Chao H, Liu Y, Lin C, Xu X, Li Z, Bao Z et al (2018) Activation of bradykinin B2 receptor induced the inflammatory responses of cytosolic phospholipase A2 after the early traumatic brain injury. Biochim Biophys Acta Mol basis Dis 1864:2957–2971
269. Imamura J, Suzuki Y, Gonda K, Roy CN, Gatanaga H, Ohuchi N et al (2011) Single particle tracking confirms that multivalent Tat protein transduction domain-induced heparan sulfate proteoglycan cross-linkage activates Rac1 for internalization. J Biol Chem 286:10581–10592
270. West MA, Prescott AR, Eskelinen EL, Ridley AJ, Watts C (2000) Rac is required for constitutive macropinocytosis by dendritic cells but does not control its downregulation. Curr Biol 10:839–848
271. Ciesielski-Treska J, Guerold B, Aunis D (1982) Immunofluorescence study on the organization of actin in astroglial cells in primary cultures. Neuroscience 7:509–522
272. Kwon KW, Park H, Doh J (2013) Migration of T cells on surfaces containing complex nanotopography. LoS One 8:e73960
273. Jordan MA, Thrower D, Wilson L (1992) Effects of vinblastine, podophyllotoxin and nocodazole on mitotic spindles. Implications for the role of microtubule dynamics in mitosis. J Cell Sci 102:401–416
274. Jordan MA, Wilson L (2004) Microtubules as a target for anticancer drugs. Nat Rev Cancer 4:253–265
275. Chen X, Khambu B, Zhang H, Gao W, Li M, Chen X et al (2014) Autophagy induced by calcium phosphate precipitates targets damaged endosomes. J Biol Chem 289:11162–11174

Chapter 13

Protocol for Assessing the Procoagulant and Prothrombotic Potential of Extracellular Vesicles

Kimberly A. Thomas, Baron K. Osborn, and Jessica C. Cardenas

Abstract

Extracellular vesicles (EVs) are small (<1 μm in diameter) particles surrounded by bilipid membranes that carry DNA, RNA, proteins, and other intracellular molecules from the cell that released the EV to a recipient cell—facilitating intercellular communication. EVs are naturally released by all types of cells in response to injury or cell death, with the aim of eliciting help from their surroundings to aid in cellular debris removal and tissue repair. The type of signals that induce EV release (i.e., injury vs apoptosis) can change the makeup of the EV membrane surface, sometimes resulting in EVs that have a more procoagulant phenotype, which is often manifested by phosphatidylserine (PS) exposure and/or tissue factor (TF) expression. These procoagulant EVs potentiate thrombosis and are associated with an increased risk of both arterial and venous thromboembolism. The goal of this chapter is to provide information and protocols with which to evaluate the procoagulant potential of EVs derived from cells in the vasculature, where prothrombotic activity can cause the most harm.

Key words Extracellular vesicles, Procoagulant, Phosphatidylserine, Tissue factor, Thrombin generation

1 Introduction

1.1 Procoagulant Extracellular Vesicles in Blood

The production of extracellular vesicles (EVs) is a process inherent to almost all cell types; however, we will be restricting this chapter to the assessment of the most commonly found EVs circulating in blood—where procoagulant activity has the most significant effects. EV production is a naturally occurring phenomenon during cell injury and death. EVs released during these processes carry important cargo as messages to neighboring cells, promoting inflammation and tissue repair pathways to help with the removal of cellular debris from the injured/dying cell [1]. The phenotype of the EV membrane can provide context on the cellular source of the membrane, the activity that led to EV release, and the ability of the EV to interact with other cells and molecules within the local environment. Key to the prothrombotic potential of EVs is their

Amit K. Srivastava (ed.), *Extracellular Vesicles*, Neuromethods, vol. 228, https://doi.org/10.1007/978-1-0716-4905-3_13,

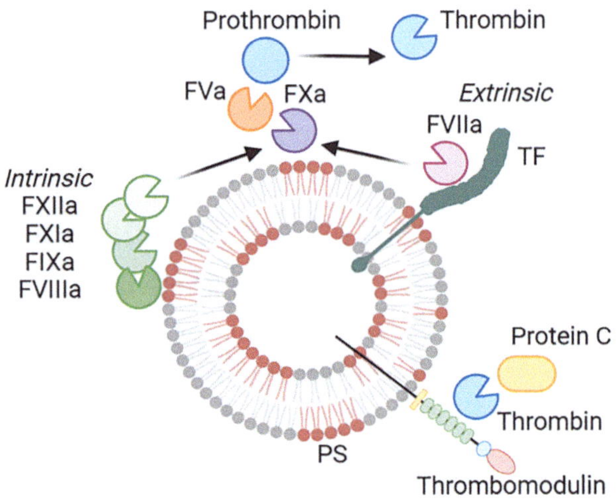

Fig. 1 Procoagulant Activity on EV Membranes. Both PS and TF play important roles in promoting coagulation on the surface of EVs. Of note, thrombomodulin expressed on endothelial-derived EVs plays a key role in dampening this coagulant phenotype through Protein C activation. (Created in BioRender)

ability to support coagulation at the EV surface—the procoagulant capacity of the EV—which is largely dictated by phosphatidylserine (PS) exposure [2] and tissue factor (TF) expression [3] (Fig. 1). Not all EVs are procoagulant by nature, as not all express PS and TF. In fact, some EVs express receptors that play anticoagulant roles, such as thrombomodulin expressed by endothelial cell-derived EVs [4] and the ability of red blood cell (RBC)-derived EVs to activate protein S and C [5].

Early assessments of EV origin and quantity in the circulation of healthy individuals identified that >50% of EVs in the blood are derived from platelets (PEVs) [6]. With the advent of improved flow cytometric detection of EVs in blood, reports again found that 50–80% of circulating EVs were PEVs [7, 8]. The next most abundant sources of EVs in blood are those derived from red blood cells (RBC; RBC-EV), followed by leukocyte-derived EVs, and lastly endothelial cell-derived EV (EC-EV) which are extremely low in number in the absence of notable pathophysiology. PEVs are the most well characterized with respect to their procoagulant function and capacity mediated by PS and TF, and their overall role in thrombosis [9] and hemostasis [2, 10]. Importantly, not all EVs express PS or TF on their surface, highlighting that not all EVs are procoagulant by nature. RBCs do not express TF; therefore, their procoagulant capacity is tightly associated with PS exposure and contact activation (FXII mediated) of coagulation [11–13]. In contrast, EVs derived from monocytes mostly support thrombin generation through TF-dependent mechanisms with little contribution from PS and FXII activation [12, 14, 15]. Endothelial-derived EVs can be TF+ or PS+, and levels of these molecules are heavily dictated by cytokine activation and shear conditions [16–18].

While PS and TF expression play key roles in determining the intrinsic procoagulant capacity of EVs, the ability of EVs to bind to and interact with other cells in the blood can also enhance pro-thrombotic potential. This “extrinsic” procoagulant capacity is normally mediated through receptor interactions that promote binding to and activation of leukocytes in a manner that potentiates coagulation. EV binding of monocytes can result in increased TF expression on the surface of the monocyte, which can support coagulation. Additionally, EVs have been shown to enhance NETosis [19, 20], which can potentiate coagulation. Thus, to have a true picture of the procoagulant capacity of an EV, one needs to assess both the intrinsic and extrinsic features that promote coagulation: PS expression, TF expression, and receptor expression that would promote binding and activation of monocytes and neutrophils.

EVs arise as components of cell injury and/or death; therefore, disease pathophysiologies that result in increased inflammation and cell damage have a marked effect on the quantity and quality of circulating EVs. EVs with increased TF expression/activity have been found in various infectious diseases, cancers, and cardiovascular disease (reviewed in Ref. [21]). The recent SARS-CoV-2/COVID-19 pandemic highlighted the contribution of TF+ PS+ EVs to disease severity through promoting thrombin generation leading to a hypercoagulable state [22–24]. Sickle cell disease and beta thalassemia can lead to increased PS exposure and RBC EV production, resulting in enhanced thrombin-generating capacity of RBC-derived EVs [25, 26]. As RBCs make up the majority of cells in the blood, even small increases in PS exposure can significantly augment thrombotic potential [27]. EVs with increased TF activity were found to be significantly elevated in patients with cancers that have an increased risk of venous thromboembolism (VTE), such as brain and pancreatic cancers [21]. Patients with acute coronary syndromes were found to have elevated levels of endothelial cell EVs that had increased procoagulant activity [28]. Lastly, traumatic injury leads to a significant increase in PEVs that have enhanced thrombin-generating capacity [29]. Of note, temporal analysis of EVs post-traumatic injury highlighted that EVs generated in the first 3 hours after injury have the greatest procoagulant capacity [30], with a steady decline back to baseline values by 12 h post injury. Collectively, these studies highlight the importance of being able to accurately measure procoagulant capacity of EVs from different cellular origins, as this functionality contributes to disease burden, irrespective of etiology.

1.2 Methods to Assay Procoagulant EV Surface Phenotype

With respect to the procoagulant capacity of the EV surface, the most important features are the phospholipid content [31, 32]—procoagulant phospholipids (PPL), to include phosphatidylserine (PS)—and TF expression, as these features most readily support coagulation. PPL are normally found on the inner leaflet of the plasma membrane, but upon cellular activation, these lipids are flipped and exposed on the outer membrane, providing the negatively charged surfaces required to support coagulation [33]. While assays to measure PPL content have traditionally been engineered to measure PPL content in plasma [34], these same platforms can be adapted to measure PPL content in EVs. The most commonly used assays employ the clotting time method (STA-Procoag-PPL assay, Diagnostica Stago; XACT assay, Haematex), where activated FXa and calcium chloride are added to a sample containing PPL and clotting time is measured. Increased PPL content decreases the amount of time it takes for the clot to form, resulting in reduced clotting times in the presence of abundant PPL [34].

The most common method used to evaluate PS expression is flow cytometry. The two most common labels used to bind exposed PS are Annexin V and Lactadherin, both of which can be conjugated to a fluorophore of the user's choosing, using standard approaches. Importantly, Annexin V can only bind PS in the presence of Ca^{2+}, whereas Lactadherin can bind in the presence or absence of Ca^{2+}. This needs to be taken into consideration depending on the EV sample source, as many plasma samples are derived from blood collected into sodium citrate (anticoagulant) and adding Ca^{2+} to this type of sample would induce clotting. Thus, Annexin V may be optimal for PS measurement in isolated EV samples, whereas Lactadherin may be better for PS measurement in biological specimen samples containing Ca^{2+} sensitive anticoagulants.

For flow cytometry acquisition, specialized high resolution small particle cytometers are needed to accurately detect EVs due to their small size. If these platforms are unavailable, an alternative is to use beads coated with streptavidin bound to biotinylated capture antibodies to essentially pull down EVs, then stain with additional antibodies/dyes to determine specificities as needed. The downside to this latter approach is the loss of single EV resolution. Lastly, the most rigorous way to measure the surface phenotype of individual EVs is to use platforms that allow for single particle resolution, such as a nanoparticle tracking analyzer (NTA, e.g., nanosight platform [35]) or an imaging flow cytometer (e.g., Amnis Cell Stream).

While TF expression on EVs can be evaluated using TF-specific antibodies via flow cytometry, or via ELISA, these antigen-based methods are not able to discern encrypted versus active TF [36–40]. There are a handful of commercial TF activity kits; however, these were found to be less specific and less sensitive than in-house assays developed by Hisada and Mackman. These in-house assays use a two-step approach wherein first FVIIa binds TF allowing for conversion of FX to FXa. Then, in the second step, a chromogenic substrate of FXa is added, then absorbance read. This assay allows for high throughput due to the 96-well format; however the 2-h long incubation of the first step has been hypothesized to allow for TF-independent activity of FX, thereby having less specificity. An alternative to this assay has been developed to assess only TF-dependent activity by adding additional factors (prothrombin, FV, and a chromogenic thrombin substrate) and decreasing the incubation period to less than 10 min as longer than 10 min leads to TF-independent activation of FX (most likely by phospholipids) [41]. While this platform also uses a similar two-step approach and commercially available coagulation factors, the readout of this assay is thrombin generation instead of FXa activity.

As mentioned above, another key surface receptor that can modulate the procoagulant capacity of EVs is thrombomodulin (TM)—an important regulator of the anticoagulant phenotype of endothelium. To fully assess the procoagulant potential of EVs derived from endothelial cells, it is necessary to measure TM expression and activity. TM expression can be evaluated using flow cytometry, whereas TM activity can be read out indirectly through protein C activation (via chromogenic or clot-based methods) or thrombin generation. In the case of the former, the most common protein C activation assays are chromogen based and can be run on multiple different hemostasis analyzers (e.g., Stago and Seimens platforms). In the case of the latter, thrombin generation assays (as outlined below) can be used. In both of these methodologies, the addition of a TM blocking antibody to the EV containing sample can block thrombomodulin activity, thereby reducing activation of protein C and increasing thrombin generation [42].

1.3 Methods to Assay EV Procoagulant Function In Vitro

The assays above highlight methods to evaluate PS, TF, and TM expression on the surface of EVs. However, they do not account for the actual generation of thrombin, which is a central feature of procoagulant capacity. Therefore, assays that measure the ability of EVs to generate thrombin and fibrin, in turn promoting coagulation, are an important functional readout of the procoagulant capacity of EVs. There are two main methods that fall under this category—thrombin generation assays (TGAs) and clotting assays.

TGAs come in many different varieties, but perhaps the most well-established assay is the calibrated automated thrombogram (CAT) [43], which evaluates thrombin generation in platelet-rich and platelet-poor plasma and can be adapted to measure thrombin generation by EVs. While there are many in-house/custom TGAs that are of a similar premise to the CAT assay, we will be discussing the CAT as the reagents are commercially available and standardized. Thrombin generation by EVs in platelet-poor plasma

[40, 44–46] can be measured using the CAT in conjunction with proprietary reagents from Diagnostica Stago, specifically the MP Reagent. This reagent contains phospholipids (4 μM) alone with a composition of 20% phosphatidylserine (PS), 20% phosphatidylethanolamine (PE), and 60% phosphatidylcholine (PC), so as to mimic the surface of activated platelets. When the MP reagent is mixed with Ca^{2+} in the presence of TF+ EVs, thrombin generation occurs. As described above, not all EVs express TF; therefore, this assay can be performed with the addition of TF (commonly Dade Innovin) to assess the maximal procoagulant potential of EVs. Moreover, the use of TF blocking antibodies can be used to determine the EV-expressed TF contributions to thrombin generation [30, 40]. Lastly, others have proposed the use of tissue factor pathway inhibitor (TFPI)-specific antibodies to "unmask" the true thrombin generating potential of EVs in PPP samples [30]. The methods outlined below (Subheading 3) provide a step-by-step protocol to use the CAT to measure thrombin generation in PPP containing EVs.

An alternative to TGAs is to use a PPL clotting assay, which evaluates clot formation induced by the presence of PPLs. As discussed above, one of the most well-known PPL clotting assays is the Stago Procoag-PPL assay, where activated FXa and calcium chloride are added to a sample containing PPL and clotting time is measured. A higher concentration of EVs in a sample increases the amount of PPL in the reaction, which in turn accelerates clot formation [34, 47].

1.4 Methods to Assay EV Procoagulant Function In Vivo

Demonstrating that the procoagulant capacity observed in vitro (in assays as described above) in fact translates to functional procoagulant activity in vivo requires animal models which recapitulate pathophysiologies associated with bleeding and a need for procoagulant activity. Models of traumatic injury and hemorrhage allow for testing of different therapeutic strategies for resuscitation and therefore are useful for evaluating therapies that are associated with procoagulant capacity. Lopez et al. demonstrated, using a murine liver trauma model, that intravenous administration of PEV reduced abdominal blood loss by 24% compared to control (sham) treated mice [48]. These PEVs were evaluated in vitro as well and showed a dose-dependent increase in clot strength and thrombin generation, thus demonstrating translatability between in vitro and in vivo procoagulant function. In another study, Dyer et al. isolated PEVs from mice with traumatic injury and found they were prothrombotic due to increased thrombin generation as assessed in vitro. Upon transfusion, these trauma PEVs reduced bleeding time in a tail bleed model and also reduced blood loss after adoptive transfer into recipient mice with polytrauma [29]. An interesting finding in this model was that in addition to reduced bleeding, there was also an increase in thrombus weight after PEV administration, further highlighting their prothrombotic potential. Collectively, these studies demonstrate the utility of murine trauma models for evaluating procoagulant capacity in vivo.

1.5 Advanced Specialized Methodologies to Evaluate Procoagulant EV

Many of the assays outlined above for evaluating TF expression of EVs are highly sensitive and often require purified EVs as sample material. An assay that could be used without the need of EV purification and correlated strongly with functional assays would be extremely useful when determining the contributions of procoagulant EVs to clinical outcomes. The ExoView R100 platform attempts to fill this niche using a TF antibody-based EV capture approach [40]. While this platform does not require EV purification prior to detection, it does not have a rapid turnaround time and it is contingent upon the specificity of the antibodies used for capture.

Another advanced approach is the use of mass spectrometry to evaluate procoagulant EV phenotype and cargo [42]. Highly purified EVs can be used in proteomics approaches to evaluate the relative quantity of key proteins such as TF and TM, while simultaneously identifying additional proteins associated with prothrombotic potential and cellular origins. Lipidomics approaches may provide insight on the quality and quantity of lipids in addition to PS that are important for driving procoagulant function.

Lastly, there is a growing interest in visualizing EV engagement with cells in the vascular space and under physiologically relevant flow conditions. For both capturing EV release from cells and for measuring EV interaction in the vascular space in vivo, high-resolution intravital microscopy is necessary. These approaches require labeling of EVs (e.g., genetically, dyes, antibodies, gold particle labeling) and real-time imaging capacity (e.g., fluorescent, bioluminescent, nuclear, photoacoustic) and specific skills [49]. Despite these hurdles, information generated from these types of experiments is both illuminating and highly mechanistic in nature. For example, Zifkos et al. found that procoagulant (PS+) EVs were far more likely to engage with leukocytes under physiological shear and adhere to injured vessel walls than non-PS expressing EVs [50]. In another study, Vats et al. demonstrated that inflammasome activation induced PEVs in a mouse model of sickle cell disease (SCD) and that these PEVs directly engaged with neutrophils to cause lung vasoocclusion in vivo [51]. These findings were recapitulated in vitro using human SCD samples and microfluidic devices. While these more advanced methodologies provide unique insight into EV phenotype and function, they do not necessarily provide quantitative metrics on thrombin generating capacity.

1.6 Conclusions/Future perspectives

The therapeutic potential of EVs as a drug delivery system has long been pursued due to their ability to store multiple types of cargo, ease of passing communication between cells, and relative immune quiescence [52–54]. However, using EVs to modulate coagulation is a newer area of therapeutic EV research. As outlined above (Sect. 1.4), procoagulant PEVs have been used to enhance hemostasis in murine models of trauma. PEVs have also been shown to promote vascular maintenance in a VEGF-induced permeability model [55] and reduce blood loss in a tail bleed model [55, 56]. PEVs specifically have the potential for streamlined manufacture and use, as the source material (platelet concentrates) is already an established, licensed therapy, and its collection and production practices are regulated by national regulatory entities. PEV production does not require expansion, as EVs from other cell sources may, and pathogen reduction technologies are already in use for platelet concentrates, ensuring minimal risk of pathogen transmission to end users. Heterogeneity in platelet donors may lead to issues with ensuring sustained similar quality across PEV products/lots and thus should be considered as PEV are developed into a clinically approved therapy.

While we have focused on the procoagulant function of EV within the vasculature, it is of note that PEV have also been demonstrated to promote angiogenesis [57–59] and improve wound repair [58, 59]. Furthermore, PEVs have been found to play important roles in megakaryocyte reprogramming in the bone marrow [60], and they may play a role in lymphatic development and maintenance akin to platelets [59]. Endothelial cell-derived EVs have also been demonstrated to help with re-endothelialization and vascular repair after injury [61, 62]. Thus, the therapeutic potential of blood cell-derived EV beyond procoagulant function is just beginning to be fully elucidated and is an area for further development.

2 Materials

2.1 Samples

1. PPP containing EVs *OR*
2. Isolated EVs of known concentration.

2.2 Reagents

2.2.1 Diagnostica Stago Reagents

1. MP Reagent (# 86222, box of 20 × 1 mL vials).
2. FluCa-Kit (# 86197, box of 20 × 1.6 mL vials of Fluo-Buffer and 1 × 0.9 mL vials of Fluo-Substrate).
3. Thrombin Calibrator (# 86192, box of 20 × 1 mL vials).

2.2.2 Additional Reagents

1. Normal pooled plasma (NPP) devoid of EV.
   1.1 In house produced or commercial NPP will work.
   1.2. If NPP already EV depleted, thaw directly for use as outlined in Subheading 3.

1.3. If NPP not already EV depleted, thaw NPP and spin at 20,000 *g* for 30 min at 4° [46], then use the supernatant as NPP devoid of EV.

2. Normal saline (0.9% NaCl in H_2O).
3. diH_2O.

2.3 Consumables

1. 96-well plates (Diagnostica Stago, #86715).
2. 1.5 mL microcentrifuge tubes.
3. 15 mL conical tubes.
4. 50 mL conical tubes.
5. P10 micropipette and tips.
6. P20 micropipette and tips.
7. P100 micropipette and tips.
8. P1000 micropipette and tips.

2.4 Equipment

1. 37 °C water bath.
2. Two beakers (one 25 mL, one large enough to hold a 50 mL conical tube; both should be in the CAT device).
3. Fluoroskan Ascent (CAT device).
4. Thrombinoscope Software.

3 Methods

3.1 Equipment and Reagent Preparation

1. Ensure the water bath is on and set to 37 °C.
2. Turn on the CAT device and log into the Thrombinoscope software. Once logged in, click "Measure," then accept the message to shut the sample drawer.

3.1.2 If you do not click "Measure" and shut the sample drawer, the software will not initiate the warming protocol, meaning the device will not start raising the internal temperature to 37 °C, delaying the assay start time (*see* **Note 1**).

3.1.3. On the main CAT window, click "Settings," followed by "Set calibrator." Click "Yes" to confirm you want to change the calibrator settings. You should do this for every run to verify you are using the correct calibrator lot. The batch number can be found on the Thrombin Calibrator reagent.

3.1.4. Enter the Batch number and activity value and press "Save."

3.1.5. Some versions of the software do not have the option to add the batch number for the thrombin calibrator (*see* **Note 2**).

3.1.6 Fill two 50 mL conical tubes with diH_2O, then place one tube in the 37 °C water bath and allow to incubate for at least 30 min. Use the other conical tube for reconstituting reagents as described below in Subheading 3.4.

3.2 Design Plate Layout

1. The manufacturer suggests that samples are assayed using triplicates (*see* **Note 3**). While this is good lab practice, in the case of small sample volumes or quantities, the end user can decide to perform the assay in duplicate if needed.
2. Each experimental sample must be paired with a calibrator sample (in duplicate or triplicate) (Fig. 2). Designate which wells will be experimental (contains the MP reagent, red assignment, "Thrombin Generation") and which wells will be the standard (contains the Thrombin Calibrator, yellow assignment, "Thrombin Calibrator"). These two sets are termed a "group." Feel free to assign new group names.
3. Once the plate layout is set, under Patterns, click "Create." Save the pattern accordingly. Use this plate layout to determine the number of wells that will receive the MP reagent, the number of wells that will receive the Thrombin Calibrator, and the total number of wells (equal to those that need FluCa).
4. Ensure the duration is set for 60 min, and the Interval is set for 20 s.
5. The software is now ready. Go prepare the samples and reagents.

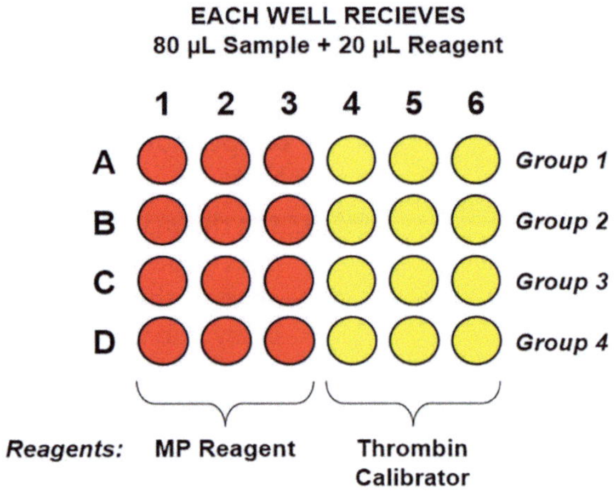

Fig. 2 Plate Layout Design

3.3 Prepare Samples

1. For samples consisting of isolated EVs of a known quantity (concentration), use the steps below to prepare samples.

   3.3.1.1. Isolated EV samples are often prepared and stored in minimal buffers that do not include the coagulation factors needed to support thrombin generation (*see* **Note 4**). Thus, isolated EV samples need to be diluted, or spiked, into a normal pooled plasma devoid of EV, in order to provide the factors needed to support thrombin generation. Previous publications have demonstrated that thrombin generation can be measured at 7–8 × 10^9 PEV.

   3.3.1.2. Determine the desired EV count per well to be tested, e.g., 1 × 10^9/well would be 1 × 10^{10}/mL as each well has a final reaction volume of 100 μL. Using $C_1V_1 = C_2V_2$, where C_1 is the [EV stock] and C_2 is the [desired EV final] and V_2 is (100 μL {final volume per well} * n, where n is the total number of wells required for the given sample), calculate V_1—the volume needed of PEV stock. Next, subtract V_1 from V_2 to determine the volume of NPP to add to V_1 to create the working sample solution to be added to the wells.

   3.3.1.3. Label clean, empty 1.5 mL microcentrifuge tubes for every PEV sample to be tested.

   3.3.1.4. Fill an ice bucket large enough to hold freezer sample boxes with ice.

   3.3.1.5. Retrieve the relevant sample boxes from −80 °C storage and place onto the prepared bed of ice. Remove EV samples to be tested, place them on ice to thaw, and return the sample boxes to −80 °C storage.

   3.3.1.6. Prepare NPP.

      3.3.1.6.1 Retrieve NPP from −80 °C storage and thaw at 37 °C for 5 min.

      3.3.1.6.2 Dilute NPP from 100% to 85% using normal saline.

   3.3.1.7. After EV sample and NPP thawing, prepare each sample to be tested using the volumes calculated above. Specifically, add the calculated V_1 of PEV stock to the appropriately labeled tube. To the same tube, add the calculated V_2 of NPP and gently mix. Leave the mixture at RT until addition to the TGA plate.

      3.3.1.7.1 If the amount of PEV solution sample to be added to the 1.5 mL microcentrifuge tube is lower than 1 μL, perform a 10× dilution of the PEV sample in normal saline (i.e.,

10 μL of PEV stock into 90 μL of normal saline). After this has been done, recalculate the amount of PEV stock (C_1) needed by the method outlined above in Subheading 3.c.i.2.

3.3.1.8. After samples are prepared, return the PEV stocks to −80 °C storage.

3.3.1.9. Use PEV spiked NPP samples as outlined below in the "Prepare Plate" section.

3.3.2 For samples that contain PEV of unknown quantity, such as platelet poor plasma generated from patient whole blood samples collected in sodium citrate, use the steps below to prepare samples.

3.3.2.1 As the quantity of PEV is unknown in these samples, consider using other methods to enumerate PEV in the samples prior to thrombin generation measurements (*see* **Note 5**).

3.3.2.2 Identify whether PEV samples are stored in plasma or a solution that does not contain coagulation factors.

3.3.2.2.1 If stored in plasma, use 80 μL of sample per well.

3.3.2.2.2 If stored in a solution that does not contain coagulation factors, add 15 μL of PEV containing sample to 85 μL of NPP for each well needed. Therefore, if 6 wells are needed, add 90 μL (6 wells × 15 μL per well) of sample to 510 μL of NPP (6 wells × 85 μL per well). Use 80 μL of this mixture as the sample for each well.

3.4 Prepare Reagents

1. Retrieve the Fluo-Substrate vial from 4 °C and allow to come to room temperature (RT) on the benchtop, protected from light.
2. Retrieve the appropriate number of vials of Fluo-Buffer from 4 °C. For 48 wells or less, one vial of Fluo-Buffer is needed; for 49–96 wells, two vials of Fluo-Buffer will be needed. Transfer the Fluo-Buffer to a 15 mL conical tube, and place Fluo-Buffer in a 37 °C water bath until use in Subheading 3.6, **step 4**.
3. Retrieve the appropriate number of vials of the MP reagent and Thrombin Calibrator from 4 °C, and allow to come to RT on the benchtop for 10 min. Each vial will provide enough reagent for 50 wells.

4. Once the MP Reagent and Thrombin Calibrator are at RT, remove the metal safety tab, then carefully remove the rubber stopper of each vial. Add 1 mL of RT diH_2O to each vial and wait 10 min. Swirl gently, and use within 30 min. Do not store or freeze the reconstituted MP Reagent or Thrombin Calibrator.

3.5 Prepare Plate

1. Retrieve the 96-well clear round bottom plate.
2. Pipet 20 μL of reconstituted reagents to the sample-assigned (red) wells and 20 μL of reconstituted thrombin calibrator to the calibrator-assigned (yellow) wells. Take care to minimize bubbles in the wells by placing the reagent directly at the bottom of the well.
3. Add 80 μL of PEV samples to the assigned wells. Take care to minimize bubbles in the wells.

3.6 Run Plate

1. Click "Plate Out" in the software, place the plate in the machine, then click "Plate In." Press "Start," which will begin a 10 min incubation/warm-up.
2. When 5 min remain in the incubation period, open the plastic hardshell cover on top of the CAT device. Ensure an empty small 25 mL beaker is placed under the dispensing tube, and a larger empty beaker is placed nearby.
3. Retrieve the 50 mL conical tube of diH_2O from the water bath and place in the larger beaker near the dispenser set up.
4. When the incubation period has 2 min remaining, prepare the FluCa reagent (Fluo-buffer + Fluo-Substrate). For 48 wells or less, add 40 μL of the Fluo-Substrate to the 15 mL conical tube containing the pre-warmed Fluo-Buffer. For 49–96 wells, add 80 μL of the Fluo-Substrate. Immediately vortex for 10–30 s until solution clears. Set 15 mL conical tube in the large beaker near the dispenser set up, next to the warm diH_2O.
5. Follow on-screen instructions for washing the machine with warm diH_2O, followed by FluCa loading (Fig. 3). The time-sensitive step is after filling the loading syringe with a FluCa reagent and placing the dispensing tube in the appropriate channel.
6. Pay close attention to where the aspirate tube and the dispensing tube are placed (Fig. 3). Ensure the aspirate tube is completely submerged in either the diH_2O or the FluCa reagent so as to not aspirate up bubbles. Ensure the tip of the dispensing tube is pressed against the side of the 15 mL conical tube containing FluCa so as to prevent bubbling (*see* **Note 6**).
7. Continue to follow on-screen instructions until the dispensing tube is removed, the peg has been replaced in the hole marked "M," and the assay has begun.

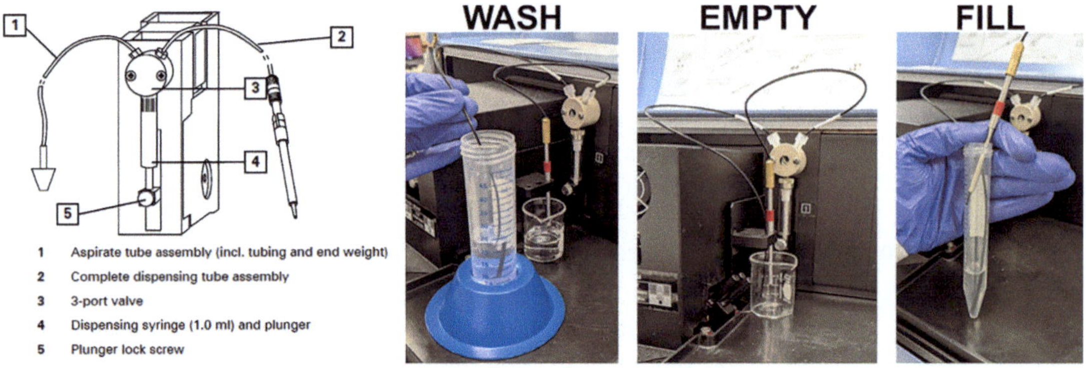

Fig. 3 Left: CAT Device Syringe pump assembly (adapted from Stago Thrombinoscope User Manual). Right: Visuals of washing, emptying, and filling procedures. Notice the location of the aspirate tube and the dispensing tube (one with the pointy tip)

8. Place the dispensing tube back in the mounted holder above the small beaker, and ensure the tip is submerged in water for the duration of the run; add more warm diH_2O if necessary. Close the CAT device cover, and place the 50 mL conical tube of warm diH_2O back in the 37 °C water bath to use for cleaning when the run is over.
9. As the run commences, a "Results" window will open, showing the fluorescence curves as they are generated over time. Verify that the calibrator curves (in green) are being generated in each window within the first 2–3 reads—this validates that the reaction is occurring correctly.

3.7 End Run

1. When the run is completed, the CAT device will automatically open the plate drawer. Remove the plate and dispose of in the appropriate biohazard receptacle.
2. Click "Plate In" to close the drawer.
3. Clean the fluidics as follows:
   3.1 Place the black aspirate tube into the conical tube of warm diH_2O.
   3.2 Ensure the small beaker is empty and placed under the dispenser tip.
   3.3 Select "Instrument," then "Prime," and change the volume to 10,000, and select "OK."
   3.4 The machine will now prime 10 × 1 mL in order to clean all of the FluCa buffer out of the lines.
   3.5 After the priming, remove the black aspirate tube from the conical tube and place in the mounted holder above the small beaker.
   3.6 Select "Instrument," then "Empty."

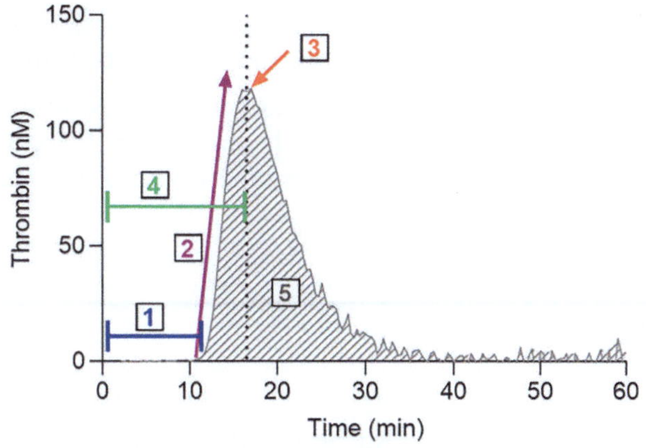

Fig. 4 Thrombogram Parameters. (1) Lagtime, minutes—time until initial thrombin formation. (2) Velocity index—rate of thrombin generation. (3) Peak, nM—maximal amount of thrombin generated at a single time point. (4) Time to peak, minutes—the time it takes to reach peak thrombin formation. (5) Endogenous thrombin potential (ETP), nM*min—total potential thrombin generated in the sample

   3.7 The machine will now pull air through the lines to dry them out.

4. After the fluidics are cleaned, turn the device off using the power button.

3.8 Review and Extract Data

1. At the end of the run, the "Results" window will display the fluorescence kinetics for each well.
2. The data generated during the run are automatically saved in a. dat file (*see* **Note 7**).
3. To export the data, click "File," then select "Export to xls." Save the file as needed. The xls file can be opened in Microsoft Excel for further analysis.

3.9 Data Analysis

1. Analysis of thrombogram data (Fig. 4) is reviewed in depth both in Thrombinoscope manuals as well as in other publications. Key metrics are as follows:
   1.1. Lag time—this is the amount of time, in minutes, it takes for initial thrombin to form. Smaller values indicate rapid thrombin production, whereas larger values suggest delays in initial thrombin formation.
   1.2. Velocity Index—this is the rate of thrombin generation (nM/min), and is essentially the slope of the line from the lagtime to the time to peak thrombin generation.
   1.3. Peak—this is the maximal amount (nM) of thrombin generated at a single time point over the course of the assay.

1.4. Time to peak—this is the time, in minutes, at which the maximal amount of thrombin was generated.

1.5. Endogenous Thrombin Potential (ETP, nM*min)—this is the area under the curve which accounts for the entire amount of thrombin generated over the course of the assay.

4 Notes

1. If you do not click "Measure" and shut the sample drawer, the software will not initiate the warming protocol—meaning the device will not start raising the internal temperature to 37 °C, delaying the assay start time. As the assay is time-sensitive once key reagents and samples are mixed in the 96-well plate, it is beneficial to ensure the device has reached temperature prior to adding samples to the plate.
2. Some versions of the software do not have the option to add the batch number for the thrombin calibrator. It is important to check this, especially if using proprietary reagents from Stago where the thrombin activity is set.
3. The manufacturer suggests that samples are assayed using triplicates. While this is good lab practice, in the case of small sample volumes or quantities, the end user can decide to perform the assay in duplicate if needed. Performing in a singlet is not recommended.
4. Isolated EV samples are often prepared and stored in minimal buffers that do not include the coagulation factors needed to support thrombin generation. Thus, isolated EV samples need to be diluted, or spiked, into a normal pooled plasma devoid of EV, in order to provide the factors needed to support thrombin generation.
5. As the quantity of PEV is unknown in these samples, consider using other methods to enumerate PEV in the samples prior to thrombin generation measurements, such as flow cytometry, nanoparticle tracking analysis, or dynamic light scattering, to enumerate EVs prior to use in this assay.
6. Pay close attention to where the aspirate tube and the dispensing tube are placed (Fig. 3). Ensure the aspirate tube is completely submerged in either the diH_2O or the FluCa reagent so as to not aspirate up bubbles. Ensure the tip of the dispensing tube is pressed against the side of the 15 mL conical tube containing FluCa so as to prevent bubbling. Bubbling will prevent the accurate volume from being loaded into the syringe and require restarting the entire assay.

7. The data generated during the run are automatically saved in a. dat file. This can be confusing as you do not save the data file before the run, and there is no prompt to save the data file after the run. Additionally, this file cannot by manipulated, and can be used for help troubleshooting issues with runs.

Acknowledgments

This study was funded by the Department of Defense (W81XWH2110683) to JCC.

References

1. Sanwlani R, Gangoda L (2021) Role of extracellular vesicles in cell death and inflammation. Cells 10:2663. https://doi.org/10.3390/cells10102663
2. Reddy EC, Rand ML (2020) Procoagulant phosphatidylserine-exposing platelets in vitro and in vivo. Front Cardiovasc Med 7. https://doi.org/10.3389/fcvm.2020.00015
3. Blood-borne tissue factor: Another view of thrombosis | PNAS. https://www-pnas-org.ucsf.idm.oclc.org/doi/full/10.1073/pnas.96.5.2311. Accessed 3 Dec 2024
4. Whelan C, Burnley-Hall N, Morris K et al (2022) The procoagulant effects of extracellular vesicles derived from hypoxic endothelial cells can be selectively inhibited by inorganic nitrite. Nitric Oxide 122–123:6–18. https://doi.org/10.1016/j.niox.2022.02.002
5. Livaja Koshiar R, Somajo S, Norström E, Dahlbäck B (2014) Erythrocyte-derived microparticles supporting activated protein C-mediated regulation of blood coagulation. PLoS One 9:e104200. https://doi.org/10.1371/journal.pone.0104200
6. Berckmans RJ, Nieuwland R, Böing AN et al (2001) Cell-derived microparticles circulate in healthy humans and support low grade thrombin generation. Thromb Haemost 85:639–646
7. Weiss R, Eichhorn T, Spittler A et al (2017) Release and cellular origin of extracellular vesicles during circulation of whole blood over adsorbent polymers for lipid apheresis. J Biomed Mater Res B Appl Biomater 105: 636–646. https://doi.org/10.1002/jbm.b.33588
8. Berckmans RJ, Lacroix R, Hau CM et al (2019) Extracellular vesicles and coagulation in blood from healthy humans revisited. J Extracell Vesicles 8:1688936. https://doi.org/10.1080/20013078.2019.1688936
9. He Y, Wu Q (2023) The effect of extracellular vesicles on thrombosis. J Cardiovasc Trans Res 16:682–697. https://doi.org/10.1007/s12265-022-10342-w
10. Mabrouk M, Guessous F, Naya A et al (2022) The pathophysiological role of platelet-derived extracellular vesicles. Semin Thromb Hemost 49:279–283. https://doi.org/10.1055/s-0042-1756705
11. Jy W, Johansen ME, Bidot C et al (2013) Red cell-derived microparticles (RMP) as haemostatic agent. Thromb Haemost 110:751–760. https://doi.org/10.1160/TH12-12-0941
12. Van Der Meijden PEJ, Van Schilfgaarde M, Van Oerle R et al (2012) Platelet- and erythrocyte-derived microparticles trigger thrombin generation via factor XIIa. J Thromb Haemost 10: 1355–1362. https://doi.org/10.1111/j.1538-7836.2012.04758.x
13. Noubouossie DF, Henderson MW, Mooberry M et al (2020) Red blood cell microvesicles activate the contact system, leading to factor IX activation via 2 independent pathways. Blood 135:755–765. https://doi.org/10.1182/blood.2019001643
14. Tripisciano C, Weiss R, Eichhorn T et al (2017) Different potential of extracellular vesicles to support thrombin generation: contributions of phosphatidylserine, tissue factor, and cellular origin. Sci Rep 7:6522. https://doi.org/10.1038/s41598-017-03262-2
15. Tripisciano C, Weiss R, Karuthedom George S et al (2020) Extracellular vesicles derived from platelets, red blood cells, and monocyte-like cells differ regarding their ability to induce factor XII-dependent thrombin generation. Front Cell Dev Biol 8. https://doi.org/10.3389/fcell.2020.00298
16. Dignat-George F, Boulanger CM (2011) The many faces of endothelial microparticles.

Arterioscler Thromb Vasc Biol 31:27–33. https://doi.org/10.1161/ATVBAHA.110.218123

17. Shet AS, Aras O, Gupta K et al (2003) Sickle blood contains tissue factor–positive microparticles derived from endothelial cells and monocytes. Blood 102:2678–2683. https://doi.org/10.1182/blood-2003-03-0693
18. Markiewicz M, Richard E, Marks N, Ludwicka-Bradley A (2013) Impact of endothelial microparticles on coagulation, inflammation, and angiogenesis in age-related vascular diseases. J Aging Res 2013:734509. https://doi.org/10.1155/2013/734509
19. Sung P-S, Hsieh S-L (2021) C-type lectins and extracellular vesicles in virus-induced NETosis. J Biomed Sci 28:46. https://doi.org/10.1186/s12929-021-00741-7
20. Jiao Y, Li W, Wang W et al (2020) Platelet-derived exosomes promote neutrophil extracellular trap formation during septic shock. Crit Care 24:380. https://doi.org/10.1186/s13054-020-03082-3
21. Hisada Y, Alexander W, Kasthuri R et al (2016) Measurement of microparticle tissue factor activity in clinical samples: a summary of two tissue factor-dependent FXa generation assays. Thromb Res 139:90–97. https://doi.org/10.1016/j.thromres.2016.01.011
22. Guervilly C, Bonifay A, Burtey S et al (2021) Dissemination of extreme levels of extracellular vesicles: tissue factor activity in patients with severe COVID-19. Blood Adv 5:628–634. https://doi.org/10.1182/bloodadvances.2020003308
23. Rosell A, Havervall S, von Meijenfeldt F et al (2021) Patients with COVID-19 have elevated levels of circulating extracellular vesicle tissue factor activity that is associated with severity and mortality-brief report. Arterioscler Thromb Vasc Biol 41:878–882. https://doi.org/10.1161/ATVBAHA.120.315547
24. Zahran AM, El-Badawy O, Ali WA et al (2021) Circulating microparticles and activated platelets as novel prognostic biomarkers in COVID-19; relation to cancer. PLoS One 16: e0246806. https://doi.org/10.1371/journal.pone.0246806
25. van Beers EJ, Schaap MCL, Berckmans RJ et al (2009) Circulating erythrocyte-derived microparticles are associated with coagulation activation in sickle cell disease. Haematologica 94: 1513–1519. https://doi.org/10.3324/haematol.2009.008938
26. Thangaraju K, Neerukonda SN, Katneni U, Buehler PW (2020) Extracellular vesicles from red blood cells and their evolving roles in health, coagulopathy and therapy. Int J Mol Sci 22:153. https://doi.org/10.3390/ijms22010153
27. Whelihan MF, Zachary V, Orfeo T, Mann KG (2012) Prothrombin activation in blood coagulation: the erythrocyte contribution to thrombin generation. Blood 120:3837–3845. https://doi.org/10.1182/blood-2012-05-427856
28. Mallat Z, Benamer H, Hugel B et al (2000) Elevated levels of shed membrane microparticles with procoagulant potential in the peripheral circulating blood of patients with acute coronary syndromes. Circulation 101:841–843. https://doi.org/10.1161/01.cir.101.8.841
29. Dyer MR, Alexander W, Hassoune A et al (2019) Platelet-derived extracellular vesicles released after trauma promote hemostasis and contribute to DVT in mice. J Thromb Haemost 17:1733–1745. https://doi.org/10.1111/jth.14563
30. Rognes IN, Hellum M, Ottestad W et al (2021) Extracellular vesicle–associated procoagulant activity is highest the first 3 hours after trauma and thereafter declines substantially: a prospective observational pilot study. J Trauma Acute Care Surg 91:681–691. https://doi.org/10.1097/TA.0000000000003333
31. Ghadami S, Dellinger K (2023) The lipid composition of extracellular vesicles: applications in diagnostics and therapeutic delivery. Front Mol Biosci 10. https://doi.org/10.3389/fmolb.2023.1198044
32. Morita S, Tsuji T, Terada T (2020) Protocols for enzymatic Fluorometric assays to quantify phospholipid classes. Int J Mol Sci 21:1032. https://doi.org/10.3390/ijms21031032
33. Protty MB, Jenkins PV, Collins PW, O'Donnell VB. The role of procoagulant phospholipids on the surface of circulating blood cells in thrombosis and haemostasis. Open Biol 12:210318. https://doi.org/10.1098/rsob.210318
34. Ramberg C, Jamaly S, Latysheva N et al (2021) A modified clot-based assay to measure negatively charged procoagulant phospholipids. Sci Rep 11:9341. https://doi.org/10.1038/s41598-021-88835-y
35. Bachurski D, Schuldner M, Nguyen P-H et al (2019) Extracellular vesicle measurements with nanoparticle tracking analysis—an accuracy and repeatability comparison between NanoSight NS300 and ZetaView. J Extracell Vesicles 8: 1596016. https://doi.org/10.1080/20013078.2019.1596016
36. Sachetto ATA, Archibald SJ, Hisada Y et al (2023) Tissue factor activity of small and large

extracellular vesicles in different diseases. Res Pract Thromb Haemost 7:100124. https://doi.org/10.1016/j.rpth.2023.100124
37. Mackman N, Sachetto ATA (2024) Challenges with measuring tissue factor antigen and activity in human plasma. Blood Vessel Thromb Hemost 1:100022. https://doi.org/10.1016/j.bvth.2024.100022
38. Hisada Y, Mackman N (2018) Measurement of tissue factor activity in extracellular vesicles from human plasma samples. Res Pract Thromb Haemost 3:44–48. https://doi.org/10.1002/rth2.12165
39. Key NS, Mackman N (2010) Tissue factor and its measurement in whole blood, plasma, and microparticles. Semin Thromb Hemost 36:865–875. https://doi.org/10.1055/s-0030-1267040
40. Price JMJ, Hisada Y, Hazeldine J et al (2023) Detection of tissue factor–positive extracellular vesicles using the ExoView R100 system. Res Pract Thromb Haemost 7:100177. https://doi.org/10.1016/j.rpth.2023.100177
41. Østerud B, Latysheva N, Schoergenhofer C et al (2022) A rapid, sensitive, and specific assay to measure TF activity based on chromogenic determination of thrombin generation. J Thromb Haemost 20:866–876. https://doi.org/10.1111/jth.15606
42. Setua S, Thangaraju K, Dzieciatkowska M et al (2022) Coagulation potential and the integrated omics of extracellular vesicles from COVID-19 positive patient plasma. Sci Rep 12:22191. https://doi.org/10.1038/s41598-022-26473-8
43. Hemker HC, Giesen P, AlDieri R et al (2002) The calibrated automated thrombogram (CAT): a universal routine test for hyper- and hypocoagulability. Pathophysiol Haemost Thromb 32:249–253. https://doi.org/10.1159/000073575
44. Bidot L, Jy W, Bidot C et al (2008) Microparticle-mediated thrombin generation assay: increased activity in patients with recurrent thrombosis. J Thromb Haemost 6:913–919. https://doi.org/10.1111/j.1538-7836.2008.02963.x
45. Gheldof D, Hardij J, Cecchet F et al (2013) Thrombin generation assay and transmission electron microscopy: a useful combination to study tissue factor-bearing microvesicles. J Extracell Vesicles 2. https://doi.org/10.3402/jev.v2i0.19728
46. Nielsen T, Kristensen AF, Pedersen S et al (2018) Investigation of procoagulant activity in extracellular vesicles isolated by differential ultracentrifugation. J Extracell Vesicles 7:1454777. https://doi.org/10.1080/20013078.2018.1454777
47. Guerreiro EM, Kruglik SG, Swamy S et al (2024) Extracellular vesicles from activated platelets possess a phospholipid-rich biomolecular profile and enhance prothrombinase activity. J Thromb Haemost 22:1463–1474. https://doi.org/10.1016/j.jtha.2024.01.004
48. Lopez E, Srivastava AK, Burchfield J et al (2019) Platelet-derived- extracellular vesicles promote hemostasis and prevent the development of hemorrhagic shock. Sci Rep 9:17676. https://doi.org/10.1038/s41598-019-53724-y
49. Liu Q, Huang J, Xia J et al (2022) Tracking tools of extracellular vesicles for biomedical research. Front Bioeng Biotechnol 10. https://doi.org/10.3389/fbioe.2022.943712
50. Zifkos K, Bochenek ML, Gogiraju R et al (2024) Endothelial PTP1B deletion promotes VWF exocytosis and venous thromboinflammation. Circ Res 134:e93–e111. https://doi.org/10.1161/CIRCRESAHA.124.324214
51. Vats R, Brzoska T, Bennewitz MF et al (2020) Platelet extracellular vesicles drive inflammasome–IL-1β–dependent lung injury in sickle cell disease. Am J Respir Crit Care Med 201:33–46. https://doi.org/10.1164/rccm.201807-1370OC
52. Kumar MA, Baba SK, Sadida HQ et al (2024) Extracellular vesicles as tools and targets in therapy for diseases. Sig Transduct Target Ther 9:1–41. https://doi.org/10.1038/s41392-024-01735-1
53. Murphy DE, de Jong OG, Brouwer M et al (2019) Extracellular vesicle-based therapeutics: natural versus engineered targeting and trafficking. Exp Mol Med 51:1–12. https://doi.org/10.1038/s12276-019-0223-5
54. Johnson J, Wu Y-W, Blyth C et al (2021) Prospective therapeutic applications of platelet extracellular vesicles. Trends Biotechnol 39:598–612. https://doi.org/10.1016/j.tibtech.2020.10.004
55. Miyazawa B, Trivedi A, Togarrati PP et al (2019) Regulation of endothelial cell permeability by platelet-derived extracellular vesicles. J Trauma Acute Care Surg 86:931–942. https://doi.org/10.1097/TA.0000000000002230
56. Lee JH, Jung H, Song J et al (2020) Activated platelet-derived vesicles for efficient hemostatic activity. Macromol Biosci 20:e1900338. https://doi.org/10.1002/mabi.201900338
57. Hayon Y, Dashevsky O, Shai E et al (2012) Platelet microparticles induce angiogenesis

and neurogenesis after cerebral ischemia. Curr Neurovasc Res 9:185–192. https://doi.org/10.2174/156720212801619018

58. Guo S-C, Tao S-C, Yin W-J et al (2017) Exosomes derived from platelet-rich plasma promote the re-epithelization of chronic cutaneous wounds via activation of YAP in a diabetic rat model. Theranostics 7:81–96. https://doi.org/10.7150/thno.16803
59. Puhm F, Boilard E, Machlus KR (2021) Platelet Extracellular Vesicles. Arterioscler Thromb Vasc Biol 41:87–96. https://doi.org/10.1161/ATVBAHA.120.314644
60. French SL, Butov KR, Allaeys I et al (2020) Platelet-derived extracellular vesicles infiltrate and modify the bone marrow during inflammation. Blood Adv 4:3011–3023. https://doi.org/10.1182/bloodadvances.2020001758
61. Bedair TM, ElNaggar MA, Joung YK, Han DK (2017) Recent advances to accelerate re-endothelialization for vascular stents. J Tissue Eng 8:2041731417731546. https://doi.org/10.1177/2041731417731546
62. Li B, Hong J, Hong M et al (2019) piRNA-823 delivered by multiple myeloma-derived extracellular vesicles promoted tumorigenesis through re-educating endothelial cells in the tumor environment. Oncogene 38:5227–5238. https://doi.org/10.1038/s41388-019-0788-4

Chapter 14

Electron Tomography-Based Reconstruction of Extracellular Vesicles in Tissue

Massimo Micaroni, Jens Berndtsson, Ornella Urzì, and Rossella Crescitelli

Abstract

Extracellular vesicles (EVs) are lipid bilayer nanoparticles involved in cell–cell communication, playing roles in both physiological and pathological processes. EVs have recently been isolated directly from solid tissues, which could provide insights into how EVs mediate communication between cells in vivo. Electron microscopy, including three-dimensional (3D) electron tomography, enables visualization of EVs within their native environments, overcoming limitations of traditional bidimensional imaging. This chapter outlines a detailed protocol for visualizing EVs in human tissues using 3D electron tomography, emphasizing the preparation of samples, including fixation, sectioning, and image acquisition. It provides details on the challenges of accurate imaging of EVs through 3D electron tomography and offers insights to improve the resolution and quality of EV visualization in tissue samples. This method allows the visualization of EVs in their native environment, thus preserving their characteristics, which can be altered by the isolation process.

Key words Extracellular vesicles, Imaging, 3D reconstruction, Electron tomography

1 Introduction

Extracellular vesicles (EVs) are nanosized (50–1000 nm) lipid-bound particles released by almost all cell types into the extracellular space [1]. Their complex content reflects the cell of origin and includes proteins [2, 3], lipids [4, 5], metabolites [6, 7], and nucleic acids, such as DNA [8, 9] and RNA [10]. Given that EVs can transfer their functional cargo into recipient cells, they are involved in cell–cell communication, mediating also cross-species interactions [11]. In the past two decades, several studies have demonstrated how EVs take part in both physiological and patho-

---

Supplementary Information The online version contains supplementary material available at https://doi.org/10.1007/978-1-0716-4905-3_14.

Amit K. Srivastava (ed.), *Extracellular Vesicles*, Neuromethods, vol. 228, https://doi.org/10.1007/978-1-0716-4905-3_14,

logical processes, playing a key role in many diseases, including cancer [12]. EVs have been isolated from several biological fluids, including blood [13], saliva [14], urine [15], cerebrospinal fluid [16], seminal fluid [17], and breast milk [18]. Although EVs isolated from biological fluids provide promising biomarker candidates, their functions remain unclear. The lack of knowledge about how EVs mediate cell–cell communication in intact tissues of the human body has led scientists to study how EVs are trafficked in tissues [19–22].

EVs are usually visualized after isolation and not in their native environment. The method used for EV isolation can alter their native structure, leading to the creation of contamination-derived artifacts. A common method used for EV isolation is high-speed centrifugation, but it has been demonstrated that this technique induces the formation of EV aggregates and the coprecipitation of proteins [23, 24]. Other methods, such as size-exclusion chromatography and ultrafiltration, can improve EV yield and purity by reducing contaminants coisolated with EVs and retaining their original morphology [25–27].

Artifacts may also be caused by the sample preparation for imaging. For instance, the "cup shape" morphology of EVs observed using transmission electron microscopy (TEM) with negative stain is an artifact caused by the high surface tension of water [28, 29]. Recently, thin sectioning of the tissue after fixation, dehydration, and embedding in plastic was performed to visualize the EVs in their native environment [19, 30]. This approach still carries limitations since it provides a bidimensional (2D) image, which makes it impossible to visualize the end of each structure, leading to the difficulty in distinguishing between EVs and cell membrane protrusions. The three-dimensional (3D) reconstruction of tissues may help to overcome this limitation and allow for better visualization of EVs within different types of tissues.

The term "tomography" was proposed in 1917 by Radon [31]; it comes from the mathematical concepts of tomography to reconstruct a 3D volume from several 2D projections [32]. Some years later, this concept was applied to TEM [33]. In electron tomography, the sample is tilted within the TEM to produce 2D images from many different perspectives to reconstruct a 3D visualization [34]. A tilt series of images of a 3D object is recorded as projections and then back-projected to retrieve the 3D volume. This technique enabled the determination of the structure of large biological complexes [35] and the structure of whole cells [36]. Crescitelli R. and coworkers performed, for the first time, the 3D tomography reconstructions with the aim of visualizing EVs in tissues [37].

In this chapter, we describe a step-by-step protocol for a visualization of EVs in human tissues using 3D tomography. An overview of the protocol is described in Fig. 1.

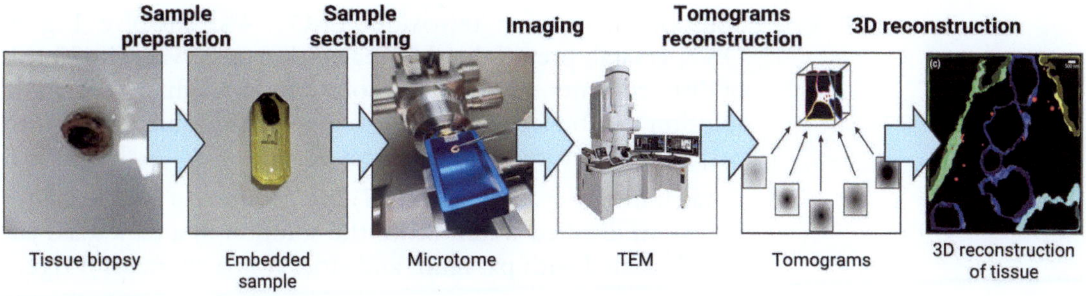

Fig. 1 Schematic overview of electron tomography-based reconstruction of EVs in tissue. (Images are adapted from tomography reconstruction published in Olofsson Bagge [37])

2 Materials

All the chemicals should have a high purity grade, and all the prepared solutions should be filtered through a 0.2 μm filter before use.

1. Na-cacodylate (Na-Caco) buffer 0.2 M. Dissolve 21.4 g of sodium cacodylate trihydrate (MW: 214.03, AGAR Sci. #R1103) in 350 mL of water. Use a plastic pipette for mixing. Adjust to pH 7.2–7.4 with 42 mL of HCl 0.2 M (690 μL of HCl 37% + 41.31 mL water). Use a plastic pipette for mixing. Aliquots (20 mL each in 50 mL Falcon tubes) are prepared and stocked in the freezer. An equal volume of distilled water is added to the vial, bringing the solution to the final concentration of 0.1 M, and then used as a washing buffer.
2. Glutaraldehyde (EMS, USA) is prepared from a stock water solution of 20%.
3. Formaldehyde (EMS, USA) is prepared from a stock water solution of 16%.
4. Sodium azide (BDH Chemicals, UK) is prepared from a stock water solution of 1%.
5. Karnovsky's fixative. In the fume hood, pipette 12.5 mL of glutaraldehyde 20% (final concentration 1.5%), 12.5 mL of formaldehyde 16% (final concentration 2%), 1 mL of sodium azide 1% (final concentration 0.02%), and 12.5 mL of Na-Caco buffer (final concentration 0.05 M). Fill up with PBS (1X) to reach the final volume of 50 mL. This solution should be freshly made.
6. Glycine (Sigma, USA) 50 mM in Na-Caco 0.1 M. Dissolve 38 mg of glycine in 10 mL of sodium cacodylate 0.1 M. NB: the solution should be prepared fresh every time before use, shaking vigorously until the powder is dissolved.

7. Solution of osmium tetroxide (OsO_4) 4%. Dissolve 1 g of OsO_4 crystal, preweighted in sealed dark glass vials, in 25 mL of distilled water. Sonicate the solution overnight at 4 °C. The solution can be stored in a dark glass jar, double sealed with parafilm, and stored in a chemical fridge.
8. Potassium ferricyanide (KFeCN) 2%. Dilute 1 g of KFeCN in 50 mL of distilled water. Store the solution in a dark glass jar, double sealed with parafilm, and store it in a chemical fridge.
9. Solution osmium tetroxide (OsO_4) 1%: potassium ferrocyanide (KFeCN) 1% (v:v) in Na-Caco 0.05 M (OsO_4/KFeCN). Mix 2 vol. KFeCN 2%, 1 vol. Na-Caco 0.2 M, and 1 vol. OsO_4 4% to reach a final concentration of 1%/1% (v:v).
10. Epoxy resin kit (EMS, US) contains 50 g Hard-Plus Resin-812, 50 g Hardener Hard-Plus, 2.5 mL Accelerator. It is prepared by mixing equal volumes of the Hard-plus resin-812 and the hardener on a magnetic stirrer for a couple of hours.
11. Aqueous solution of uranyl acetate (UAc) 2%. Dilute 1 g of UAc (Merck, USA) in 50 mL of distilled water into a dark bottle. Stir the solution overnight at room temperature (R.T.). The day after, the solution is passed through a 0.22 μm filter and stored in the fridge.
12. 10 nm gold conjugated protein A (Aurion, Netherlands).

    NB: The size of the beads is dependent on the magnification used during image acquisition. Lower magnification requires larger beads, while higher magnification necessitates smaller beads.

    The tools needed for preparing the EM grids are indicated in Fig. 2.

3 Methods

3.1 Resin Embedding

The biopsies were taken from the liver of a patient affected by metastatic uveal melanoma. Both nontumor liver and uveal melanoma metastatic tissues were collected. After removal, the biopsies were immediately moved into Karnovsky's fixative. Incubation of the biopsies is performed at RT and in the dark, unless stated otherwise. The incubation steps on a rotor wheel were all performed at 10 rpm/min at RT (this speed applies to all further steps where specified).

The samples are immediately processed for TEM.

1. In a chemical fume hood, using a single-edged X100 razor blade (Fisher Sc., USA), reshape the samples in $1 \times 1 \times 1$ mm^3 to be consistent between different biopsies; this process also guarantees efficient penetration for the fixative

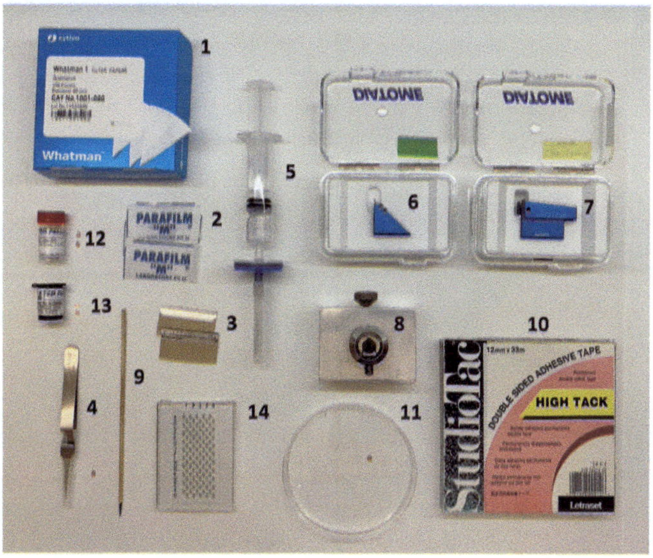

Fig. 2 Tools needed to prepare sections (1) Whatmann filter paper, (2) parafilm, (3) razor blades, (4) inverted tweezer, (5) syringe water with 0.22 μm filters, (6) Trim90 diamond knive (Diatome, Switzerland), (7) Ultra Maxi 35° diamond knive (Diatome), (8) support for ultramicrotome holder samples, (9) homemade eyelash, (10) double side tape, (11) double side tape loading station for sitting section onto slot grids, (12) 1 × 2 mm slot Cu grids, (13) 150 mesh Cu grids, and (14) grid box

and following chemicals. Once prepared, transfer the samples into 2 mL conical vials.

2. Wash the sample with 0.1 M Na-Caco two times for 5–10 min each in a rotor wheel.
3. Perform an additional wash with 50 mM glycine in 0.1 M Na-Caco for 10 min in a rotor wheel.
4. Perform two more washes with 0.1 M Na-Caco for 5–10 min each in a rotor wheel.
5. Stain the samples to fix the lipids as well as to make them electron dense, incubating the biopsies in OsO_4/KFeCN for 1 h at RT.

   NB: OsO_4/KFeCN. Prepare this mix a few minutes before using. Generally, this must be prepared during the last wash with Na-Caco 0.1 M.

   NB: Both OsO_4 and KFeCN solutions are extremely light sensitive. Handle the chemicals, keeping them protected from direct light as much as possible. The vials must be light safe (or covered by aluminum foil during the above mentioned staining).
6. Wash with distilled water, 3 times, 10 min each step in a rotor wheel.

7. Stain the biopsies in UAc 1% (diluting the UAc 2% with an equal volume of distilled water) for 30 min at R.T., in a light-shielded tube.

   NB: The biopsies stained with UAc 1% make the protein electrodense.
8. Wash with distilled water, 3 times, 10 min each step in a rotor wheel.
9. Dehydrate the samples to replace the water with solvent. Dehydrate the sample with increasing concentrations of ethanol—30%, 50%, 75%, 85%, and 100% ethanol incubation. Each incubation step lasts 10 min.
10. Incubate the biopsies in 100% acetone, followed by two more washes, 10 min each.

    All the dehydration steps are performed at R.T. in a rotor wheel.
11. Embed the sample in plastic resin using an epoxy resin kit (EMS, USA).
12. Incubate the sample in increasing concentration (25%, 50%, 75%) of complete Hard-plus Resin-812 (EMS, USA) diluted in pure acetone for 20 min in a rotor wheel.
13. Replace the acetone-resin mix with pure epoxy resin. Leave the vials containing the biopsies in a fume hood with open lids to facilitate the evaporation of any remaining traces of solvents. Each incubation with pure epoxy resin is performed 3 times for 20 min each step.
14. Finally, add 240 μL of accelerator (EMS, USA) to 10 mL epoxy resin and mix for 5–10 min in a rotor wheel.
15. Incubate the samples containing the epoxy resin-accelerator mix for 10 min in a rotor wheel.
16. Move the samples into the oven at 60 °C ON. This will let the epoxy resin solidify.
17. The day after, remove the samples from the oven and remove the plastic tubes with a scalpel.

3.2 Sectioning

The samples are then processed for sectioning using an ultramicrotome UC-6 (Leica, Austria).

1. Place the block in the support for ultramicrotome holder samples.
2. Trim the sample block to create a block face 1 × 1 mm in size. To do that, use a Trim90 diamond knife (Diatome, Switzerland) previously filled with filtered water using a syringe with a 0.22 μm filter.
3. Once the block is trimmed, cut a 70 nm thick section using Ultra Maxi 35° diamond knife to find the area of interest,

looking for intercellular spaces where the EVs could be located, and to check the quality of the preparation. The sections are loaded onto copper hexagonal 150 mesh grids.

4. As soon as the area of interest is chosen, cut 3–4 ultrathin 300 nm thick sections using an Ultra Maxi 35° diamond knife (Diatome).
5. Load the sections onto a formvar/carbon-coated copper slot grid (EMS, USA) with a hole 2 × 1 mm (Fig. 3).
6. To collect the sections on the grid, with the help of a home-made eyelash, group the sections on the water surface.
7. Use a bendt empty slot grid as a loop to pick the sections up. The practical description on how to bend the grid using a razor blade is shown in Video 1.
8. Collect the section filter after cutting using the bended grid previously prepared (Video 1). Slowly dip the grid into the water where the section is (Fig. 3a). Grab the grid from the bended part (Fig. 3b). Using tweezer, move the grids with the section on top of another grid (not bended grid previously placed on a double-sided tape sit in a 10 cm petri dish) (Fig. 3c, d). Using filter paper, let the water be sucked and flip the bended grid (Fig. 3e). Remove the bended grid (Fig. 3f) and move the slot grid (with section on top) in a grid box (Fig. 3g, h).
9. Handle the gride with inverted tweezer, each slot is submerged for 2 min into a 50 μL drop solution of 10 nm gold conjugated protein A diluted 1:55 in water on top of a cleaned parafilm (Fig. 4). This procedure allows to randomly deposit gold beads on both sides of the sections, useful as reference markers during the deconvolution of the 3D-TEM tomography (see below).

3.3 Acquisition by Transmission Electron Microscopy (TEM)

The acquisition is performed using a TEM Talos L120C (Thermo-Fisher Sc., USA) operating at 120 keV, nominal magnification of 11,500× (1.31 nm/pixel), using a CMOS 4kx4k BM-Ceta camera.

1. Load the slot grid on the single-tilt tomography holder with the section side face down and the major axis of the slot perpendicular to the holder.

   N.B. Once the holder is loaded onto the microscope, the side of the grid with the section will face the electron beam.
2. Align the optics of the microscope as normal, centering the beam, and the condenser and objective aperture around it.
3. When a region of interest has been identified, set the beam to a low magnification and high dose (low spot size) with a field of view that covers the area and its surroundings for 5 min.

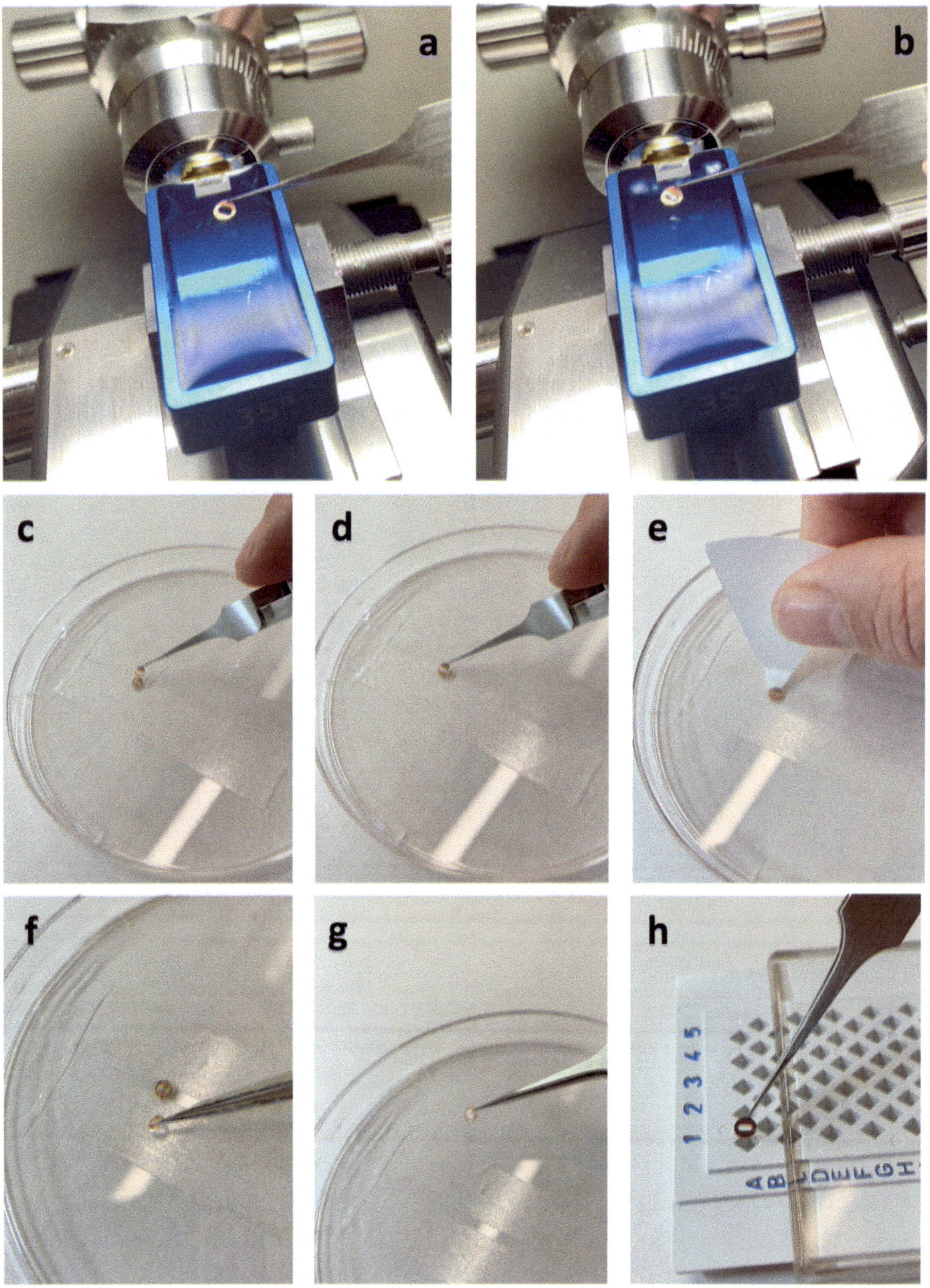

Fig. 3 Overview of the passages to collect the tissue section. Section floating onto the pool of the diamond knife (**a**) was collected holding an "empty" slot grid (without carbon-formvar layer) as a loop using an inverted tweezer (**b**). The slot carrying the section was loaded onto a carbon-formvar coated slot grid, sitting in between two layers of bi-adhesive tape (**c**). The slot with the section was left on top of it (**d**) before removing the water in between the two slots using a filter paper (**e**). This allows the transfer of the sections onto the carbon-coated slot grid. Then, the empty slot grid is removed (**f**) and the slot with the section is moved from the biadhesive tape using the inverted tweezers (**g**) and to a grid box for storage (**h**)

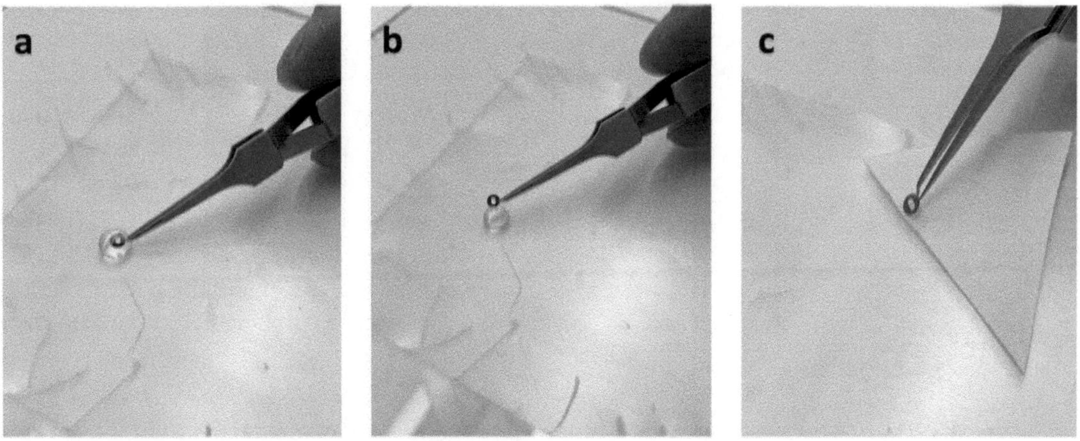

Fig. 4 Key passages to stain the section with 10 nm gold beads. Holding the slot grids with an inverted anticapillary tweezer into a 100 μL drop of protein-A 10 nm-gold conjugated, the beads are deposited on both faces of the sections (**a**). After 1 min, pull the slot out of the drop (**b**). The slot must be handled vertically above a filter paper Whatmann (**c**) to let it dry, before storing it in a grid box

NB: This overexposure, which is called "baking of the section," will thin it out a little bit and make the region more stable to the beam.

4. Decide which magnification to acquire images at. Choose the right magnification to visualize both cells and the extracellular space. It is important to set the dose toward the higher end of what the detector can handle when the sample is at 0°, as the sample will be thicker with higher degrees of tilt.

5. Perform calibrations in the *Tomography 5* software (Thermo-Fisher Sc., USA) so that the autofocus function and the image shift are calibrated. Additionally, the stage-tilt needs to be aligned using the function *Image Filter Settings.* The presets: *Thon Ring, Drift, Focus, and Tracking* can be set to the same as *Exposure*, whereas the *Search/Template* preset should be set to at least two magnification steps below the other presets (8500× if *Exposure* is set to 11,500×). For the low-resolution imaging presets, *Atlas* and *Overview/Positioning*, the magnification is not as important; set them to 155× and 380×, respectively.

   NB: Make sure to impose a small amount of under focus (−1 to −3 μm, depending on chosen magnification) for the *Exposure* preset as this will prevent any accidental overfocusing of your images.

6. When imaging the tilt series, set the acquisition of serial projection images of the selected region of interest of the 300 nm thick section, rotating 1° from −65° to +65°. Also, wait 3 s after tilting for each image acquisition to allow CompuStage to settle. Set the autofocus occurrence to every 5°. This collection will yield a 131-frame tilt series that can be processed further.

3.4 Image Analysis: From Tilt-Series to Tomogram

1. The generated 131 projection images result in a stack file tilt-series, which is then deconvolved using the *IMOD/eTomo* package software (University of Colorado, USA).
2. For extra information regarding the image tomographic processing and reconstruction via *eTomo*, visit the IMOD4.11 homepage (https://bio3d.colorado.edu/imod/doc/etomoTutorial.html) [38]. In brief, the following steps are taken:
   - Import the tilt series into *eTomo* and select the plastic section template, make sure that single tilt is selected for the data type, and that the bead size of the gold beads is correct.
   - Very few modifications to the prefilled protocol need to be made. The X-ray erasing and coarse alignment of the tilt series can be made as per the tutorial.
   - When creating a fiducial model for the fine alignments of the tilt series, two approaches are used: (1) gold bead tracking (Fig. 5a) and (2) patch tracking (Fig. 5b).
3. Gold bead tracking approach: 100 beads are randomly selected within a bounding box covering a square ~90% of the 0° tilt image. These are then tracked throughout the tilt series, filling all the gaps that the automatic filling does not find.
4. A patch-tracked model is generated only in areas where the sample is present, avoiding empty spaces.
   - Fine alignment of the tilt series based on the fiducial model is done as in the tutorial. Care is taken to track the gold beads throughout the tilt series so that it never switches target in case of crossing beads.
   - Next is to position the tomogram in the z-direction (Fig. 5c). This is done by drawing out the boundary model either manually or automatically. Choosing a tomogram thickness that will fit the tomogram and keep it in a mostly horizontal position is important. Starting at 1000 μm is recommended as the tomogram will be trimmed in later steps.
   - Create the full aligned stack according to the available tutorial (https://bio3d.colorado.edu/imod/doc/etomoTutorial.html).
   - The last step is generating the tomogram and trimming it. Based on previous steps, all values should be filled in correctly. Open the final tomogram and identify where in the z-stack the sample is first and last seen. Trim off the excess frames.

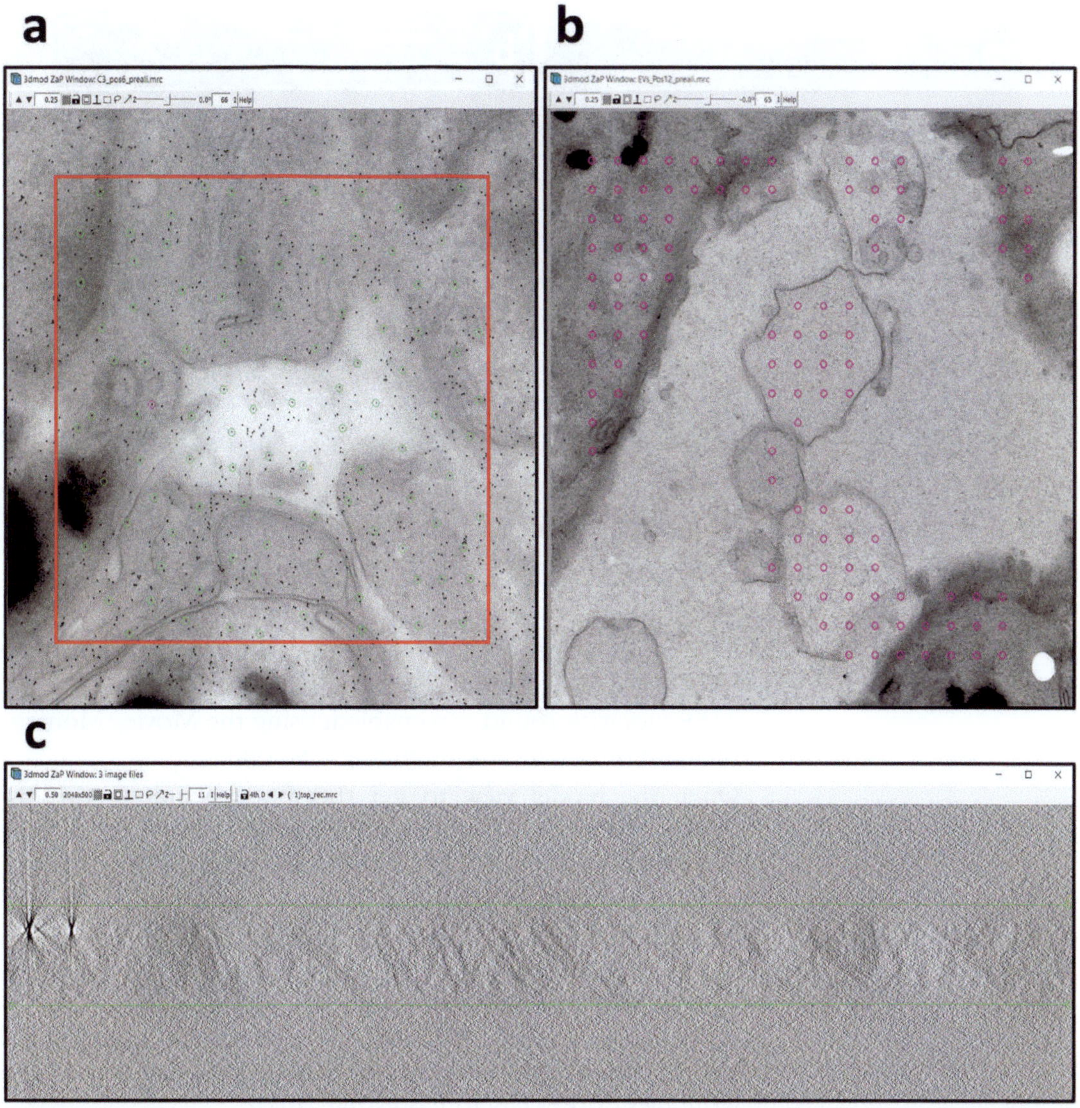

Fig. 5 Alignment of tilt series and tomogram positioning. Alignment of tilt-series done by either gold beads (**a**) or patch tracking (**b**). Tomogram positioning in the z-dimension (**c**) is done by marking out the top and bottom of the stack for all the frames. (Images are adapted from tomography reconstruction published in Olofsson Bagge [37])

3.5 3D-Modeling of Tomogram and Video Rendering

1. Modeling of the tomogram is done in *IMOD* using the sculpture tool available to shape objects in the zap window (Fig. 6a). Make sure to outline the object every 10–20 frames of the z-stack and interpolate the missing frames using linear interpolation. Enable the scale bar and set it to 500 nm.

   NB: It is recommended to increase the outline thickness of the objects from 1 to 3 so that it is easier to see the line.

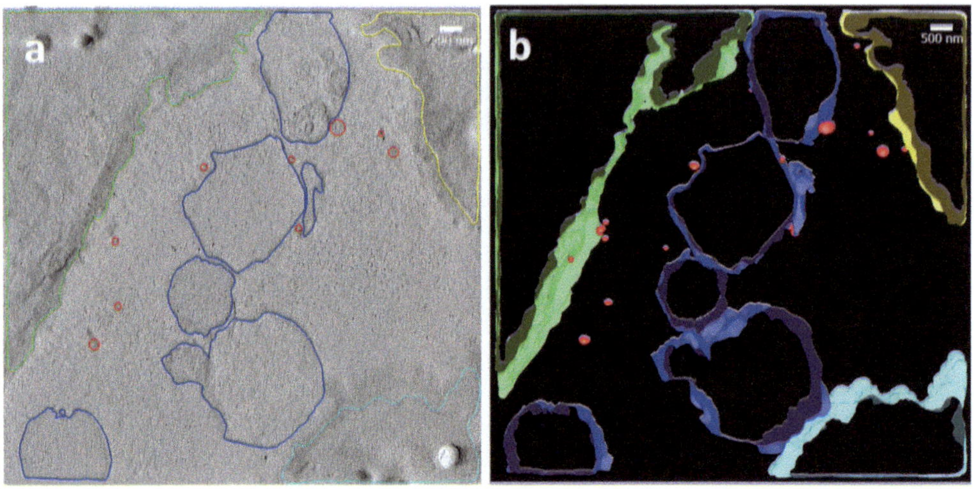

Fig. 6 IMOD modeling of tomograms. Modeling of tomograms done in IMOD using the sculpture function (**a**), and its corresponding model view (**b**) after sculpting. (Data adapted from Olofsson Bagge [37])

2. Make sure the mode is set to movie and not model, and that the scale of the Zap window is 0.25. Then export the full z-stack as TIFF files with round trip enabled, using the Movie/Montage function of *IMOD* at 25 frames per second.
3. Open the model view to get the model window with the meshed 3D model. This is made into a solid surface by meshing the objects with the "style of fill" enabled (Fig. 6b).
4. In the Movie/Montage panel of the model window, align the image with the Zap window image by setting all the rotation (X, Y, and Z) values to 0, the X, and Y translation values to −2000 (Z translation can be set to 1), and the zoom factor to 2000. Export a 360° rotation of the model by selecting full 360 Y, changing the number of frames to 360, and selecting write files (TIFFs), before pressing "make."

   NB: The scale bar needs to be enabled in the Zap window for it to appear in the model-window.
5. A 3D video of the resulting model is obtained using images taken from *IMOD* and compiled onto a timeline using *OpenShot Video Editor* (OpenShot studios, LLC, USA). For the first and last frame of the movie, choose 1 s image length, for all other choose 0.04 s [corresponding to 25 frames per second (fps)].
6. Export the video using the target MP4 (h.264), video profile HD square 25 fps, and high quality.

4 Notes

1. Single-sided sharp blades are recommended to dissect the tissue biopsy rather than using surgery blades. This will ensure a smoother cut.
2. In our experience, inverted tweezers are preferable to the classical direct tweezers, because once the grid is held, they guarantee a better grip and stability.
3. Buffers and chemicals are preferably prepared fresh just before use, from higher concentration stocks or frozen aliquots, to minimize the risk of oxidation as well as precipitates.
4. The trick to use an empty slot grid should be favored for the perfect loop (or any other kind of homemade loops) because it already gives an orientation of the section along the hole of the slot grid, making the transfer to the carbon-formvar coated grids easier.
5. The V shape of the TEM tomography holder allows for reaching higher tilt angles compared to a routine holder, providing a better locking system and a better view of the sections during the acquisition.

Acknowledgments

The authors acknowledge the Centre for Cellular Imaging (CCI) at the University of Gothenburg and the National Microscopy Infrastructure, NMI (VR-RFI2019-00217), for providing assistance in microscopy.

Conflict of Interest Rossella Crescitelli has received a research grant from Vetenskapsrådet Etableringsbidrag (Starting Grant from the Swedish Research Council) (Grant # 2023-02239), the Assar Gabrielsson's Foundation (Grant # FB23-01), the Serena Ehrenström foundation, the Ann-Lisa och Bror Björnssons Foundation, Wilhelm och Martina Lundgrens Vetenskapsfond (Grant # 2025-SA-4821 and 2023-SA-4142), and Magnus Bergvalls Foundation (2024-1009).

Rossella Crescitelli has developed multiple EV-associated patents for putative clinical utilization. Rossella Crescitelli owns equity in Exocure Bioscience Inc.

References

1. van Niel G, D'Angelo G, Raposo G (2018) Shedding light on the cell biology of extracellular vesicles. Nat Rev Mol Cell Biol 19(4): 213–228. https://doi.org/10.1038/nrm.2017.125
2. Vallejo MC, Sarkar S, Elliott EC, Henry HR, Powell SM, Diaz Ludovico I et al (2023) A proteomic meta-analysis refinement of plasma extracellular vesicles. Sci Data 10(1):837. https://doi.org/10.1038/s41597-023-02748-1
3. Lischnig A, Bergqvist M, Ochiya T, Lässer C (2022) Quantitative proteomics identifies proteins enriched in large and small extracellular vesicles. Mol Cell Proteomics 21(9):100273. https://doi.org/10.1016/j.mcpro.2022.100273
4. Cianciaruso C, Beltraminelli T, Duval F, Nassiri S, Hamelin R, Mozes A et al (2019) Molecular profiling and functional analysis of macrophage-derived tumor extracellular vesicles. Cell Rep 27(10):3062–80.e11. https://doi.org/10.1016/j.celrep.2019.05.008
5. Blandin A, Dugail I, Hilairet G, Ponnaiah M, Ghesquière V, Froger J et al (2023) Lipidomic analysis of adipose-derived extracellular vesicles reveals specific EV lipid sorting informative of the obesity metabolic state. Cell Rep 42(3): 112169. https://doi.org/10.1016/j.celrep.2023.112169
6. Puhka M, Takatalo M, Nordberg ME, Valkonen S, Nandania J, Aatonen M et al (2017) Metabolomic profiling of extracellular vesicles and alternative normalization methods reveal enriched metabolites and strategies to study prostate cancer-related changes. Theranostics 7(16):3824–3841. https://doi.org/10.7150/thno.19890
7. Hayasaka R, Tabata S, Hasebe M, Ikeda S, Hikita T, Oneyama C et al (2022) Metabolomics of small extracellular vesicles derived from isocitrate dehydrogenase 1-mutant HCT116 cells collected by semi-automated size exclusion chromatography. Front Mol Biosci 9: 1049402. https://doi.org/10.3389/fmolb.2022.1049402
8. Crescitelli R, Filges S, Karimi N, Urzì O, Alonso-Agudo T, Ståhlberg A et al (2022) Extracellular vesicle DNA from human melanoma tissues contains cancer-specific mutations. Front Cell Dev Biol 10:1028854. https://doi.org/10.3389/fcell.2022.1028854
9. Jiao Y, Gao L, Zhang T, He Z, Zheng SY, Liu W (2024) Profiling DNA cargos in single extracellular vesicles via hydrogel-based droplet digital multiple displacement amplification. Anal Chem 96(3):1293–1300. https://doi.org/10.1021/acs.analchem.3c04666
10. Valadi H, Ekström K, Bossios A, Sjöstrand M, Lee JJ, Lötvall JO (2007) Exosome-mediated transfer of mRNAs and microRNAs is a novel mechanism of genetic exchange between cells. Nat Cell Biol 9(6):654–659. https://doi.org/10.1038/ncb1596
11. Mathieu M, Martin-Jaular L, Lavieu G, Théry C (2019) Specificities of secretion and uptake of exosomes and other extracellular vesicles for cell-to-cell communication. Nat Cell Biol 21(1):9–17. https://doi.org/10.1038/s41556-018-0250-9
12. Bebelman MP, Smit MJ, Pegtel DM, Baglio SR (2018) Biogenesis and function of extracellular vesicles in cancer. Pharmacol Ther 188:1–11. https://doi.org/10.1016/j.pharmthera.2018.02.013
13. Xu L, Liang Y, Xu X, Xia J, Wen C, Zhang P et al (2021) Blood cell-derived extracellular vesicles: diagnostic biomarkers and smart delivery systems. Bioengineered 12(1):7929–7940. https://doi.org/10.1080/21655979.2021.1982320
14. Comfort N, Bloomquist TR, Shephard AP, Petty CR, Cunningham A, Hauptman M et al (2021) Isolation and characterization of extracellular vesicles in saliva of children with asthma. Extracell Vesicles Circ Nucl Acids 2: 29–48. https://doi.org/10.20517/evcna.2020.09
15. Merchant ML, Rood IM, Deegens JKJ, Klein JB (2017) Isolation and characterization of urinary extracellular vesicles: implications for biomarker discovery. Nat Rev Nephrol 13(12):731–749. https://doi.org/10.1038/nrneph.2017.148
16. Muraoka S, Jedrychowski MP, Yanamandra K, Ikezu S, Gygi SP, Ikezu T (2020) Proteomic profiling of extracellular vesicles derived from cerebrospinal fluid of Alzheimer's disease patients: a pilot study. Cells 9(9). https://doi.org/10.3390/cells9091959
17. Roca J, Rodriguez-Martinez H, Padilla L, Lucas X, Barranco I (2022) Extracellular vesicles in seminal fluid and effects on male reproduction. An overview in farm animals and pets. Anim Reprod Sci 246:106853. https://doi.org/10.1016/j.anireprosci.2021.106853

18. Admyre C, Johansson SM, Qazi KR, Filén JJ, Lahesmaa R, Norman M et al (2007) Exosomes with immune modulatory features are present in human breast milk. J Immunol 179(3):1969–1978. https://doi.org/10.4049/jimmunol.179.3.1969
19. Crescitelli R, Lässer C, Jang SC, Cvjetkovic A, Malmhäll C, Karimi N et al (2020) Subpopulations of extracellular vesicles from human metastatic melanoma tissue identified by quantitative proteomics after optimized isolation. J Extracell Vesicles 9(1):1722433. https://doi.org/10.1080/20013078.2020.1722433
20. Jang SC, Crescitelli R, Cvjetkovic A, Belgrano V, Olofsson Bagge R, Sundfeldt K et al (2019) Mitochondrial protein enriched extracellular vesicles discovered in human melanoma tissues can be detected in patient plasma. J Extracell Vesicles 8(1):1635420. https://doi.org/10.1080/20013078.2019.1635420
21. Vella LJ, Scicluna BJ, Cheng L, Bawden EG, Masters CL, Ang CS et al (2017) A rigorous method to enrich for exosomes from brain tissue. J Extracell Vesicles 6(1):1348885. https://doi.org/10.1080/20013078.2017.1348885
22. Liang W, Sagar S, Ravindran R, Najor RH, Quiles JM, Chi L et al (2023) Mitochondria are secreted in extracellular vesicles when lysosomal function is impaired. Nat Commun 14(1):5031. https://doi.org/10.1038/s41467-023-40680-5
23. Yuana Y, Böing AN, Grootemaat AE, van der Pol E, Hau CM, Cizmar P et al (2015) Handling and storage of human body fluids for analysis of extracellular vesicles. J Extracell Vesicles 4:29260. https://doi.org/10.3402/jev.v4.29260
24. Linares R, Tan S, Gounou C, Arraud N, Brisson AR (2015) High-speed centrifugation induces aggregation of extracellular vesicles. J Extracell Vesicles 4:29509. https://doi.org/10.3402/jev.v4.29509
25. Böing AN, van der Pol E, Grootemaat AE, Coumans FA, Sturk A, Nieuwland R (2014) Single-step isolation of extracellular vesicles by size-exclusion chromatography. J Extracell Vesicles:3. https://doi.org/10.3402/jev.v3.23430
26. Nordin JZ, Lee Y, Vader P, Mäger I, Johansson HJ, Heusermann W et al (2015) Ultrafiltration with size-exclusion liquid chromatography for high yield isolation of extracellular vesicles preserving intact biophysical and functional properties. Nanomedicine 11(4):879–883. https://doi.org/10.1016/j.nano.2015.01.003
27. Lozano-Ramos I, Bancu I, Oliveira-Tercero A, Armengol MP, Menezes-Neto A, Del Portillo HA et al (2015) Size-exclusion chromatography-based enrichment of extracellular vesicles from urine samples. J Extracell Vesicles 4: 27369. https://doi.org/10.3402/jev.v4.27369
28. Kurtjak M, Kereïche S, Klepac D, Križan H, Perčić M, Krušić Alić V et al (2022) Unveiling the native morphology of extracellular vesicles from human cerebrospinal fluid by atomic force and cryogenic electron microscopy. Biomedicine 10(6). https://doi.org/10.3390/biomedicines10061251
29. Rikkert LG, Nieuwland R, Terstappen L, Coumans FAW (2019) Quality of extracellular vesicle images by transmission electron microscopy is operator and protocol dependent. J Extracell Vesicles 8(1):1555419. https://doi.org/10.1080/20013078.2018.1555419
30. Crescitelli R, Lässer C, Lötvall J (2021) Isolation and characterization of extracellular vesicle subpopulations from tissues. Nat Protoc 16(3): 1548–1580. https://doi.org/10.1038/s41596-020-00466-1
31. Radon J (1986) On the determination of functions from their integral values along certain manifolds. IEEE Trans Med Imaging 5(4): 170–176. https://doi.org/10.1109/tmi.1986.4307775
32. Penczek PA (2010) Fundamentals of three-dimensional reconstruction from projections. Methods Enzymol 482:1–33. https://doi.org/10.1016/s0076-6879(10)82001-4
33. De Rosier DJ, Klug A (1968) Reconstruction of three dimensional structures from electron micrographs. Nature 217(5124):130–134. https://doi.org/10.1038/217130a0
34. Ercius P, Alaidi O, Rames MJ, Ren G (2015) Electron tomography: a three-dimensional analytic tool for hard and soft materials research. Adv Mater 27(38):5638–5663. https://doi.org/10.1002/adma.201501015
35. Beck M, Lucić V, Förster F, Baumeister W, Medalia O (2007) Snapshots of nuclear pore complexes in action captured by cryo-electron tomography. Nature 449(7162):611–615. https://doi.org/10.1038/nature06170

36. Grimm R, Singh H, Rachel R, Typke D, Zillig W, Baumeister W (1998) Electron tomography of ice-embedded prokaryotic cells. Biophys J 74(2 Pt 1):1031–1042. https://doi.org/10.1016/s0006-3495(98)74028-7
37. Olofsson Bagge R, Berndtsson J, Urzì O, Lötvall J, Micaroni M, Crescitelli R (2023) Three-dimensional reconstruction of interstitial extracellular vesicles in human liver as determined by electron tomography. J Extracell Vesicles 12(12):e12380. https://doi.org/10.1002/jev2.12380
38. Kremer JR, Mastronarde DN, McIntosh JR (1996) Computer visualization of three-dimensional image data using IMOD. J Struct Biol 116(1):71–76. https://doi.org/10.1006/jsbi.1996.0013

Chapter 15

Strategies for Loading miRNA Cargo into Extracellular Vesicles

Nico G. Menjivar, Camila Azzolin de Souza, Ahmed Gad, Luca A. Souza, Samuel Gebremedhn, Juliano C. da Silveira, and Dawit Tesfaye

Abstract

Naturally occurring lipid-enveloped nanoparticles, coined extracellular vesicles (EVs), bear exponential promise as functional therapeutic delivery vehicles. Recent studies have evidenced that the RNA cargo (particularly microRNAs; miRNAs) of EVs participates in altering recipient cell function, which has led researchers to investigate mechanisms to engineer/modify EVs miRNA cargo for mediation and functional delivery by means of regulating the gene expression of target cells. Here, we describe selected experimental techniques that have been used to investigate the dynamics and efficiency of miRNA loading into EVs and their subsequent capacity to deliver highly specialized cargo to targeted recipient cells. Notably, following the determination of identifying candidate miRNAs, we selected both "passive" and "active" loading measures to modify the basal miRNA cargo level found within the EVs. We intend to provide a comprehensive overview highlighting both the advantages and limitations of the specified methodical practices, as detailed in the outcome of a specified workflow on the generation of miRNA-modified EVs for downstream assessment using both quantitative and qualitative techniques to determine loading (EVs) and delivery (target cells) efficiencies. This chapter focuses on describing the methods and protocols for a subset of exogenous and endogenous strategies currently used for the targeted loading of miRNAs into EVs. Additionally, approaches used for determining their efficiencies and the mechanisms of targeted delivery of the proposed miRNAs into recipient cells will be discussed in detail.

Key words Extracellular vesicles, miRNAs, Cargo modifications, Delivery vehicles

1 Introduction: Decoding the Regulation of MicroRNAs into Extracellular Vesicles

The rapid advancements in RNA therapeutics have prompted pressing research in the field of nanotechnology-based targeted drug delivery, particularly lipid-based nanocarriers [1, 2]. Contrasting with the popularity of synthetic nanocarriers and liposomes, cell-

Authors Nico G. Menjivar and Camila Azzolin de Souza have equally contributed to this chapter.

Amit K. Srivastava (ed.), *Extracellular Vesicles*, Neuromethods, vol. 228, https://doi.org/10.1007/978-1-0716-4905-3_15,

derived extracellular vesicle (EV)-based carrier systems exude a more realistic and natural role in mediating cellular communication, with the structural composition of a plausible and versatile platform for RNA encapsulation and delivery [3–5]. Their intrinsic capability to interact with target cells is known as an evolutionarily conserved process [6]. Extracellular vesicles are a heterogeneous group of small, lipid-bound nanoparticles that exhibit low toxicity and immunogenicity profiles [7], which is markedly useful for their potential clinical application in vivo.

Constitutively present in both prokaryotic and eukaryotic organisms, EVs are enriched with several subclasses of noncoding RNAs, including microRNAs (miRNAs), transfer RNAs (tRNAs), ribosomal RNAs (rRNAs), mitochondrial RNAs (mtRNAs), circular RNAs (circRNAs), and long noncoding RNAs (lncRNAs) [8]. Of particular interest, miRNAs represent a class of short noncoding RNAs (approximately 18–22 nucleotides in length) [9, 10] that are stable in circulation when bound to Argonaute (AGO) proteins, high-density lipoproteins (HDLs), and/or enveloped within EVs [11–13]. Given their attraction as a credible delivery vehicle, miRNAs encapsulated within EVs have been extensively characterized in the context of different reproductive processes [14–18]. Previous reports have indicated that miRNAs are exclusively secreted within and protected by membranous EVs, given that circulating miRNAs exist in an extremely stable and RNase-resistant state, detected within various preparations of purified EVs [19, 20]. Others have independently disclosed that, aside from effectively being contained within EVs, extracellular miRNAs are predominantly associated with RNA-binding proteins (RBPs), such as AGO2, that protect extracellular miRNAs from degradation by ubiquitous RNases [21–24]. Although the mechanisms by which miRNAs are packaged into EVs are not fully understood, multiple reports indicate that it is highly dependent and selectively regulated through both protein-bound and protein-independent forms, of which, RBPs are the first class of protein molecules involved in the binding of specific RNA molecules (miRNA sequence motifs) to facilitate its sorting endeavor into EVs. Loss of function and targeted mutation studies have revealed that the packaging and release of miRNAs into and through EVs in response to specific stimuli are regulated by the interplay between specific sequence motifs amidst the miRNAs and RNA-binding proteins [25]. Although various conditions of entropy prompt such functional undertakings that orchestrate the often-characterized EV-miRNA profiles, a plethora of strategies exist to offset selective miRNA contents for the subsequent delivery of specified molecules to target cells [26].

1.1 Mechanisms to Enrich Selective MicroRNAs into Extracellular Vesicles

Loading functionally significant miRNAs into EVs in large enough quantities, yet still preserving their activity and the structural integrity of the EV-carrying vessel, remains a major challenge. Largely, EVs are modified using active strategies post-secretion [7, 27] or by genetic alterations made to the secreting cells to produce modified EVs in surplus [28, 29]. Traditionally, the passive loading of miRNAs into EVs includes a directed interaction of incubating donor cells with the target molecule, allowing cells to secrete EVs that naturally possess an abundant load of the target molecule [30]. However, this is not the only mechanism for the passive loading of miRNAs to occur. The co-incubation of EVs with a target molecule at room temperature in a conventional culture medium or phosphate buffered saline (PBS) for a fixed period of time is also a plausible approach in this regard [31]. The diffusion of miRNAs into EVs occurs along the concentration gradient, with the efficiency dependent on the hydrophobicity of the target molecules [32].

Contrastingly, methods to actively load miRNAs into EVs include electroporation, saponin treatment, sonication, hypotonic dialysis, freeze–thaw cycles, and the use of specific catalyzing reagents (Exo-Fect™). Electroporation involves the use of electrical pulses to create temporary micropores in the EV-lipid bilayer, allowing biomolecules to easily pass through the previously less transmissible barrier of protection [33]. Saponins, on the other hand, are surfactants that cause disruptions in the EV membrane upon contact with EVs, as they antagonistically interact with cholesterol in the lipid bilayer. They are typically composed of D-glucose, L-rhamnose, D-galactose, D-glucuronic acid, L-arabinose, D-xylose, or D-fucose [34]. Sonication is carried out through a device that generates high-frequency sound waves that agitate particles in a solution, inducing deformations in the EV membrane that enable the entry of targeted molecules for incorporation [35]. Hypotonic dialysis, another method of active loading, induces osmotic shock, causing EVs to swell and become porous without inducing complete lysis. Upon restoration of normal osmotic pressure gradients, pores then close, entrapping the target biomolecule within its contents [36]. The freeze–thaw method subjects EVs to thermal shock, damaging the lipid membrane structure and creating temporary pores. This process involves incubating EVs with the target molecule at room temperature, followed by freezing at −80 °C or in liquid nitrogen, and then thawing back to room temperature, repeated at least three times to ensure effective incorporation [37]. Concerning specific catalyzing reagents, *System Biosciences Co.* has recently developed a commercial kit for the incorporation of bioactive molecules using their proprietary Exo-Fect™ reagent, which has shown considerable implications with regard to cargo loading capacity and subsequent delivery [38].

Collectively, our groups use a combinatorial approach (filtration, ultracentrifugation, and size exclusion chromatography) to isolate, purify, and characterize EVs from sources including biological fluids and conditioned cell culture medium. In this chapter, we describe in detail, strategies to incorporate miRNAs into EVs with a focus on both endogenous (cellular overexpression) and exogenous measures (electroporation and Exo-Fect™) (Table 1). The schematic approach, mechanism of cellular action, and the potential biodistribution of the target miRNAs are indicated in Fig. 1. Moreover, we provide plausible and reliable mechanisms to determine cargo loading efficiencies and subsequent delivery to target cells, with the in silico application of predicting miRNA functional targets for downstream validation (Fig. 2 and Table 4).

Table 1
Different strategies used to load miRNAs into EVs

miRNA enrichment	Incorporation method	Advantages	Disadvantages	Principle description	Source
Endogenous	HiPerfect® transfection	Easily operated	Compromised cellular homeostasis; toxicity; low loading efficiency	Method incorporating a proprietary reagent (ideally suited for miRNA mimics or inhibitors) composed of a blend of cationic and neutral lipids propagating miRNA uptake and release inside of EV-secreting cells.	[39]
Exogenous	Electroporation	High loading efficiency	Impacted structural integrity; EV/cargo aggregation	Method involving applying an electric current to suspended EVs in a conductive solution, creating a provisionally porous membrane, allowing cargo to diffuse into its luminal cavity prior to the resealing of its structural binding.	[7, 40, 41]
Exogenous	Exo-Fect™	High loading efficiency	Cargo aggregation/ transfection complex bias	Method composing a two-step system involving a proprietary modified cell-penetrating peptide (CPP) (1) responsible in complexing miRNAs entry into EVs, (2) followed by the removal of free CPP.	[40, 42–46]

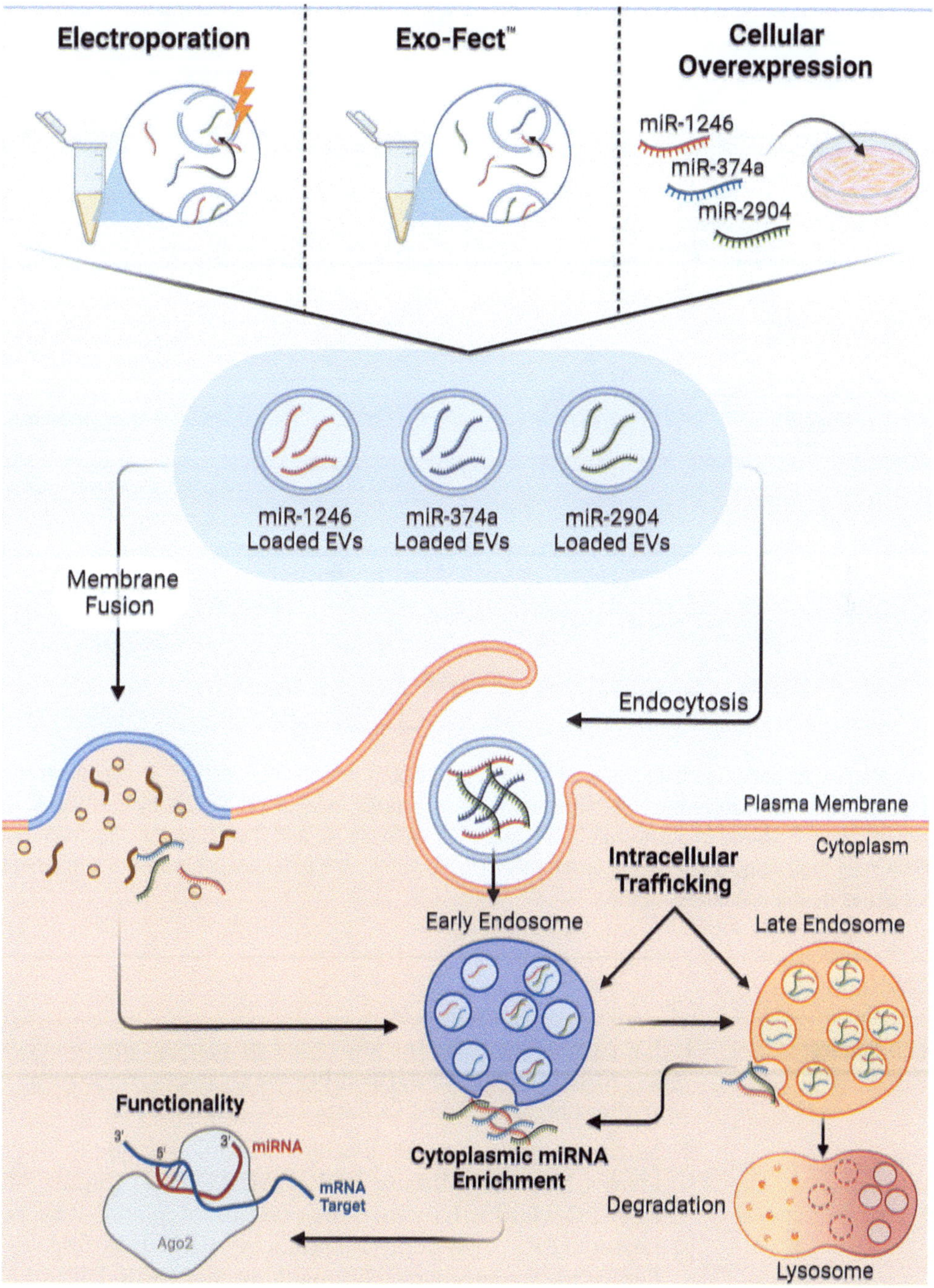

Fig. 1 Schematic illustration depicting the transfection of EVs modified using electroporation, Exo-Fect™, or overexpression of candidate microRNAs (miRNAs) into recipient cells. The miRNA-modified EVs can be delivered to recipient cells through either membrane fusion or endocytosis. Once inside the cell, these EVs may be integrated into early endosomes, facilitating the cytoplasmic enrichment of the target miRNAs, or alternatively, EVs may be incorporated into late endosomes, where part of the EVs and their cargo could undergo lysosomal degradation

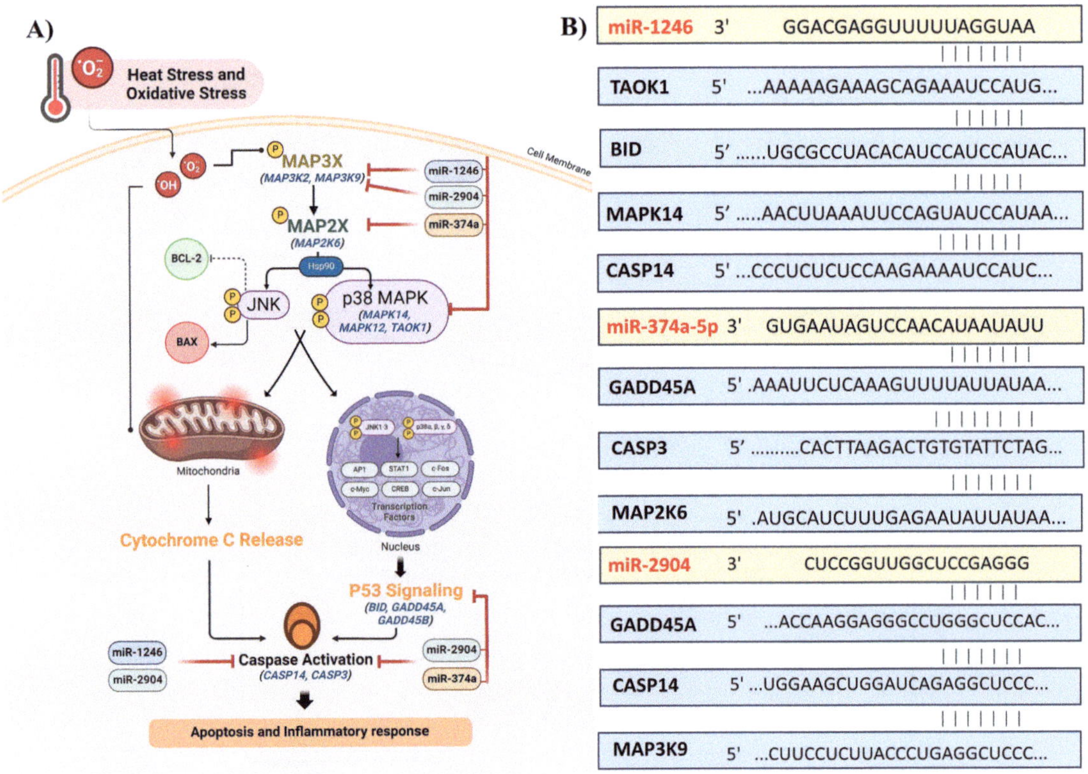

Fig. 2 Strategies to understand the role of candidate miRNAs in cellular function. A hypothetical model depicting the potential involvement of target miRNAs in various cellular processes by posttranscriptional regulation of genes involved in JUNK-p38 MAPK signaling pathway (**A**) and the sequence alignment between the miRNA seed region and 3'UTR of the target mRNA (**B**) to predict the potential posttranscriptional regulation of target genes by the candidate miRNAs

2 Materials

2.1 Cell Culture

1. EV-producing cells: Intrafollicular primary granulosa cells collected from abattoir-derived bovine ovaries (JBS Foods, USA) at the discretion of BPO Parts LLC (La Salle, CO).
2. Media composition: Dulbecco's Modified Eagle's Medium/Ham's Nutrient Mixture F12 (DMEM/F12) (Sigma-Aldrich 51445C-1000 mL) with 10% exosome-depleted fetal bovine serum (FBS) (System Biosciences EXO-FBS-50A-1), 1% Penicillin–Streptomycin (Sigma-Aldrich P4333-100 mL), and 1% Amphotericin B solution (Sigma-Aldrich A2941-20 mL) as final concentrations.

2.2 Extracellular Vesicle Isolation and Characterization

1. Refrigerated centrifuge (Eppendorf™ Centrifuge 5804 R), microcentrifuge (Labnet Prism R™ Refrigerated MicroCentrifuge C2500-R), and an ultracentrifuge (Beckman Coulter Optima XE-90 Ultracentrifuge A94471 with the SW 55Ti

rotor) with the capacity to spin at 500/3000× *g*, 17,200× *g* for 30 min, and 120,000× *g*, respectively.

2. 0.22 μm Sterile filters (Millex® SLGPR33RB).
3. Exo-spin™ mini columns (CELL guidance systems EX03-50).
4. Cold, filtered, and sterile Dulbecco's Phosphate Buffered Saline without calcium and magnesium chloride (DPBS (1X) (ThermoFisher Scientific 14190-144).
5. Transmission electron microscope: FEI/TFS Tecnai T12 Spirit TEM (FEI Company; Hillsboro, OR, USA), operating at 100 kV, with an AMT CCD.
6. Nanoparticle imaging apparatus equipped with 405/488/520/640 nm laser sensitivity: ZetaView® QUATT 4 Nanosight Instrument (Particle Metrix; Ammersee, Germany).

2.3 Cargo Preparation

1. Synthetic miRNA sequences:
   (a) hsa-miR-1246: AAUGGAUUUUUGGAGCAGG (Qiagen; miRCURY LNA miRNA Mimic).
   (b) hsa-miR-374a-5p: UUAUAAUACAACCUGAUAAGUG (Qiagen; miRCURY LNA miRNA Mimic).
   (c) Syn-bta-miR-2904: GGGAGCCUCGGUUGGCCUC (Qiagen; miScript miRNA Mimic).

   Resuspend the dried down mimic (5.0 nmol) in 75 μL of nuclease-free, sterile water to yield a final concentration of 66.7 μM.

2.4 Production of miRNA-Modified Extracellular Vesicles

1. Synthetic miRNA mimic overexpression in EV-producing cells:
   (a) HiPerFect® Transfection Reagent (Qiagen 301705).
2. Electroporation:
   (a) Gene Pulser® Electroporation Buffer (BioRad 1652677).
   (b) Gene Pulser® Cuvette (BioRad 1652086).
   (c) Gene Pulser Xcell™ Electroporation System (BioRad 1652660) with the Gene Pulser Xcell™ ShockPod™ Cuvette Chamber (BioRad 1652669).
3. Exo-Fect™ siRNA/miRNA Transfection Kit:
   (a) Exo-Fect™ siRNA/miRNA Transfection Kit (System Biosciences EXFT200A-1).
4. Exo-Fect™ Exosome Transfection Kit:
   (a) Exo-Fect™ Exosome Transfection Kit (System Biosciences EXFT20A-1).
5. RNase I (10 U/μL) (Thermo Scientific™ EN0601).

2.5 RNA Isolation

1. EVs (loading):
   (a) Exosomal RNA Isolation Kit (Norgen Biotek Corp. 58000).
   (b) Ethyl Alcohol 200 Proof (PHARMCO 111000200).
2. Cells (delivery):
   (a) miRNeasy® Mini Kit (Qiagen 1038703).
   (b) Ethyl Alcohol 200 Proof (PHARMCO 111000200).
   (c) Chloroform (Sigma C2432-1L).
   (d) RNase-Free DNase (Qiagen 79254).
3. NanoDrop™ 2000 Spectrophotometer (ThermoFisher Scientific ND-2000).

2.6 miRNA Conversion to CDNA

1. TaqMan™ MicroRNA Reverse Transcription Kit (Applied Biosystems™ 4,366,596).
   TaqMan™ MicroRNA Assays (U6 snRNA, bta-miR-1246, hsa-miR-374, bta-miR-2904).
2. VWR® PCR Tubes (VWR 732-0545).
3. Mastercycler® nexus—PCR Thermal Cycler (Eppendorf 6331000025).

2.7 miRNA Cargo Detection

1. Droplet Digital PCR (ddPCR):
   (a) QX200 Droplet Digital PCR System (BioRad) includes: QX200™ Droplet Generator (1864002), QX200™ Droplet Reader (1864003), PX1 PCR Plate Sealer (1814000), and C1000 Touch™ Thermal Cycler (1851197).
   (b) ddPCR Supermix for Probes (No dUTP) (BioRad 1863023).
   (c) DG8 Cartridges for QX100/QX200 Droplet Generator (BioRad 1864008).
   (d) Droplet Generator DG8™ Gasket (BioRad 1863009).
   (e) Droplet Generation Oil for Probes (BioRad 1863005).
   (f) ddPCR Plates 96-Well, Semi-Skirted (BioRad 12001925).
   (g) Pierceable Foil Heat Seal (BioRad 1814040).
2. Nanoflow cytometry:
   (a) CytoFLEX, Flow Cytometer (Beckman Coulter B84130).
   (b) Megamix-Plus SSC beads (BioCytex 7803).
   (c) Megamix-Plus FSC beads (BioCytex 7802).
3. Fluorescence quantification:
   (a) Paraformaldehyde (Sigma-Aldrich 158127-5G).
   (b) Glycine (Sigma-Aldrich G7126).

(c) Hoechst (Sigma-Aldrich B2261-25MG).
(d) ProLong™ Gold (Thermo Fisher Scientific P10144).
(e) Thunder Imager 3D Assay (Leica Microsystems 11889113).
(f) ImageJ software: https://imagej.net/ij/download.html

3 Methods

3.1 Establish Source of Extracellular Vesicles

1. Grow cells in EV-depleted media (*see* **Note 1**) in 24-well plate with lid (Fisher Scientific CC7682-7524).
2. Isolate EVs using a combinatorial approach adapted and described previously [17, 47]. Briefly, EV-containing conditioned media were centrifuged in sequential order at 500× *g* for 10 min, 3000× *g* for 10 min, and 17,200× *g* for 30 min and filtered through a 0.22 μm sterile filter to remove residual cellular contaminants, large particles, and protein aggregates. The resulting fluid was then subjected to two consecutive rounds of ultracentrifugation at 120,000× *g* for 70 min (*see* **Note 2**), followed by further standardization by elution through SEC columns (*see* **Note 3**). All centrifugation steps were performed at 4 °C.
3. Store at −80 °C (long term) or at 4 °C (short term) until further use.

3.2 Passive and Active Loading

1. Prepare passive loading steps:
   (a) Prepare cell culture ~24 h before transfection to an optimal density of 60–70% confluency, followed by a media change prior to transfection.
   (b) Cellular overexpression: On ice, combine 1.2 mL of F-12 w/o serum, 36 μL of HiPerFect® Transfection Reagent, and 1.08 μL (66.7 μM) synthetic miRNA mimics. Incubate for 10 min.
2. Prepare active loading steps:
   (a) Electroporation: On ice, combine 800 μL of Gene Pulser® Electroporation Buffer, 200 μL of isolated EVs, and 1.5 μL (66.7 μM) synthetic miRNA mimics.
   (b) Exo-Fect™ siRNA/miRNA Transfection Kit: According to the manufacturer's recommendation, combine the following on ice: 4 μL Transfection Reagent, 102.25 μL Transfection Buffer, and 0.75 μL (66.7 μM) synthetic miRNA mimics. Incubate at room temperature for 15 min.

(c) Exo-Fect™ Exosome Transfection Kit (EXFT20A-1): According to the manufacturer's recommendation, constitute the working reagents as follows: 1.5 μL Transfection Reagent, 20 μL Nucleic Acids (20 pMol of si/miRNA), 70 μL Sterile DPBS (1X), and 50 μL EVs suspended in DPBS (1X) (~50–300 μg of EVs). Incubate at 37 °C for 10 min.

3. Perform passive loading steps:

(a) Cellular overexpression: Distribute 51.54 μL of the combined reagents from **step 1(b)** to each well.

4. Perform active loading steps:

(a) Electroporation: Equally distribute the combined reagents from **step 2(a)** into three Gene Pulser® Cuvettes (0.2 cm gap), place them into the Gene Pulser Xcell™ ShockPod™ Cuvette Chamber, and within the Gene Pulser Xcell™ Electroporation System, run a "Time constant protocol" for each aliquoted sample: Voltage (V): 500, TC (ms): 10, Cuvette (mm): 2.

(b) Exo-Fect™ siRNA/miRNA Transfection Kit: According to the manufacturer's recommendation, add 100 μL of isolated EVs to the reaction in **step 2(b)** and incubate for 1 h at 38.5 °C. Finally, elute the mixture using the provided resin columns (*see* **Note 4**).

(c) Exo-Fect™ Exosome Transfection Kit (EXFT20A-1): Following the initial period of incubation, immediately place this solution on ice, add 30 μL ExoQuick-TC, and incubate on ice for 30 min to stop the reaction. Then, centrifuge for 3 min at 13,000–14,000 rpm, remove the supernatant, and resuspend the pellet in sterile DPBS (1X).

3.3 EV Characterization Post-Modification

After purifying the miRNA-modified EVs (*see* **Note 5**), we aimed in part to verify their presence by abiding by the detailed methods of characterization set forth by the International Society for Extracellular Vesicles [48, 49], including nanoparticle concentration/size distribution among samples and their ultrastructural morphological visualization (*see* Fig. 3).

3.3.1 NTA and TEM

1. Determine ultrastructural morphology (*see* **Note 6**), size, and relative concentration of EVs (*see* **Note 7**) (Fig. 3).

3.4 Total RNA Isolation and Reverse Transcription

1. EVs (*see* **Note 8**):

(a) Add 300 μL of Lysis Buffer A and 37.5 μL of Lysis Additive B to the 200 μL of purified exosomes.

(b) Mix well by vortexing for 10 s, then incubate at room temperature for 10 min.

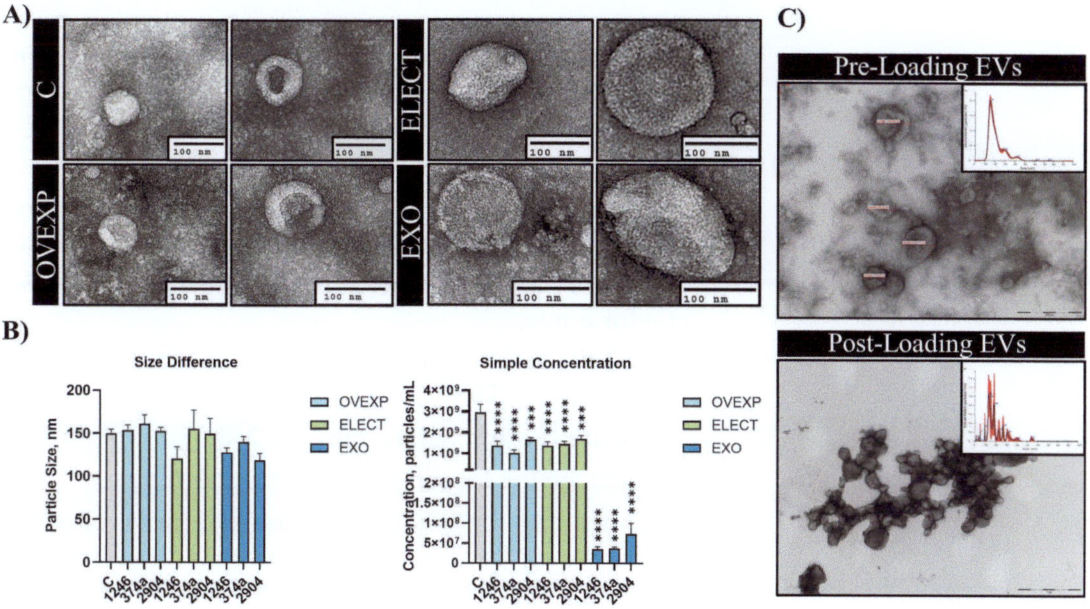

Fig. 3 Morphological and nano tracking analysis of modified EVs. Characterization of modified EVs was performed using transmission electron microscopy (TEM) (**A**) and the ZetaView NTA system (**B** and **C**). TEM images revealed an increase in EV size for those modified by electroporation and the Exo-Fect™ method compared to non-modified control EVs and those derived from overexpression of target miRNAs. However, no significant differences in average particle size were observed between the different modification methods. A significant reduction in EV concentration was detected across all three modification methods, with the Exo-Fect™ method showing the greatest reduction compared to the non-modified controls

(c) After incubation, add 500 μL of 96–100% ethanol to the mixture from **step 2** and mix well by vortexing for 10 s.

(d) Transfer 500 μL of the mixture from **step 3** into a Mini Spin column. Centrifuge for 1 min at 3300× *g* (~6000 RPM). Discard the flowthrough and reassemble the spin column with its collection tube.

(e) Repeat **step 4** one more time to transfer the remaining mixture from **step 3** into the Mini Spin Column.

(f) Apply 600 μL of Wash Solution A to the column and centrifuge for 30 s at 3300× *g* (~6000 RPM). Discard the flowthrough and reassemble the spin column with its collection tube.

(g) Repeat **step 6** one more time, for a total of two washes.

(h) Spin the column, empty, for 1 min at 13,000× *g* (~14,000 RPM). Discard the collection tube.

(i) Transfer the spin column to a fresh 1.7 mL elution tube. Apply 30–50 μL (24 μL) of Elution Solution A to the column and centrifuge for 1 min at 2000 RPM, followed by 2 min at 8000 RPM.

(j) For maximum recovery, transfer the eluted buffer back to the column and let it stand at room temperature for 2 min. Centrifuge for 1 min at 400× *g* (~2000 RPM), followed by 2 min at 5800× *g* (~8000 RPM).

2. Cells (*see* **Note 9**).
   (i) Add 700 μL QIAzol Lysis Reagent to the sample and disrupt and homogenize using an appropriate method.
   (ii) Incubate the homogenate at room temperature (15–25 °C) for 5 min.
   (iii) Add 140 μL chloroform and cap the tube securely. Shake vigorously for 15 s.
   (iv) Incubate at room temperature for 2–3 min.
   (v) Centrifuge for 15 min at 12,000× *g* at 4 °C.
   (vi) Transfer the upper aqueous phase to a new collection tube. Avoid transferring any interphase. Add 1.5 volumes (usually 525 μL) of 100% ethanol and mix thoroughly by pipetting.
   (vii) Pipette up to 700 μL of the sample, including any precipitate, into an RNeasy® Mini column in a 2 mL collection tube. Close the lid and centrifuge at ≥8000× *g* for 15 s at room temperature. Discard the flowthrough.
   (viii) Repeat **step** 7 using the remainder of the sample.
   (ix) Perform DNase digest: Pipet 80 μL of DNaseI incubation mix (10 μL of DNase +70 μL of RDD Buffer) directly onto the RNeasy® Mini column membrane and incubate at room temperature for 15 min.
   (x) Add 700 μL Buffer RWT to the RNeasy Mini column. Close the lid and centrifuge for 15 s at ≥8000× *g*. Discard the flowthrough. This step is optional if working with cultured cells.
   (xi) Pipette 500 μL Buffer RPE onto the RNeasy Mini column. Close the lid and centrifuge for 15 s at ≥8000× *g*. Discard the flowthrough.
   (xii) Add 500 μL Buffer RPE to the RNeasy Mini column. Close the lid and centrifuge for 2 min at ≥8000× *g*.
   (xiii) Optional: Place the RNeasy Mini column into a new 2 mL collection tube. Centrifuge at full speed for 1 min to further dry the membrane.
   (xiv) Transfer the RNeasy Mini column to a new 1.5 mL collection tube. Pipette 14–16 μL RNase-free water directly onto the RNeasy Mini column membrane. Close the lid, and centrifuge for 1 min at ≥8000 × *g* to elute.

Table 2
Reverse-transcription reaction components and volume per reaction

Components	Volume/reaction
10X RT buffer	0.75 μL
dNTP mix w/dTTp	0.075 μL
RNase inhibitor	0.095 μL
MultiScribe™ RT enzyme	0.5 μL
Rnase-free water	2.08 μL
Target-specific stem-loop reverse transcription primer	1.5 μL
Total RNA	2.5 μL (5 ng/μL)
Total volume	7.5 μL

(xv) If the expected RNA yield is >30 μg, repeat **step 14** using an additional 14–16 μL RNase-free water or using the eluate from **step 14** (if high RNA concentration is required). Reuse the collection tube from **step 14**.

(a) miRNA Reverse Transcription to cDNA (*see* **Note 10**),

(b) Mix and briefly centrifuge each component before use (Table 2).

(c) Combine the following in a 0.2-mL thin-walled PCR tube on ice: 10× RT buffer, dNTP mix w/dTTp, RNase inhibitor, MultiScribe™ RT enzyme, and H_2O.

(d) Aliquot 3.5 μL per tube.

(e) Add 1.5 μL of primer specific for miRNA (box contains two tubes for each miRNA, use the tube indicated with "RT").

(f) Add 2.5 μL of diluted RNA (5 ng/μL).

(g) Collect the contents of the tube by brief centrifugation, and then place on ice for 5 min before placing in a thermal cycler: 16 °C for 30 min, 42 °C for 30 min, 85 °C for 5 min, and 4 °C ∞.

3.5 Procedures to Quantify Target microRNAs Loaded into Extracellular Vesicles/Delivered microRNAs into Target Cells

1. Droplet Digital PCR (ddPCR) (*see* **Note 11 and** Figs. 4 and 5):

(a) Prepare a 20 μL reaction mixture for each sample (Table 3). Mix thoroughly by vortexing and briefly centrifuge. Allow the reaction tubes to equilibrate at room temperature for about 3 min.

(b) Load the reaction mixtures into the bottom of the sample wells of the DG8 droplet generator cartridges.

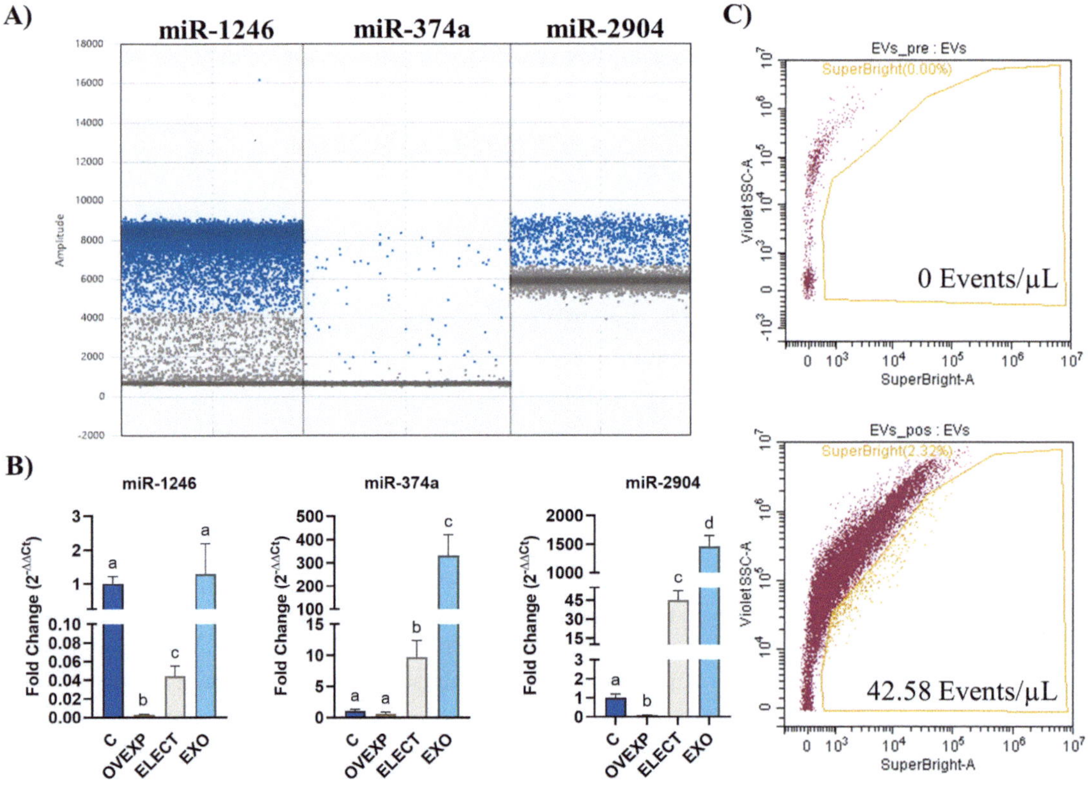

Fig. 4 Loading efficiency validation of the target miRNAs into EVs using passive (overexpression) and active (electroporation and Exo-Fect™) approaches. Droplet digital PCR results of the target miRNAs in control EVs before modification with the target miRNAs (**A**). Relative abundance of the target miRNAs in EVs modified by via overexpression (OVEXP), electroporation (ELECT), and Exo-Fect™ (EXO) compared to the unmodified control (**C**) EVs (**B**). Flow cytometry results of EVs before (EVs_pre) and after (EVs_pos) modification (**C**)

(c) Dispense 70 μL oil for probes into the oil wells of the DG8 droplet generator cartridges. Begin droplet generation within 2 min of oil loading.

(d) Subject the loaded cartridges to the droplet generator. Approximately 10,000–20,000 monodispersed droplets for each sample will be generated in the droplet wells.

(e) Slowly draw the generated droplets (~ 40 μL) into the pipet tip over ~5 s. Position the pipette tip near the bottom of the well in a 96-well PCR plate and slowly dispense the droplets. Seal the plate using a PCR Plate Sealer at 180 °C for 5 s.

(f) Proceed to thermal cycling using the following conditions: 95 °C for 10 min, 40 cycles of 94 °C for 30 s, 60 °C for 1 min, and 98 °C for 10 min.

(g) Place the plate containing the droplets into the Droplet reader and follow the instructions in the QuantaSoft

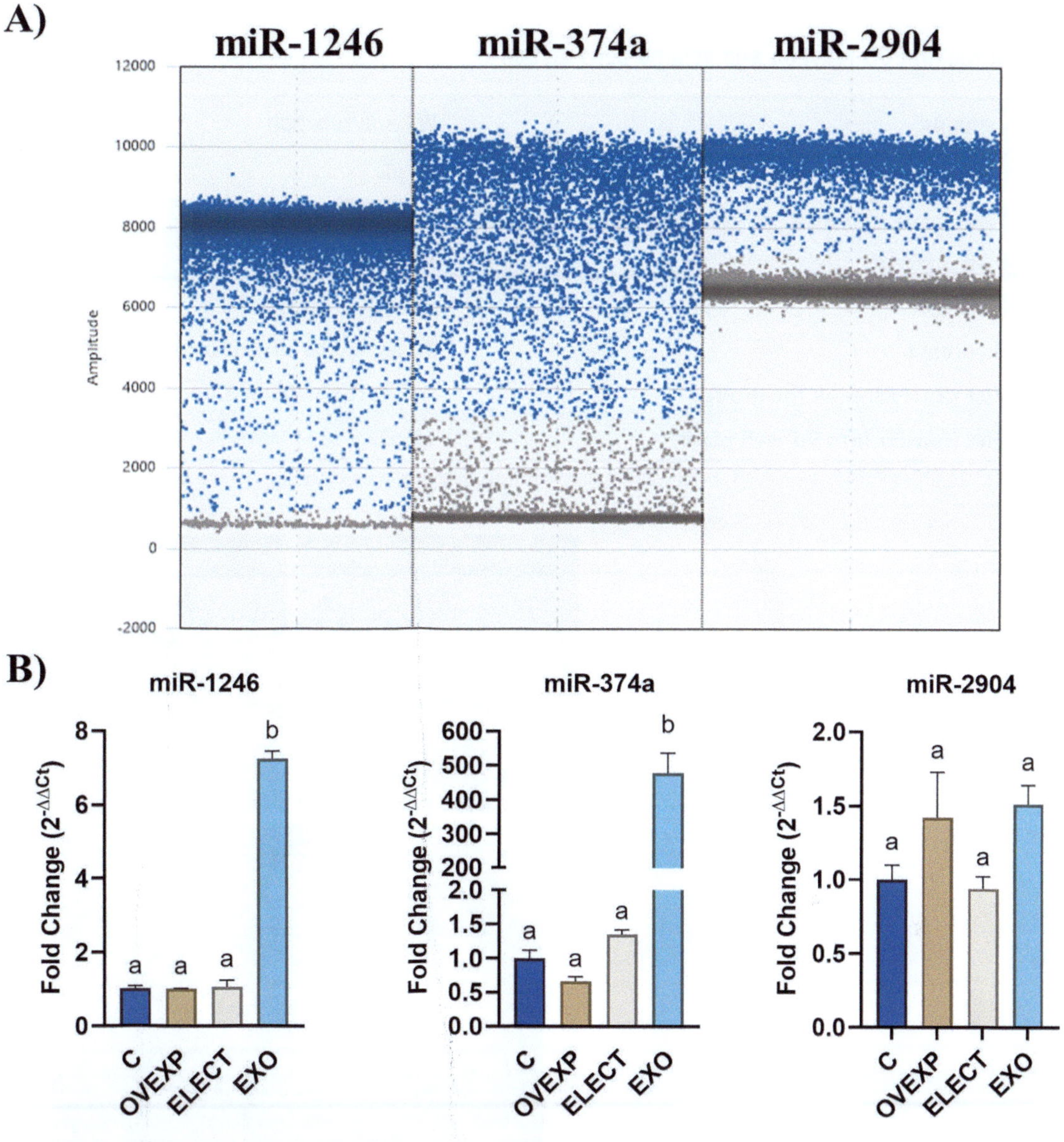

Fig. 5 Delivery efficiency among targeted recipient cells. Droplet digital PCR results indicating the enrichment of target miRNAs in recipient cells transfected with modified EVs (**A**). Relative expression of target miRNAs among recipient cells transfected with EVs modified via overexpression (OVEXP), electroporation (ELECT), and Exo-Fect™ (EXO) compared to unmodified control (labeled as C in the graph) EVs (**B**)

software to determine the absolute quantification (copy numbers) of the targets.

2. Nanoflow cytometry (Fig. 6):
   (a) Turn on the CytoFLEX/computer and proceed with the daily start-up procedure.

Table 3
Droplet digital PCR reaction and volume per reaction

Components	Volume/reaction
ddPCR Supermix for Probes (No dUTP)	10 μL
TaqMan® Assay (TaqMan® probe and PCR primer set)	1 μL
Rnase -free water	Variable
cDNA sample	Variable
Total volume	**20 μL (to be loaded into a sample well)**
Droplet generation oil for probes	**70 μL (to be loaded into an oil well)**
Droplet transfer into 96-well plate	**40 μL**

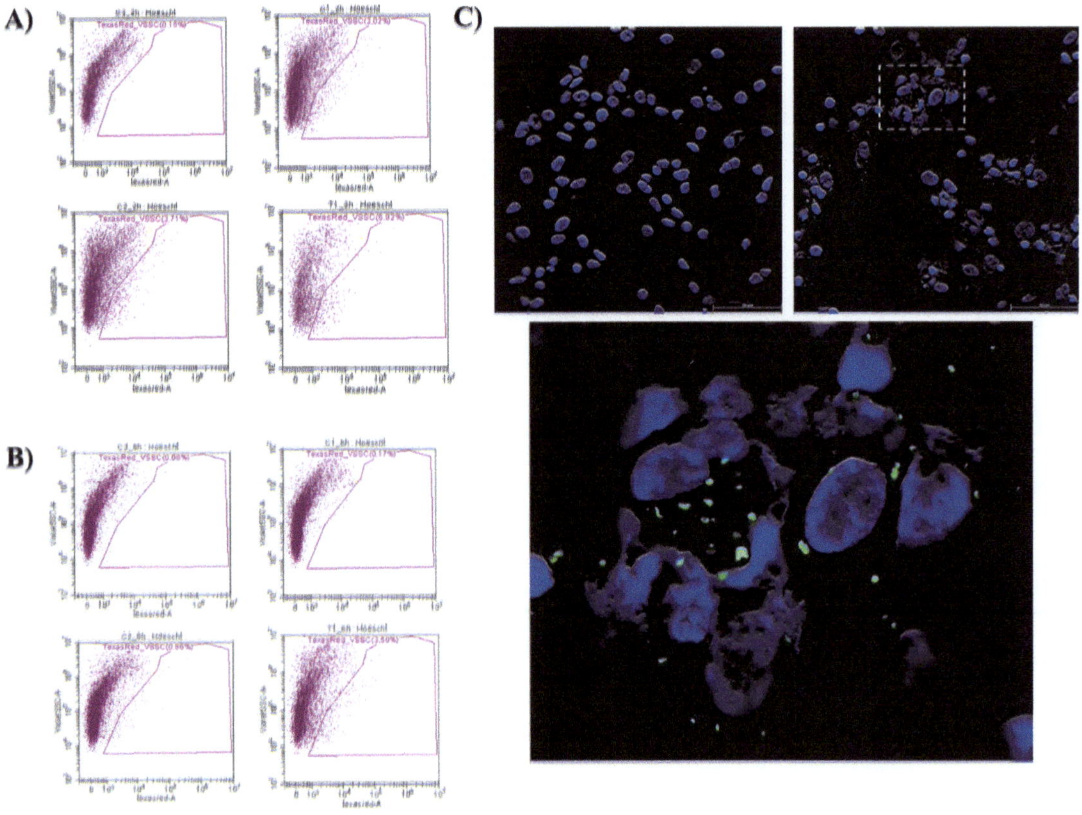

Fig. 6 Evaluation of EV-miRNA scramble delivery among granulosa cells using nanoflow cytometry and fluorescence microscopy. Increased number of AF 488-positive events in granulosa cells supplemented with EVs enriched with miRNA scramble, 2 (**A**) and 6 h (**B**) post-supplementation. Visible intracellular punctate and perinuclear structures among granulosa cells supplemented with EVs enriched with miRNA scramble following 6 h of supplementation (**C**)

Table 4
In silico tools for the prediction/validation of miRNA targets

Database	Description	Website
miRWalk	Comprehensive resource, hosting 'predicted' and 'experimentally' validated miRNA-target interaction pairs	http://mirwalk.umm.uni-heidelberg.de/
miRTarBase	Curated database of accumulated experimentally verified miRNA-target interactions	https://mirtarbase.cuhk.edu.cn/
TargetScan	Platform for the prediction of biological targets of miRNAs by searching for the presence of conserved 8mer, 7mer, and 6mer sites that match the seed region of each miRNA	https://www.targetscan.org/
miRDB	Server utilizing custom miRNA or gene target sequences for transcriptome-wide prediction of gene targets or miRNA regulators in one of the five species: human, mouse, rat, dog, or chicken	https://mirdb.org/

(b) Change the filter configuration of the Violet laser (VSSC-405 nm).

(c) Execute the QC measurement using the VSSC filter configuration.

(d) Start a new experiment and set the dot plots and histograms as follows: Dot plot 1: VSSC 405 nm, log-FL1 488 nm, log: detects the Gigamix beads. Dot plot 2: VSSC 405 nm, log-FSC 488 nm, log: determines the region for size. Dot plot 3: Time (120 s)-VSSC 405 nm, log: follow events during washing steps. Prepare additional dot plots and histograms according to your fluorescence staining needs.

(e) Adjust the gain and the threshold so that the number of events per second is ≤ 2000, and the abort rate is ≤ 5%, followed by measuring the Gigamix beads and determining the size that will be used for the experiment.

(f) Measure all samples at the flow rate *slow* (10 μL/min) and record at least 50 μL of each sample.

(g) Start by reading the negative controls and then determine the gate that will be used for the EVs loaded with fluorescence-conjugated miRNAs.

(h) Proceed with the reading of all samples, verify the number of events per μL, and compare between groups.

(i) To perform nanoflow cytometry for EV identification/incorporation, the process is very similar to the aforementioned process (step(**d**)). Instead, use nuclei markers, such as Hoechst, to determine the population of interest and the presence of the miRNA -conjugated fluorescence that was encapsulated within the EVs.

3. Fluorescence Quantification.
    (a) Following the co-incubation of fluorescence-conjugated miRNA loaded EVs, fix the cells in a 4% paraformaldehyde solution for 20 min. and subsequently treat the cells with a glycine solution (0.1 M) for 15 min. to avoid a green background noise and counterstain them with Hoechst for nuclei visualization for 15 min.
    (b) Mount the slides using ProLong ™ Gold and wait at least 48 h to allow optimal antifade and optical properties to develop.
    (c) Perform slide readings using the inverted biological microscope mentioned in Subheading 2.7 (3).
    (d) Capture ten distinct fields from each group and each biological replicate.
    (e) In the ImageJ software, download the merged image and separate the channels (Image>Color>Split Channels).
    (f) With the channels separated, use the photo that has the fluorescence channel of interest: Edit > Invert.
    (g) Adjust the threshold so that the red dots cover only the signal: Image > Adjust > Threshold > Apply.
    (h) Select the area where the signal is located: Analyze > Set Measurements > check the options: Area, Mean gray value, Standard deviation, Minimum and Maximum gray values; Analyze > Measure.
    (i) Perform normalization by dividing the mean fluorescence intensity of the channel of interest from each image by the mean fluorescence intensity of the cell nuclei channel (Hoechst). Then, calculate the average of the ten images taken for each treatment. Compare the averages between the biological replicates.

4 Notes

1. Initially, seed the granulosa cells at a rate of 2.5×10^5 (250,000) cells per well, containing 600 μL of the EV-depleted F-12 mixture, and culture at 38.5 °C in a 95% air humidity and 5% CO_2 atmospheric chamber.
2. Following the initial spin, remove the supernatant, leaving behind approximately 500 μL of the suspended EV-containing pellet. Resuspend the pellet in 3 mL of chilled DPBS (1X), and vigorously pipette up and down to "wash" the resulting suspension. Repeat the centrifugation an additional time, remove the supernatant, and proceed to SEC.

3. According to the manufacturer's protocol, initially precipitate the EV-containing pellet in a 2:1 ratio (sample: buffer) overnight at 4 °C. Centrifuge the precipitate at 16,000× *g* for 60 min. Apply the EV-containing supernatant to the column for purification, followed by 180 μL of chilled DPBS (1X) to elute the EVs.
4. Centrifuge the clean-up column at 1000× *g* for 30 s to remove the storage buffer. Apply 500 μL of column buffer on top of the resin and centrifuge for an additional 30 s at 1000× *g*. Add the entire content of the sample from **step 2(b)** to the resin in the column and mix at room temperature on a rotator for 15 min. Transfer the column to a 2 mL tube, and centrifuge at 1000× *g* for 30 s to obtain transfected (loaded) EVs.
5. Post-modification, EVs were collected and pelleted as determined previously. RNase digestion of unbound miRNAs was then performed. Add 15–20 μL of RNase I per 1 μg of RNA and incubate at 37 °C for 30 min, followed by purification through a spin column.
6. Similarly, the morphology and size of EVs were determined using a particle beam of electrons for visualization using transmission electron microscopy (TEM) (Fig. 3a). For this, a small volume of isolated EV samples volume (6–8 μL) was applied to a carbon-coated copper mesh TEM grid for 1–2 min, followed by negative staining with 2% aqueous uranyl acetate. TEM imaging was done on an FEI/TFS Tecnai T12 Spirit TEM (FEI Company; Hillsboro, OR, USA), operating at 100 kV, with an AMT CCD.
7. A ZetaView® QUATT 4 Nanosight Instrument (Particle Metrix) was used to determine the particle concentration and size distribution in all EV samples (Fig. 3b). Briefly, isolated EVs were diluted in sterile DPBS (1X), loaded into the device, and analyzed. Measurements were taken at 11 positions via scatter mode using a 488 nm laser to determine the size distribution and concentration within a sample. Analysis of EV concentrations was calculated using the Zetaview Software (version 8.05.12 SP1) before and after sample dilution.
8. Before starting, prepare a working concentration of Wash Solution A by adding 42 mL of 96–100% ethanol to the supplied 18 mL of concentrated Wash Solution A. Store all solutions at room temperature prior to use.
9. This protocol is for purifying total RNA, including small RNAs, from animal cells and tissue. If necessary, redissolve any precipitate in Buffer RWT by warming. Except for phase separation (Step 5), all steps should be performed at room temperature (15–25 °C). Work quickly. Add ethanol (96–100%) to Buffer RWT and Buffer RPE concentrates before

use (see bottle label for volume). Before starting with Step 1 for the first time, select disruption and homogenization methods according to recommendations in the miRNeasy Mini Kit Handbook.

10. The following procedure is designed after dilution of total RNA to 5 ng/μL.
11. Thaw all components to room temperature. Mix thoroughly by vortexing each tube to ensure homogeneity because a concentration gradient may form during −20 °C storage. Centrifuge briefly to collect contents at the bottom of the tubes. Prepare samples at the desired concentration before setting up the reaction mix.

5 Conclusions

Loading miRNAs into EVs for delivery purposes is an attractive approach for different biomedical applications. In this chapter, we explore techniques applied to reproductive biology. In terms of reproductive biology, the use of EVs to deliver miRNAs can improve genetic gain and fertility rates. Importantly, it is critical to understand which EVs interact with which cell type, this will help to drive the therapeutic use of EVs loaded with specific miRNAs. The reason for that is that EVs can interact differently with different cell types, possibly having different efficiencies in the miRNA delivery to these cells. Another important topic to be explored in the future is the mechanism of action of the delivered miRNAs. Where do they go? Are they recycled back into EVs? What is the minimum efficient dose? Can we use these EVs for systemic treatments? Thus, EVs loaded with miRNAs synthetically are here and can be used to manipulate cells' fate and response to environmental stressors, helping cells to overcome these challenges.

Acknowledgments

NGM is financially supported by grant funds from the USDA-NIFA-AFRI Predoctoral Fellowship, Grant number 2023-67011-40511. CAS is financially supported by grant funds from the São Paulo Research Foundation, Brazil. FAPESP #21/06645-0; #22/01433-7, #22/01505-8; #22/09461-0.

References

1. O'Brien K, Breyne K, Ughetto S, Laurent LC, Breakefield XO (2020) RNA delivery by extracellular vesicles in mammalian cells and its applications. Nat Rev Mol Cell Biol 21(10): 585–606
2. Herrmann IK, Wood MJA, Fuhrmann G (2021) Extracellular vesicles as a next-generation drug delivery platform. Nat Nanotechnol 16(7):748–759

3. Valadi H, Ekström K, Bossios A, Sjöstrand M, Lee JJ, Lötvall JO (2007) Exosome-mediated transfer of mRNAs and microRNAs is a novel mechanism of genetic exchange between cells. Nat Cell Biol 9(6):654–659
4. Sohel MMH, Hoelker M, Noferesti SS, Salilew-Wondim D, Tholen E, Looft C, Rings F, Uddin MJ, Spencer TE, Schellander K (2013) Exosomal and non-exosomal transport of extra-cellular microRNAs in follicular fluid: implications for bovine oocyte developmental competence. PLoS One 8(11):e78505
5. Elsharkasy OM, Nordin JZ, Hagey DW, de Jong OG, Schiffelers RM, Andaloussi SE, Vader P (2020) Extracellular vesicles as drug delivery systems: why and how? Adv Drug Deliv Rev 159:332–343
6. Woith E, Fuhrmann G, Melzig MF (2019) Extracellular vesicles—connecting kingdoms. Int J Mol Sci 20(22):5695
7. Alvarez-Erviti L, Seow Y, Yin H, Betts C, Lakhal S, Wood MJA (2011) Delivery of siRNA to the mouse brain by systemic injection of targeted exosomes. In Nat Biotechnol 29 (4):341–345. https://doi.org/10.1038/nbt.1807
8. Schulz-Siegmund M, Aigner A (2021) Nucleic acid delivery with extracellular vesicles. Adv Drug Deliv Rev 173:89–111
9. Friedman RC, Farh KK-H, Burge CB, Bartel DP (2009) Most mammalian mRNAs are conserved targets of microRNAs. Genome Res 19(1):92–105
10. Cortez MA, Bueso-Ramos C, Ferdin J, Lopez-Berestein G, Sood AK, Calin GA (2011) MicroRNAs in body fluids—the mix of hormones and biomarkers. Nat Rev Clin Oncol 8(8):467–477
11. Wagner J, Riwanto M, Besler C, Knau A, Fichtlscherer S, Röxe T, Zeiher AM, Landmesser U, Dimmeler S (2013) Characterization of levels and cellular transfer of circulating lipoprotein-bound microRNAs. Arterioscler Thromb Vasc Biol 33(6): 1392–1400
12. Qu L, Ding J, Chen C, Wu Z-J, Liu B, Gao Y, Chen W, Liu F, Sun W, Li X-F (2016) Exosome-transmitted lncARSR promotes sunitinib resistance in renal cancer by acting as a competing endogenous RNA. Cancer Cell 29(5):653–668
13. Groot M, Lee H (2020) Sorting mechanisms for MicroRNAs into extracellular vesicles and their associated diseases. Cells 9(4):1044
14. Gebremedhn S, Gad A, Ishak GM, Menjivar NG, Gastal MO, Feugang JM, Prochazka R, Tesfaye D, Gastal EL (2023) Dynamics of extracellular vesicle-coupled microRNAs in equine follicular fluid associated with follicle selection and ovulation. Mol Hum Reprod 29(4):gaad009
15. Gebremedhn S, Gad A, Aglan HS, Laurincik J, Prochazka R, Salilew-Wondim D, Hoelker M, Schellander K, Tesfaye D (2020) Extracellular vesicles shuttle protective messages against heat stress in bovine granulosa cells. Sci Rep 10(1): 15824
16. Gad A, Menjivar NG, Felton R, Durrant B, Tesfaye D, Ruggeri E (2024) Mapping the follicle-specific regulation of extracellular vesicle-mediated microRNA transport in the southern white rhinoceros (Ceratotherium simum simum). Biol Reprod 111(2):376–390
17. Menjivar NG, Gad A, Thompson RE, Meyers MA, Hollinshead FK, Tesfaye D (2023) Bovine oviductal organoids: a multi-omics approach to capture the cellular and extracellular molecular response of the oviduct to heat stress. BMC Genomics 24(1):646
18. Gad A, Joyce K, Menjivar NG, Heredia D, Rojas CS, Tesfaye D, Gonella-Diaza A (2023) Extracellular vesicle-microRNAs mediated response of bovine ovaries to seasonal environmental changes. J Ovarian Res 16(1):101
19. Wang K, Zhang S, Weber J, Baxter D, Galas DJ (2010) Export of microRNAs and microRNA-protective protein by mammalian cells. Nucleic Acids Res 38(20):7248–7259
20. Hoy AM, Buck AH (2012) Extracellular small RNAs: what, where, why? Biochem Soc Trans 40(4):886–890
21. Turchinovich A, Weiz L, Langheinz A, Burwinkel B (2011) Characterization of extracellular circulating microRNA. Nucleic Acids Res 39(16):7223–7233
22. Arroyo JD, Chevillet JR, Kroh EM, Ruf IK, Pritchard CC, Gibson DF, Mitchell PS, Bennett CF, Pogosova-Agadjanyan EL, Stirewalt DL (2011) Argonaute2 complexes carry a population of circulating microRNAs independent of vesicles in human plasma. Proc Natl Acad Sci 108(12):5003–5008
23. Chevillet JR, Kang Q, Ruf IK, Briggs HA, Vojtech LN, Hughes SM, Cheng HH, Arroyo JD, Meredith EK, Gallichotte EN (2014) Quantitative and stoichiometric analysis of the microRNA content of exosomes. Proc Natl Acad Sci 111(41):14888–14893
24. Albanese M, Chen Y-FA, Hüls C, Gärtner K, Tagawa T, Mejias-Perez E, Keppler OT, Göbel C, Zeidler R, Shein M (2021) MicroRNAs are minor constituents of extracellular

vesicles that are rarely delivered to target cells. PLoS Genet 17(12):e1009951

25. Garcia-Martin R, Wang G, Brandão BB, Zanotto TM, Shah S, Kumar Patel S, Schilling B, Kahn CR (2022) MicroRNA sequence codes for small extracellular vesicle release and cellular retention. Nature 601(7893):446–451
26. Menjivar NG, Oropallo J, Gebremedhn S, Souza LA, Gad A, Puttlitz CM, Tesfaye D (2024) MicroRNA nano-shuttles: engineering extracellular vesicles as a cutting-edge biotechnology platform for clinical use in therapeutics. Biol Proced Online 26(1):14
27. Ohno S-i, Takanashi M, Sudo K, Ueda S, Ishikawa A, Matsuyama N, Fujita K, Mizutani T, Ohgi T, Ochiya T (2013) Systemically injected exosomes targeted to EGFR deliver antitumor microRNA to breast cancer cells. Mol Ther 21(1):185–191
28. Kojima R, Bojar D, Rizzi G, Hamri GC-E, El-Baba MD, Saxena P, Ausländer S, Tan KR, Fussenegger M (2018) Designer exosomes produced by implanted cells intracerebrally deliver therapeutic cargo for Parkinson's disease treatment. Nat Commun 9(1):1305
29. Gunnels TF, Stranford DM, Mitrut RE, Kamat NP, Leonard JN (2022) Elucidating design principles for engineering cell-derived vesicles to inhibit SARS-CoV-2 infection. Small 18(19):2200125
30. Pascucci L, Coccè V, Bonomi A, Ami D, Ceccarelli P, Ciusani E, Viganò L, Locatelli A, Sisto F, Doglia SM (2014) Paclitaxel is incorporated by mesenchymal stromal cells and released in exosomes that inhibit in vitro tumor growth: a new approach for drug delivery. J Control Release 192:262–270
31. Lamichhane TN, Jay SM (2018) Production of extracellular vesicles loaded with therapeutic cargo. In: Targeted drug delivery methods and protocols, pp 37–47
32. Luan X, Sansanaphongpricha K, Myers I, Chen H, Yuan H, Sun D (2017) Engineering exosomes as refined biological nanoplatforms for drug delivery. Acta Pharmacol Sin 38(6): 754–763
33. Gilligan KE, Dwyer RM (2017) Engineering exosomes for cancer therapy. Int J Mol Sci 18(6):1122
34. Podolak I, Galanty A, Sobolewska D (2010) Saponins as cytotoxic agents: a review. Phytochem Rev 9:425–474
35. Lamichhane TN, Jeyaram A, Patel DB, Parajuli B, Livingston NK, Arumugasaamy N, Schardt JS, Jay SM (2016) Oncogene knockdown via active loading of small RNAs into extracellular vesicles by sonication. Cell Mol Bioeng 9:315–324
36. Millán CG, Castañeda AZ, Marinero MLS, Lanao JM (2004) Factors associated with the performance of carrier erythrocytes obtained by hypotonic dialysis. Blood Cell Mol Dis 33(2):132–140
37. Haney MJ, Klyachko NL, Zhao Y, Gupta R, Plotnikova EG, He Z, Patel T, Piroyan A, Sokolsky M, Kabanov AV (2015) Exosomes as drug delivery vehicles for Parkinson's disease therapy. J Control Release 207:18–30
38. Ding Y, Cao F, Sun H, Wang Y, Liu S, Wu Y, Cui Q, Mei W, Li F (2019) Exosomes derived from human umbilical cord mesenchymal stromal cells deliver exogenous miR-145-5p to inhibit pancreatic ductal adenocarcinoma progression. Cancer Lett 442:351–361
39. Zhai L, Shen H, Sheng Y, Guan Q (2021) ADMSC Exo-MicroRNA-22 improve neurological function and neuroinflammation in mice with Alzheimer's disease. In J Cell Mol Med 25(15):7513–7523. https://doi.org/10.1111/jcmm.16787
40. Abreu RC. de, Ramos CV, Becher C, Lino M, Jesus C; da Costa M, Paula A et al. (2021) Exogenous loading of miRNAs into small extracellular vesicles. In J Extracell Vesicles 10 (10):e12111. https://doi.org/10.1002/jev2.12111
41. Momen-Heravi, F, Bala S, Bukong T, Szabo G (2014) Exosome-mediated delivery of functionally active miRNA-155 inhibitor to macrophages. In Nanomedicine: nanotechnology, biology, and medicine 10(7), pp. 1517–1527. https://doi.org/10.1016/j.nano.2014.03.014
42. An M, K K, Park J, Ryu DR, Shin JA, Lee K, Jihee et al. (2017) Extracellular matrix-derived extracellular vesicles promote cardiomyocyte growth and electrical activity in engineered cardiac atria. In Biomaterials 146:49–59. https://doi.org/10.1016/j.biomaterials.2017.09.001
43. Aqil F, Munagala R, Jeyabalan J, Agrawal AK, Kyakulaga Al-H, Wilcher SA, Gupta RC (2019) Milk exosomes - Natural nanoparticles for siRNA delivery. In Cancer Lett 449:186–195. https://doi.org/10.1016/j.canlet.2019.02.011
44. Castaño C, Kalko S, Novials A, Párrizas M (2018) Obesity-associated exosomal miRNAs modulate glucose and lipid metabolism in mice. In Proc Natl Acad Sci USA 115 (48):12158–12163. https://doi.org/10.1073/pnas.1808855115
45. Morton MC, Neckles VN, Seluzicki CM, Holmberg JC, Feliciano DM (2018) Neonatal

Subventricular Zone Neural Stem Cells Release Extracellular Vesicles that Act as a Microglial Morphogen. In Cell Rep 23(1), pp. 78–89. https://doi.org/10.1016/j.celrep.2018.03.037

46. Zeng Z, Li Y, Pan Y, Lan X, Song F, Sun Jingbo et al (2018) Cancer-derived exosomal miR-25-3p promotes pre-metastatic niche formation by inducing vascular permeability and angiogenesis. In Nature communications 9(1), p. 5395. https://doi.org/10.1038/s41467-018-07810-w
47. Menjivar NG, Gad A, Gebremedhn S, Ghosh S, Tesfaye D (2023) Granulosa cell-derived extracellular vesicles mitigate the detrimental impact of thermal stress on bovine oocytes and embryos. Front Cell Dev Biol 11: 1142629
48. Théry C, Witwer KW, Aikawa E, Alcaraz MJ, Anderson JD, Andriantsitohaina R, Antoniou A, Arab T, Archer F, Atkin-Smith GK (2018) Minimal information for studies of extracellular vesicles 2018 (MISEV2018): a position statement of the International Society for Extracellular Vesicles and update of the MISEV2014 guidelines. J Extracell Vesicles 7(1):1535750
49. Welsh JA, Goberdhan DC, O'Driscoll L, Buzas EI, Blenkiron C, Bussolati B, Cai H, Di Vizio D, Driedonks TA, Erdbrügger U (2024) Minimal information for studies of extracellular vesicles (MISEV2023): from basic to advanced approaches. J Extracell Vesicles 13(2):e12404

Chapter 16

Bioinformatics Tools and Databases for Extracellular Vesicles and Particles Research

Sriram Gummadi, Sai V. Chitti, Suresh Mathivanan, Sanjay Shahi, and Pamali Fonseka

Abstract

Extracellular vesicles (EVs) are membrane-bound structures released by all cell types that carry diverse cargo, including proteins, lipids, nucleic acids, and metabolites. Cells also secrete protein–nucleic acid complexes that lack a membrane and are called extracellular particles (EPs). EVs play pivotal roles in intercellular and intracellular communication, for the regulation of physiological and pathological processes determined by the cargo content. The molecular content of EVs varies depending on the originating cell type and its physiological state; thus, EVs are emerging as potential biomarkers for various pathological conditions. In this regard, profiling EVs enables a deeper understanding of their biological roles. Several publicly accessible, web-based repositories—such as ExoCarta, EVpedia, and Vesiclepedia—catalogued multiple datasets related to EV cargo to aid the research community. These resources help gain insights into molecular mechanisms and pathophysiological processes associated with different disease conditions from which these EVs are isolated. This chapter discusses the current EV databases that are mostly used in EV research and bioinformatics tools available to perform integrated analyses for the identification of key functional components and gain a comprehensive understanding of the roles of EVs in health and disease.

Key words Extracellular vesicles, Extracellular particles, Exosomes, Microvesicles, EV databases

1 Introduction

Extracellular vesicles (EVs) are phospholipid bilayer structures secreted by all cell types into the extracellular environment [1–3]. Likewise, cells also secrete small protein-nucleic-acid complexes that are not membrane bound and are termed as extracellular particles (EPs) [2]. EVs and EPs are increasingly recognized for their critical roles in intercellular and intracellular communication [4, 5]. EVs carry a diverse range of biomolecules, including proteins, lipids, and nucleic acids [6]. Based on their size, biogenesis, and release mechanisms, EVs are broadly classified into exosomes, shedding microvesicles (also known as ectosomes), apoptotic bodies, migrasomes, exophers, and large oncosomes (Fig. 1) [1–3,

Amit K. Srivastava (ed.), *Extracellular Vesicles*, Neuromethods, vol. 228, https://doi.org/10.1007/978-1-0716-4905-3_16,

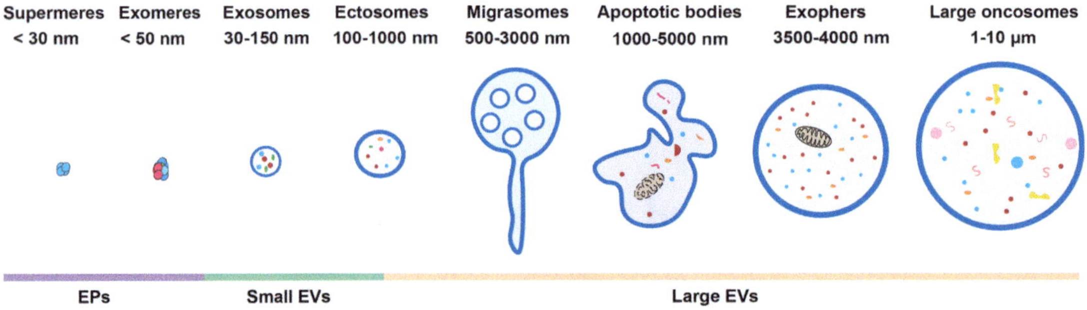

Fig. 1 Subtypes of extracellular vesicles (EVs) and extracellular particles (EPs). Based on their operational terms and size, EVs and particles are generally classified into large EVs, small EVs, and EPs

7]. Exosomes, typically ranging from 30 to 150 nm in diameter, originate from inward budding of the endosomal membrane, forming multivesicular bodies (MVBs). These MVBs subsequently fuse with the plasma membrane, releasing exosomes into the extracellular space [6–9]. Microvesicles (100–1000 nm) are shed directly from the plasma membrane through a process involving outward budding [1, 6, 8, 10]. Apoptotic bodies (up to 5 μm) are released during programmed cell death [1, 6]. Exophers are a type of large EVs ranging from 3.5 to 4 μm, are released through plasma membrane budding and contain protein aggregates [11]. Large oncosomes, ranging from 1 to 10 μm, are exclusively shed by cancer cells [12]. In 2015, Ma et al. discovered the novel class of EVs known as migrasomes (500–3000 nm), whose formation depends on cell migration (mostly in immune cells and metastatic cancer cells) [13]. EVs are pivotal in various physiological and pathological processes, including immune response, tissue repair, and tumor progression. They facilitate cell-to-cell communication by transferring biomolecules that can significantly influence the behavior of recipient cells [1, 6, 7]. In pathological conditions such as cancer, EVs derived from tumors are known to contribute to all hallmarks of cancer, including but not limited to immune evasion, proliferation, resisting cell death, modulation of the extracellular matrix, and vasculature contributing to metastasis and cachexia [14, 15]. For instance, EVs can promote tumorigenesis by transferring oncogenic factors to neighboring cells, thereby creating a supportive microenvironment for tumor growth [16, 17]. Additionally, EVs have been implicated in neurodegenerative diseases [18], cardiovascular diseases [19, 20], and microbial and viral infections [21–24].

The potential of EVs as diagnostic biomarkers and therapeutic agents has acquired significant attention in recent years [25, 26]. Since they are found in bodily fluids such as blood, urine, and saliva, they can be accessed for noninvasive diagnostic applications [27]. Specific EVs have been identified as a source of biomarkers for various cancers, aiding in early detection and monitoring of disease progression [25]. For instance, CA19-9 is a biomarker for cancer-associated venous thromboembolism in pancreatic cancer and has been detected in EVs [28]. Furthermore, the ability to engineer EVs for targeted drug delivery presents a promising avenue for therapeutic interventions, particularly in oncology and regenerative medicine [18, 25, 27]. However, potential to utilize EVs as targets or drug delivery vehicles for management of diseases requires cautious and extensive preliminary studies as well as rigorous clinical trials [29–30].

2 EV Cargo

EVs carry a variety of biomolecules, often referred to as cargo, including proteins, lipids, RNAs such as messenger RNA (mRNA), micro RNA (miRNA), long noncoding RNA (lncRNA), and DNA, which reflect the functional state of the parent cell [6, 32].

3 Proteins

The protein composition of EVs is diverse and reflects the physiological state of the parent cell [33–35]. Proteins associated with EVs can be broadly categorized into membrane proteins, cytosolic proteins, and proteins involved in vesicle trafficking [36, 37]. Membrane proteins are crucial for the structural integrity and functionality of EVs [38]. Tetraspanins, such as CD63, CD81, and CD9, are the most common membrane proteins found in exosomes (small EVs). These proteins play vital roles in facilitating cell–cell interactions, mediating signaling pathways which are essential for various biological functions, including cell adhesion, migration, and fusion with target cells [38, 39].

Cytosolic proteins can also be incorporated into EVs through selective packaging mechanisms [9, 32, 40]. Heat shock proteins (HSPs), which are involved in protein folding and cellular stress responses, are frequently found in EVs [41]. Additionally, enzymes such as protein kinases and phosphatases can be present in EVs, influencing signaling pathways in recipient cells [39]. Proteins associated with endosomal origins, such as Alix and Tsg101, engage in the biogenesis and sorting of exosomal cargo, playing a pivotal role in ensuring the proper packaging of proteins and lipids into the vesicles [9, 42]. The selective incorporation of specific proteins into EVs is critical for their functionality, allowing them to exert distinct biological effects on recipient cells [42]. The study of the protein composition of EVs is vital for understanding their natural properties as well as their roles in intercellular communication, disease progression, and potential therapeutic applications [6, 31, 37, 42, 43].

4 Nucleic Acids

Extracellular vesicles play a crucial role in the transfer of nucleic acids, including mRNA, miRNA, and DNA [36, 37]. The presence of these nucleic acids in EVs has significant implications for intercellular communication and regulation of gene expression [37]. mRNA found in EVs can be transferred to recipient cells, horizontal gene transfer, where it can be translated into proteins, influencing the behavior and function of these cells [37]. The mRNA content of EVs reflects the transcriptional state of the parent cell, making it a valuable tool for studying cellular responses to various stimuli and pathological conditions [27, 37].

Similar to mRNAs, miRNAs can be selectively packaged into EVs through specific sorting mechanisms, often reflecting the physiological state of the donor cell [44]. miRNAs are a class of small noncoding RNAs that play crucial roles in posttranscriptional regulation of gene expression [45]. Thus, EVs-mediated transfer of these cargo components allows for the modulation of gene expression in recipient cells, thereby impacting various biological processes such as proliferation, differentiation, and apoptosis [46]. The transfer of miRNAs via EVs has been implicated in numerous diseases, including cancer, where tumor-derived exosomal miRNAs can influence the tumor microenvironment and promote metastasis [44, 45, 47].

While the concept of EVs or exosomes as carrier of DNA remained controversial for a significant amount of time, several studies have validated the fact that DNA is indeed detected in EVs [2, 48–50]. The functional implications of DNA transfer via EVs are still being explored, but it suggests a novel mechanism for genetic regulation and cellular communication [2, 48]. The nucleic acid content of EVs offers a wealth of information about the cellular environment and can serve as biomarkers for various diseases [51]. Analyzing the nucleic acids carried by EVs via bioinformatics tools provides insights into disease mechanisms and potential therapeutic targets, making them a focal point in current biomedical research.

5 Lipids

Lipids are one of the abundant components of the cell membrane, and as EVs are membrane bound vesicles, lipids form a significant proportion of the EV structure and cargo [6, 52]. The lipid bilayer of EVs consists primarily of phospholipids, cholesterol, and sphingolipids, each contributing to the overall properties of the vesicles [53]. Common phospholipids found in EVs include phosphatidylcholine, phosphatidylserine, and phosphatidylethanolamine [54, 55]. The presence of phosphatidylserine is particularly noteworthy; it is usually located on the inner leaflet of the plasma membrane but translocates to the outer leaflet during apoptosis and microvesicle formation, serving as an "eat me" signal for phagocytic cells [52]. This translocation plays a crucial role in mediating immune responses and facilitating the clearance of apoptotic cells [53, 54].

Cholesterol is essential for maintaining the structural integrity of EVs and influences their interactions with other cells [56]. Cholesterol-rich regions in the membrane, known as lipid rafts, serve as platforms for signaling molecules, enhancing the vesicles' ability to mediate intercellular communication [56]. Sphingolipids are involved in the formation of lipid rafts and can modulate membrane fluidity, impacting the sorting and release of EVs [54]. Similar to other cargo components, the specific lipid composition of EVs can vary depending on the cell type and physiological conditions of the parent cell [53]. Understanding the lipid composition of EVs is crucial for deciphering their roles in biological processes and their potential as therapeutic vehicles [53, 54]. Furthermore, literature suggests the ability to manipulate lipid composition may enhance the efficacy of EV-based therapies, allowing for targeted delivery of drugs and genetic materials [54].

6 Metabolites

Metabolites in EVs include a wide variety of small molecules like ATP, cAMP, and reactive oxygen species (ROS), that reflect the metabolic state of the parent cell [57, 58]. EVs can transfer metabolites to recipient cells, influencing processes such as energy metabolism, cellular signaling, immune responses, and even disease progression. For example, fatty acids and amino acids carried in EVs can provide energy to neighboring cells, while signaling molecules like eicosanoids can modulate inflammation [59]. The presence of metabolites in EVs makes them valuable as carriers of metabolic signals for diseases like cancers, neurodegenerative conditions, and microbial infections as tuberculosis [59, 60].

7 Databases for EV Research

To further benefit the scientific community worldwide, specialized databases have been developed to store, curate, and analyze EV-related data. These databases serve as valuable resources for researchers by providing organized access to various aspects of EV biology, including molecular composition, functional characteristics, EV-associated diseases, and experimental protocols.

- *ExoCarta* (http://www.exocarta.org)

  ExoCarta is a web-based and freely accessible database that hosts exosome specific proteins, RNAs, and lipid contents from 10 different species. It is manually curated and contains both published and unpublished exosomal studies. It hosts protein–protein interaction networks and biological pathways of exosomal proteins [61]. With the association of scientific community annotations, this database will develop into a comprehensive resource featuring high-quality exosomal research.
- *Vesiclepedia* (http://www.microvesicles.org/)

  Vesiclepedia is a free web-based compendium of DNA, RNA, proteins, lipids, and metabolites that are associated with EVs and EPs [62]. The website is manually curated and allows users to query a biomolecule of interest to retrieve the experimental details of EV studies, such as sample source, identification and isolation methods, and the subclass of the identified EV [62, 63]. The latest Vesiclepedia is updated as per the MISEV 2024 guidelines [1, 62]. In this update, the EV-QUANT feature was added, which allows for relative quantification between EV protein or RNA, or lipid cargo samples within one specific study. This feature helps EV researchers to identify what biomolecule is identified in which experiment and in samples, and can compare between conditions in real time [62]. This compendium is extremely useful for understanding the diversity of EV cargo and the cellular pathways involved in EV biogenesis, secretion, and function.
- *EVpedia* (http://evpedia.info)

  EVpedia is an integrated database that catalogues the proteome, transcriptome, and lipidome of EVs derived from archaea, bacteria, and eukaryotes, including humans [64]. It offers a variety of analytical tools for researchers, such as search and browse functionalities for vesicular proteins, comparison of datasets through ortholog identification, gene ontology enrichment analyses, and network analyses of vesicular proteins [64, 65]. This platform serves as a vital community resource aimed at advancing systematic studies of EVs, thereby enhancing research in the field of EVs.
- *EV-TRACK* (http://evtrack.org)

  EV-TRACK is a community-driven, open-access database designed to standardize the reporting of EV-related studies. It includes data about the methodology, sample preparation, isolation, characterization, and use of EVs in scientific studies [66]. EV-TRACK allows researchers to find studies on EVs based on experimental design, and it emphasizes the importance of reproducibility and transparency in EV research.

- *EV-ADD* (www.evdnadatabase.com)

  EV-associated DNA database (EV-ADD) is a manually curated EV-DNA data derived from human biofluids, such as liquid biopsies, as well as in vitro studies. Users can easily query the database to find information on various aspects, including EV isolation methods and characterization techniques, and protocols related to EV-DNA isolation [48]. EV-ADD aims to advance EV-DNA research forward and serves as a vital resource for researchers in the field of EVs.

- *EV-COMM* (http://sdc.iue.ac.cn/evs/list/)

  EV-COMM (EV-mediated communication) is a database designed to facilitate the study of intercellular and interspecies interactions mediated by EVs. This innovative resource compiles detailed records of interactions between donor and recipient cells or species, categorized by cell type, and tissue or organ [67]. EV-COMM provides users with a web interface that allows for easy browsing, searching, visualization, and downloading of interaction records. Researchers can filter interactions based on specific contexts, cell types, species, or functional cargoes. This resource aims to enhance understanding of EV-mediated communication in physiological and pathological conditions.

8 Software Tools for Analysis

EVs carry a complex cargo that includes proteins, RNA, lipids, and other biomolecules, making them valuable sources of information for understanding cellular communication and various physiological processes [6, 7, 37, 68, 69]. Given the complexity and richness of EV cargo, high-throughput analysis methods are essential for deciphering the functions of EVs (Fig. 2). To gain insights into the precise roles of EVs, most studies, whether hypothesis-driven or data-driven, rely on comprehensive global analyses. A variety of software tools, both free-to-use and commercial, are currently available for analyzing EV-based omics datasets (Table 1). These tools commonly employ a strategy that involves mapping the provided gene or protein list to custom biological databases. The outcome of this analysis is a set of statistically enriched biological annotations, which allow researchers to identify key biological pathways and processes that govern the function of EVs. These enrichment analyses are critical for elucidating the molecular mechanisms underlying the biological roles of EVs.

- *DAVID (Database for Annotation, Visualization, and Integrated Discovery)*

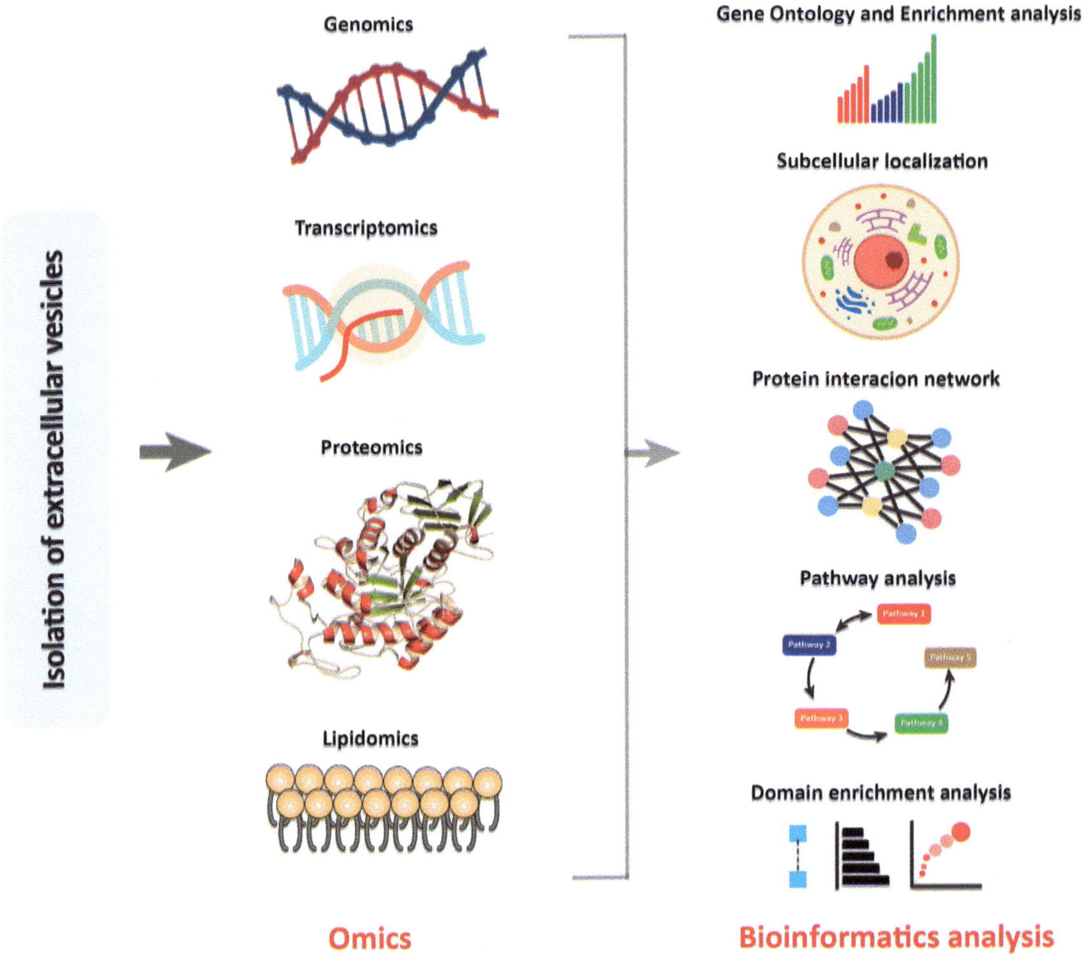

Fig. 2 A flowchart illustrating the integrated bioinformatics analysis that can be carried out to identify vital components in EV dataset analysis. Isolated EVs can be analyzed to profile their genomic, transcriptomic, proteomic, and lipidomic compositions. Various integrated analyses can be performed on the resulting datasets, including gene ontology (GO) enrichment for biological processes and molecular functions, protein interaction network mapping, pathway analysis, domain enrichment, and subcellular localization studies

DAVID is a web-based tool widely used for functional annotation and enrichment analysis of gene lists. It provides users with a variety of databases for gene ontology (GO), pathways (e.g., KEGG, Reactome), and protein domains [71]. In the context of EV research, DAVID is useful for identifying significant biological processes and pathways such as exosome formation, vesicle-mediated transport, and other extracellular activities. It offers a comprehensive, user-friendly interface for gene set analysis, and its broad database of gene annotations makes it a go-to tool for discovering the functional implications of EVs.

- *Cytoscape*

Table 1
Online tools for functional enrichment analysis

Tool name	URL	Analysis	Platform
DAVID [70]	https://david.ncifcrf.gov/	GO, pathway enrichment, functional annotation, protein domain analysis	Web
Enrichr [71]	https://maayanlab.cloud/Enrichr/	GO, KEGG, DisGeNET, drug-targets, disease-gene associations	Web
GSEA [72]	https://www.gsea-msigdb.org/gsea/index.jsp	Gene set enrichment analysis, MSigDB gene sets	Web, R, Python
Cytoscape [73]	https://cytoscape.org/	Pathway analysis, protein interaction networks, custom enrichment, network analysis	Web, Desktop (with plugins)
FunRich [74, 75]	http://www.funrich.org/	GO, KEGG, Reactome, Reactome pathway analysis	Windows (Standalone)
IPA®	http://www.ingenuity.com/products/ipa	Pathway analysis, molecular networks, gene expression analysis, regulatory networks, disease pathway	Web (Paid)
MetaCor™	https://www.clarivate.com/products/metacore/	Pathway analysis, network analysis, functional annotation, disease pathways, biomarker discovery	Web (Paid)

Cytoscape is an open-source software platform widely used for visualizing, analyzing, and integrating complex biological networks. It excels at representing interactions such as protein–protein interactions, gene regulatory networks, and metabolic pathways. With its extensive collection of plugins, Cytoscape enhances network visualization, analysis, and data integration [73]. This platform enables users to upload biological interaction data in multiple formats and visualize intricate networks of molecular interactions. Researchers can explore relationships between EV cargo proteins, identify network clusters, and perform enrichment analyses using powerful plugins such as BiNGO and EnrichmentMap [76, 77]. These capabilities make Cytoscape an indispensable tool for studying the molecular networks.

- *FunRich (Functional Enrichment Analysis Tool)*

FunRich is a standalone software tool designed for functional enrichment and pathway analysis of EV data, providing users with the ability to perform enrichment analyses using various background databases while allowing for updates and modifications [74]. Users can easily download and refresh the background database from UniProt [78]. Recent updates include miRNA enrichment analysis, a plugin for analyzing EV datasets from the Vesiclepedia database, oncogene comparison using the COSMIC database, ID conversion tools, and plot customizable graphs of publication quality [74, 75]. These features collectively make FunRich an invaluable resource for researchers in the EV field.

- *Enrichr*

  Enrichr is an interactive, user-friendly web-based tool designed for gene list enrichment analysis, widely utilized in bioinformatics for interpreting large genomic datasets [71, 79]. It provides access to a vast collection of over 400,000 annotated gene sets from approximately 300 libraries, encompassing various biological processes, diseases, and drug interactions [71]. A more advanced version, Enrichr-KG, integrates selected gene set libraries into a knowledge graph database, facilitating comprehensive analysis across multiple libraries and enhancing the visualization of relationships between genes and enriched terms [80]. This graphical representation allows users to explore hidden associations between genes and annotated enriched terms from various datasets. Additionally, Enrichr offers convenient result visualizations, making it an excellent resource for gaining rapid insights into biological data.

- *GSEA (Gene Set Enrichment Analysis)*

  Gene set enrichment analysis (GSEA) is a computational method used to determine whether a predefined set of genes shows statistically significant differences in expression levels between two biological states [72]. It is particularly effective when applied to RNA-Seq or microarray data, as it identifies subtle but consistent changes in pathways and gene sets that may be overlooked when analyzing individual genes [77]. The analysis involves ranking all genes based on their expression levels and then calculating an enrichment score (ES) for each gene set, which reflects the degree to which the genes in that set are concentrated at the extremes of the ranked list. GSEA utilizes predefined gene sets from databases like the molecular signatures database (MSigDB), allowing researchers to leverage curated pathways and functional annotations for their analyses [81]. Its robust statistical framework, which includes permutation testing for significance assessment, and its ability to detect subtle changes make it a valuable tool in bioinformatics.

- *Ingenuity Pathway Analysis® (IPA) and MetaCore™*

  IPA® and MetaCore™ are powerful commercial software platforms designed to analyze large-scale OMICS datasets, such as genomics, transcriptomics, proteomics, and metabolomics, by identifying key biological pathways, molecular interactions, and gene networks. IPA®, developed by QIAGEN, offers a broad range of analysis capabilities and integrates extensive public biological data from various sources, such as gene expression profiles and disease associations, to provide deep functional insights. MetaCore™, developed by Clarivate, focuses on systems biology and functional genomics, emphasizing the construction and interpretation of biological networks within the context of disease mechanisms and therapeutic interventions. Both platforms rely on curated datasets enriched by manual curation of the scientific literature, leveraging publicly available databases to ensure comprehensive, up-to-date data integration. While both tools offer significant advantages in terms of data analysis and network construction, they also have certain limitations, such as the reliance on curated content, which can introduce bias or limit the scope of emerging biological findings [68, 69, 82].

9 Conclusion

The rapid advancements in high-throughput experiments have significantly accelerated our ability to study complex biological processes underlying both physiological and pathological conditions [83]. In this context, bioinformatics analyses play a crucial role in interpreting the vast amounts of omics data and identifying key functional factors that drive these processes. Bioinformatics tools are constantly evolving, and having a deep understanding of how these tools, along with the underlying databases, are created is crucial for making informed interpretations and avoiding potential pitfalls. One of the major challenges in the EV field remains identifying the cellular origins of EVs in complex biological samples such as blood, urine, and plasma due to their heterogeneous nature and the complexity of the sources. Addressing this challenge by developing a robust bioinformatics tool to predict the cellular origin of EVs would greatly enhance our understanding of EV biology and accelerate the translation of EV research into clinical applications, unlocking their full potential in diagnostics and therapeutic applications.

Conflict of Interest

The authors have no conflicts of interest to declare that are relevant to the content of this chapter.

References

1. Welsh JA, Goberdhan DCI, O'Driscoll L, Buzas EI, Blenkiron C, Bussolati B, Cai H, Di Vizio D, Driedonks TAP, Erdbrügger U, Falcon-Perez JM, Fu QL, Hill AF, Lenassi M, Lim SK, Mahoney MG, Mohanty S, Möller A, Nieuwland R, Ochiya T, Sahoo S, Torrecilhas

AC, Zheng L, Zijlstra A, Abuelreich S, Bagabas R, Bergese P, Bridges EM, Brucale M, Burger D, Carney RP, Cocucci E, Crescitelli R, Hanser E, Harris AL, Haughey NJ, Hendrix A, Ivanov AR, Jovanovic-Talisman T, Kruh-Garcia NA, Ku'ulei-Lyn Faustino V, Kyburz D, Lässer C, Lennon KM, Lötvall J, Maddox AL, Martens-Uzunova ES, Mizenko RR, Newman LA, Ridolfi A, Rohde E, Rojalin T, Rowland A, Saftics A, Sandau US, Saugstad JA, Shekari F, Swift S, Ter-Ovanesyan D, Tosar JP, Useckaite Z, Valle F, Varga Z, van der Pol E, van Herwijnen MJC, Wauben MHM, Wehman AM, Williams S, Zendrini A, Zimmerman AJ, Théry C, Witwer KW (2024) Minimal information for studies of extracellular vesicles (MISEV2023): from basic to advanced approaches. J Extracell Vesicles 13(2):e12404. https://doi.org/10.1002/jev2.12404

2. Malkin EZ, Bratman SV (2020) Bioactive DNA from extracellular vesicles and particles. Cell Death Dis 11(7):584. https://doi.org/10.1038/s41419-020-02803-4
3. Liu YJ, Wang C (2023) A review of the regulatory mechanisms of extracellular vesicles-mediated intercellular communication. Cell Commun Signal 21(1):77. https://doi.org/10.1186/s12964-023-01103-6
4. Asare-Werehene M, Nakka K, Reunov A, Chiu C-T, Lee W-T, Abedini MR, Wang P-W, Shieh D-B, Dilworth FJ, Carmona E, Le T, Mes-Masson A-M, Burger D, Tsang BK (2020) The exosome-mediated autocrine and paracrine actions of plasma gelsolin in ovarian cancer chemoresistance. Oncogene 39(7): 1600–1616. https://doi.org/10.1038/s41388-019-1087-9
5. Raposo G, Stahl PD (2019) Extracellular vesicles: a new communication paradigm? Nat Rev Mol Cell Biol 20(9):509–510. https://doi.org/10.1038/s41580-019-0158-7
6. Kalra H, Drummen GP, Mathivanan S (2016) Focus on extracellular vesicles: introducing the next small big thing. Int J Mol Sci 17(2):170. https://doi.org/10.3390/ijms17020170
7. Kalluri R, LeBleu VS (2020) The biology, function, and biomedical applications of exosomes. Science 367(6478). https://doi.org/10.1126/science.aau6977
8. Xu M, Ji J, Jin D, Wu Y, Wu T, Lin R, Zhu S, Jiang F, Ji Y, Bao B, Li M, Xu W, Xiao M (2023) The biogenesis and secretion of exosomes and multivesicular bodies (MVBs): intercellular shuttles and implications in human diseases. Genes Dis 10(5):1894–1907. https://doi.org/10.1016/j.gendis.2022.03.021
9. Kang T, Atukorala I, Mathivanan S (2021) Biogenesis of extracellular vesicles. Subcell Biochem 97:19–43. https://doi.org/10.1007/978-3-030-67171-6_2
10. Théry C, Witwer KW, Aikawa E, Alcaraz MJ, Anderson JD, Andriantsitohaina R, Antoniou A, Arab T, Archer F, Atkin-Smith GK, Ayre DC, Bach JM, Bachurski D, Baharvand H, Balaj L, Baldacchino S, Bauer NN, Baxter AA, Bebawy M, Beckham C, Bedina Zavec A, Benmoussa A, Berardi AC, Bergese P, Bielska E, Blenkiron C, Bobis-Wozowicz S, Boilard E, Boireau W, Bongiovanni A, Borràs FE, Bosch S, Boulanger CM, Breakefield X, Breglio AM, Brennan M, Brigstock DR, Brisson A, Broekman ML, Bromberg JF, Bryl-Górecka P, Buch S, Buck AH, Burger D, Busatto S, Buschmann D, Bussolati B, Buzás EI, Byrd JB, Camussi G, Carter DR, Caruso S, Chamley LW, Chang YT, Chen C, Chen S, Cheng L, Chin AR, Clayton A, Clerici SP, Cocks A, Cocucci E, Coffey RJ, Cordeiro-da-Silva A, Couch Y, Coumans FA, Coyle B, Crescitelli R, Criado MF, D'Souza-Schorey C, Das S, Datta Chaudhuri A, de Candia P, De Santana EF, De Wever O, Del Portillo HA, Demaret T, Deville S, Devitt A, Dhondt B, Di Vizio D, Dieterich LC, Dolo V, Dominguez Rubio AP, Dominici M, Dourado MR, Driedonks TA, Duarte FV, Duncan HM, Eichenberger RM, Ekström K, El Andaloussi S, Elie-Caille C, Erdbrügger U, Falcón-Pérez JM, Fatima F, Fish JE, Flores-Bellver M, Försönits A, Frelet-Barrand A, Fricke F, Fuhrmann G, Gabrielsson S, Gámez-Valero A, Gardiner C, Gärtner K, Gaudin R, Gho YS, Giebel B, Gilbert C, Gimona M, Giusti I, Goberdhan DC, Görgens A, Gorski SM, Greening DW, Gross JC, Gualerzi A, Gupta GN, Gustafson D, Handberg A, Haraszti RA, Harrison P, Hegyesi H, Hendrix A, Hill AF, Hochberg FH, Hoffmann KF, Holder B, Holthofer H, Hosseinkhani B, Hu G, Huang Y, Huber V, Hunt S, Ibrahim AG, Ikezu T, Inal JM, Isin M, Ivanova A, Jackson HK, Jacobsen S, Jay SM, Jayachandran M, Jenster G, Jiang L, Johnson SM, Jones JC, Jong A, Jovanovic-Talisman T, Jung S, Kalluri R, Kano SI, Kaur S, Kawamura Y, Keller ET, Khamari D, Khomyakova E, Khvorova A, Kierulf P, Kim KP, Kislinger T, Klingeborn M, Klinke DJ, 2nd, Kornek M, Kosanović MM, Kovács Á F, Krämer-Albers EM, Krasemann S, Krause M, Kurochkin IV, Kusuma GD, Kuypers S, Laitinen S, Langevin SM, Languino LR, Lannigan J, Lässer C, Laurent LC, Lavieu G, Lázaro-Ibáñez E, Le Lay S, Lee MS, Lee YXF, Lemos DS, Lenassi M,

Leszczynska A, Li IT, Liao K, Libregts SF, Ligeti E, Lim R, Lim SK, Linē A, Linnemannstöns K, Llorente A, Lombard CA, Lorenowicz MJ, Lörincz Á M, Lötvall J, Lovett J, Lowry MC, Loyer X, Lu Q, Lukomska B, Lunavat TR, Maas SL, Malhi H, Marcilla A, Mariani J, Mariscal J, Martens-Uzunova ES, Martin-Jaular L, Martinez MC, Martins VR, Mathieu M, Mathivanan S, Maugeri M, McGinnis LK, McVey MJ, Meckes DG, Jr., Meehan KL, Mertens I, Minciacchi VR, Möller A, Møller Jørgensen M, Morales-Kastresana A, Morhayim J, Mullier F, Muraca M, Musante L, Mussack V, Muth DC, Myburgh KH, Najrana T, Nawaz M, Nazarenko I, Nejsum P, Neri C, Neri T, Nieuwland R, Nimrichter L, Nolan JP, Nolte-'t Hoen EN, Noren Hooten N, O'Driscoll L, O'Grady T, O'Loghlen A, Ochiya T, Olivier M, Ortiz A, Ortiz LA, Osteikoetxea X, Østergaard O, Ostrowski M, Park J, Pegtel DM, Peinado H, Perut F, Pfaffl MW, Phinney DG, Pieters BC, Pink RC, Pisetsky DS, Pogge von Strandmann E, Polakovicova I, Poon IK, Powell BH, Prada I, Pulliam L, Quesenberry P, Radeghieri A, Raffai RL, Raimondo S, Rak J, Ramirez MI, Raposo G, Rayyan MS, Regev-Rudzki N, Ricklefs FL, Robbins PD, Roberts DD, Rodrigues SC, Rohde E, Rome S, Rouschop KM, Rughetti A, Russell AE, Saá P, Sahoo S, Salas-Huenuleo E, Sánchez C, Saugstad JA, Saul MJ, Schiffelers RM, Schneider R, Schøyen TH, Scott A, Shahaj E, Sharma S, Shatnyeva O, Shekari F, Shelke GV, Shetty AK, Shiba K, Siljander PR, Silva AM, Skowronek A, Snyder OL, 2nd, Soares RP, Sódar BW, Soekmadji C, Sotillo J, Stahl PD, Stoorvogel W, Stott SL, Strasser EF, Swift S, Tahara H, Tewari M, Timms K, Tiwari S, Tixeira R, Tkach M, Toh WS, Tomasini R, Torrecilhas AC, Tosar JP, Toxavidis V, Urbanelli L, Vader P, van Balkom BW, van der Grein SG, Van Deun J, van Herwijnen MJ, Van Keuren-Jensen K, van Niel G, van Royen ME, van Wijnen AJ, Vasconcelos MH, Vechetti IJ, Jr., Veit TD, Vella LJ, Velot É, Verweij FJ, Vestad B, Viñas JL, Visnovitz T, Vukman KV, Wahlgren J, Watson DC, Wauben MH, Weaver A, Webber JP, Weber V, Wehman AM, Weiss DJ, Welsh JA, Wendt S, Wheelock AM, Wiener Z, Witte L, Wolfram J, Xagorari A, Xander P, Xu J, Yan X, Yáñez-Mó M, Yin H, Yuana Y, Zappulli V, Zarubova J, Žėkas V, Zhang JY, Zhao Z, Zheng L, Zheutlin AR, Zickler AM, Zimmermann P, Zivkovic AM, Zocco D, Zuba-Surma EK (2018) Minimal information for studies of extracellular vesicles 2018 (MISEV2018): a position statement of the International Society for Extracellular Vesicles and update of the MISEV2014 guidelines. J Extracell Vesicles 7(1):1535750. https://doi.org/10.1080/20013078.2018.1535750

11. Sohal IS, Kasinski AL (2023) Emerging diversity in extracellular vesicles and their roles in cancer. Front Oncol 13:1167717. https://doi.org/10.3389/fonc.2023.1167717
12. Minciacchi VR, You S, Spinelli C, Morley S, Zandian M, Aspuria PJ, Cavallini L, Ciardiello C, Reis Sobreiro M, Morello M, Kharmate G, Jang SC, Kim DK, Hosseini-Beheshti E, Tomlinson Guns E, Gleave M, Gho YS, Mathivanan S, Yang W, Freeman MR, Di Vizio D (2015) Large oncosomes contain distinct protein cargo and represent a separate functional class of tumor-derived extracellular vesicles. Oncotarget 6(13): 11327–11341. https://doi.org/10.18632/oncotarget.3598
13. Ma L, Li Y, Peng J, Wu D, Zhao X, Cui Y, Chen L, Yan X, Du Y, Yu L (2015) Discovery of the migrasome, an organelle mediating release of cytoplasmic contents during cell migration. Cell Res 25(1):24–38. https://doi.org/10.1038/cr.2014.135
14. Chitti SV, Fonseka P, Mathivanan S (2018) Emerging role of extracellular vesicles in mediating cancer cachexia. Biochem Soc Trans 46(5):1129–1136. https://doi.org/10.1042/BST20180213
15. Shahi S, Cianciarulo C, Nedeva C, Mathivanan S (2021) Extracellular Vesicles Regulate Cancer Metastasis. Subcell Biochem 97:275–296. https://doi.org/10.1007/978-3-030-67171-6_11
16. Anand S, Samuel M, Kumar S, Mathivanan S (2019) Ticket to a bubble ride: cargo sorting into exosomes and extracellular vesicles. Biochim Biophys Acta Proteins Proteom 1867(12):140203. https://doi.org/10.1016/j.bbapap.2019.02.005
17. Zhang Q, Higginbotham JN, Jeppesen DK, Yang YP, Li W, McKinley ET, Graves-Deal R, Ping J, Britain CM, Dorsett KA, Hartman CL, Ford DA, Allen RM, Vickers KC, Liu Q, Franklin JL, Bellis SL, Coffey RJ (2019) Transfer of functional cargo in exomeres. Cell Rep 27(3):940–954.e946. https://doi.org/10.1016/j.celrep.2019.01.009
18. Sigdel S, Swenson S, Wang J (2023) Extracellular vesicles in neurodegenerative diseases: an update. Int J Mol Sci 24(17). https://doi.org/10.3390/ijms241713161
19. Fu S, Zhang Y, Li Y, Luo L, Zhao Y, Yao Y (2020) Extracellular vesicles in cardiovascular diseases. Cell Death Discovery 6(1):68. https://doi.org/10.1038/s41420-020-00305-y
20. Kovácsházi C, Hambalkó S, Sayour NV, Gergely TG, Brenner GB, Pelyhe C, Kapui D,

Weber BY, Hültenschmidt AL, Pállinger É, Buzás EI, Zolcsák Á, Kiss B, Bozó T, Csányi C, Kósa N, Kellermayer M, Farkas R, Karvaly GB, Wynne K, Matallanas D, Ferdinandy P, Giricz Z (2024) Effect of hypercholesterolemia on circulating and cardiomyocyte-derived extracellular vesicles. Sci Rep 14(1):12016. https://doi.org/10.1038/s41598-024-62689-6
21. Kumari S, Bandyopadhyay B, Singh A, Aggarwal S, Yadav AK, Vikram NK, Guchhait P, Banerjee A (2023) Extracellular vesicles recovered from plasma of severe dengue patients induce CD4+ T cell suppression through PD-L1/PD-1 interaction. MBio 14(6):e0182323. https://doi.org/10.1128/mbio.01823-23
22. Lane S, White TLA, Walsh EE, Cattley RT, Cumberland R, Hawse WF, Delgoffe GM, Badylak SF, Bomberger JM (2023) Antiviral epithelial-macrophage crosstalk permits secondary bacterial infections. MBio 14(5): e0086323. https://doi.org/10.1128/mbio.00863-23
23. Martin C, Ligat G, Malnou CE (2023) The Yin and the Yang of extracellular vesicles during viral infections. Biomed J 47:100659. https://doi.org/10.1016/j.bj.2023.100659
24. Prados-Rosales R, Weinrick BC, Piqué DG, Jacobs WR Jr, Casadevall A, Rodriguez GM (2014) Role for mycobacterium tuberculosis membrane vesicles in iron acquisition. J Bacteriol 196(6):1250–1256. https://doi.org/10.1128/jb.01090-13
25. Mathew M, Zade M, Mezghani N, Patel R, Wang Y, Momen-Heravi F (2020) Extracellular vesicles as biomarkers in cancer immunotherapy. Cancers (Basel) 12(10). https://doi.org/10.3390/cancers12102825
26. Chitti SV, Nedeva C, Manickam R, Fonseka P, Mathivanan S (2022) Extracellular vesicles as drug targets and delivery vehicles for cancer therapy. Pharmaceutics 14(12):2822
27. Takahashi Y, Takakura Y (2023) Extracellular vesicle-based therapeutics: extracellular vesicles as therapeutic targets and agents. Pharmacol Ther 242:108352. https://doi.org/10.1016/j.pharmthera.2023.108352
28. Shibata C, Otsuka M, Ishigaki K, Seimiya T, Kishikawa T, Fujishiro M (2024) CA19-9-positive extracellular vesicle is a risk factor for cancer-associated thrombosis in pancreatic cancer. Gastro Hep Adv 3(4):551–561. https://doi.org/10.1016/j.gastha.2024.02.005
29. Duong A, Parmar G, Kirkham AM, Burger D, Allan DS (2023) Registered clinical trials investigating treatment with cell-derived extracellular vesicles: a scoping review. Cytotherapy 25(9):939–945. https://doi.org/10.1016/j.jcyt.2023.04.007
30. Herrmann IK, Wood MJA, Fuhrmann G (2021) Extracellular vesicles as a next-generation drug delivery platform. Nat Nanotechnol 16(7):748–759. https://doi.org/10.1038/s41565-021-00931-2
31. Shahi S, Kang T, Fonseka P (2024) Extracellular vesicles in pathophysiology: a prudent target that requires careful consideration. Cells 13(9):754
32. Lee YJ, Shin KJ, Chae YC (2024) Regulation of cargo selection in exosome biogenesis and its biomedical applications in cancer. Exp Mol Med 56(4):877–889. https://doi.org/10.1038/s12276-024-01209-y
33. Cappe B, Vadi M, Sack E, Wacheul L, Verstraeten B, Dufour S, Franck J, Xie W, Impens F, Hendrix A, Lafontaine DLJ, Vandenabeele P, Riquet FB (2023) Systematic compositional analysis of exosomal extracellular vesicles produced by cells undergoing apoptosis, necroptosis and ferroptosis. J Extracell Vesicles 12(10):e12365. https://doi.org/10.1002/jev2.12365
34. Garza AP, Wider-Eberspächer E, Morton L, van Ham M, Pállinger É, Buzás EI, Jänsch L, Dunay IR (2024) Proteomic analysis of plasma-derived extracellular vesicles: pre- and postprandial comparisons. Sci Rep 14(1): 23032. https://doi.org/10.1038/s41598-024-74228-4
35. Whitham M, Parker BL, Friedrichsen M, Hingst JR, Hjorth M, Hughes WE, Egan CL, Cron L, Watt KI, Kuchel RP, Jayasooriah N, Estevez E, Petzold T, Suter CM, Gregorevic P, Kiens B, Richter EA, James DE, Wojtaszewski JFP, Febbraio MA (2018) Extracellular vesicles provide a means for tissue crosstalk during exercise. Cell Metab 27(1):237–251.e234. https://doi.org/10.1016/j.cmet.2017.12.001
36. Doyle LM, Wang MZ (2019) Overview of extracellular vesicles, their origin, composition, purpose, and methods for exosome isolation and analysis. Cells 8(7). https://doi.org/10.3390/cells8070727
37. Kumar MA, Baba SK, Sadida HQ, Marzooqi SA, Jerobin J, Altemani FH, Algehainy N, Alanazi MA, Abou-Samra AB, Kumar R, Al-Shabeeb Akil AS, Macha MA, Mir R, Bhat AA (2024) Extracellular vesicles as tools and targets in therapy for diseases. Signal Transduct Target Ther 9(1):27. https://doi.org/10.1038/s41392-024-01735-1
38. Hallal S, Tűzesi Á, Grau GE, Buckland ME, Alexander KL (2022) Understanding the extracellular vesicle surface for clinical

molecular biology. J Extracell Vesicles 11(10): e12260. https://doi.org/10.1002/jev2.12260

39. Jankovičová J, Sečová P, Michalková K, Antalíková J (2020) Tetraspanins, more than markers of extracellular vesicles in reproduction. Int J Mol Sci 21(20). https://doi.org/10.3390/ijms21207568
40. Raposo G, Stoorvogel W (2013) Extracellular vesicles: exosomes, microvesicles, and friends. J Cell Biol 200(4):373–383. https://doi.org/10.1083/jcb.201211138
41. Wu Q, Zhang H, Sun S, Wang L, Sun S (2021) Extracellular vesicles and immunogenic stress in cancer. Cell Death Dis 12(10):894. https://doi.org/10.1038/s41419-021-04171-z
42. Li J, Zhang Y, Dong PY, Yang GM, Gurunathan S (2023) A comprehensive review on the composition, biogenesis, purification, and multifunctional role of exosome as delivery vehicles for cancer therapy. Biomed Pharmacother 165:115087. https://doi.org/10.1016/j.biopha.2023.115087
43. Gangoda L, Liem M, Ang CS, Keerthikumar S, Adda CG, Parker BS, Mathivanan S (2017) Proteomic profiling of exosomes secreted by breast cancer cells with varying metastatic potential. Proteomics 17:23–24. https://doi.org/10.1002/pmic.201600370
44. Xu D, Di K, Fan B, Wu J, Gu X, Sun Y, Khan A, Li P, Li Z (2022) MicroRNAs in extracellular vesicles: sorting mechanisms, diagnostic value, isolation, and detection technology. Front Bioeng Biotechnol 10:948959. https://doi.org/10.3389/fbioe.2022.948959
45. O'Brien J, Hayder H, Zayed Y, Peng C (2018) Overview of MicroRNA biogenesis, mechanisms of actions, and circulation. Front Endocrinol (Lausanne) 9:402. https://doi.org/10.3389/fendo.2018.00402
46. Cecchin R, Troyer Z, Witwer K, Morris KV (2023) Extracellular vesicles: the next generation in gene therapy delivery. Mol Ther 31(5): 1225–1230. https://doi.org/10.1016/j.ymthe.2023.01.021
47. Sepúlveda F, Mayorga-Lobos C, Guzmán K, Durán-Jara E, Lobos-González L (2023) EV-miRNA-mediated intercellular communication in the breast tumor microenvironment. Int J Mol Sci 24(17). https://doi.org/10.3390/ijms241713085
48. Tsering T, Li M, Chen Y, Nadeau A, Laskaris A, Abdouh M, Bustamante P, Burnier JV (2022) EV-ADD, a database for EV-associated DNA in human liquid biopsy samples. J Extracell Vesicles 11(10):e12270. https://doi.org/10.1002/jev2.12270
49. Thakur BK, Zhang H, Becker A, Matei I, Huang Y, Costa-Silva B, Zheng Y, Hoshino A, Brazier H, Xiang J, Williams C, Rodriguez-Barrueco R, Silva JM, Zhang W, Hearn S, Elemento O, Paknejad N, Manova-Todorova K, Welte K, Bromberg J, Peinado H, Lyden D (2014) Double-stranded DNA in exosomes: a novel biomarker in cancer detection. Cell Res 24(6):766–769. https://doi.org/10.1038/cr.2014.44
50. Lázaro-Ibáñez E, Sanz-Garcia A, Visakorpi T, Escobedo-Lucea C, Siljander P, Ayuso-Sacido Á, Yliperttula M (2014) Different gDNA content in the subpopulations of prostate cancer extracellular vesicles: apoptotic bodies, microvesicles, and exosomes. Prostate 74(14): 1379–1390. https://doi.org/10.1002/pros.22853
51. Chitti SV, Nedeva C (2021) Extracellular vesicles contain putative cancer biomarkers. Subcell Biochem 97:363–374. https://doi.org/10.1007/978-3-030-67171-6_13
52. Sakuragi T, Nagata S (2023) Regulation of phospholipid distribution in the lipid bilayer by flippases and scramblases. Nat Rev Mol Cell Biol 24(8):576–596. https://doi.org/10.1038/s41580-023-00604-z
53. Fyfe J, Casari I, Manfredi M, Falasca M (2023) Role of lipid signalling in extracellular vesicles-mediated cell-to-cell communication. Cytokine Growth Factor Rev 73:20–26. https://doi.org/10.1016/j.cytogfr.2023.08.006
54. Ghadami S, Dellinger K (2023) The lipid composition of extracellular vesicles: applications in diagnostics and therapeutic delivery. Front Mol Biosci 10:1198044. https://doi.org/10.3389/fmolb.2023.1198044
55. Zhang H, Freitas D, Kim HS, Fabijanic K, Li Z, Chen H, Mark MT, Molina H, Martin AB, Bojmar L, Fang J, Rampersaud S, Hoshino A, Matei I, Kenific CM, Nakajima M, Mutvei AP, Sansone P, Buehring W, Wang H, Jimenez JP, Cohen-Gould L, Paknejad N, Brendel M, Manova-Todorova K, Magalhães A, Ferreira JA, Osório H, Silva AM, Massey A, Cubillos-Ruiz JR, Galletti G, Giannakakou P, Cuervo AM, Blenis J, Schwartz R, Brady MS, Peinado H, Bromberg J, Matsui H, Reis CA, Lyden D (2018) Identification of distinct nanoparticles and subsets of extracellular vesicles by asymmetric flow field-flow fractionation. Nat Cell Biol 20(3):332–343. https://doi.org/10.1038/s41556-018-0040-4
56. Pfrieger FW, Vitale N (2018) Cholesterol and the journey of extracellular vesicles. J Lipid Res 59(12):2255–2261. https://doi.org/10.1194/jlr.R084210

57. Harmati M, Bukva M, Böröczky T, Buzás K, Gyukity-Sebestyén E (2021) The role of the metabolite cargo of extracellular vesicles in tumor progression. Cancer Metastasis Rev 40(4):1203–1221. https://doi.org/10.1007/s10555-021-10014-2
58. Wu Y, Chen W, Guo M, Tan Q, Zhou E, Deng J, Li M, Chen J, Yang Z, Jin Y (2022) Metabolomics of extracellular vesicles: a future promise of multiple clinical applications. Int J Nanomedicine 17:6113–6129. https://doi.org/10.2147/ijn.S390378
59. Zhang Y, Liang F, Zhang D, Qi S, Liu Y (2023) Metabolites as extracellular vesicle cargo in health, cancer, pleural effusion, and cardiovascular diseases: an emerging field of study to diagnostic and therapeutic purposes. Biomed Pharmacother 157:114046. https://doi.org/10.1016/j.biopha.2022.114046
60. Luo P, Mao K, Xu J, Wu F, Wang X, Wang S, Zhou M, Duan L, Tan Q, Ma G, Yang G, Du R, Huang H, Huang Q, Li Y, Guo M, Jin Y (2020) Metabolic characteristics of large and small extracellular vesicles from pleural effusion reveal biomarker candidates for the diagnosis of tuberculosis and malignancy. J Extracell Vesicles 9(1):1790158. https://doi.org/10.1080/20013078.2020.1790158
61. Keerthikumar S, Chisanga D, Ariyaratne D, Al Saffar H, Anand S, Zhao K, Samuel M, Pathan M, Jois M, Chilamkurti N, Gangoda L, Mathivanan S (2016) ExoCarta: a web-based compendium of Exosomal cargo. J Mol Biol 428(4):688–692. https://doi.org/10.1016/j.jmb.2015.09.019
62. Chitti SV, Gummadi S, Kang T, Shahi S, Marzan AL, Nedeva C, Sanwlani R, Bramich K, Stewart S, Petrovska M, Sen B, Ozkan A, Akinfenwa M, Fonseka P, Mathivanan S (2024) Vesiclepedia 2024: an extracellular vesicles and extracellular particles repository. Nucleic Acids Res 52(D1):D1694–d1698. https://doi.org/10.1093/nar/gkad1007
63. Pathan M, Fonseka P, Chitti SV, Kang T, Sanwlani R, Van Deun J, Hendrix A, Mathivanan S (2019) Vesiclepedia 2019: a compendium of RNA, proteins, lipids and metabolites in extracellular vesicles. Nucleic Acids Res 47 (D1):D516–d519. https://doi.org/10.1093/nar/gky1029
64. Kim DK, Kang B, Kim OY, Choi DS, Lee J, Kim SR, Go G, Yoon YJ, Kim JH, Jang SC, Park KS, Choi EJ, Kim KP, Desiderio DM, Kim YK, Lötvall J, Hwang D, Gho YS (2013) EVpedia: an integrated database of high-throughput data for systemic analyses of extracellular vesicles. J Extracell Vesicles 2. https://doi.org/10.3402/jev.v2i0.20384
65. Kim DK, Lee J, Kim SR, Choi DS, Yoon YJ, Kim JH, Go G, Nhung D, Hong K, Jang SC, Kim SH, Park KS, Kim OY, Park HT, Seo JH, Aikawa E, Baj-Krzyworzeka M, van Balkom BW, Belting M, Blanc L, Bond V, Bongiovanni A, Borràs FE, Buée L, Buzás EI, Cheng L, Clayton A, Cocucci E, Dela Cruz CS, Desiderio DM, Di Vizio D, Ekström K, Falcon-Perez JM, Gardiner C, Giebel B, Greening DW, Gross JC, Gupta D, Hendrix A, Hill AF, Hill MM, Nolte-'t Hoen E, Hwang DW, Inal J, Jagannadham MV, Jayachandran M, Jee YK, Jørgensen M, Kim KP, Kim YK, Kislinger T, Lässer C, Lee DS, Lee H, van Leeuwen J, Lener T, Liu ML, Lötvall J, Marcilla A, Mathivanan S, Möller A, Morhayim J, Mullier F, Nazarenko I, Nieuwland R, Nunes DN, Pang K, Park J, Patel T, Pocsfalvi G, Del Portillo H, Putz U, Ramirez MI, Rodrigues ML, Roh TY, Royo F, Sahoo S, Schiffelers R, Sharma S, Siljander P, Simpson RJ, Soekmadji C, Stahl P, Stensballe A, Stępień E, Tahara H, Trummer A, Valadi H, Vella LJ, Wai SN, Witwer K, Yáñez-Mó M, Youn H, Zeidler R, Gho YS (2015) EVpedia: a community web portal for extracellular vesicles research. Bioinformatics 31(6):933–939. https://doi.org/10.1093/bioinformatics/btu741
66. Van Deun J, Mestdagh P, Agostinis P, Akay Ö, Anand S, Anckaert J, Martinez ZA, Baetens T, Beghein E, Bertier L, Berx G, Boere J, Boukouris S, Bremer M, Buschmann D, Byrd JB, Casert C, Cheng L, Cmoch A, Daveloose D, De Smedt E, Demirsoy S, Depoorter V, Dhondt B, Driedonks TA, Dudek A, Elsharawy A, Floris I, Foers AD, Gärtner K, Garg AD, Geeurickx E, Gettemans J, Ghazavi F, Giebel B, Kormelink TG, Hancock G, Helsmoortel H, Hill AF, Hyenne V, Kalra H, Kim D, Kowal J, Kraemer S, Leidinger P, Leonelli C, Liang Y, Lippens L, Liu S, Lo Cicero A, Martin S, Mathivanan S, Mathiyalagan P, Matusek T, Milani G, Monguió-Tortajada M, Mus LM, Muth DC, Németh A, Nolte-'t Hoen EN, O'Driscoll L, Palmulli R, Pfaffl MW, Primdal-Bengtson B, Romano E, Rousseau Q, Sahoo S, Sampaio N, Samuel M, Scicluna B, Soen B, Steels A, Swinnen JV, Takatalo M, Thaminy S, Théry C, Tulkens J, Van Audenhove I, van der Grein S, Van Goethem A, van Herwijnen MJ, Van Niel G, Van Roy N, Van Vliet AR, Vandamme N, Vanhauwaert S, Vergauwen G, Verweij F, Wallaert A, Wauben M, Witwer KW, Zonneveld MI, De Wever O, Vandesompele J, Hendrix A (2017) EV-TRACK: transparent reporting and centralizing knowledge in extracellular vesicle research. Nat Methods 14(3):

228–232. https://doi.org/10.1038/nmeth.4185

67. Chen J, Lin JJ, Wang W, Huang H, Pan Z, Ye G, Dong S, Lin Y, Lin C, Huang Q (2024) EV-COMM: a database of interspecies and intercellular interactions mediated by extracellular vesicles. J Extracell Vesicles 13(4): e12442. https://doi.org/10.1002/jev2.12442
68. Müller T, Schrötter A, Loosse C, Helling S, Stephan C, Ahrens M, Uszkoreit J, Eisenacher M, Meyer HE, Marcus K (2011) Sense and nonsense of pathway analysis software in proteomics. J Proteome Res 10(12): 5398–5408. https://doi.org/10.1021/pr200654k
69. Henderson-Maclennan NK, Papp JC, Talbot CC Jr, McCabe ER, Presson AP (2010) Pathway analysis software: annotation errors and solutions. Mol Genet Metab 101(2–3): 134–140. https://doi.org/10.1016/j.ymgme.2010.06.005
70. Dennis G, Sherman BT, Hosack DA, Yang J, Gao W, Lane HC, Lempicki RA (2003) DAVID: database for annotation, visualization, and integrated discovery. Genome Biol 4(5): P3. https://doi.org/10.1186/gb-2003-4-5-p3
71. Kuleshov MV, Jones MR, Rouillard AD, Fernandez NF, Duan Q, Wang Z, Koplev S, Jenkins SL, Jagodnik KM, Lachmann A, McDermott MG, Monteiro CD, Gundersen GW, Ma'ayan A (2016) Enrichr: a comprehensive gene set enrichment analysis web server 2016 update. Nucleic Acids Res 44(W1): W90–W97. https://doi.org/10.1093/nar/gkw377
72. Subramanian A, Tamayo P, Mootha VK, Mukherjee S, Ebert BL, Gillette MA, Paulovich A, Pomeroy SL, Golub TR, Lander ES, Mesirov JP (2005) Gene set enrichment analysis: a knowledge-based approach for interpreting genome-wide expression profiles. Proc Natl Acad Sci USA 102(43):15545–15550. https://doi.org/10.1073/pnas.0506580102
73. Shannon P, Markiel A, Ozier O, Baliga NS, Wang JT, Ramage D, Amin N, Schwikowski B, Ideker T (2003) Cytoscape: a software environment for integrated models of biomolecular interaction networks. Genome Res 13(11):2498–2504. https://doi.org/10.1101/gr.1239303
74. Pathan M, Keerthikumar S, Ang CS, Gangoda L, Quek CY, Williamson NA, Mouradov D, Sieber OM, Simpson RJ, Salim A, Bacic A, Hill AF, Stroud DA, Ryan MT, Agbinya JI, Mariadason JM, Burgess AW, Mathivanan S (2015) FunRich: an open access standalone functional enrichment and interaction network analysis tool. Proteomics 15(15): 2597–2601. https://doi.org/10.1002/pmic.201400515
75. Fonseka P, Pathan M, Chitti SV, Kang T, Mathivanan S (2021) FunRich enables enrichment analysis of OMICs datasets. J Mol Biol 433(11):166747. https://doi.org/10.1016/j.jmb.2020.166747
76. Maere S, Heymans K, Kuiper M (2005) BiNGO: a Cytoscape plugin to assess overrepresentation of gene ontology categories in biological networks. Bioinformatics 21(16): 3448–3449. https://doi.org/10.1093/bioinformatics/bti551
77. Reimand J, Isserlin R, Voisin V, Kucera M, Tannus-Lopes C, Rostamianfar A, Wadi L, Meyer M, Wong J, Xu C, Merico D, Bader GD (2019) Pathway enrichment analysis and visualization of omics data using g:Profiler, GSEA, Cytoscape and EnrichmentMap. Nat Protoc 14(2):482–517. https://doi.org/10.1038/s41596-018-0103-9
78. UniProt Consortium (2021) UniProt: the universal protein knowledgebase in 2021. Nucleic Acids Res 49(D1):D480–D489. https://doi.org/10.1093/nar/gkaa1100
79. Chen EY, Tan CM, Kou Y, Duan Q, Wang Z, Meirelles GV, Clark NR, Ma'ayan A (2013) Enrichr: interactive and collaborative HTML5 gene list enrichment analysis tool. BMC Bioinformatics 14:128. https://doi.org/10.1186/1471-2105-14-128
80. Evangelista JE, Xie Z, Marino GB, Nguyen N, Clarke DJB, Ma'ayan A (2023) Enrichr-KG: bridging enrichment analysis across multiple libraries. Nucleic Acids Res 51(W1):W168–W179. https://doi.org/10.1093/nar/gkad393
81. Liberzon A, Birger C, Thorvaldsdóttir H, Ghandi M, Mesirov JP, Tamayo P (2015) The molecular signatures database (MSigDB) hallmark gene set collection. Cell Syst 1(6): 417–425. https://doi.org/10.1016/j.cels.2015.12.004
82. Cirillo E, Parnell LD, Evelo CT (2017) A review of pathway-based analysis tools that visualize genetic variants. Front Genet 8:174. https://doi.org/10.3389/fgene.2017.00174
83. Satam H, Joshi K, Mangrolia U, Waghoo S, Zaidi G, Rawool S, Thakare RP, Banday S, Mishra AK, Das G, Malonia SK (2023) Next-generation sequencing technology: current trends and advancements. Biology (Basel) 12(7). https://doi.org/10.3390/biology12070997

INDEX

Amit K. Srivastava (ed.), *Extracellular Vesicles*, Neuromethods, vol. 228, https://doi.org/10.1007/978-1-0716-4905-3,

N

P

R

S

T

U

V

W

FSC
www.fsc.org

MIX
Papier aus verantwortungsvollen Quellen
Paper from responsible sources
FSC® C105338

If you have any concerns about our products,
you can contact us on
ProductSafety@springernature.com

In case Publisher is established outside the EU,
the EU authorized representative is:
Springer Nature Customer Service Center GmbH
Europaplatz 3, 69115 Heidelberg, Germany

Printed by Libri Plureos GmbH
in Hamburg, Germany